Behavioral Ecology of Insect Parasitoids

To Maura, Emilio, and Esther

To Danielito, Andresuchi, and Esteli

To Frietson and Joris

To John Maynard-Smith and Bill Hamilton

Behavioral Ecology of Insect Parasitoids

From Theoretical Approaches to Field Applications

Edited by Éric Wajnberg,
Carlos Bernstein,
and Jacques van Alphen

Blackwell
Publishing

© 2008 by Blackwell Publishing Ltd

BLACKWELL PUBLISHING
350 Main Street, Malden, MA 02148-5020, USA
9600 Garsington Road, Oxford OX4 2DQ, UK
550 Swanston Street, Carlton, Victoria 3053, Australia

First published 2008 by Blackwell Publishing Ltd

1   2008

Library of Congress Cataloging-in-Publication Data

Behavioral ecology of insect parasitoids : from theoretical approaches to field applications / edited by Eric Wajnberg, Carlos Bernstein, and Jacques J.M. van Alphen.
       p. cm.
    Includes bibliographical references and index.
    ISBN 978-1-4051-6347-7 (hardcover : alk. paper) 1. Parasitic insects—Behavior. 2. Parasitoids—Behavior. 3. Parasitic insects—Ecology. 4. Parasitoids—Ecology. 5. Insect pests—Biological control. I. Wajnberg, E. II. Bernstein, Carlos. III. Alphen, Jacques van.

      QL496.B384 2008
      595.717′857—dc22

                                                                           2007025136

A catalogue record for this title is available from the British Library

Set in 10/12.5pt Minion
by Graphicraft Limited, Hong Kong
Printed and bound in Singapore
by Markono Print Media Pte Ltd

The publisher's policy is to use permanent paper from mills that operate a sustainable forestry policy, and which has been manufactured from pulp processed using acid-free and elementary chlorine-free practices. Furthermore, the publisher ensures that the text paper and cover board used have met acceptable environmental accreditation standards.

For further information on
Blackwell Publishing, visit our website:
www.blackwellpublishing.com

Contents

Contributors

Pierre Bernhard
Polytech'Nice Sophia Antipolis
930, Route des Colles
BP 145
06903 Sophia Antipolis Cedex
France
Tel: +33 4 92 96 51 52
Fax: +33 4 92 96 51 55
e-mail: pierre.bernhard@polytech.unice.fr

Carlos Bernstein
Biométrie et Biologie Évolutive
Université de Lyon; Université Lyon I
43 BD du 11 Novembre 1918
69622 Villeurbanne Cedex
France
Tel: +33 4 72 43 14 38
Fax: +33 4 72 43 13 88
e-mail: carlosbe@biomserv.univ-lyon1.fr

Guy Boivin
Centre de Recherche et de Développement en Horticulture
Agriculture et Agroalimentaire Canada
430, boulevard Gouin
Saint-Jean-sur-Richelieu
Québec J3B 3E6
Canada
Tel: +1 450 346 4494 ext. 210
Fax: +1 450 346 7740
e-mail: boiving@agr.gc.ca

Michael B. Bonsall
Department of Zoology

University of Oxford
South Parks Road
Oxford OX1 3PS
UK
Tel: +44 1 865 281064
Fax: +44 1 865 310447
e-mail: michael.bonsall@zoo.ox.ac.uk

Jérôme Casas
University of Tours
Institut de Recherche sur la Biologie de l'Insecte
IRBI-CNRS UMR6035
Av. Monge
37200 Tours
France
Tel: +33 2 47 36 69 78
Fax: +33 2 47 36 69 66
e-mail: casas@univ-tours.fr

Luc-Alain Giraldeau
Université du Québec à Montréal
Case postale 8888
Succursale Centre-ville
Montréal (Québec) H3C 3P8
Canada
Tel: +1 514 987 3000 ext. 3244
Fax: +1 514 987 4647
e-mail: giraldeau.luc-alain@uqam.ca

H. Charles J. Godfray
Department of Zoology
University of Oxford
South Parks Road
Oxford OX1 3PS
UK
Tel: +44 1865 271176
Fax: +44 1865 310447
e-mail: charles.godfray@zoo.ox.ac.uk

Richard F. Green
Department of Mathematics and Statistics
University of Minnesota Duluth
Duluth, MN 55812
USA
Tel: +1 218 726 7229
Fax: +1 218 726 8399
e-mail: rgreen@d.umn.edu

Patsy Haccou
Section Theoretical Biology
Institute of Biology
Leiden University
P.O. Box 9516
2300 RA Leiden
The Netherlands
Tel: +31 71 527 4917
Fax: +31 71 527 4900
e-mail: p.haccou@biology.leidenuniv.nl

Ian C.W. Hardy
School of Biosciences
University of Nottingham
Sutton Bonington Campus
Loughborough LE12 5RD
UK
Tel: +44 115 951 6052
Fax: +44 115 951 6060
e-mail: ian.hardy@nottingham.ac.uk

George E. Heimpel
Department of Entomology
University of Minnesota
1980 Folwell Ave
St Paul, MN 55108
USA
Tel: +1 612 624 3480
Fax: +1 612 625 5299
e-mail: heimp001@umn.edu

Lia Hemerik
Wageningen University
Biometris, Department of Mathematical and Statistical Methods
P.O. Box 100
6700AC Wageningen
The Netherlands
Tel: +31 317 482083
Fax: +31 317 483554
e-mail: Lia.Hemerik@wur.nl

Monika Hilker
Department of Applied Zoology/Animal Ecology
Institute of Biology
Freie Universität Berlin
Haderslebener Str. 9
12163 Berlin

Germany
Tel: +49 30 838 559 13
Fax: +49 30 838 538 97
e-mail: hilker@zedat.fu-berlin.de

Thomas S. Hoffmeister
Institute of Ecology and Evolutionary Biology
FB 2, Biology
University of Bremen
Leobener Str./NW2
28359 Bremen
Germany
Tel: +49 421 218 4290
Fax: +49 421 218 4504
e-mail: hoffmeister@uni-bremen.de

Anthony R. Ives
Department of Zoology
University of Wisconsin – Madison
Madison, WI 53706
USA
Tel: +1 608 262 1519
Fax: +1 608 265 6320
e-mail: arives@wisc.edu

Mark Jervis
Cardiff School of Biosciences
Cardiff University
Cardiff CF10 3TL
UK
Tel: +44 29 20 874948
Fax: +44 29 20 874305
e-mail: jervis@cardiff.ac.uk

Alex R. Kraaijeveld
School of Biological Sciences
University of Southampton
Bassett Crescent East
Southampton SO16 7PX
UK
Tel: +44 2380 593436
Fax: +44 2380 594459
e-mail: arkraa@soton.ac.uk

Jeremy McNeil
Department of Biology
The University of Western Ontario

London, ON N6A 5B7
Canada
Tel: +1 519 661 3487
Fax: +1 519 661 3935
e-mail: jnmcneil@gmail.com

Nick J. Mills
University of California
Department of Environmental Science, Policy, and Management
Mulford Hall
Berkeley, CA 94720-3114
USA
Tel: +1 510 642 1711
Fax: +1 510-643 5438
e-mail: nmills@nature.berkeley.edu

Paul J. Ode
North Dakota State University
Department of Entomology
270 Hultz Hall
Fargo, ND
USA
Tel: +1 701 231 5934
Fax: +1 701 231 8557
e-mail: Paul.Ode@ndsu.edu

Jean-Sébastien Pierre
INRA
Biologie des organismes et des populations appliquée à la protection des plantes
Domaine de la Motte
BP 35327
35653 Le Rheu Cedex
France
Tel: +33 2 23 48 70 83
Fax: +33 2 23 48 51 50
e-mail: jean-sebastien.pierre@rennes.inra.fr, jean-sebastien.pierre@univ-rennes1.fr

Bernard Roitberg
Behavioral Ecology Research Group and
Centre for Pest Management
Department of Biology
Simon Fraser University
8888 University Drive
Burnaby, BC V5A 1S6
Canada
Tel: +1 604 291 3585
Fax: +1 604 291 3496
e-mail: roitberg@sfu.ca

William E. Snyder
Department of Entomology
Washington State University
Pullman, WA 99164-6382
USA
Tel: +1 509 335 3724
Fax: +1 509 335 1009
e-mail: wesnyder@wsu.edu

Michael R. Strand
Department of Entomology
420 Biological Sciences
University of Georgia
Athens, GA 30602-2603
USA
Tel: +1 706 583 8237
Fax: +1 706 542 2279
e-mail: mrstrand@bugs.ent.uga.edu

Jacques J.M. van Alphen
Institute of Evolutionary and Ecological Sciences
Leiden University
PO Box 9516
2300 RA Leiden
The Netherlands
Tel: +31 71 527 4992
Fax: +31 71 527 4900
e-mail: J.J.M.van.Alphen@biology.leidenuniv.nl

Minus van Baalen
UMR 7625 « Fonctionnement et Évolution des Systèmes Écologiques »
Université Pierre et Marie Curie
Bât. A, 7ème Étage Case 237
7, quai St.-Bernard
75252 Paris Cedex 05
France
Tel: +33 1 44 27 25 45
Fax: +33 1 44 27 35 16
e-mail: minus.van.baalen@ens.fr

Louise E.M. Vet
Netherlands Institute of Ecology (NIOO-KNAW)
P.O. Box 1299
3600 BG Maarssen
The Netherlands
Tel: +31 294 239 312
Fax: +31 294 239 078
e-mail: l.vet@nioo.knaw.nl

Éric Wajnberg
INRA
400 Route des Chappes
BP 167
06903 Sophia Antipolis Cedex
France
Tel: +33 4 92 38 64 47
Fax: +33 4 92 38 65 57
e-mail: wajnberg@sophia.inra.fr

Preface

Parasitoids are fascinating insects, whose adult females lay their eggs in or on other insects. The parasitoid larvae develop by feeding on the host bodies, resulting in the death of the host. Parasitoids are found in nearly all terrestrial ecosystems and show a vast biological and ecological diversity and a wide array of specific adaptations, making them ideal subjects for comparative research. For their reproduction, they depend on finding and attacking hosts and, as a consequence, they are under strong natural selection to develop efficient host search and attack strategies, often through elaborate behavioral mechanisms. This makes these insects superb models for testing evolutionary hypotheses, also because a direct link exists between host search and attack behavior of a parasitoid and its fitness, as the number of hosts parasitized is proportional to the number of offspring produced. Further, since their reproduction results in the death of their hosts, parasitoids are often important factors in the natural control of insect populations, thus preventing insect pests. Certain species are mass-produced and released on a large scale to limit or suppress insect pests attacking different crops. 'Biological control', as this technique is known, can lead to a highly significant reduction in the use of toxic chemical pesticides, thus reducing the impact on non-target organisms.

This book originated from the European scientific program 'Behavioural Ecology of Insect Parasitoids', financially supported by the European Science Foundation (ESF). Behavioral ecology is a scientific discipline that strives to understand animal behavior under natural conditions in evolutionary terms, i.e. by asking what the adaptive advantages of a particular behavior are. This is done by comparing the actual behavior of animals with predictions of theoretical models of how animals should behave so as to optimize their fitness, given a realistic set of constraints. The aim of the models, frequently expressed in mathematical terms, is to better define the questions and to help in designing experiments. Experimental work puts the hypothesis to the test, and the differences between experimental results and theory helps in identifying any weakness in our understanding. This suggests aspects of the problem that might have been overlooked and that would be subsequently incorporated into new models or tested in new experiments. The scope of behavioral ecology extends to the population level. Incorporating optimal behaviors and deviations from these archetypes, into models of population dynamics, allows increasing their realism by putting them on a firm evolutionary footing. In recent decades, this approach has been developed with ample success by using different animals (e.g. mammals, birds, fishes, and also insects).

Biological control has resulted both in remarkable successes and in definitive failures. Biological control programs follow in general an empirical approach and practitioners often have a limited understanding of the reasons for the different fates (success or otherwise) of control attempts. There is, as a consequence, a clear need to base pest control practises on a firm, formal scientific basis. As all living beings have been shaped by natural selection, evolutionary thinking is the key to the understanding of the workings of nature ('Nothing in Biology makes sense except in the light of evolution,' Dobzhansky (1973) *The American Biology Teacher* **35**: 125–9), on which sound species management should be based.

In parasitoids, parasitism behavior is central to the success of biological control, because the death of the host results from the production of progeny. In spite of this, the theoretical and experimental achievements in the understanding of the evolution of parasitoid behavior and life history traits have been seldom put to use as a means of improving the efficacy of biological control programs. However, it has been successfully applied in the selection of the most promising candidates for release, in improving parasitoid mass rearing efficiency, and in the evaluation of success and failure of parasitoids as biocontrol agents. As a consequence, in this volume, we combine the study of fundamental aspects of parasitoid behavior with a discussion of their possible consequences for the efficacy of selective pest control.

This book contains 18 chapters, each of them written by two distinguished specialists, covering virtually all the key aspects of parasitoid behavior and their relevance for efficient biological control programs.

The first part presents current issues in behavioral ecology of insect parasitoids. It starts with Chapter 1 linking optimal foraging behavior to efficient biological control. Chapter 2 proposes an accurate definition of fitness, how it translates in these particular insects, and how fitness should be estimated. Then, since behavioral ecology addresses the behavior of parasitoids under natural conditions, the same conditions under which biological control takes place, Chapter 3 contributes to the understanding of parasitoid decision-making under field situations. After Chapter 4 has addressed the important issue of competition between parasitoids and other species foraging for hosts at the same trophic level, Chapters 5 to 9 discuss the 'classical' questions of parasitoid behavior, namely responses to chemical cues for finding hosts (Chapter 5), the physiological mechanisms involved in behavioral decisions (Chapter 6), food searching strategies (Chapter 7), patch time allocation (Chapter 8), and competition between foraging females on patches of hosts (Chapter 9). Finally, Chapter 10 deals with potential risk assessment strategies adopted by these insects.

The second part of the volume addresses the extension of the evolutionary approach of behavioral ecology to other related scientific questions. To start with, Chapter 11 looks at consequences of parasitoid behavior in a multitrophic context. Then, Chapter 12 discusses sex ratio control and Chapter 13 considers the consequences of parasitoid behavioral ecology for population dynamics. Finally, Chapter 14 raises the potential link between parasitoid behavior and the development of resistance/virulence physiological mechanisms in host-parasitoid associations.

Since a behavioral ecology approach of such tiny animals cannot be developed without the use of specific technical tools, the last section of the volume addresses some methodological issues, especially those developed for the study of insect parasitoids. Chapter 15 presents how state-dependent problems should be addressed, while Chapter 16 discusses more specifically Bayesian approaches. Finally, Chapter 17 presents how genetic algorithms can be used to find optimal behavioral decisions under different environmental conditions and Chapter 18 summarizes the most recent statistical methods that should be used for a sound analysis of behavioral data.

We hope that this volume is timely, and that it will foster research on the behavioral ecology of insect parasitoids and propose new and interesting venues for future research. We hope that it will rapidly become an important reference for both scientists and students working on parasitoid biology and for everyone involved in using parasitoids in biological control programs.

We want to thank several referees that read and commented critically on one or more chapters. They include Pierre Bernhard, Carlos Bernstein, Jérôme Casas, Patrick Coquillard, Anne-Marie Cortesero, Christine Curty, René Feyereisen, Luc-Alain Giraldeau, Patsy Haccou, Thomas Hoffmeister, Mark Jervis, Finn Kjellberg, Nick Mills, Franco Pennacchio, Jean-Sébastien Pierre, Marylène Poirié, Odile Pons, Geneviève Prévost, Bernie Roitberg, Brigitte Tenhumberg, Jacques van Alphen, Minus van Baalen, Brad Vinson, and Éric Wajnberg.

Much editing work has been done in order to homogenize the content of the book, but all information, results, and discussion provided in each chapter are under the single responsibility of their corresponding authors.

We finally want to express our sincere thanks to the ESF and to the people at Blackwell Publishing for their efficient help and support in the production of this book.

Éric Wajnberg
Carlos Bernstein
Jacques van Alphen

Part 1

Current issues in behavioral ecology of insect parasitoids

1

Optimal foraging behavior and efficient biological control methods

Nick J. Mills and Éric Wajnberg

Abstract

Insect parasitoids have been used for the biological control of insect pests through classical importations for the control of invasive phytophagous species, through seasonal or inundative releases for short-term suppression of indigenous or invasive pests, and through conservation of parasitoid activity by the provisioning of resource subsidies and alteration of management practices. In all cases, success in the suppression of a pest is dependent upon the behavioral decisions made by the parasitoid in searching for and parasitizing hosts. For example, in the case of classical biological control, patch choice decisions that maximize parasitoid fitness will tend to increase its regional impact, leading to greater suppression of the pest and success in biological control. In contrast, for augmentative biological control, the goal is to provide local pest suppression and behavior that maximizes fitness might, for example, lead parasitoids to abandon local patches of their hosts before the pest has been suppressed to the desired level. Thus, the behavioral ecology of insect parasitoids is central to the successful implementation of biological control programs.

We explore how optimal foraging effects the suppression of global pest densities in a metapopulation context and to what extent the physiological condition and behavioral decisions of foraging parasitoids are likely to influence establishment and impact in classical biological control. In the case of inundative biological control, we discuss the trade-off between optimal foraging behavior and the level of pest suppression at a local scale and consider the use of chemical attractants and arrestants to increase parasitoid activity and patch time allocation. We also discuss the influence of host size and quality, and sex ratio (*Wolbachia* infection) on parasitoid mass rearing. Finally, the influence of nectar subsidies on parasitoid foraging behavior and host suppression is considered in the context of developing more efficient methods for conservation biological control.

1.1 Introduction

Biological control represents the action of living natural enemies in suppressing the abundance or activity of pests. As a naturally occurring ecosystem service, globally, biological control has been loosely valued at $400 billion per year (Costanza et al. 1997); while a more conservative estimate of $4.5 billion per year has been attributed to the services provided by indigenous predators and parasitoids of native agricultural pests in the USA (Losey & Vaughan 2006). Although natural enemies include predators (that must consume many prey individuals to complete their development), pathogens (bacteria, fungi, and viruses), parasites (soil-inhabiting entomopathogenic nematodes), and antagonists (competitors) in addition to parasitoids, the latter are the most important group in the context of biological control of insect pests.

There are three broad categories that describe how parasitoids can be used in biological control: importation, augmentation, and conservation. Importation or classical biological control makes use of host-specific parasitoids imported from the region of origin of invasive pests and has received the greatest amount of attention (Mills 2000, Hoddle 2004). The introduction of exotic parasitoids for the control of invasive pests continues to fascinate ecologists, fuel theoretical models of host–parasitoid interactions, and yet defy a simple and unified mechanistic explanation. Since the initial success of the introduction of the vedalia beetle (*Rodolia cardinalis*) from Australia for the control of cottony cushion scale (*Icerya purchasi*) in California in 1886 (Caltagirone & Doutt 1989), biological control practitioners have continued to implement biological control as an effective strategy for the management of invasive insect pests, while ecologists have struggled to find a consistent explanation for the success or failure of these programmes (Murdoch et al. 2003).

When parasitoids of invasive or indigenous pests are unable to persist year round or to increase in numbers quickly enough to suppress pest damage, augmentative biological control, involving the periodic release of insectary-produced parasitoids, can be effective. Augmentation has been used most effectively in protected or semi-protected environments such as glasshouses and cattle or poultry houses (Daane et al. 2002, Heinz et al. 2004), with rather less success under open field conditions (Collier & Van Steenwyk 2004). Augmentative biological control can be approached through inoculation or inundation. Inoculation of small numbers of parasitoids can be used to improve colonization at critical periods for season-long pest suppression, as practised under certain conditions for the control of greenhouse whitefly (*Trialeurodes vaporariorum*) on tomato by *Encarsia formosa* (Hoddle et al. 1998). Alternatively, inundation of large numbers of parasitoids can be used for immediate suppression, but often without a lasting impact, as used for control of house flies (*Musca domestica*) by *Spalangia cameroni* (Skovgard & Nachman 2004). In contrast, conservation biological control focuses on the enhancement of both introduced and indigenous parasitoid populations through provisioning of limiting resources or alteration of crop production practices. Parasitoids are often limited by the availability of essential resources such as nectar or overwintering sites and are excluded from crops by use of incompatible pesticides. Thus, success in conservation biological control can result, for example, from perimeter planting of annual buckwheat as a nectar subsidy for the aphid parasitoid *Aphidius rhopalosiphi* (Tylianakis et al. 2004) or from removal of incompatible insecticides as demonstrated in the effective suppression of the brown planthopper in rice in Indonesia (Kenmore 1996).

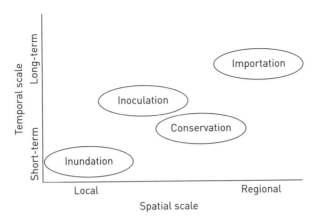

Fig. 1.1 A schematic representation of the four main approaches to applied biological control to reflect the differential spatial and temporal scales of the processes involved.

For each of these categories, success in the suppression of an insect pest is dependent upon the behavioral decisions made by parasitoids in both searching for and parasitizing hosts. Thus, the behavioral ecology of insect parasitoids is central to the successful implementation of biological control programs. However, linkages between variation in parasitoid behavior and its consequences for population dynamics remain few and have proved to be an elusive and difficult goal (Ives 1995, Vet 2001). In the context of parasitoid foraging behavior, there are important distinctions between the different approaches to applied biological control based on both the spatial and temporal scale of the processes involved (Fig. 1.1). The aim of importation differs from all other categories of biological control in that success requires regional suppression of a pest population, extending in some cases to a substantial part of whole continents, as in the successful control of the cassava mealybug through importation of *Anagyrus lopezi* (Neuenschwander 2003). In addition, as relatively small numbers of parasitoids are introduced into spatially and numerically extensive populations of the pest, success in importation biological control takes longer to achieve and may span several years. Bellows (2001) estimated that the average time taken to achieve suppression of an invasive pest through importation of parasitoids is six to 13 generations, but complete suppression on a large regional scale can take up to 12 years as in the case of the cassava mealybug (Neuenschwander 2003). In contrast, inundation, as an approach to biological control, is used for immediate impact, within a single generation of a pest and tends to be confined to a very local scale such as an individual field or orchard. Although the implementation of inundation can extend to large areas, such as releases of *Trichogramma brassicae* in 2002 on 77,000 hectares for suppression of European corn borer in France (Wajnberg & Hassan 1994), the process itself still operates at a very local scale. Intermediate between importation and inundation, both in terms of spatial and temporal scale, are inoculation and conservation. In both cases, the aim is generally to provide season-long control and the hope is that the impact of the intervention might extend on a spatial scale beyond the points of implementation. Inoculation is based on the notion that released parasitoids will continue to affect the pest population over several generations, as in the early season releases of *E. formosa* for control of greenhouse whitefly

(Hoddle et al. 1998) and so can be considered to operate at a slightly greater temporal scale. In contrast, conservation, through enhancement of the suitability of the environment for parasitoids, might be considered to operate on a slightly broader spatial scale as in the provision of nectar subsidies, where parasitoids able to use such subsidies have greater mobility and can be found more distantly from the source (Heimpel & Jervis 2005).

In this chapter, we explore how the distinct spatial and temporal scales of the four main approaches to applied biological control are influenced by different aspects of parasitoid foraging behavior. In considering each of the approaches, we begin with a brief discussion of the pertinent foraging decisions and subsequently consider practical applications and future opportunities. First, we consider optimal patch choice and the extent to which it might affect the success of importation biological control. In the context of augmentative biological control, we discuss the trade-off between optimal foraging behavior and level of pest suppression at a local scale and consider the use of infochemicals to increase patch residence time (see also Chapter 5 by Hilker and McNeill). We also discuss the influence of host size and quality, and sex ratio (*Wolbachia* infection) (see also Chapter 12 by Ode and Hardy) on the efficiency of parasitoid mass rearing. Finally, we focus on foraging decisions that affect current versus future reproduction in the context of nectar subsidies as a component of conservation biological control.

1.2 Importation biological control

For importation biological control, host-specific parasitoids are imported from the region of origin of an exotic invasive pest. The goal of this approach is for the introduced parasitoid to become established throughout the region colonized by the invasive pest and to provide long-term suppression at low pest densities. The introduction of *Aphytis paramaculicornis* and *Coccophagoides utilis* into California in the 1950s to control olive scale (*Parlatoria oleae*) provides a stellar example with olive scale remaining a scarce insect in California some 50 years after the initial parasitoid introductions (Huffaker et al. 1986, Rochat & Gutierrez 2001). While there has been a series of important successes against invasive insect pests in many different regions of the world, there remains an even greater list of failures in which the introduced parasitoids either did not become established in the target region or, if they did so, there was no notable impact on the abundance of the target pest. Using the historical record of classical biological control introductions worldwide, only 38% of 1450 unique pest-introduced parasitoid combinations have resulted in establishment and 44% of 551 established parasitoids have provided partial to complete control of the pest, corresponding to a 17% overall rate of success (Mills 1994, 2000).

An important question that arises from the historical record is to what extent the overall rate of success in classical biological control can be improved. In this context, as noted by Mills (2000), it is important to distinguish between establishment, i.e. the colonization of a new environment by an introduced parasitoid and impact, i.e. the reduction of pest population abundance by the action of an established parasitoid. From studies of invasive species, it is apparent that there are no widely applicable characteristics of successful invaders (Mack et al. 2000, Sakai et al. 2001), with establishment being determined by the ability of a small founder population to survive and reproduce in a novel environment. Thus, the establishment phase of importation biological control seems less likely to be influenced by foraging decisions than by more general population processes, such as Allee effects, genetic bottlenecks, and demographic stochasticity, and by the favorability of the environment, as determined by species richness, disturbance, and environmental stochasticity. Nonetheless, Mills (2000) suggested that the success of parasitoid establishment could be enhanced through manipulation of the holding conditions to maximize the fitness of parasitoids destined for field release.

In contrast, pest suppression results from the impact of an exotic parasitoid that does become established in a favorable environment and this process typically occurs over a period of 6 to 13 generations (Bellows 2001) and extends over a broad geographic scale (Fig. 1.1). Taking this into account, it is not unreasonable to assume that larger-scale processes will dominate smaller-scale processes and there is some supporting evidence for this assumption from field studies of parasitoids (Thies et al. 2003, Cronin 2004). In addition, there is growing evidence that parasitoids can assess variation in host densities among patches from a distance using volatile infochemical signals (Geervliet et al. 1998, Vet 2001). Thus, patch choice decisions by parasitoids that determine the distribution of parasitoid foraging effort among host patches are more likely to influence the impact of classical biological control than foraging decisions made within host patches.

1.2.1 Behavioral context – optimal patch choice

Phytophagous hosts occur in discrete patches in the environment (Godfray 1994, Wajnberg 2006) and parasitoids seldom exist as isolated individuals within host patches and almost certainly interact with conspecific individuals (see also Chapter 9 by Haccou and van Alphen), if not with competing species or enemies, requiring them to make decisions with regard to patch choice. The optimal strategy for patch choice for a population of foragers is frequently represented by the ideal free distribution (IFD) in which individual foragers are distributed among patches such that each has an equal rate of gain from the patches that they occupy (Fretwell & Lucas 1970, Kacelnik et al. 1992, Tregenza 1995). This simple representation of patch choice includes the simplifying assumptions of instant movement among patches at no cost, equal competitive ability of foragers, and perfect knowledge of the variation in resources among patches. Although experimental evidence suggests that few foragers exactly match the simple model of an IFD (Tregenza 1995), it has nonetheless become one of the most widely applied theoretical concepts in behavioral ecology.

An IFD can be generated by both exploitative and interference competition between animals foraging among patches in which resources are depletable and thus, decline in suitability over time (Tregenza 1995, Sutherland 1996). Foragers will tend to favor patches with the highest resource densities for ease of resource acquisition, but at the same time will experience interference competition that will tend to reduce the rate of gain of resources. Thus, one particularly interesting interpretation of the IFD is that it represents the point at which acquisition and interference balance out to generate an equal rate of gain among patches (Sutherland 1983, 1996). In this way, the IFD can be defined by the interference coefficient m of Hassell and Varley (1969), such that when $m = 1$ (exact matching), foragers match the distribution of resources and the impact on the resource population is spatially density independent. However, when $m < 1$, foragers aggregate in patches of higher resource density, such that when $m = 0$, all individuals forage in the patch with the highest resource density and this generates a spatially density-dependent pattern of mortality among patches.

Walde and Murdoch (1988) assembled a set of 75 previous studies of spatial patterns of parasitism in the field. Although parasitism does not necessarily represent the distribution of adult parasitoids, being confounded by per capita rates of attack within patches, it provides a preliminary picture of the possible patterns of mortality that result from parasitoids foraging among host patches. From the 75 studies, 49% showed density independence, 23% showed positive density dependence, and 28% inverse density dependence. This evidence suggests that, while an IFD with $m = 1$ is consistent with approximately half of the studies, both higher and lower levels of interference would be necessary to account for the full range of spatial density dependence observed from patterns of parasitism among patches. Aside from this broad approach, only two studies have more specifically addressed IFD for parasitoids, a laboratory study with *Venturia canescens* (Tregenza et al. 1996) and a greenhouse study of the foraging behavior of *Lysiphlebus testaceipes* (Fauvergue et al. 2006), although several studies have monitored aggregation by parasitoids in the field (Waage 1983, Wang et al. 2004, Legaspi & Legaspi 2005). In each case, in common with more extensive tests using vertebrates, these studies show a greater level of foraging by parasitoids at lower density host patches, or under-matching, than expected.

A number of theoretical studies have examined the consequences of relaxing the basic assumptions of an IFD to include factors such as learning ability, travel costs, unequal competitive ability, and speed of patch quality assessment that might account for deviations from exact matching (Bernstein et al. 1988, 1991, Tregenza 1995). Each of these factors can influence the distribution of foragers among patches, indicating that simple individual behaviors can lead to complex distributions of competitors. More recently, Jackson et al. (2004a) suggested that under-matching of foragers to resources can readily be resolved by incorporating simple random movements into an individual-based model of the IFD. Of course, one of the other factors that could influence the distribution of foragers among patches is the risk of predation and observations of foraging under field conditions, highlighting the importance of predation for adult parasitoids (Rosenheim 1998). In this regard, it is interesting to note that Jackson et al. (2004b) developed a model in which foragers minimize the risk of predation per unit of resource gain. This model leads to perfect matching of foragers and resources when there is perfect knowledge at both trophic levels, but results in under-matching if the level of knowledge or movement of the resource population is greater than that of the forager population.

Although the spatial ecology of host–parasitoid interactions has received increasing attention in recent years (Hassell 2000, Murdoch et al. 2003, Cronin & Reeve 2005), reflecting a more general awareness of the importance of spatial processes in population and community ecology, the link between patch choice and population dynamics has yet to be explored in detail. For simplicity, many host–parasitoid models that incorporate spatial heterogeneity are based on just two host patches, but spatial structure and the population consequences of patch choice decisions by parasitoids can be more explicitly developed through lattice models (Rohani & Miramontes 1995, Kean & Barlow 2000, Childs et al. 2004). Despite the proliferation of spatial host–parasitoid models, the prime focus of these studies has been on mechanisms for the persistence of metapopulations that are locally unstable (Bernstein et al. 1999, Briggs & Hoopes 2004). In the context of classical biological control, although metapopulation persistence is one of the two characteristics of success, it is the degree of suppression of host abundance that is of greater importance.

The only metapopulation model to have addressed host suppression is that of Rohani and Miramontes (1995), in which parasitoids respond to the distribution of host densities

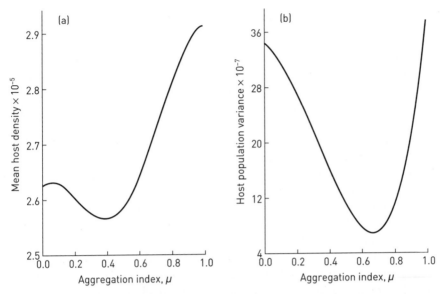

Fig. 1.2 The influence of parasitoid aggregation on (a) the global mean host density and (b) the variance in patch densities for a lattice metapopulation model of a Nicholson–Bailey parasitoid (adapted from Rohani & Miramontes 1995). An aggregation index of 1 represents exact matching, while an index of less than 1 reflects under matching or the inefficiency of the parasitoid in responding to the distribution of hosts among patches.

among neighborhood patches as defined by an aggregation parameter μ (Hassell & May 1973), such that $\mu = 1$ represents exact matching and $\mu < 1$ represents under-matching. It is important to note that there are two aspects of host suppression in a spatially structured environment, one being the mean host density among patches and the other being the variance in host densities among patches. The Rohani and Miramontes (1995) model indicates that the lowest mean host metapopulation densities are achieved at a relatively low aggregation index ($\mu = 0.4$) representing a high level of under-matching by the parasitoid (Fig. 1.2a). This suggests that parasitoids that are less than perfect in their distribution of foraging effort in relation to host densities among patches could still play an important role in biological control. On the other hand, it may be more valuable in the context of classical biological control to ensure that no host patches experience damaging host densities, in which case variance in host densities may be more important than the overall mean. The Rohani and Miramontes (1995) model indicates that the lowest variance in host metapopulation density can only be achieved at a higher aggregation index ($\mu = 0.7$), suggesting that more optimal parasitoid behavior (Fig. 1.2b) may be needed to prevent damaging host densities in all patches, albeit at the expense of a greater mean host metapopulation density. To what extent these results would change through incorporation of host density dependence within patches, a saturating functional response, or density-dependent dispersal into the model remains to be explored, but this study does provide an initial indication that parasitoid inefficiencies in responding to the patchiness of hosts may not be incompatible with biological control and the success of host suppression.

1.2.2 Optimal foraging and importation biological control

The aim of importation biological control is the long-term suppression of pest populations through the introduction of exotic specialist parasitoids from the region of origin of
an invasive species. As noted above, the two phases of an introduction are establishment
and impact and little attention has been paid to improving the success of establishment.
In this respect, it is interesting to note, from the biological control record, that 63% of
the phytophagous insects introduced for the control of weeds become established (Syrett
et al. 2000) in comparison to 36% of the insect parasitoids introduced against insect pests
(Mills 2000). This suggests that there could be opportunities for improving establishment
rates of parasitoids and one that has been explored to a limited extent is the influence
of holding conditions on the subsequent reproductive capacity and behavioral characteristics of parasitoids that are being processed for field release (Hougardy et al. 2005,
Hougardy & Mills 2006, 2007).

In any introduction program it is necessary to hold adult parasitoids for a period of
time in rearing cages to accumulate sufficient emergence to justify effective field releases.
During this holding period, which can typically last several days and can sometimes represent up to 25% of the adult life span, parasitoids are mated and given sugar-rich food
and will experience either host deprivation if hosts are withheld or egg depletion if hosts
are provided. In the absence of hosts, parasitoids might be expected to accumulate mature
eggs, which could increase their motivation for foraging once released, but could also
experience egg resorption and become temporarily unable to oviposit. The absence of hosts
might also reduce egg maturation rates and prevent the acquisition of host-associated cues
for host finding. In contrast, in the presence of hosts, although parasitoids would learn
to find hosts, the expenditure of eggs would necessarily lead to a reduction in future
reproduction and may also reduce the motivation for host finding.

In conjunction with parasitoid introductions for classical biological control of the
codling moth in California (Mills 2005), we considered the effects of both host deprivation
and egg expenditure on the reproductive capacity and behavior of a cocoon (prepupal)
parasitoid, *Mastrus ridibundus* (Hymenoptera, Ichneumonidae). When deprived of hosts,
M. ridibundus maintained a maximal egg load for up to 7 days and showed a peak of
oviposition on the first day that hosts became available, although daily rates of host attack
fell to a lower level subsequently, with a relatively low lifetime fecundity that was independent of the duration of deprivation (Hougardy et al. 2005).

In contrast, although egg expenditure led to declining egg loads, daily attack rates, and
lifetime fecundity with increasing duration of holding, only egg load was lower than the
comparable values for host-deprived parasitoids. When parasitoids that had experienced
1–9 days of host deprivation or egg expenditure were released into a field cage to estimate
the success of patch and host finding, those that had experienced host deprivation showed
no reduction in foraging success even after 9 days in the absence of hosts, whereas those
that had experienced egg expenditure showed a progressive decline in both patch and host
finding (Hougardy & Mills 2007). In addition, using mark-release-recapture experiments
with immunological markers in the field, *M. ridibundus* females showed a dispersal rate
of 81.5 m²/h after experiencing four or more days of host deprivation, as compared to a
rate of 2.1 m²/h for those that experienced either a lower level of host deprivation or all
levels of egg expenditure (Hougardy & Mills 2006). Thus, pre-release conditions can have

a marked influence on post-release performance, both in terms of reproductive potential and foraging behavior and deserves closer attention in the future. For *M. ridibundus*, if large numbers of parasitoids are available and the aim is to establish the parasitoid over a broad area as quickly as possible, then depriving parasitoids of hosts for 4 days before release would enhance their dispersal through the release region. A more likely aim, however, would be to establish the parasitoid in a more localized area and, in this case, parasitoids should be exposed to hosts, but should not be held for more than 2 days prior to release to avoid any reduction in foraging ability.

While there are no options available to manipulate the host patch choice decisions made by parasitoids introduced as classical biological controls, it is nonetheless valuable to know whether the impact of established parasitoids is influenced by their ability to match their foraging effort to the heterogeneity in host densities among patches. We know of no studies that have directly addressed host patch choice by parasitoid species used in classical biological control, but there is some indirect evidence from aggregative distributions of parasitism. For example, Hassell (1980) showed that there is a positive relationship between parasitism by the tachinid *Cyzenis albicans* and winter moth density among trees. Although the success of the biological control programs against the winter moth in Nova Scotia and British Columbia seems likely to involve the indirect influence of predation of winter moth pupae in the soil (Roland 1988), the role of *C. albicans* continues to be disputed (Bonsall & Hassell 1995, Roland 1995). In contrast to the winter moth example, however, there is no evidence of a relationship between parasitism and host density among trees for parasitoids that have proved to be successful in the biological control of diaspid scales: *Aphytis melinus* for California red scale (Reeve & Murdoch 1985, Smith & Maelzer 1986), *A. paramaculicornis* and *C. utilis* for olive scale (Murdoch et al. 1984), and *Aphytis yanonensis* and *Coccobius fulvus* for arrowhead scale (Matsumoto et al. 2004). The lack of response of parasitism to host density could result either from under-matching in the spatial distribution of adult parasitoids or from a reduction in the per capita performance of parasitoids that do effectively orient toward the higher host density patches. However, in the absence of any direct evidence of the distribution of foraging adults for parasitoids that have become established in biological control programs, it remains unclear to what extent host patch choice decisions are likely to support or constrain the impact of introduced parasitoids and if this is an aspect of biological control that deserves closer attention in the future.

1.3 Augmentative biological control

Besides introductions of natural enemies from the region of origin of invasive pests, biological control also includes the periodic release of individuals for immediate or season-long suppression of pests. Augmentative releases of mass-reared parasitoids have resulted in the development of small-scale commercial insectaries in many regions of the world over the last 30 years and it is estimated that more than 125 natural enemy species are commercially available and used on about 16 million ha globally each year (van Lenteren 2000). Inoculative releases are most frequently used early in the season to create a reproducing population of natural enemies in the crop or target environment, with the founder population initiating a series of generations that persist throughout the growing season. The best example of inoculative augmentation for parasitoids is the use of *E. formosa* for control of greenhouse whitefly in Europe (Hoddle et al. 1998). For inoculative releases of *E. formosa* to be successful, the crop must be able to tolerate a sufficient whitefly population to allow the parasitoid to persist through reproduction. As *E. formosa* is a host feeder, too low a host density will lead to hosts being used more frequently for host feeding than for reproduction, which often leads to extinction. Thus, inoculative releases have worked most effectively in vegetable crops, such as cucumber and tomato, which are able to tolerate some honeydew production, whereas in floral crops, where control requirements are more stringent, inundative releases are necessary, which do no allow for sustained reproduction by *E. formosa*. van Lenteren (2000) estimated that biological control is used on 14,000 of a total 300,000 ha of protected crops globally, with *E. formosa* being the most frequently used natural enemy representing 33% of the monetary sales of natural enemies used in glasshouses.

Inoculative augmentation of parasitoids in biological control is functionally similar to the introduction of parasitoids in classical biological control. The difference is one of temporal and spatial scale (Fig. 1.1) but, in both cases, success is dependent upon a reproducing parasitoid population suppressing the density of a pest population. Thus, in common with classical biological control, spatial processes and patch choice is expected to be among the most important aspects of the foraging behavior of *E. formosa* with regard to seasonal control. In this respect, it is of interest to note that suppression of greenhouse whitefly is less stable in small greenhouses (van Lenteren et al. 1996). This suggests that, in the absence of sufficient spatial scale, a small glasshouse acts more like a local patch in which Nicholson–Bailey dynamics dominate (Nicholson & Bailey 1935) and extensive host feeding and superparasitism by *E. formosa* can accelerate the likelihood of extinction.

Inundative releases of insect parasitoids are used for immediate impact on the pest population, often with no expectation of successful reproduction and carry over to subsequent generations. In this way, the use of parasitoids in inundative augmentation can be likened to the use of a biological insecticide. Thus, inundation is based on maximizing the immediate killing power of the released parasitoids rather than on the dynamics of interacting host–parasitoid populations over a series of generations. The most frequently used parasitoids in inundative release programmes include *Trichogramma* species for control of lepidopteran pests in cereals, cotton, and field vegetables worldwide (Wajnberg & Hassan 1994), *Cotesia flavipes* for sugarcane borer control in South America, *A. melinus* for control of California red scale in citrus in the USA, and *Muscdifurax* and *Spalangia* species for control of filth flies in North America and Europe (van Lenteren 2000). A particularly interesting success story is the development of inundative releases of *A. melinus* in citrus. Mass production of *A. melinus* was initiated by the Fillmore Insectary, USA in 1960 as part of the biological control focus of the 9000 acres of citrus grown by Fillmore Citrus Protective District, a grower cooperative. In 1986, releases of *A. melinus* from the Fillmore Insectary were estimated at 190 million parasitoids (Carpenter 2005). Not only have inundative releases worked well in coastal citrus in southern California, they have also proved effective and commercially viable in the San Joaquin Valley where higher temperatures result in the production of smaller California red scale that are less preferred by *A. melinus* (Moreno & Luck 1992, Luck et al. 1996).

1.3.1 Behavioral context – optimal patch and host use

Patch use decisions and optimal foraging theory have been studied extensively, if somewhat sporadically, since the first appearance of the marginal value theorem (Charnov 1976, Houston & McNamara 1999, Green 2006). The latter predicts how long an individual forager should stay in a patch in order to maximize its long-term rate of gain or fitness and that the patch becomes successively depleted with time (see also Chapter 8 by van Alphen and Bernstein). The optimal time at which to leave the patch is when the current rate of gain falls to the overall rate for the environment.

Patch use decisions have also been studied experimentally for a variety of parasitoid species with an emphasis on the range of factors that can influence patch time allocation and the informational cues that are used by parasitoids in developing patch-leaving rules (van Alphen et al. 2003, Burger et al. 2006, Wajnberg 2006). While patch residence time for parasitoids is primarily determined by the rate of successful oviposition events (or host encounters) within patches, it is now well known that parasitoids can adapt their strategy of patch use in response to experience and information gained while foraging. Patch residence time in parasitoids has been shown to be influenced by genetic variability (Wajnberg et al. 1999, 2004), seasonality (Roitberg et al. 1992), physiological status (Outreman et al. 2005), adult food (Stapel et al. 1997), experience (Keasar et al. 2001, van Baaren et al. 2005), by the presence of competitors (Bernstein & Driessen 1996, Wajnberg et al. 2004, Goubault et al. 2005) or enemies (E. Hougardy & N. J. Mills, unpublished), and by chemical cues associated with host plant damage (Wang & Keller 2004, Tentelier et al. 2005), hosts (Waage 1979, Shaltiel & Ayal 1998), or enemies (Petersen et al. 2000). The mechanism used by parasitoids for optimal patch use appears to be incremental when hosts are aggregated, such that each oviposition increases the probability of staying, but decremental when hosts are regularly distributed, such that each oviposition decreases the probability of staying. In addition, recent evidence suggests that parasitoids may also have the flexibility to switch between the two mechanisms as circumstances change (Driessen & Bernstein 1999, Outreman et al. 2005, Burger et al. 2006). From a biological control perspective, perhaps the most important influences on patch residence time are chemical cues (kairomones or synomones), which are believed to provide an initial evaluation of patch quality (Shaltiel & Ayal 1998, Tentelier et al. 2005) and competition, which can lead either to increased or to decreased patch residence time (Wajnberg et al. 2004, Goubault et al. 2005).

In addition to patch use decisions, parasitoids must also make choices between host individuals and make decisions about host acceptance, sex allocation, and clutch size. As solitary parasitoids seldom exploit host patches alone, both direct competition with other foraging females and indirect competition through encounters with chemical markers or previously parasitized hosts can influence host acceptance and the tendency to super-parasitize (Visser et al. 1992, Plantegenest et al. 2004). Once a host has been accepted, the optimal strategy of host use, assuming that host quality affects female fitness more than male fitness, is to allocate daughters to higher quality hosts and males to lower quality hosts (Charnov et al. 1981, see also Chapter 12 by Ode and Hardy). For parasitoids, host quality is often equated with size and there is good evidence from solitary species that the primary sex ratio is generally correlated with host size (Godfray 1994). Of course, size is not the only component of host quality, as host plant, host species, host age, and previous parasitism can also influence the primary sex ratio (King 1987, Campan & Benrey 2004, Shuker & West 2004, Ueno 2005, see also Chapter 12 by Ode and Hardy).

For gregarious species, the optimal clutch size has frequently been considered to be the number of eggs that maximizes the parent females' fitness gain from the whole clutch, often referred to as the Lack clutch size (Lack 1947, Godfray 1994). However, the majority of experimental laboratory studies have observed clutch sizes that are smaller than the Lack clutch size (Godfray 1994, Zaviezo & Mills 2000). This suggests that the lifetime reproductive success of gregarious parasitoids is not always determined by the size of a single clutch, but can be modified by environmental conditions that influence the likelihood of future reproduction. Thus, the optimal strategy, when the expectation of future reproduction is high, is more toward maximization of the fitness gain per egg (a reduction from the Lack clutch size), a situation that may well apply to parasitoids under laboratory conditions. However, when the expectation of future reproduction is low, the optimal strategy is to maximize the fitness gain per clutch (Lack clutch size), a situation that may be more applicable to insect parasitoids under field conditions. Environmental factors that can influence the expectation of future reproduction include host encounter rates, parasitoid survivorship, and competition (Iwasa et al. 1984, Visser & Rosenheim 1998).

Finally, sex allocation within parasitoid clutches is influenced by local mate competition (LMC) which generates female-biased offspring sex ratios, regardless of whether mothers use a patch of hosts simultaneously or sequentially (Hamilton 1967, Werren 1980, see also Chapter 12 by Ode and Hardy). Hamilton's (1967) theory of LMC predicts that, when mating takes place between the offspring generated by one of a few mothers, sex ratios should be female biased to limit the competition between brothers for mates and that, as the number of mothers increases, the female bias declines. There is considerable experimental evidence that LMC can account for the variation in sex ratios of parasitoids (Godfray 1994). However, in an interesting recent extension of LMC, Shuker et al. (2005, 2006) pointed out the importance of asynchronous emergence of offspring from sequential females visiting the same patch of hosts. Under such circumstances, optimal parasitoid sex ratios can vary for different hosts in the same patch due to differential levels of competition between males that emerge asynchronously within the patch. Thus, optimal sex allocation presents a complex problem for gregarious parasitoids involving host quality, clutch size, and asymmetrical LMC, suggesting that an absence of perfect information may at times constrain their ability to respond accurately (Shuker & West 2004).

1.3.2 Optimal foraging and inundative biological control

The aim of inundative biological control is to release as many insectary-produced parasitoids as needed to generate sufficient mortality to suppress pest densities on a localized scale and prevent crop damage. At such a localized scale, patch use and host use decisions by the parasitoids are of much greater importance than patch choice, both in the context of mass production and impact following release. In the context of mass production, the goal is to produce vast numbers of selected parasitoids in insectaries without compromising their ability to function as intended after release (van Lenteren 2003). The emphasis here is primarily on production, with quality control serving not so much to optimize the fitness of the individuals produced as to maintain an acceptable level of field performance.

One particularly interesting practical application of the host quality model of sex allocation in parasitoids (Charnov et al. 1981) to mass production concerns the manipulation of host size to reduce male-biased sex ratios in *Diglyphus isaea* (Ode & Heinz 2002, Chow & Heinz 2005, 2006). *Diglyphus isaea* is commercially produced for inundative releases against *Liriomyza* leafminers in glasshouses, but the cost of production often prohibits greater adoption of this approach in comparison to insecticides. As is typical for a solitary idiobiont parasitoid, *D. isaea* produces more daughters on larger host larvae and bases its assessment of host size on recent experience of the distribution of host sizes in a patch (Ode & Heinz 2002). *Diglyphus isaea* was found to produce about 60% male offspring when presented with hosts of an intermediate size over 3 days, whereas a sequence of increasing host sizes each day over the same time period reduced male production to 40% and a sequence of decreasing host sizes increased male production to 74% (Ode & Heinz 2002). Taking this to a more practical level for parasitoid production, Chow and Heinz (2005) showed that *D. isaea* produces 60% male offspring when presented simultaneously with small host and large hosts on separate plants in rearing cages but, when both host sizes were present on the same plants, male bias could be reduced to 48%. Over an 8-week period of simulated mass rearing (Fig. 1.3a), the combination of host sizes on plants produced an equal number of wasps, but with a significantly lower male bias (10% reduction with

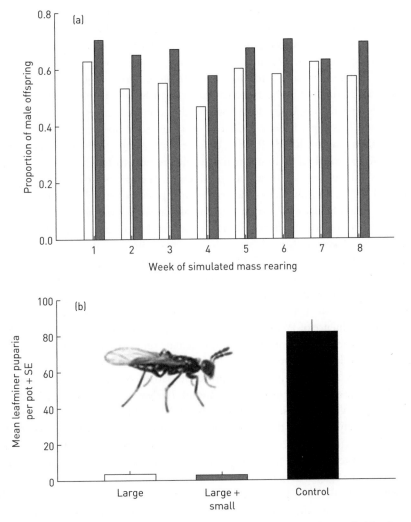

Fig. 1.3 Exposing *D. isaea* to a combination of large and small hosts (white bars) rather than to high-quality large hosts alone (gray bars) can (a) reduce the proportion of male offspring in mass production and (b) produce female wasps that are equally effective in suppressing leafminer densities in relation to untreated controls (black bar) (adapted from Chow & Heinz 2005, 2006).

no reduction in female size) than the standard insectary production procedure of providing large hosts alone (Chow & Heinz 2005). Further, in greenhouse trails, parasitoids produced from the novel host size combination approach to production of *D. isaea* were as effective in reducing survivorship of the leafminer *Liriomyza langei* (Fig. 1.3b) and damage to chrysanthemums as parasitoids produced from large hosts (Chow & Heinz 2006). This example provides a clear indication that the cost-effectiveness of mass production of an idiobiont parasitoid can be enhanced through manipulation of the foraging behavior of the parasitoid. A similar potential has been identified for the production of *Catolaccus grandis* (Heinz 1998), a parasitoid of cotton boll weevil and the approach may be more broadly applicable to idiobiont parasitoids that share the same host-size-based sex allocation behavior.

Mass production of gregarious parasitoids necessarily involves the use of rearing cages in which multiple females simultaneously parasitize hosts in close proximity. Under these conditions, it would be expected that LMC might lead to a reduction in the production of female offspring. However, in a study of mass rearing protocols for two soft scale parasitoids *Metaphycus flavus* and *Metaphycus stanleyi*, Bernal et al. (1999) found that the sex ratio was dominated by host quality rather than by interactions with other females. Contrary to expectations, larger-scale hosts produced more females and larger broods than smaller hosts. The larger broods produced on larger-scale hosts not only produced constant sex ratios, but produced offspring of larger size. Although these two parasitoids did not respond to crowding in an optimal way with respect to LMC, their lack of response clearly does not compromise their mass production. The observation that host quality may dominate LMC as an influence on sex ratios in captive parasitoid rearings may be more general, as a similar lack of response to female crowding was found for *Parallorhogas pyralophagus*, a gregarious ectoparasitoid of the stemborer *Eoreuma loftini* (Bernal et al. 2001) and for *Anagyrus kamali*, a solitary endoparasitoid of the colony forming pink hibiscus mealybug (Sagarra et al. 2000). Such a scenario would also be consistent with the previous observation that sex ratios in *D. isaea* can readily be improved through manipulation of host quality despite the multiple foundresses of a mass-rearing environment.

Of interest to note here, in the context of maximizing female production in mass rearing, is the influence of *Wolbachia* infection in *Trichogramma* species. Stouthamer (2003) suggested that the selection of unisexual (i.e. female only) strains of *Trichogramma* could benefit mass production as no hosts would be wasted on the production of males and thus, production costs could be reduced. In a direct comparison of unisexual and sexual (through antibiotic treatment) forms of the same line of *Trichogramma deion* and *Trichogramma cordubensis* in a glasshouse setting, it was found that both forms found host egg patches equally effectively but that the sexual form parasitized more hosts per patch than the unisexual form (Silva et al. 2000). The latter effect is probably due to the lower offspring production of the unisexual form, suggesting that the use of unisexual parasitoids would be most effective against solitary hosts.

In considering the field performance of insectary-produced parasitoids, arguably the most important constraint is that the constancy and simplicity of an insectary environment inevitably selects for a limited set of genotypes that proliferate under rearing conditions, but that are not so well adapted to function effectively under field conditions (Nunney 2003, Wajnberg 2004). While this has led to the development of some valuable recommendations for the maintenance of genetic diversity in the captive rearing of parasitoids (Roush & Hopper 1995, Nunney 2003, Wajnberg 2004), there is less information on the extent to which captive rearing influences the foraging behavior of parasitoids. Captive rearing did not prevent *T. brassicae* from showing optimal behavior in the exploitation of localized patches of hosts (Wajnberg et al. 2000), but patch-leaving rules that result in parasitoids abandoning a patch before all potential hosts have been attacked are not optimal for inundative biological control. The ideal outcome for biological control would be to maximize host attack in every patch irrespective of patch host density and the marginal gain with respect to other patches. In this way, optimal foraging runs counter to the goal of inundative biological control. Although patch residence times might be increased through use of natural flightless mutants, such as known for the coccinellid *Harmonia axyridis* (Tourniaire et al. 2000), a more widely applicable approach is through use of behavior-modifying infochemicals.

It is well known that parasitoids are responsive to infochemical cues and that learning of such cues plays an important role in parasitoid foraging (Vet et al. 2003, see also Chapter 5 by Hilker and McNeil). This has led to consideration of applications such as the priming of insectary-reared parasitoids with infochemicals prior to field release (Hare & Morgan 1997) and the spraying of crops with compounds that will either stimulate parasitoid search or retain parasitoids in patches where extended periods of search are desired.

The potential for priming is well illustrated by the oviposition behavior of *A. melinus*, a parasitoid of California red scale, for which host recognition is mediated by the presence and quantity of the contact chemical O-caffeoyltyrosine present in scale covers (Hare & Morgan 1997). The contact chemical is highest in concentration in the covers of third instar scale, the preferred host stage and the threshold concentration that stimulates ovipositor probing can be reduced either through experience with California red scale covers or with the chemical itself. For mass production, *A. melinus* is reared on an alternative host, the oleander scale on squash and, as this scale lacks the host recognition chemical, mass-reared parasitoids do not have any experience of this contact cue when field released. Hare and Morgan (1997) showed that it is feasible to prime mass-reared parasitoids and that primed parasitoids do show an increased level of probing of California red scale. Subsequently, Hare et al. (1997) showed that this can lead to a 6–11% enhancement of parasitism rates in sleeve cages in the field. Although the recognition chemical can be produced synthetically, commercial application of *Aphytis* priming awaits the development of a mechanical procedure for priming thousands of wasps, the concentration of chemical necessary to ensure effective priming, and verification that such a system would be effective for parasitoid releases in commercial orchards.

In addition to initial priming of wasps prior to field release, consideration has also been given to spraying crops directly with host recognition chemicals (Prokopy & Lewis 1993), particularly for *Trichogramma* releases (Lewis et al. 1979). However, this approach has met with more variable success and it remains unclear whether a uniform coating of plant surfaces with kairomones would stimulate or disrupt parasitoid foraging behavior. In a laboratory study of the aphid parasitoid *Aphelinus asychis*, Li et al. (1997) showed that the presence of aphid honeydew on leaves could at least double patch residence times, but that this increase applied only to parasitoids with no or limited (1 day) experience with hosts and was not apparent for parasitoids that were more fully experienced (3–4 days with hosts). In many cases, parasitoid mass-rearing protocols do produce naïve female wasps and thus, uniform coatings of inexpensive contact kairomones could lead to foraging patterns that are not optimal for the individuals released, but more effective in terms of suppression of pest densities and crop damage.

1.4 Conservation biological control

Conservation biological control focuses on the enhancement of both introduced and indigenous parasitoid populations through the enhancement of limiting resources or the removal of incompatible pesticides. The potential impact of synthetic pesticides on parasitoids is well documented and has given rise to the well-known phenomena of pest resurgence (Hardin et al. 1995) and secondary pest outbreaks. Although removal of excessive pesticide use or the adoption of more selective pesticide products can lead to effective conservation of parasitoid populations, this aspect of conservation biological control concerns the survivorship rather than foraging behavior of parasitoids in crop production systems and will not be discussed further. In contrast, the provisioning of limiting resources as an approach to the conservation of parasitoids in cropping systems is the least well understood and implemented component of biological control with little documentation of the elements of success (Ehler 1998, Landis et al. 2000). As pointed out by Gurr et al. (2000), many studies of conservation biological control have focused on habitat manipulation, such as crop diversification. Under such circumstances, it becomes difficult to separate the relative importance of the bottom-up influence of resource concentration from the top-down influence of enemies when pest populations change in abundance. While natural enemy abundance often increases in response to crop diversification, there is limited verification that increased enemy abundance leads to greater pest population suppression.

More recently, there has been renewed interest in the provisioning of nectar subsidies as a more specific limiting resource for parasitoids in cropping systems (Heimpel & Jervis 2005, Wäckers et al. 2005). Although extra-floral nectar and honeydew can also be important sugar sources for parasitoids, floral nectar is more readily manipulated in farmer fields in the context of the implementation of conservation biological control. Not only does this approach more specifically target the foraging behavior of parasitoids, it also provides a more focused direction for field-based studies in conservation biological control. This then raises the question of optimal patch choice between hosts and adult food, the distance or ease of access of nectar sources from host patches and the extent to which foraging for food could reduce the time available to search for hosts (see also Chapter 7 by Bernstein and Jervis).

1.4.1 Behavioral context – optimal use of nectar subsidies

As a variant on patch choice, foraging parasitoids also face decisions of whether to stay in a patch of hosts or to select an alternative patch containing plant-provided food (see also Chapter 7 by Bernstein and Jervis). The most important form of plant-provided food for parasitoids is nectar. Not only have parasitoids frequently been observed feeding from flowers (Jervis et al. 1993), but in many cases the longevity and realized fecundity of parasitoids are known to be greatly enhanced in the presence of floral nectar (Wäckers 2004, 2005). Although parasitoids vary in the frequency with which they require carbohydrate sources to sustain survivorship and flight, foraging females must make important decisions of whether to search for nectar subsidies to support future reproduction or for hosts to maximize current reproduction (see also Chapter 6 by Strand and Casas and Chapter 7 by Bernstein and Jervis). Nectar subsidies come with both direct and indirect costs. Direct costs are associated with the potential of increased mortality while foraging on flowers (Rosenheim 1998) and indirect opportunity costs are associated with the time lost while feeding rather than ovipositing (Sirot & Bernstein 1996). As an initial step, using a stochastic dynamic programming model, Sirot and Bernstein (1996) determined that the optimal solution for the distribution of parasitoids between patches of hosts and food is influenced by both the availability of food sources and the dependence of survivorship on energy reserves (see also Chapter 7 by Bernstein and Jervis). More recently, Tenhumberg et al. (2006) extended this approach to relax some of the assumptions and to include an energy cost for host searching. They found that, in contrast to the Sirot and Bernstein (1996) model, parasitoids should always search for food rather than hosts when energy reserves drop to a low level, even if food availability and rewards are low. However, Bernstein and Jervis (Chapter 7) show that the reason for the contradiction between the two models is more a matter of the parameter values chosen than the assumptions of the model *per se*.

While there have been numerous laboratory studies on the impact of floral nectar on the performance of individual parasitoids (Wäckers 2005) and an increasing number of field studies on the influence of nectar subsidies on parasitism (Gurr et al. 2005, Heimpel & Jervis 2005), the consequences of floral nectar for the dynamics of host–parasitoid inter- actions at a population level are poorly understood. Křivan and Sirot (1997) confirmed the suggestion of Sirot and Bernstein (1996) that the inclusion of floral subsidies can stabilize a host–parasitoid model, but provided no indication of the consequences for host population suppression. Kean et al. (2003) addressed this problem by asking specifi- cally how an increase in parasitoid longevity or fecundity, through provisioning of nectar subsidies, would affect the equilibrium density of a host population. By including para- sitoid longevity and fecundity (maximum number of attacks) into a simple extension of a Lotka–Volterra host–parasitoid model, they were able to show that increased fecundity is of less importance than increased longevity and that the effect of increased parasitoid longevity in suppressing a host population depends upon whether a parasitoid is primarily egg or time limited and whether time spent on nectar subsidies is likely to result in a reduc- tion in the search rate for hosts (Fig. 1.4). In other words, parasitoids that are more pro-ovigenic (with a high ovigeny index *sensu* Jervis et al. 2001) are less likely to benefit from increased longevity, whereas those that are more synovigenic (low ovigeny index) could provide enhanced pest suppression in the presence of a nectar subsidy as time spent searching for food should not limit the daily number of hosts attacked for a time-limited parasitoid (see also Chapter 7 by Bernstein & Jervis for further details).

1.4.2 Optimal foraging and conservation biological control

Infochemicals originating from a damaged host plant (synomones) or from the host itself (kairomones) are well-known signals that aid parasitoids in the location of suitable hosts. Although less well known, there is also increasing evidence that parasitoids also respond to floral odors in their search for sugars to support maintenance and flight (Wäckers 1994, Jacob & Evans 2001). The responsiveness of parasitoids to host-related versus food-related cues then depends upon the level of hunger, with starved females responding preferentially to food odors and well-fed females responding preferentially to host-associated odors (Jervis et al. 1996, Lewis et al. 1998, Desouhant et al. 2005). This responsiveness can lead to para- sitoids maintaining a fairly constant level of energy under field conditions in the presence of an abundant adult food supply, as shown for *Venturia canescens* (Casas et al. 2003).

The concept of using floral nectar subsidies to enhance the abundance or activity of parasitoids is based on three important observations: (i) crop monocultures are often devoid of sugars; (ii) parasitoid longevity is often greatly enhanced when fed on sugars;

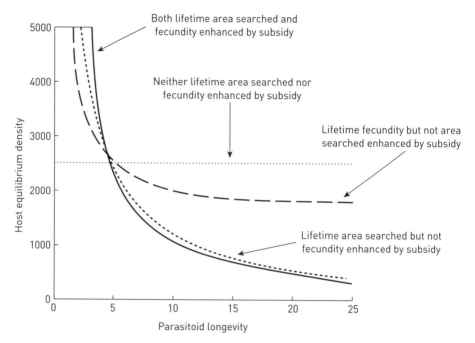

Fig. 1.4 The potential influence of nectar subsidies on the ability of a parasitoid to reduce the equilibrium density of a host population based on an extended Lotka–Volterra model in relation to parasitoid longevity (adapted from Kean et al. 2003). Four possible scenarios are presented, depending upon whether the subsidy has an influence on lifetime fecundity or area searched by the parasitoid (see text for details).

and (iii) parasitoids often use floral nectar under natural conditions. In reviewing the experimental evidence for improved parasitoid performance in the presence of floral nectar, Heimpel and Jervis (2005) noted that there was evidence of increased parasitism in 7 of 20 field studies, but that only one of these 7 showed a simultaneous reduction in pest density, while 2 did not and 4 did not monitor host density. Since this review, several other studies have shown enhanced rates of parasitism under field conditions in the presence of floral nectar (Tylianakis et al. 2004, Lavandero et al. 2005, Berndt et al. 2006, Winkler et al. 2006), but there have been no further reports of a reduction in pest densities. An increase in parasitism in the presence of floral nectar can result from a combination of two effects: an increase in parasitoid density due to greater attraction or retention of parasitoid females and an increase in the per capita performance of the parasitoids. It is not clear which of these factors may have been more important, but it is interesting to note that increased rates of parasitism were reported for both host-feeding and non-host-feeding parasitoids, suggesting that host feeding alone may not compensate for a parasitoid's need for sugars to fuel flight and support longevity (but see Giron et al. 2004). It should also be noted here that sucrose sprays have been shown to be sufficient to increase the abundance of the alfalfa weevil parasitoid *Bathyplectes curculionis* and weevil parasitism during the first crop of alfalfa in fields where aphids are not abundant (Jacob & Evans 1998). Nonetheless, despite increasing evidence for the importance of flower feeding for parasitoids (Wäckers

2005), it appears that there is far less evidence that the presence of floral nectar will translate to improved biological control.

Observations of increased parasitism in the presence of floral subsidies has been sufficient, however, to generate considerable interest in the possibility to enhance parasitoid populations and their performance in agricultural crops that tend to lack natural sources of suitable sugars (Landis et al. 2000, Gurr et al. 2005). In the context of conservation biological control, two important questions arise: which floral subsidies to use and how close they need to be to the crop? To answer the first question it is important to consider five different features of the flowers of a particular plant species: availability in space and time, apparency in terms of olfactory and visual cues, accessibility in relation to parasitoid mouthpart morphology, chemical composition with respect to sugars, stimulants and deterrents, and specificity in enhancing parasitoids rather than pests or higher order predators (Gurr et al. 2005, Wäckers 2005). Although annual buckwheat (*Fagopyrum esculentum*) has become something of a model plant for floral subsidy studies, Wäckers (2004) has shown that the flowers of different plant species can differ considerably in both olfactory attractiveness and accessibility for three different species of parasitoid. Some nectar constituents can also either act as deterrents or be toxic to parasitoids (Wäckers 2001, 2005). Thus, the selection of flowering plant species as insectary mixes for use in conservation biological control, not only needs to take these features into consideration, but also needs to be tailored for variation among parasitoid species.

The question of how close a floral subsidy needs to be to a crop, to be readily found and used by adult parasitoids, remains largely unknown. In an interesting study of parasitism of grain aphids by *A. rhopalosiphi* in the presence of annual buckwheat, Tylianakis et al. (2004) showed that parasitism declined exponentially from 36% immediately adjacent to a floral patch to zero beyond a distance of 14 m (Fig. 1.5), suggesting that the foraging distance of this parasitoid may be relatively small. While foraging distance is likely to increase

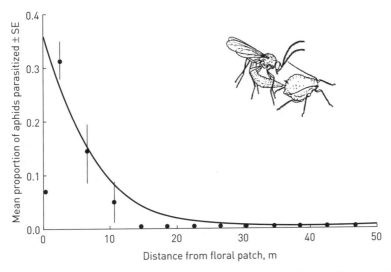

Fig. 1.5 Parasitism of grain aphids by *A. rhopalosiphi* in relation to distance into the crop from a patch of buckwheat flowers that acted as a source of floral nectar (adapted from Tylianakis et al. 2004).

with parasitoid size, there have been few studies of parasitoid foraging distance and the majority of these are concerned with movement in relation to hosts rather than food sources (Desouhant et al. 2003). However, Lavandero et al. (2005) found that *Diadegma semiclausum* could be trapped at distances of 80 m from a source of floral nectar marked with rubidium. Nonetheless, there remains insufficient data from which to base any assessment of the necessary proximity of floral subsidies. Similarly, the question of how much floral nectar is needed to support a suitable population of parasitoids in an agricultural crop has yet to be addressed. However, these examples suggest that proximity might be more important than quantity and that, to reach to the middle of agricultural fields, floral subsidies may need to be integrated into a crop in the form of headland plantings or strips rather than being confined to perimeter plantings.

1.5 Conclusion

Behavioral ecology and optimal foraging theory provide a valuable basis for developing improvements in the application of biological control. As the four main approaches to biological control differ in both temporal and spatial scale, however, it is important to consider the corresponding scale of foraging decisions that are most applicable to each approach, with larger-scale processes likely dominating smaller-scale processes. A behavioral approach also appears promising in more formally linking measurements at the individual level to processes at the population level and helping to shift the implementation of biological control away from its traditional roots of trail-and-error toward a more exact science in which success can be more readily predicted.

Host patch choice is the appropriate scale of behavior for biological control importations and inoculations and, while such behavior offers little opportunity for manipulation to improve success, it could help to clarify the differences between success and failure. Additional theoretical studies that focus on host suppression rather than stability would provide a better basis for understanding to what extent inefficiencies in the distribution of foraging effort could enhance or compromise the success of biological control. In this respect, initiation of field observations on the spatial distribution of foraging effort for established parasitoids would fill a current vacuum and inform the continued debate that aggregation to patches of higher host density is a beneficial aspect of biological control. Patch and host use decisions are the appropriate scales of behavior for inundative biological control, with opportunities to improve both mass production of parasitoids and their subsequent impact following field release. Manipulating sex ratios through host quality and use of uniparental (i.e. female only) strains can reduce the production costs for mass-reared parasitoids. Similarly, patch residence times and consequent parasitism can be increased for inundative releases using infochemicals as crop sprays or for priming parasitoids with host recognition cues prior to release. Although not yet adopted by insectaries and biological control practitioners due to insufficient practical development, these techniques offer considerable potential that could readily lead to commercial application. In the case of conservation biological control, the provision of nectar subsidies for adult parasitoids has attracted considerable attention in recent years, involving patch choice decisions that influence the trade-off between current and future reproduction. While nectar subsidies are used by adult parasitoids in the field and have been shown to increase parasitism, at least locally, there is little experimental evidence that this has translated to reduced pest densities, indicating the need for additional and more critical field tests.

References

Bellows, T.S. (2001) Restoring population balance through natural enemy introductions. *Biological Control* **21**: 199–205.

Bernal, J.S., Luck, R.F. and Morse, J.G. (1999) Host influences on sex ratio, longevity, and egg load of two *Metaphycus* species parasitic on soft scales: implications for insectary rearing. *Entomologia Experimentalis et Applicata* **92**: 191–204.

Bernal, J.S., Gillogly, P.O. and Griset, J. (2001) Family planning in a stemborer parasitoid: sex ratio, brood size and size-fitness relationships in *Parallorhogas pyralophagus* (Hymenoptera: Braconidae), and implications for biological control. *Bulletin of Entomological Research* **91**: 255–64.

Berndt, L.A., Wratten, S.D. and Scarratt, S.L. (2006) The influence of floral resource subsidies on parasitism rates of leafrollers (Lepidoptera: Tortricidae) in New Zealand vineyards. *Biological Control* **37**: 50–5.

Bernstein, C. and Driessen, G. (1996) Patch-marking and optimal search patterns in the parasitoid *Venturia canescens*. *Journal of Animal Ecology* **65**: 211–19.

Bernstein, C., Kacelnik, A. and Krebs, J.R. (1988) Individual decisions and the distribution of predators in a patchy environment. *Journal of Animal Ecology* **57**: 1007–26.

Bernstein, C., Kacelnik, A. and Krebs, J.R. (1991) Individual decisions and the distribution of predators in a patchy environment. II. The influence of travel costs and structure of the environment. *Journal of Animal Ecology* **60**: 205–26.

Bernstein, C., Auger, P. and Poggiale, J.C. (1999) Predator migration decisions, the ideal distribution, and predator–prey dynamics. *American Naturalist* **153**: 267–81.

Bonsall, M.B. and Hassell, M.P. (1995) Identifying density-dependent processes: a comment on the regulation of winter moth. *Journal of Animal Ecology* **64**: 781–4.

Briggs, C.J. and Hoopes, M.F. (2004) Stabilizing effects in spatial parasitoid–host and predator–prey models: a review. *Theoretical Population Biology* **65**: 299–315.

Burger, J.M.S., Huang, Y., Hemerik, L., van Lenteren, J.C. and Vet, L.E.M. (2006) Flexible use of patch-leaving mechanism in a parasitoid wasp. *Journal of Insect Behavior* **19**: 155–70.

Caltagirone, L.E. and Doutt, R.L. (1989) The history of the vedalia importation to California and its impact on the development of biological control. *Annual Review of Entomology* **34**: 1–16.

Campan, E. and Benrey, B. (2004) Behavior and performance of a specialist and a generalist parasitoid of bruchids on wild and cultivated beans. *Biological Control* **30**: 220–8.

Carpenter, M. (2005) Fillmore Citrus Protective District. http://www.rinconvitova.com/history%20BC.htm#Fillmore%20Citrus%20Protective%20District

Casas, J., Driessen, G., Mandon, N. et al. (2003) Energy dynamics in a parasitoid foraging in the wild. *Journal of Animal Ecology* **72**: 691–7.

Charnov, E.L. (1976) Optimal foraging: the marginal value theorem. *Theoretical Population Biology* **9**: 129–36.

Charnov, E.L., Hartogh, R.L.L-D., Jones, W.T. and van den Assem, J. (1981) Sex ratio evolution in a variable environment. *Nature* **289**: 27–33.

Childs, D.Z., Bonsall, M.B. and Rees, M. (2004) Periodic local disturbance in host–parasitoid metapopulations: host suppression and parasitoid persistence. *Journal of Theoretical Biology* **227**: 13–23.

Chow, A. and Heinz, K.M. (2005) Using hosts of mixed sizes to reduce male-biased sex ratio in the parasitoid wasp, *Diglyphus isaea*. *Entomologia Experimentalis et Applicata* **117**: 193–9.

Chow, A. and Heinz, K.M. (2006) Control of *Liriomyza langei* on chrysanthemum by *Diglyphus isaea* produced with a standard or modified parasitoid rearing technique. *Journal of Applied Entomology* **130**: 113–21.

Collier, T. and Van Steenwyk, R. (2004) A critical evaluation of augmentative biological control. *Biological Control* **31**: 245–56.

Costanza, R., d'Arge, R., de Groot, R. et al. (1997) The value of the world's ecosystem services and natural capital. *Nature* **387**: 253–60.

Cronin, J.T. (2004) Host–parasitoid extinction and colonization in a fragmented prairie landscape. *Oecologia* **139**: 503–14.

Cronin, J.T. and Reeve, J.D. (2005) Host–parasitoid spatial ecology: a plea for a landscape-level synthesis. *Proceedings of the Royal Society of London Series B Biological Science* **272**: 2225–35.

Daane, K.M., Mills, N.J. and Tauber, M.J. (2002) Biological pest controls: augmentative controls. In: Pimentel, D. (ed.) *Encyclopedia of Pest Management*. Marcel Dekker, New York, pp. 36–8.

Desouhant, E., Driessen, G., Lapchin, L., Wielaard, S. and Bernstein, C. (2003) Dispersal between host populations in field conditions: navigation rules in the parasitoid *Venturia canescens*. *Ecological Entomology* **28**: 257–67.

Desouhant, E., Driessen, G., Amat, I. and Bernstein, C. (2005) Host and food searching in a parasitic wasp *Venturia canescens*: a trade-off between current and future reproduction? *Animal Behaviour* **70**: 145–52.

Driessen, G. and Bernstein, C. (1999) Patch departure mechanisms and optimal host exploitation in an insect parasitoid. *Journal of Animal Ecology* **68**: 445–59.

Ehler, L.E. (1998) Conservation biological control: past, present and future. In: Barbosa, P. (ed.) *Conservation Biological Control*. Academic Press, San Diego, pp. 1–8.

Fauvergue, X., Boll, R., Rochat, J., Wajnberg, É., Bernstein, C. and Lapchin, L. (2006) Habitat assessment by parasitoids: consequences for population distribution. *Behavioral Ecology* **17**: 522–31.

Fretwell, S.D. and Lucas, H.L. (1970) On territorial behaviour and other factors influencing habitat distribution in birds. *Acta Biotheoretica* **19**: 16–36.

Geervliet, J.B.F., Ariens, S., Dicke, M. and Vet, L.E.M. (1998) Long-distance assessment of patch profitability through volatile infochemicals by the parasitoids *Cotesia glomerata* and *C. rubecula* (Hymenoptera: Braconidae). *Biological Control* **11**: 113–21.

Giron, D., Pincebourde, S. and Casas, J. (2004) Lifetime gains of host-feeding in a synovigenic parasitic wasp. *Physiological Entomology* **29**: 436–42.

Godfray, H.C.J. (1994) *Parasitoids. Behaviour and Evolutionary Ecology*. Princeton University Press, Princeton.

Goubault, M., Outreman, Y., Poinsot, D. and Cortesero, A.M. (2005) Patch exploitation strategies of parasitic wasps under intraspecific competition. *Behavioral Ecology* **16**: 693–701.

Green, R.F. (2006) A simpler, more general method of finding the optimal foraging strategy for Bayesian birds. *Oikos* **112**: 274–84.

Gurr, G.M., Wratten, S.D. and Barbosa, P. (2000) Success in conservation biological control of arthropods. In: Gurr, G. and Wratten, S. (eds.) *Measures of Success in Biological Control*. Kluwer Academic Publishers, Dordrecht, pp. 105–32.

Gurr, G.M., Wratten, S.D., Tylianakis, J., Kean, J. and Keller, M. (2005) Providing plant foods for natural enemies in farming systems: balancing practicalities and theory. In: Wäckers, F.L., van Rijn, P.C.J. and Bruin, J. (eds.) *Plant-provided Food for Carnivorous Insects*. Cambridge University Press, Cambridge, pp. 326–47.

Hamilton, W.D. (1967) Extraordinary sex ratios. *Science* **156**: 477–88.

Hardin, M.R., Benrey, B., Coll, M., Lampe, W.O., Roderick G.K. and Barbosa, P. (1995) Arthropod pest resurgence: an overview of potential mechanisms. *Crop Protection* **14**: 3–18.

Hare, J.D. and Morgan, D.J.W. (1997) Mass-priming *Aphytis*: behavioral improvement of insectary-reared biological control agents. *Biological Control* **10**: 207–14.

Hare, J.D., Morgan, D.J.W. and Nguyun, T. (1997) Increased parasitization of California red scale in the field after exposing its parasitoid, *Aphytis melinus*, to a synthetic kairomone. *Entomologia Experimentalis et Applicata* **82**: 73–81.

Hassell, M.P. (1980) Foraging strategies, population models, and biological control: a case study. *Journal of Animal Ecology* **49**: 603–28.

Hassell, M.P. (2000) *The Spatial and Temporal Dynamics of Host–Parasitoid Interactions*. Oxford University Press, Oxford.

Hassell, M.P. and May R.M. (1973) Stability in insect host–parasitoid models. *Journal of Animal Ecology* **42**: 693–726.

Hassell, M.P. and Varley, G.C. (1969) New inductive population model for insect parasites and its bearing on biological control. *Nature* **223**: 1133–6.

Heimpel, G.E. and Jervis, M.A. (2005) Does floral nectar improve biological control by parasitoids? In: Wäckers, F.L., van Rijn, P.C.J. and Bruin, J. (eds.) *Plant-provided Food for Carnivorous Insects*. Cambridge University Press, Cambridge, pp. 267–304.

Heinz, K.M. (1998) Host size-dependent sex allocation behaviour in a parasitoid: implications for *Catolaccus grandis* (Hymenoptera: Pteromalidae) mass rearing programmes. *Bulletin of Entomological Research* **88**: 37–45.

Heinz, K.M., van Driesche, R.G. and Parrella, M.P. (2004) *Biocontrol in Protected Culture*. Ball Publishing, Batavia.

Hoddle, M.S. (2004) Restoring balance: using exotic species to control invasive exotic species. *Conservation Biology* **18**: 38–49.

Hoddle, M.S., van Driesche, R.G. and Sanderson, J.P. (1998) Biology and use of the whitefly parasitoid *Encarsia formosa*. *Annual Review of Entomology* **43**: 645–69.

Hougardy, E. and Mills, N.J. (2006) The influence of host deprivation and egg expenditure on the rate of dispersal of a parasitoid following field release. *Biological Control* **37**: 206–13.

Hougardy, E. and Mills, N.J. (2007) Influence of host deprivation and egg expenditure on the patch and host-finding behavior of the parasitoid wasp *Mastrus ridibundus*. *Journal of Insect Behavior* **20**: 229–46.

Hougardy, E., Bezemer, T.M. and Mills, N.J. (2005) Effects of host deprivation and egg expenditure on the reproductive capacity of *Mastrus ridibundus*, an introduced parasitoid for the biological control of codling moth in California. *Biological Control* **33**: 96–106.

Houston, A.I. and McNamara, J.M. (1999) *Models of Adaptive Behaviour*. Cambridge University Press, Cambridge.

Huffaker, C.B., Kennett, C.E. and Tassan, R.L. (1986) Comparison of parasitism and densities of *Parlatoria oleae* 1952–1982 in relation to ecological theory. *American Naturalist* **128**: 379–93.

Ives, A.R. (1995) Spatial heterogeneity and host–parasitoid population dynamics: do we need to study behavior? *Oikos* **74**: 366–76.

Iwasa, Y., Suzuki, Y. and Matsuda, H. (1984) Theory of oviposition strategies of parasitoids. I. Effect of mortality and limited egg number. *Theoretical Population Biology* **26**: 205–27.

Jackson, A.L., Humphries, S. and Ruxton, G.D. (2004a) Resolving the departures of observed results from the ideal free distribution with simple random movements. *Journal of Animal Ecology* **73**: 612–22.

Jackson, A.L., Ranta, E., Lundberg, P., Kaitala, V. and Ruxton, G.D. (2004b) Consumer–resource matching in a food chain when both predators and prey are free to move. *Oikos* **106**: 445–50.

Jacob, H.S. and Evans, E.W. (1998) Effects of sugar spray and aphid honeydew on field populations of the parasitoid *Bathyplectes curculionis* (Hymenoptera: Ichneumonidae). *Environmental Entomology* **27**: 1563–8.

Jacob, H.S. and Evans, E.W. (2001) Influence of food deprivation on foraging decisions of the parasitoid *Bathyplectes curculionis* (Hymenoptera: Ichneumonidae). *Annals of the Entomological Society of America* **94**: 605–11.

Jervis, M.A., Kidd, N.A.C., Fitton, M.G., Huddleston, T. and Dawah, H.A. (1993) Flower-visiting by hymenopteran parasitoids. *Journal of Natural History* 27: 67–105.

Jervis, M.A., Kidd, N.A.C. and Heimpel, G.E. (1996) Parasitoid adult feeding behaviour and biocontrol: a review. *Biocontrol News & Information* 17: 11–26.

Jervis, M.A., Heimpel, G.E., Ferns, P.N., Harvey, J.A. and Kidd, N.A.C. (2001) Life-history strategies in parasitoid wasps: a comparative analysis of 'ovigeny'. *Journal of Animal Ecology* 70: 442–58.

Kacelnik, A., Krebs, J.R. and Bernstein, C. (1992) The ideal free distribution and predator–prey populations. *Trends in Ecology & Evolution* 7: 50–5.

Kean, J.M. and Barlow, N.D. (2000) Can host–parasitoid metapopulations explain successful biological control? *Ecology* 81: 2188–97.

Kean, J., Wratten, S., Tylianakis, J. and Barlow, N. (2003) The population consequences of natural enemy enhancement, and implications for conservation biological control. *Ecology Letters* 6: 604–12.

Keasar, T., Ney-Nifle, M., Mangel, M. and Swezey, S. (2001) Early oviposition experience affects patch residence time in a foraging parasitoid. *Entomologia Experimentalis et Applicata* 98: 123–32.

Kenmore, P.E. (1996) Integrated pest management in rice. In: Persley, G.J. (ed.) *Biotechnology and Integrated Pest Management*. CABI Publishing, Wallingford, pp. 76–97.

King, B.H. (1987) Offspring sex ratios in parasitoid wasps. *Quarterly Review of Biology* 62: 367–96.

Křivan, V. and Sirot, E. (1997) Searching for food or hosts: the influence of parasitoids behavior on host–parasitoid dynamics. *Theoretical Population Biology* 51, 201–9.

Lack, D. (1947) The significance of clutch size. *Ibis* 89: 309–352.

Landis, D.A., Wratten, S.D. and Gurr, G.M. (2000) Habitat management to conserve natural enemies of arthropod pests in agriculture. *Annual Review of Entomology* 45: 175–201.

Lavandero, B., Wratten, S., Shishehbor, P. and Worner, S. (2005) Enhancing the effectiveness of the parasitoid *Diadegma semiclausum* (Helen): movement after use of nectar in the field. *Biological Control* 34: 152–8.

Legaspi, B.C. and Legaspi, J.C. (2005) Foraging behavior of field populations of *Diadegma* spp. (Hymenoptera : Ichneumonidae): testing for density-dependence at two spatial scales. *Journal of Entomological Science* 40: 295–306.

Lewis, W.J., Beevers, M., Nordlund, D.A., Gross, H.R. and Hagen, K.S. (1979) Kairomones and their use for management of entomophagous insects. IX. Investigations of various kairomone treatment patterns for *Trichogramma* spp. *Journal of Chemical Ecology* 5: 673–80.

Lewis, W.J., Stapel, J.O., Cortesero, A.M. and Takasu, K. (1998) Understanding how parasitoids balance food and host needs: importance to biological control. *Biological Control* 11: 175–83.

Li, C., Roitberg, B.D. and Mackauer, M. (1997) Effects of contact kairomone and experience on initial giving-up time. *Entomologia Experimentalis et Applicata* 84: 101–4.

Losey, J.E. and Vaughan, M. (2006) The economic value of ecological services provided by insects. *BioScience* 56: 311–23.

Luck, R.F., Forster, L.D. and Morse, J.G. (1996) An ecologically based IPM program for citrus in California's San Joaquin Valley using augmentative biological control. International Citrus Congress, Sun City, South Africa. *International Society of Citriculture* 1: 499–503.

Mack, R.N., Simberloff, D., Lonsdale, W.M., Evans, H., Clout, M. and Bazzaz, F.A. (2000) Biotic invasions: causes, epidemiology, global consequences, and control. *Ecological Applications* 10: 689–710.

Matsumoto, T., Itioka, T., Nishida, T. and Inoue, T. (2004) A test of temporal and spatial density dependence in the parasitism rates of introduced parasitoids on host, the arrowhead scale (*Unaspis yanonensis*) in stable host–parasitoids system. *Journal of Applied Entomology* 128: 267–72.

Mills, N.J. (1994) Biological control: some emerging trends. In: Leather, S.R., Watt, A.D., Mills, N.J. and Walters, K.F.A. (eds.) *Individuals, Populations and Patterns in Ecology*. Intercept, Andover, pp. 213–22.

Mills, N.J. (2000) Biological control: the need for realistic models and experimental approaches to parasitoid introductions. In: Hochberg, M.E. and Ives, A.R. (eds.) *Parasitoid Population Biology*. Princeton University Press, Princeton, pp. 217–34.

Mills, N.J. (2005) Selecting effective parasitoids for biological control introductions: codling moth as a case study. *Biological Control* **34**: 274–82.

Moreno, D.S. and Luck, R.F. (1992) Augmentative releases of *Aphytis melinus* (Hymenoptera: Aphelinidae) to suppress California red scale (Homoptera: Diaspididae) in southern California lemon orchards. *Journal of Economic Entomology* **85**: 1112–19.

Murdoch, W.W., Reeve, J.D., Huffaker, C.B. and Kennett, C.E. (1984) Biological control of olive scale *Parlatoria oleae* and its relevance to ecological theory. *American Naturalist* **123**: 371–92.

Murdoch, W.W., Briggs, C.J. and Nisbet, R.M. (2003) *Consumer–Resource Dynamics*. Princeton University Press, Princeton.

Neuenschwander, P. (2003) Biological control of cassava and mango mealybugs in Africa. In: Neuenschwander, P., Borgemeister, C. and Langewald, J. (eds.) *Biological Control in IPM Systems in Africa*. CABI Publishing, Wallingford, pp. 45–59.

Nicholson, A.J. and Bailey, V.A. (1935) The balance of animal populations. *Proceedings of the Zoological Society of London* **1**: 551–98.

Nunney, L. (2003) Managing captive populations for release: a population-genetic perspective. In: van Lenteren, J.C. (ed.) *Quality Control and Production of Biological Control Agents: Theory and Testing Procedures*. CABI Publishing, Wallingford, pp. 73–87.

Ode, P.J. and Heinz, K.M. (2002) Host-size-dependent sex ratio theory and improving mass-reared parasitoid sex ratios. *Biological Control* **24**: 31–41.

Outreman, Y., Le Ralec, A., Wajnberg, É. and Pierre, J-S. (2005) Effects of within- and among-patch experiences on the patch-leaving decision rules in an insect parasitoid. *Behavioral Ecology & Sociobiology* **58**: 208–17.

Petersen, G., Matthiesen, C., Francke, W. and Wyss, U. (2000) Hyperparasitoid volatiles as possible foraging behaviour determinants in the aphid parasitoid *Aphidius uzbekistanicus* (Hymenoptera: Aphidiidae). *European Journal of Entomology* **97**: 545–50.

Plantegenest, M., Outreman, Y., Goubault, M. and Wajnberg, É. (2004) Parasitoids flip a coin before deciding to superparasitize. *Journal of Animal Ecology* **73**: 802–6.

Prokopy, R.J. and Lewis, W.J. (1993) Application of learning to pest management. In: Papaj, D.R. and Lewis, A.C. (eds.) *Insect Learning: Ecological and Evolutionary Perspectives*. Chapman & Hall, New York, pp. 308–42.

Reeve, J.D. and Murdoch, W.W. (1985) Aggregation by parasitoids in the successful control of the California USA red scale: a test of theory. *Journal of Animal Ecology* **54**: 797–816.

Rochat, J. and Gutierrez, A.P. (2001) Weather-mediated regulation of olive scale by two parasitoids. *Journal of Animal Ecology* **70**: 476–90.

Rohani, P. and Miramontes, O. (1995) Host–parasitoid metapopulations: the consequences of parasitoid aggregation on spatial dynamics and searching efficiency. *Proceedings of the Royal Society of London Series B Biological Science* **260**: 35–42.

Roitberg, B.D., Mangel, M., Lalonde, R.G., Roitberg, C.A., van Alphen, J.J.M. and Vet, L. (1992) Seasonal dynamic shifts in patch exploitation by parasitic wasps. *Behavioral Ecology* **3**: 156–65.

Roland, J. (1988) Decline in winter moth populations in North America, direct versus indirect effect of introduced parasites. *Journal of Animal Ecology* **57**: 523–32.

Roland, J. (1995) Response to Bonsall and Hassell 'Identifying density-dependent processes: A comment on the regulation of winter moth'. *Journal of Animal Ecology* **64**: 785–6.

Rosenheim, J.A. (1998) Higher-order predators and the regulation of insect herbivore populations. *Annual Review of Entomology* **43**: 421–47.

Roush, R.T. and Hopper, K.R. (1995) Use of single family lines to preserve genetic variation in laboratory colonies. *Annals of the Entomological Society of America* **88**; 713–17.

Sagarra, L.A., Vincent, C. and Stewart, R.K. (2000) Mutual interference among female *Anagyrus kamali* Moursi (Hymenoptera: Encyrtidae) and its impact on fecundity, progeny production and sex ratio. *Biocontrol Science & Technology* **10**: 239–44.

Sakai, A.K., Allendorf, F.W., Holt, J.S. et al. (2001) The population biology of invasive species. *Annual Review of Ecology & Systematics* **32**: 305–32.

Shaltiel, L. and Ayal, Y. (1998) The use of kairomones for foraging decisions by an aphid parasitoid in small host aggregations. *Ecological Entomology* **23**: 319–29.

Shuker, D.M. and West, S.A. (2004) Information constraints and the precision of adaptation: sex ratio manipulation in wasps. *Proceedings of the National Academy of Sciences USA* **101**: 10363–7.

Shuker, D.M., Pen, I., Duncan, A.B., Reece, S.E. and West, S.A. (2005) Sex ratios under asymmetrical local mate competition: theory and a test with parasitoid wasps. *American Naturalist* **166**: 301–16.

Shuker, D.M., Pen, I. and West, S.A. (2006) Sex ratios under asymmetrical local mate competition in the parasitoid wasp *Nasonia vitripennis*. *Behavioral Ecology* **17**: 345–52.

Silva, I.M.M.S., van Meer, M.M.M., Roskam, M.M., Hoogenboom, A., Gort, G. and Stouthamer, R. (2000) Biological control potential of *Wolbachia*-infected versus uninfected wasps: Laboratory and greenhouse evaluation of *Trichogramma cordubensis* and *T. deion* strains. *Biocontrol Science & Technology* **10**: 223–38.

Sirot, E. and Bernstein, C. (1996) Time sharing between host searching and food searching in parasitoids: state-dependent optimal strategies. *Behavioral Ecology* **7**: 189–94.

Skovgard, H. and Nachman, G. (2004) Biological control of house flies *Musca domestica* and stable flies *Stomoxys calcitrans* (Diptera: Muscidae) by means of inundative releases of *Spalangia cameroni* (Hymenoptera: Pteromalidae). *Bulletin of Entomological Research* **94**: 555–67.

Smith, A.D.M. and Maelzer, D.A. (1986) Aggregation of parasitoids and density-independence of parasitism in field population of the wasp *Aphytis melinus* and its host the red scale *Aonidiella aurantii*. *Ecological Entomology* **11**: 425–34.

Stapel, J.O., Cortesero, A.M., De Moraes, C.M., Tumlinson, J.H. and Lewis, W.J. (1997) Extrafloral nectar, honeydew, and sucrose effects on searching behavior and efficiency of *Microplitis croceipes* (Hymenoptera: Braconidae) in cotton. *Environmental Entomology* **26**: 617–23.

Stouthamer, R. (2003) The use of unisexual wasps in biological control. In: van Lenteren, J.C. (ed.) *Quality Control and Production of Biological Control Agents: Theory and Testing Procedures.* CABI Publishing, Wallingford, pp. 93–113.

Sutherland, W.J. (1983) Aggregation and the ideal free distribution. *Journal of Animal Ecology* **52**: 821–8.

Sutherland, W.J. (1996) *From Individual Behaviour to Population Ecology.* Oxford University Press, Oxford.

Syrett, P., Briese, D.T. and Hoffmann, J.H. (2000) Success in biological control of terrestrial weeds by arthropods. In: Gurr, G. and Wratten, S. (eds.) *Measures of Success in Biological Control.* Kluwer Academic Publishers, Dordrecht, pp. 189–230.

Tenhumberg, B., Siekmann, G. and Keller, M.A. (2006) Optimal time allocation in parasitic wasps searching for hosts and food. *Oikos* **113**: 121–31.

Tentelier, C., Wajnberg, É. and Fauvergue, X. (2005) Parasitoids use herbivore-induced information to adapt patch exploitation behaviour. *Ecological Entomology* **30**: 739–44.

Thies, C., Steffan-Dewenter, I. and Tscharntke, T. (2003) Effects of landscape context on herbivory and parasitism at different spatial scales. *Oikos* **101**: 18–25.

Tourniaire, R., Ferran, A., Giuge, L., Piotte, C. and Gambier, J. (2000) A natural flightless mutation in the ladybird, *Harmonia axyridis*. *Entomologia Experimentalis et Applicata* **96**: 33–8.

Tregenza, T. (1995) Building on the ideal free distribution. *Advances in Ecological Research* **26**: 253–307.

Tregenza, T., Thompson, D.J. and Parker, G.A. (1996) Interference and the ideal free distribution: oviposition in a parasitoid wasp. *Behavioral Ecology* **7**: 387–94.

Tylianakis, J.M., Didham, R.K. and Wratten, S.D. (2004) Improved fitness of aphid parasitoids receiving resource subsidies. *Ecology* **85**: 658–66.

Ueno, T. (2005) Effect of host age and size on offspring sex ratio in the pupal parasitoid *Pimpla* (=*Coccygomimus*) *luctuosa* (Hymenoptera: Ichneumonidae). *Journal of the Faculty of Agriculture Kyushu University* **50**: 399–405.

van Alphen, J.J.M., Bernstein, C. and Driessen, G. (2003) Information acquisition and time allocation in insect parasitoids. *Trends in Ecology & Evolution* **18**: 81–7.

van Baaren, J., Boivin, G. and Outreman, Y. (2005) Patch exploitation strategy by an egg parasitoid in constant or variable environment. *Ecological Entomology* **30**: 502–9.

van Lenteren, J.C. (2000) Measures of success in biological control of arthropods by augmentation of natural enemies. In: Gurr, G. and Wratten, S. (eds.) *Measures of Success in Biological Control*. Kluwer Academic Publishers, Dordrecht, pp. 77–103.

van Lenteren, J.C. (2003) *Quality Control and Production of Biological Control Agents: Theory and Testing Procedures*. CABI Publishing, Wallingford.

van Lenteren, J.C., van Roermund, H.J.W. and Sütterlin, S. (1996) Biological control of greenhouse whitefly (*Trialeurodes vaporariorum*) with the parasitoid *Encarsia formosa*: How does it work? *Biological Control* **6**: 1–10.

Vet, L.E.M. (2001) Parasitoid searching efficiency links behaviour to population processes. *Applied Entomology & Zoology* **36**: 399–408.

Vet, L.E.M., Lewis, W.J., Papaj, D.R. and van Lenteren, J.C. (2003) A variable-response model for parasitoid foraging behaviour. In: van Lenteren, J.C. (ed.) *Quality Control and Production of Biological Control Agents: Theory and Testing Procedures*. CABI Publishing, Wallingford, pp. 25–39.

Visser, M.E. and Rosenheim, J.A. (1998) The influence of competition between foragers on clutch size decisions in insect parasitoids. *Biological Control* **11**: 169–74.

Visser, M.E., van Alphen, J.J.M. and Hemerik, L. (1992) Adaptive superparasitism and patch time allocation in solitary parasitoids: an ESS model. *Journal of Animal Ecology* **61**: 93–101.

Waage, J.K. (1979) Foraging for patchily distributed hosts by the parasitoid *Nemeritis canescens*. *Journal of Animal Ecology* **48**: 353–71.

Waage, J.K. (1983) Aggregation in field parasitoid populations: foraging time allocation by a population of *Diadegma* (Hymenoptera: Ichneumonidae). *Ecological Entomology* **8**: 447–54.

Wäckers, F.L. (1994) The effect of food deprivation on the innate visual and olfactory preferences in the parasitoid *Cotesia rubecula*. *Journal of Insect Physiology* **40**: 641–9.

Wäckers, F.L. (2001) A comparison of nectar- and honeydew sugars with respect to their utilization by the hymenopteran parasitoid *Cotesia glomerata*. *Journal of Insect Physiology* **47**: 1077–84.

Wäckers, F.L. (2004) Assessing the suitability of flowering herbs as parasitoid food sources: flower attractiveness and nectar accessibility. *Biological Control* **29**: 307–14.

Wäckers, F.L. (2005) Suitability of (extra-)floral nectar, pollen, and honeydew as insect food sources. In: Wäckers, F.L., van Rijn, P.C.J. and Bruin, J. (eds.) *Plant-provided Food for Carnivorous Insects*. Cambridge University Press, Cambridge, pp. 17–74.

Wäckers, F.L., van Rijn, P.C.J. and Bruin, J. (2005) *Plant-provided Food for Carnivorous Insects*. Cambridge University Press, Cambridge.

Wajnberg, É. (2004) Measuring genetic variation in natural enemies used for biological control: Why and how? In: Ehler, L., Sforza, R. and Mateille T. (eds.) *Genetics, Evolution and Biological Control*. CABI Publishing, Wallingford, pp. 19–37.

Wajnberg, É. (2006) Time-allocation strategies in insect parasitoids: from ultimate predictions to proximate behavioural mechanisms. *Behavioral Ecology and Sociobiology* **60**: 589–611.

Wajnberg, É. and Hassan, S.A. (1994) *Biological Control with Egg Parasitoids*. CABI Publishing, Wallingford.

Wajnberg, É., Rosi, M.C. and Colazza, S. (1999) Genetic variation in patch time allocation in a parasitic wasp. *Journal of Animal Ecology* **68**: 121–33.

Wajnberg, É., Fauvergue, X. and Pons, O. (2000) Patch leaving decision rules and the marginal value theorem: an experimental analysis and a simulation model. *Behavioral Ecology* **11**: 577–86.

Wajnberg, É., Curty, C. and Colazza, S. (2004) Genetic variation in the mechanisms of direct mutual interference in a parasitic wasp: consequences in terms of patch-time allocation. *Journal of Animal Ecology* **73**: 1179–89.

Walde, S.J. and Murdoch, W.W. (1988) Spatial density dependence in parasitoids. *Annual Review of Entomology* **33**: 441–66.

Wang, X.G. and Keller, M.A. (2004) Patch time allocation by the parasitoid *Diadegma semiclausum* (Hymenoptera: Ichneumonidae). III. Effects of kairomone sources and previous parasitism. *Journal of Insect Behavior* **17**: 761–76.

Wang, X.G., Duff, J., Keller, M.A., Zalucki, M.P., Liu, S.S. and Bailey, P. (2004) Role of *Diadegma semiclausum* (Hymenoptera: Ichneumonidae) in controlling *Plutella xylostella* (Lepidoptera: Plutellidae): cage exclusion experiments and direct observation. *Biocontrol Science & Technology* **14**: 571–86.

Werren, J.H. (1980) Sex ratio adaptations to local mate competition in a parasitic wasp. *Science* **208**: 1157–9.

Winkler, K., Wäckers, F.L., Bukovinszkine-Kiss, G. and van Lenteren, J. (2006) Sugar resources are vital for *Diadegma semiclausum* fecundity under field conditions. *Basic & Applied Ecology* **7**: 133–40.

Zaviezo, T. and Mills, N.J. (2000) Factors influencing the evolution of clutch size in a gregarious insect parasitoid. *Journal of Animal Ecology* **69**: 1047–57.

2

Parasitoid fitness: from a simple idea to an intricate concept

Minus van Baalen and Lia Hemerik

Abstract

In behavioral ecology it has to be demonstrated that a given behavioral strategy maximizes fitness. For most organisms the link between behavior and fitness (lifetime reproductive success) is complex and difficult to assess experimentally. Parasitoids have long been among the favorite organisms for studies on the evolution of behavior because, for these species, all behavior related to oviposition can be directly translated into contributions to lifetime reproductive success. Because parasitoids have to locate their hosts in order to oviposit, the fitness gains of searching strategies are at first sight easy to establish. However, as reviewed in this chapter, it does not generally suffice to count the number of parasitized hosts to obtain a fitness measure for parasitoids. This fitness measure disregards the effect of differing host distributions and time allocation of the parasitoid, of parasitoid longevity, and of the cost of an egg, as reflected in the occurrence of host feeding, superparasitism, and the need to evade the host's immune system (encapsulation) and other defenses.

In this chapter, we discuss the use of mathematical methods to derive the appropriate fitness measure instead of merely hypothesizing it. Two ways of doing this exist. One method is the bottom-up approach to determine how a focal individual maximizes its expected number of surviving offspring. This method has the advantage of permitting an assessment of the stochasticity in how a parasitoid experiences its environment. The other method derives fitness from the capacity of a mutant strategy to invade a resident system. This approach cannot easily incorporate stochasticity but it brings to light the consequences of eco-evolutionary feedback loops (via resource exploitation, interference, and so on). Both methods show that parasitoid fitness depends in potentially complicated ways on their life history strategies. Analyzing these fitness functions gives us insight into the selective pressures that parasitoids are subject to and allow us to explore new venues of research.

2.1 Introduction

Natural selection shapes behavior in the same way as it shapes other traits, that is, if there is a relationship between the studied behavior and fitness and if the behavior has a genetic basis (Krebs & Davies 1997). This leads to the prediction that, given the possibilities of an organism and the constraints it is subject to, its behavioral goals are achieved as efficiently as possible. Behavioral ecology is based on this insight and has led to many theories about what kind of behavior is expected to evolve under all kinds of conditions for a wide variety of organisms.

When optimal foraging theory (Stephens & Krebs 1986) was developed in the late 1970s and early 1980s, parasitoids were among the favorite organisms to test its predictions. It was argued that, for this class of organisms, the link between behavior (foraging strategies) and fitness was particularly clear, because an egg can be deposited in every successfully located host and therefore implies a fitness increment. For the same reason, many tests of sex allocation theory have been carried out using parasitoids as model organisms (Hardy 1992, Godfray 1994). For the simplest of tests, the close link between fitness and encountering hosts is true enough, but certain aspects specific to parasitoid biology exist that confuse or obscure this straightforward reasoning.

Host distributions are typically clumpy (Atkinson & Shorrocks 1984, Driessen & Hemerik 1991), which may give rise to aggregation of the parasitoids with various population dynamical (Hassell & May 1973, 1974, May 1978, Chesson & Murdoch 1986) and evolutionary consequences (van Baalen & Sabelis 1993, 1999). However, to come up with a simple fitness measure that incorporates this heterogeneity, for instance in the form of a weighed sum, is nearly impossible.

With marginal value theories (MVTs) we can calculate the optimal residence time in a host patch as a function of the mean rate of oviposition in the environment (Charnov 1976, see also Chapter 8 by van Alphen and Bernstein), but typically this value cannot be assessed analytically or even easily observed empirically. Since the early developments, the scope of parasitoid research has widened. There is empirical support for some of the basic predictions of the MVT (Lei & Camara 1999, Wajnberg et al. 2006), but there are still many loose ends as the theory depends on many simplifying assumptions. A significant effort continues to be devoted to a better understanding of host search and oviposition decisions and their population dynamical consequences (see also Chapter 13 by Bonsall and Bernstein).

Another difficulty arises when we realize that, in contrast to what underlies the 'host encounter equals fitness increment' assumption, eggs do not come cheap to parasitoids. Some pro-ovigenic parasitoid species have their lifetime supply of eggs ready at eclosion, but other synovigenic species have to mature eggs during their adult lifetime. In both types, the decision to oviposit in or on a given host may therefore depend on how many eggs the parasitoid has and whether better opportunities can be expected in the near future (Mangel & Clark 1988, Clark & Mangel 2000). For synovigenic species, the decision between oviposition and host feeding of a host upon encounter is based on weighing current expected reproductive output against future expected offspring as a result of current host feeding (Burger et al. 2004).

Some parasitoid species, called solitary species, can realize only one adult offspring per host. For these species superparasitism, i.e. the deposition of an extra conspecific egg in an already parasitized host (van Alphen & Visser 1990), might be advantageous under given circumstances (van der Hoeven & Hemerik 1990, Visser et al. 1992a). There generally exists a period, just after the host is parasitized for the first time, in which the second or later egg has a positive probability to win the competition (Visser et al. 1992b, Gates 1993, Field et al. 1997). In addition, for so-called gregarious parasitoids that put more than a single egg in a host, superparasitism may pay as the results of competition imply different net costs to subsequent parasitoids (Ives 1989, Nagelkerke et al. 1996).

Another aspect of the parasitoid–host interaction is that the evolutionary arms race between hosts and parasitoids has often produced hosts that have attained a way of escaping parasitism. For instance, some hosts are able to encapsulate eggs of their natural enemy to a certain extent. Streams (1968) already mentioned this capability of *Drosophila melanogaster* when it was parasitized by *Leptopilina heterotoma*), while other host species have developed behavioral means to escape attack (Hochberg 1997).

The different aspects of parasitoid oviposition behavior have generated their own spheres of research to such an extent that what is currently lacking is an integrative framework. In the 1970s, the Nicholson–Bailey model (Nicholson & Bailey 1935) served as a baseline model for studying population dynamical consequences of different suggested mechanisms (Hassell 1978), while the MVT (Charnov 1976) formed the basis for studying behavioral decisions. These two approaches are difficult to combine (Visser 1991) and, since then, the problems have only grown. Indeed, one of the advantages of the Nicholson–Bailey model is that it is conceptually appealing while at the same time abundantly and obviously wrong. This tension has stimulated much of the early research into host–parasitoid population dynamics. Similarly, predictions based on the MVT have a clear interpretation but equally clearly parasitoids tend not to follow them, which pose many interesting questions about how parasitoids acquire the information that they need. Such studies have yielded much insight into how parasitoids and their interaction with hosts differ from the assumptions that went into the simpler models.

Here we discuss that there exists a large gap between practical and theoretical approaches to parasitoid fitness. First, we give some context and indications about how empirical and theoretical biologists consider fitness. Even though the precise definition of fitness has been rapidly evolving during the past decades (Brommer 2000), it has always been clear that 'fitness' is something that is difficult or even impossible to measure directly. Even at the theoretical level, definitions of fitness differ widely with respect to what aspects and mechanisms are included. For instance, the stochastic dynamic programming (SDP) approach, being related to life history theory, defines fitness as the overall number of descendants of a given strategy without taking into account eco-evolutionary feedback loops that ultimately regulate this number. Adaptive dynamics, in contrast, goes to great lengths to incorporate this feedback in its fitness measure (capacity to invade) but typically only for fixed behavioral traits or at most for simple individual behaviors. Note that it is still true that parasitoid foraging allows for better tests of behavioral ecology theories than predators, for which the link between foraging and fitness is less straightforward. However, even for parasitoids a fitness measure integrates most life history characteristics in potentially complex ways.

2.2 Practical approaches to parasitoid fitness

In the literature there exists a useful distinction between measures of true fitness (indicating a mutant's capacity to increase in frequency, often expressed as a reproduction ratio R_0 or a per capita population growth rate) and measures of indirect fitness such as

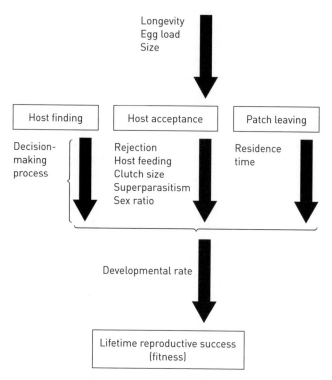

Fig. 2.1 Factors determining fitness or lifetime reproductive success.

fecundity, longevity, or size (Stephens & Krebs 1986). An indirect fitness measure is only accepted as a correct fitness measure if it changes monotonically with the true fitness measure. Roitberg et al. (2001) state that few biologists ever measure fitness directly. It should be noted that different measures of fitness have their impact in different parts of the foraging process (Fig. 2.1). For instance, the body size of parasitoids, although directly related to potential fitness (fecundity and, less strictly, egg load), certainly is not directly related to realized fitness. The latter kind of fitness is what affects population dynamics and is determined by the spatial distribution of hosts and the time allocation of the parasitoid under study (Vos & Hemerik 2003). How parasitoids decide to exploit their environment has its direct impact on realized fitness. The cues upon which a parasitoid reacts are of great importance for its realized lifetime reproductive success (a fitness measure). From here, we first derive theoretical fitness measures in different settings before returning to the question of how to deal theoretically with the fact that host feeding, superparasitism, and encapsulation all have different impacts on fitness.

2.3 Theoretical approach to parasitoid fitness

2.3.1 Adaptive dynamics

When we refer to 'true' fitness, what do we actually mean? Different scientific domains have developed different and sometimes apparent contradictory definitions. The framework of adaptive dynamics has recently been developed (Metz et al. 1992, 1996, Geritz et al. 1997, 1998) in an attempt to bridge the gap between quantitative population genetics and population dynamics. This framework explicitly takes into account ecological feedbacks, leading to frequency and density dependence. Typically, we start with a standard ecological population model, either defined in continuous or discrete time. The simplest adaptive dynamics approach, that we use in the example below, is then based on the following procedure:

1 start with a population dynamical interaction, here a host–parasitoid interaction;
2 determine which of the two populations is considered to be 'the evolving population', here the parasitoid population;
3 extend the model in such a way that it incorporates two strains of the evolving population, the so-called 'resident' versus 'mutant' strain; and
4 then focus on the dynamics of the mutant strain that is rare relative to the resident strain.

Therefore, population dynamics, including the interaction with the host population, are largely determined by the resident population. Thus, the dynamics of the mutant is essentially decoupled in the sense that mutant traits affect neither the dynamics of the resident nor those of the host. The standard approach in adaptive dynamics is that mutation occurs on a timescale that is long relative to the timescale of convergence to an ecological attractor (stable point and limit cycle). If a new (rare) mutant appears, it can either decrease or increase in numbers. This is essentially determined by the invasion fitness, that is the per capita growth rate of the new mutant in an environment consisting of only residents. If invasion fitness is negative, then the mutant eventually disappears and nothing happens. However, if invasion fitness is positive, the mutant invades and typically replaces the resident, becoming the new resident itself. In the process where a successful invading mutant parasitoid out-competes the resident, the parasitoid population has evolved. Those of the mutant have replaced the traits of the resident. Because the original mutant now has become the resident, a new mutant can arise that can give rise to a new trait substitution sequence.

As the method of adaptive dynamics explicitly incorporates the evolution–ecology feedback, it is particularly suited to address questions concerning the evolution of parasitoid–host systems. In such systems, evolutionary changes in one trophic level are bound to have consequences on the dynamics of the whole interaction. In our case, for instance, evolutionary changes in parasitoid searching behavior most likely affect the distribution of the hosts. A feedback then arises because selection on the searching strategies depends on the host distribution (van Baalen & Sabelis 1993).

Much of the theory in parasitoid biology is based on the biology of what might be called 'Nicholson–Bailey' animals, that is hosts and parasitoids that reproduce simultaneously at a single instant in time without any temporal spread whatsoever and in simple spatial structures (if any). As the research over multiple decades has shown (Godfray 1994, Hochberg & Ives 2000), parasitoids inhabit a spatio-temporally complex world instead, which requires heavy decision making. This complexity has made its way into the models that are employed to such an extent that it has become difficult to 'infer' parasitoid fitness from first principles rather than hypothesizing it. We now discuss in more detail two examples taken from our own work to illustrate this.

2.3.2 Deriving fitness from first principles

The fitness measure in SDP models is the cumulative number of surviving offspring of an average foraging parasitoid (Mangel & Clark 1988, Clark & Mangel 2000, Mangel 2006, see also Chapter 15 by Roitberg and Bernhard). This fitness measure is linked to the standard reproduction ratio, but is calculated bottom-up from first principles. In this approach, the population-dynamical consequences are considered neither on the ecological nor on the evolutionary timescales. The limitation of this approach is therefore that the consequences of the eco-evolutionary feedback are not taken into account. Its large advantage is that fitness incorporates an explicit consideration of the full stochasticity faced by a foraging parasitoid. That is, the foraging process of a parasitoid is bit by bit translated into occurrence probabilities of a particular series of events. The resulting overall fitness measure is the expected lifetime reproductive success, calculated as the sum of the fitness weights from all possible event series.

As an example, consider the host-feeding, solitary parasitoid *Encarsia formosa*. Burger et al. (2004) developed a stochastic dynamic model for this parasitoid during its full lifetime. They modeled the lifetime of *E. formosa* during the day and the night separately. As one of the building blocks, the authors derived an expression for the fitness increment of this synovigenic parasitoid during one day consisting of an active period of t_{max} time steps in which the parasitoid forages and an inactive period (night) of i time steps. This fitness measure depends on the mature egg load x, the energy level y, the time t within the day (measured in discrete time steps), and the parasitoid's age a in days (for an overview of the symbols see Table 2.1). The fitness was first bipartitioned based on the event of interruption during the foraging process. The resulting expected fitness is the probability of being interrupted during foraging p_{int} or not $(1 - p_{int})$ and the expected fitness associated with both these disjoint events:

$$F(x,y,t,a) = p_{int} \begin{bmatrix} \text{Expected fitness} \\ \text{when interrupted} \end{bmatrix} + (1 - p_{int}) \begin{bmatrix} \text{Expected fitness} \\ \text{when not interrupted} \end{bmatrix} \quad (2.1)$$

The expected fitness when interrupted is the product probability of surviving the remaining active period $t_{max} - t$, assuming survival is exponentially distributed (probability $\exp\{-(t_{max} - t)p_m\}$) and the fitness value at time step t_{max} when $m(t_{max} - t)$ egg equivalents are matured into eggs during $t_{max} - t$ time steps:

$$\begin{bmatrix} \text{Expected fitness} \\ \text{when interrupted} \end{bmatrix} = e^{-(t_{max}-t)p_m} F(x + m(t_{max} - t), y - m(t_{max} - t), t_{max}, a) \quad (2.2)$$

Thus, now the first term after the equals sign in Equation (2.1) is ready. When the parasitoid is not interrupted however, it can encounter no host or a host of type h $(= 1, \ldots, h_{max})$ in the time step. The encounter probability with host h is $p_e(h)$ and, thus, the probability of not encountering a host in the time step is $1 - \Sigma p_e(h)$. The encounter probabilities should be multiplied with the expected fitness for the disjoint events and summed (Equation 2.3) to obtain the expected fitness for not being interrupted:

Table 2.1 Explanation of the symbols in the text

Symbol	Description	Unit
a	Parasitoid age	Days
d	Foraging decision*	–
$F(x, y, t, a)$	Maximum expected reproductive output between (t, a) and death, given x eggs and y egg equivalents at (t, a)	–
$g(h)$	Gain from feeding on host type h	Egg equivalents
h	Host type	–
h_{max}	Number of different host types	–
$m(t_m)$	Number of egg equivalents matured during t_m time steps	–
$p_e(h)$	Probability of encountering host type h	Per time step
p_{int}	Probability of being interrupted during foraging	Per time step
p_m	Probability of instantaneous mortality	Per time step
t	Within day time	Time steps
$t_h(d)$	Handling time for decision d	Time steps
t_{max}	Length of foraging period within 1 day	Time steps
$W(h)$	Direct fitness increment (egg to adult survival) from oviposition in host type h	–
x	Mature egg load	Eggs
y	Energy level	Egg equivalents

*0 = rejection, 1 = host feeding, 2 = oviposition.

$$\begin{bmatrix} \text{Expected fitness} \\ \text{when not interrupted} \end{bmatrix} = \left(1 - \sum_{h=1}^{h_{max}} p_e(h)\right)\begin{bmatrix} \text{Expected fitness} \\ \text{when no host is} \\ \text{encountered} \end{bmatrix}$$
$$+ \sum_{h=1}^{h_{max}} \left(p_e(h) \begin{bmatrix} \text{Expected fitness when host type} \\ h \text{ is encountered and handled} \\ \text{with optimal decision} \end{bmatrix}\right) \quad (2.3)$$

When no host is encountered the expected fitness is the product of the probability that the parasitoid survives the time step and the fitness value in the next time step when $m(1)$ egg equivalents are matured during one time step (Equation 2.4):

$$\begin{bmatrix} \text{Expected fitness} \\ \text{when no host is} \\ \text{encountered} \end{bmatrix} = e^{-p_m} F(x + m(1), y - m(1), t + 1, a) \quad (2.4)$$

On encountering a host of type h, the parasitoid should decide what to do: decision d is defined as (0) for rejection, (1) for host feeding, or (2) for oviposition. The optimal decision is the decision from which the highest fitness increment is expected. Rejection of the encountered host implies that the expected fitness equals the probability of surviving the handling time, $t_h(0)$ time steps, needed to reject a host, multiplied by the fitness value at time step $t + t_h(0)$ associated with $m(t_h(0))$ egg equivalents matured into eggs during that period $t_h(0)$. If the parasitoid decides to host feed, survival and oogenesis occur over a period $t_h(1)$ time steps needed to feed upon a host and the fitness value at time step $t + t_h(1)$ is associated with an energy state additionally increased by $g(h)$ egg equivalents. If the parasitoid has at least one egg, it is able to oviposit. When the parasitoid decides to oviposit, the parasitoid gains an immediate fitness increment $W(h)$, which is the egg to adult survival of an oviposition in host type h. This fitness gain is paid for by a reduced egg load in the next time step $t + t_h(2)$. Solitary parasitoids do not make clutch size decisions and therefore the egg load is simply decreased by 1. Survival and oogenesis occur over a period of $t_h(2)$ time steps needed to oviposit a host. Summarizing:

$$\begin{bmatrix} \text{Expected fitness} \\ \text{when host type } h \\ \text{is encountered} \\ \text{and handled with} \\ \text{optimaldecision} \end{bmatrix} =$$

$$\max_{\text{decision } d} \begin{bmatrix} d = 0 : e^{-t_h(0)p_m} F(x + m(t_h(0)), y - m(t_h(0)), t + t_h(0), a) \\ d = 1 : e^{-t_h(1)p_m} F(x + m(t_h(1)), y - m(t_h(1)) + g(h), t + t_h(1), a) \\ d = 2 : W(h) + e^{-t_h(2)p_m} F(x + m(t_h(2)) - 1, y - m(t_h(2)), t + t_h(2), a) \end{bmatrix} \quad (2.5)$$

The total dynamic programming equation 'within days' is given by Equations (2.1)–(2.5). For the 'between days' dynamics, we refer to Burger et al. (2004).

The advantage of this approach is that it allows incorporation of the consequences of changes in internal states such as the egg complement x and energy reserves y. For instance, analysis shows that, under field conditions with low host densities, parasitoids should refrain from host feeding (Burger et al. 2004) but, under conditions where the parasitoid is used as a biological control agent, host feeding certainly increases the expected lifetime reproductive success (see also Chapter 7 by Bernstein and Jervis). Thus, the approach gives useful insight into the factors that determine optimum strategies. What this approach cannot do, however, is make predictions about whether a given optimum strategy is also evolutionarily stable, as it assumes that the focal individual lives in a fixed, given environment. For instance, in the model expressed by Equations (2.1)–(2.5), encounter rates p_h are treated as given constants, whereas in reality they depend on the dynamics of the interaction and the strategies adopted by the other members of the parasitoid population.

2.3.3 Deriving fitness from population dynamics

There are more elaborate flavors of adaptive dynamics, incorporating more realistic assumptions (regarding mutations, trait space, genetic architecture, population structure, and so on), but they are all based upon the idea that evolutionary processes depend crucially on the invasion fitness of mutants. The problem of working out the correct invasion fitness of parasitoids in deterministic models that incorporate the ecology–evolution feedback loop is not simple. On the one hand, this is a nuisance because it requires us to develop new approaches to overcome this problem. On the other hand, it renders an important result as it shows that classical fitness measures may not reflect reality. Thus, working out how invasion fitness depends on behavioral traits may teach us new things about parasitoid behavior.

Let us start, for instance, by considering the standard Nicholson–Bailey model (Nicholson & Bailey 1935). The familiar equations for the dynamics of host (N_t) and parasitoid (P_t) densities in discrete time are given by

$$\begin{cases} N_{t+1} = \lambda N_t e^{-aP_t} \\ P_{t+1} = cN_t(1 - e^{-aP_t}) \end{cases} \quad (2.6)$$

Here, λ represents the finite rate of increase of the host population, implying an exponentially growing host population without self-limitation, in the absence of parasitism. a is the searching efficiency of the parasitoids and c is the number of offspring emerging from a parasitized host. The fact that the proportion of hosts escaping parasitoid attack is expressed by the zero term of a Poisson distribution implies that hosts encounter parasitoids at random (Hemerik et al. 2002). While much of the literature on host parasitoids assesses the consequences of non-random search (parasitoid aggregation), we will not discuss this aspect in this chapter. The advantage of the current choice is that it is conceptually simple (it represents the null case of no spatial structure) and easy to analyze.

First, we observe that, even in this simple setting, it is not immediately clear how we should extend these equations to incorporate structure in the parasitoid population. For continuous-time systems like the Lotka–Volterra predator–prey model, adding a mutant is simple: just add a predation term to the prey equation and add an equation describing the dynamics of the mutant predator. For the discrete-time systems, we have to make choices about how mutant and resident parasitoids interact during one generation.

The simplest one-host, two-parasitoid extension of the Nicholson–Bailey model is the following (van Baalen & Sabelis 1993):

$$\begin{cases} N_{t+1} = \lambda N_t e^{-a_P P_t - a_Q Q_t} \\[2mm] P_{t+1} = c_P N_t \dfrac{a_P P_t}{a_P P_t + a_Q Q_t}(1 - e^{-a_P P_t - a_Q Q_t}) \\[2mm] Q_{t+1} = c_Q N_t \dfrac{a_Q Q_t}{a_P P_t + a_Q Q_t}(1 - e^{-a_P P_t - a_Q Q_t}) \end{cases} \quad (2.7)$$

Here P_t and Q_t represent the resident and mutant strains of parasitoids. In this model, the competing parasitoid strains are allowed to have different search and conversion efficiencies, $a_P \neq a_Q$ and/or $c_P \neq c_Q$. Equation (2.7) is based on the additional assumption that there is no within-host advantage of either strain. The first parasitoid to parasitize a given host always wins, although that is not the case in the real world and depends on the time between the different ovipositions (Visser et al. 1992b, Gates 1993, Field et al. 1997).

If one of the two strains (say 'mutant' Q_t) is rare, we have $Q_t \ll P_t$ and we can approximate the resident system by the Nicholson–Bailey model (Equation 2.6). In addition, the mutant Q_t is governed by an adapted version of the third term in Equation (2.7):

$$Q_{t+1} = c_Q N_t \frac{a_Q Q_t}{a_P P_t}(1 - e^{-a_P P_t}) \tag{2.8}$$

From this equation, we can easily infer the capacity for long-term invasion of the mutant given that the resident dynamics is known, because the mutant increases if

$$\lim_{n \to \infty} \sqrt[n]{\prod_{i=1}^{n} c_Q N_i \frac{a_Q}{a_P P_i}(1 - e^{-a_P P_i})} > 1 \tag{2.9}$$

Of course, the Nicholson–Bailey model is the quintessence of an unstable model, which means that the fitness measure that we derived is essentially academic. The resident system does not persist so there is nothing for mutants to invade. It is important to note, however, that our fitness measure also applies to settings that lead to some more realistic models. For instance, adding density-dependent reproduction in the host population can render the model, if not stable, then still permanent. The host and resident parasitoid persist even if they fluctuate. Then, the invasion condition (Equation 2.9) presents the correct way to calculate invasion capacity once the resident dynamics (N_t and P_t) are known. If the resident system settles at an equilibrium ($\bar{N}, \bar{P}$), then the invasion conditions simplifies into

$$c_Q \bar{N} \frac{a_Q}{a_P \bar{P}}(1 - e^{-a_P \bar{P}}) > 1 \tag{2.10}$$

The left-hand side of this condition is often called the 'basic reproduction ratio' (often denoted R_0) of the parasitoids, so

$$R_0 = c_Q \bar{N} \frac{a_Q}{a_P \bar{P}}(1 - e^{-a_P \bar{P}}) \tag{2.11}$$

R_0 is a fitness measure because any mutant with an $R_0 > 1$ has on average more than one offspring to replace itself, which implies that the mutant's strategy will invade. In the adaptive dynamics framework, a mutant that has adopted the same strategy as the resident has $R_0 = 1$, reflecting the fact that it is in population dynamical equilibrium. Note that this fitness measure depends explicitly on both the host and parasitoid equilibrium densities. If both life history and environment are built into a model, then the reproductive ratio R_0 and the per capita growth rate $r = \ln(\lambda)$ are linked and, in evolutionary–ecological equilibrium, it holds that $r = R_0 - 1 = 0$ (Mylius & Diekmann 1995).

From this simplest of models, we can conclude that the mutant parasitoid that maximizes the product $c_Q a_Q$ optimizes its invasion capacity. When adopted by the resident population, this strategy cannot be invaded by any other strategy and, hence, is an 'evolutionary stable strategy' (ESS) (Maynard Smith & Price 1973). For this example, this conclusion is not particularly enlightening. A parasitoid should maximize its searching efficiency and increase its conversion efficiency. There may be a trade-off between these two parameters (i.e. energy used for searching is no longer available for egg production) but, if not, the parasitoids are expected to be at the physiological limit that maximizes the traits simultaneously.

2.3.4 Top-down or bottom-up?

Evolution occurs because successful mutants replace their less successful ancestors. This fundamental principle underlies all modeling approaches, including the adaptive dynamics framework (Metz et al. 1992, Geritz et al. 1998). The main problem is to identify which mutations can invade under what conditions. These conditions can be derived from an extended (multi-strain) population dynamical model. Usually the invasion fitness of a new mutant is expressed in terms of its so-called 'dominant Lyapunov exponent' or long-term rate of growth when rare (i.e. the logarithm of the left-hand side of Equation 2.9). When its long-term per capita growth rate is positive, a mutant eventually invades and (except in special cases) replaces ancestral populations. Only when none of the possible mutants has a positive invasion fitness does evolution stop and an ESS has been reached. Before that happens, other interesting evolutionary phenomena such as branching can have occurred.

A potential problem is that, in general, the long-term rate of growth has no simple relationship with reproduction ratio style fitness measures. However, a careful analysis has shown that positive invasion fitness corresponds with a reproductive ratio larger than unity when a mutant is rare and the effect of density dependence is correctly incorporated (Mylius & Diekmann 1995). It should be noted that the latter proviso is necessary because some definitions of the reproduction ratio stipulate the absence of density dependence. The use of reproductive ratios as fitness measures is not limited to discrete time or discrete generation models, as they apply equally well to continuous-time settings with overlapping generations.

The two fitness measures that we have developed above are both reproduction ratios and thus, compatible with each other and lead to correct evolutionary predictions. The problem is that both approaches lead to equations that are not so easy to analyze.

2.4 Aspects affecting parasitoid fitness

The following briefly discusses a number of aspects that are pertinent to parasitoid life histories but are difficult to capture in fitness measures. The first case that we discuss is that of egg limitation. This is a conceptually simple case that nevertheless leads to unsolvable mathematical expressions. Thereafter, we discuss a number of other examples without presenting the complex mathematical expressions that should be studied to analyze them.

2.4.1 Egg limitation

Equation (2.11) for lifetime reproductive success was derived from a population-level description of the parasitoid–host interaction (i.e. by adding a mutant population to the resident system (Equation (2.6)). Here, we show that the same expression can be obtained using an individual-based way of reasoning. Assume that a mutant parasitoid has a particular probability of stochastically encountering exactly $\underline{n}$ hosts during its lifetime, with the probability $\Pr(\underline{n} = n) = f_Q(n)$. Then, the parasitoid's expected lifetime reproductive success is

$$R_0 = \sum_{n=0}^{\infty} f_Q(n) n c_Q \sigma \qquad (2.12)$$

where σ represents the probability that a located host is successfully parasitized, producing c_Q offspring. If there were no other parasitoids in the system and hosts do not employ countermeasures, we can set $\sigma = 1$ and, if we then assume that the mean number of hosts encountered is $\bar{n} = a_Q \tilde{N}$, the parasitoid's R_0 would be equal to

$$R_0 = c_Q a_Q \tilde{N} \tag{2.13}$$

However, the mutant is typically not the only parasitoid interacting with the host population and we have to take into account that hosts may also be found by other parasitoids. As this may affect the success of parasitization, σ should depend on the densities (and strategies) of other parasitoids: $\sigma = \sigma(P_t)$. It is shown in the Appendix that, if host–parasitoid encounters are random and if the first parasitoid to locate a host wins the competition for this host, then

$$\sigma(P_t) = \frac{1 - e^{-a_p P_t}}{a_p P_t} \tag{2.14}$$

and we recover Equation (2.6).

Another assumption that goes into the derivation of Equation (2.6) is that the parasitoids are able to put an egg into every host they encounter. Though some parasitoid females carry large numbers of eggs in their ovarioles, they might encounter more hosts than they can cater for. In the outline below, we show that there are reasons to expect that parasitoid egg loads should not be much larger than the expected number of oviposition occasions.

Suppose that hosts are not found by other parasitoids but that the probability an offspring develops into a parasitized host only depends upon how many eggs the parasitoid has laid before, $\sigma = \sigma(n)$, for example: then if every female has a fixed egg load of E eggs at hatching

$$\sigma(n) = \begin{cases} 1 & (n \leq E) \\ E/n & (n > E) \end{cases} \tag{2.15}$$

In fact, $\sigma(\underline{n})$ represents the average survival probability of a clutch of n eggs. Then, if the number of host encounters exceeds the egg load, the parasitoid has n times E/n, which results in E offspring. Then, the expected lifetime reproductive success can be expressed as (Ellers et al. 2000, van Baalen 2000)

$$R_0 = c_Q \left(\sum_{n=0}^{E} f_Q(n) n + \sum_{n=E+1}^{\infty} f_Q(n) E \right) \tag{2.16}$$

This equation can be rewritten in the following form:

$$R_0 = c_Q \sum_{n=0}^{E} f_Q(n) n + c_Q E \left(1 - \sum_{n=0}^{E} f_Q(n) \right) \tag{2.17}$$

where the first term sums fitness contributions over the cases where the parasitoid encounters fewer hosts than her egg load while the second term collects all cases where the parasitoid runs out of eggs before dying. The factor $1 - \sum_{n=0}^{E} f_Q(n)$ gives the probability of the parasitoid being egg limited. Unfortunately, for arbitrary distributions these summations are difficult to solve analytically.

Ellers et al. (2000) presented a numerical analysis of a similar model, which has the additional ingredient of a trade-off between the egg load and host longevity (not discussed here but which affects the expected number of hosts encountered). For a given set of parameters, we can numerically approximate the optimum egg load E relative to the expected number of hosts encountered. For a large (but relatively arbitrary) set of parameter values, they concluded that an evolved egg load tends to be larger (but not much) than the expected number of hosts encountered, suggesting that egg-limited parasitoids should be relatively rare. Ideally, the egg load matches the number of host encounters. Having too few eggs implies wasted opportunities, while having too many eggs reduces longevity (which also reduces opportunity).

One limitation of this approach is that it does not include the eco-evolutionary feedback loop. It indeed neglects the fact that an evolving parasitoid population will (more likely than not) change either the density or the distribution of their host population (or both). van Baalen (2000) carried out an analysis of a simpler model (similar to the one discussed in this section) to arrive at the conclusion that, if there is a constraint between egg load and egg size, then an ESS strategy exists that is regulated by an eco-evolutionary feedback loop. If all parasitoids produce few but competitively dominant eggs, cycles result with a relatively high mean host density. Then, parasitoids are favored to produce small eggs (i.e. many but weaker eggs), but this reduces the mean host density, favoring decreased egg loads.

As is illustrated by Ellers et al.'s (2000) model, an insight gained in recent decades is that parasitoid longevity is an important trait. In the Nicholson–Bailey framework, this trait does not appear because all parasitoids are assumed to live exactly one time step and then die. In reality, however, parasitoids live in a structured environment and their longevity relative to season length may be crucially variable (Heimpel et al. 1998).

Another approach, pioneered by Briggs et al. (1995), puts the whole host–parasitoid interaction in continuous time and traces the different classes (with different egg loads) of parasitoids using differential equations. In this way, Briggs et al. (1995) showed that host feeding may affect the equilibrium densities of host and parasitoids (see also Chapter 7 by Bernstein and Jervis) but is not expected to affect the stability of the equilibrium. This continuous time approach has the advantage of allowing explicit consideration of the longevity of the parasitoids but at the expense of an easily analyzable population dynamical model, in particular when the model involves time delays.

2.4.2 Patch exploitation

Often hosts occur in discrete clumps or patches (see also Chapter 8 by van Alphen and Bernstein). Then parasitoids face the problem of how to exploit such patches. Of course, having located a patch of hosts, they should try to exploit it, but at the same time they should avoid staying too long on exploited patches and miss better opportunities. Studies on optimal patch time allocation suggest simple behavioral rules or mechanistic models (Charnov 1976, Waage 1979, Stephens & Krebs 1986, Driessen et al. 1995, Wajnberg 2006, see also Chapter 8 by van Alphen and Bernstein) and deduce behavioral rules using statistical modeling (Haccou et al. 1991).

It is obvious that the fitness of a parasitoid in a dynamically changing patchy environment is strongly determined by how it creates and uses the opportunities for oviposition and how it reacts to encounters with healthy and parasitized hosts (Kolss et al. 2006). However, to generate accurate predictions about what behavior will be favored requires a correct, non-biased fitness measure. Unfortunately, it is almost impossible to come up with simple expressions. There are two reasons for this. First, not only the environment encountered by a parasitoid is stochastic, but the probability distribution describing for instance the spatial distribution of the host in this environment can vary independently or depending on the focal parasitoid's previous behavior or experience and that of its conspecifics that exploit the same patch. Even an approach that assumes no feedback loops, such as the SDP approach (see also Chapter 15 by Roitberg and Bernhard), faces the difficult task of working out how to calculate average lifetime reproductive success over all possible host encounter sequences.

Due to patch depletion, a parasitoid searching for hosts in a patch experiences a diminishing encounter rate with unparasitized and thus suitable hosts. To obtain a maximum lifetime reproductive success, individual parasitoids should be highly flexible in their decision-making process to face ever-changing circumstances (see also Chapter 8 by van Alphen and Bernstein). The available time should be used most efficiently. Therefore, the parasitoid constantly has to decide whether to stay in the patch and continue to search for hosts or to leave the current patch, search for, and travel to another patch in the habitat. Studies of patch time always refer to the MVT (Charnov 1976, see also Chapter 8 by van Alphen and Bernstein and Chapter 9 by Haccou and van Alphen). However, in this simple model the parasitoid is expected to leave when its gain rate in the current patch drops below the average gain rate in its full habitat. Because we cannot expect parasitoids to be omniscient with respect to the total number of hosts or conspecifics and the parasitization status of the hosts, the simple theoretical reasoning cannot be expected to apply in the real world. For in-depth discussions of these topics we refer to Chapter 8 by van Alphen and Bernstein, who review how information acquisition can affect optimal patch exploitation strategies and to Chapter 9 by Haccou and van Alphen for a discussion of the effect of the mode of competition.

Here, we merely point out what can be done to circumvent some of the mathematical problems that arise. One solution is to assume that fitness increments, instead of coming in a large discrete patch of hosts, are sufficiently small that they can be described as a continuous 'harvest'. Such a continuous fitness accumulation approach allowed Sjerps and Haccou (1994) and, more recently, Hamelin et al. (2007) to formulate and analyze differential games to work out optimum residence times for multiple parasitoids sharing a patch.

Obtaining a fitness measure for a parasitoid's lifetime involves knowing the distributions of patch residence and travel times, the relative encounter with healthy and parasitized hosts, and the distribution of the parasitoid's longevity. Taking averages over such distributions renders the calculations tedious or outright impossible, but in simple cases it might be feasible.

2.4.3 Superparasitism

Another simplifying aspect of the Nicholson–Bailey caricature is that it assumes that hosts found more than once still produce the same average number of parasitoid offspring. That is, there is a parameter for the number of parasitoids to emerge from a parasitized host but it is implicitly assumed that these are offspring from the same mother. Now, in many parasitoid species, hosts may be superparasitized (Parker & Courtney 1984, Bakker et al. 1985, Ives 1989) and even more often when parasitoids are exploiting a patch at the same time (Hemerik et al. 2002). Moreover, the developing larvae of solitary species can be aggressive and kill each other by contest competition, with only the oldest one likely to survive (Bakker et al. 1985). Many parasitoids have developed elaborate ways to mark parasitized hosts (Bakker et al. 1985, Roitberg & Prokopy 1987) though it is not obvious whether subsequent parasitoids should always avoid ovipositing in marked hosts (Nagelkerke et al. 1996, Plantegenest et al. 2004). As for patch exploitation strategies, testable predictions about the conditions and consequences of host marking require a correct fitness function. A common approach is to use SDP to work out optimal host acceptation strategies (Roitberg & Mangel 1989, see also Chapter 15 by Roitberg and Bernhard).

In addition, the decision making of parasitoids involved in the process of superparasitism depends on the time between the first and the subsequent oviposition. Gates (1993) reported a decreasing probability of superparasitism in *Antrocephalus pandens* as the time between the ovipositions increases. In addition, the probability of emerging from a superparasitized host can be smaller (Field et al. 1997). In fact, an analysis of resource sharing by juveniles (of arbitrary species) in patchy environments (Rohlfs & Hoffmeister 2003) suggested that, in some situations, there might actually be a benefit when parasitoid larvae are not alone.

Fitness functions for superparasitism involve a game theoretic component as the fitness of one individual strongly depends on the strategies adopted by others (van der Hoeven & Hemerik 1990, Visser et al. 1992a). The problems that this creates become particularly pertinent with host marking, as this allows one individual to manipulate another by influencing the other's perception of its options by providing information. Then, one female's fitness cannot be disentangled from the behavior of its conspecifics it is interacting with. This may pose problems in particular for SDP approaches, which typically consider the environment to be fixed. For instance, SDP identifies optimum strategies relatively easily because it solves equations by going 'backwards' in time (i.e. start with no future reproductive output left at the end of a parasitoid's life and work out which decision in the previous state maximizes fitness increase and so on backwards to $t = 0$, where the maximum lifetime fitness is known). However, when a decision at time t_1 modifies a demographic value at time t_2 later (for instance when the survival of a parasitoid egg depends on the decision of a parasitoid that arrives later), solving backwards in time is no longer feasible.

2.4.4 Encapsulation and resistance

Many host species, such as *D. melanogaster*, are able to defend themselves against parasitism by their parasitoids (*L. heterotoma* and *Asobara tabida* in the case of *D. melanogaster*) by encapsulation of the parasitoids' eggs (see also Chapter 14 by Kraaijeveld and Godfray). For *L. heterotoma*, it has been shown that eggs laid later are more susceptible to encapsulation (Streams 1968). This defense, however, is not perfect and the parasitoid eggs may escape from encapsulation, for instance if they are embedded in host tissue instead of floating in the hemolymph. Kraaijeveld et al. (2001) found that the parasitoid *A. tabida* could be selected to become resistant against encapsulation (see also Chapter 14 by Kraaijeveld and Godfray). The cost that has to be paid for obtaining this resistance is in a slower hatching of the parasitoid egg within the host larva.

That this system of defense and counter-defense may give rise to an arms race between host and parasitoid (Carton & Nappi 1991) is indicated by the fact that there is considerable geographic variation in the hosts' capacity to encapsulate and the parasitoids' so-called 'virulence' (Kraaijeveld & van Alphen 1994, 1995, see also Chapter 14 by Kraaijeveld and Godfray). The mere fact that encapsulation occurs implies that parasitoid eggs are costly, otherwise parasitoids would always produce maximally virulent eggs and encapsulation would not be observed. However, parasitoids have limited resources and they need to decide how to partition this into egg quantity and quality. Though this dilemma has been analyzed for seed size in plants, few models have tried to assess the consequences for parasitoids, presumably because it depends on egg limitation, which is difficult to model (van Baalen 2000).

2.5 Discussion

Though studies of parasitoid biology were greatly stimulated by the postulated direct link between behavior and fitness, subsequent developments have shown that the simplicity of this link is deceptive. In fact, working out lifetime reproductive success of parasitoids typically is a daunting task. This hampers further development in the field, because without a proper fitness measure it is difficult to infer the selective pressures that shape parasitoid life history traits. In particular, quantitatively assessing the costs and benefits associated with a given behavior becomes nearly impossible.

One of the main aspects that makes parasitoid fitness difficult to express is the fundamental role that stochasticity plays in the foraging part of their life. In contrast to consumers like those grazing herbivores that encounter an almost continuously varying resource supply, parasitoids encounter their hosts in discrete entities and at such low numbers that stochastic variation is inevitable. This implies that calculating the mean fitness involves taking averages over distributions and, in all but the simplest cases, renders the calculations tedious or outright impossible. For instance, the Nicholson–Bailey model is solvable only when simple distributions for the probability of attacking and encountering are assumed.

A first advantage of nevertheless trying to work out fitness measures is that the exercise tells us exactly what selective pressure parasitoids are subject to. New insights may thus result, for instance that natural selection should favor parasitoids that are in between the classical dichotomy of pro-ovigeny and synovigeny, which seems to be supported by the available observations (Ellers et al. 2000).

Another important insight that thus emerged is that parasitoid longevity is not a given constant but depends on a number of factors that are subject to evolution. This renders parasitoid life history dilemmas more complicated. The number of hosts a parasitoid encounters indeed not only depends on how many there are and how they are distributed in its environment but also on how long the parasitoid can expect to live. The variance in host encounters that results is a crucial variable affecting selection on egg load. Egg load thus depends potentially on all other parasitoid life history traits (Heimpel et al. 1998).

As we have discussed, there exist two different ways to derive parasitoid fitness measures (or fitness measures in general). The first way is to derive fitness from a calculation of the expected number of surviving offspring of a focal individual. The advantage of this method is that it allows incorporating the full stochasticity that is experienced by the focal individual. The disadvantage is that it does not give insight into the consequences of (inevitable) feedback mechanisms that operate in host–parasitoid dynamics. The second way is the adaptive dynamics approach, which derives fitness from the capacity of a rare mutant strategy to increase in density. In practise, it is more difficult to incorporate stochasticity into this method. The effects of ubiquitous eco-evolutionary feedbacks, on the other hand, are more easily assessed. We want to stress that the two approaches are not fundamentally different. This is illustrated by the fact that the same fitness measure can be derived from a population-level approach (such as adaptive dynamics) as from an individual-level approach (such as SDP).

Parasitoid fitness is the integral outcome of the behaviors performed by a parasitoid in its full lifetime. Recent studies show that notions like longevity, time allocation, behavioral decision making with respect to (super)parasitism, and host feeding are all important in determining lifetime reproductive success. While it remains true that locating hosts is of paramount importance to a parasitoid and, thus, a convenient model system for investigating the effects of certain behaviors, host location is not all there is to its struggle for existence. In particular, behavior that affects life history components and, therewith, realized fitness needs a more sophisticated paradigm than the original idea that fitness can straightforwardly be inferred from the number of oviposition occasions.

References

Atkinson, W.D. and Shorrocks, B. (1984) Aggregation of larval Diptera over discrete and ephemeral breeding sites: the implications for coexistence. *American Naturalist* **124**: 336–51.

Bakker, K., van Alphen, J.J.M., van Batenburg, F.H.D. et al. (1985) The function of host discrimination and superparasitization in parasitoids. *Oecologia* **67**: 572–6.

Briggs, C.J., Nisbet, R.M., Murdoch, W.W., Collier, T.R. and Metz, J.A.J. (1995) Dynamical effects of host-feeding in parasitoids. *Journal of Animal Ecology* **64**: 403–16.

Brommer, J.E. (2000) The evolution of fitness in life-history theory. *Biological Reviews* **75**: 377–404.

Burger, J.M.S., Hemerik, L., van Lenteren, J.C. and Vet, L.E.M. (2004) Reproduction now or later: optimal host-handling strategies in the whitefly parasitoid *Encarsia formosa*. *Oikos* **106**: 117–30.

Carton, Y. and Nappi, A. (1991) The *Drosophila* immune reaction and the parasitoid capacity to evade it – genetic and coevolutionary aspects. *Acta Oecologia* **12**: 89–104.

Charnov, E.L. (1976) Optimal foraging: the marginal value theorem. *Theoretical Population Biology* **9**: 129–36.

Chesson, P.L. and Murdoch, W.W. (1986) Aggregation of risk: relationships among host–parasitoid models. *American Naturalist* **127**: 696–715.

Clark, C.W. and Mangel, M. (2000) *Dynamic State Variable Models in Ecology: Methods and Applications*. Oxford University Press, Oxford.

Driessen, G. and Hemerik, L. (1991) Aggregative responses of parasitoids and parasitism in populations of *Drosophila* breeding in fungi. *Oikos* **61**: 96–107.

Driessen, G., Bernstein, C., van Alphen, J.J.M. and Kacelnik, A. (1995) A count down mechanism for host searching in the parasitoid *Venturia canescens*. *Journal of Animal Ecology* **64**: 117–25.

Ellers, J., Sevenster, J.G. and Driessen, G. (2000) Egg load evolution in parasitoids. *American Naturalist* **156**: 650–65.

Field, S.A., Keller, M.A. and Calbert, G. (1997) The pay-off from superparasitism in the egg parasitoid *Trissolcus basalis*, in relation to patch defence. *Ecological Entomology* **22**: 142–9.

Gates, S. (1993) Self and conspecific superparasitism by the solitary parasitoid *Anthrocephalus pandens*. *Ecological Entomology* **18**: 303–9.

Geritz, S.A.H., Metz, J.A.J., Kisdi, E. and Meszéna, G. (1997) The dynamics of adaptation and evolutionary branching. *Physical Review Letters* **78**: 2024–7.

Geritz, S.A.H., Kisdi, E., Meszéna, G. and Metz, J.A.J. (1998) Evolutionarily singular strategies and the adaptive growth and branching of the evolutionary tree. *Evolutionary Ecology* **12**: 35–57.

Godfray, H.C.J. (1994) *Parasitoids: Behavioural and Evolutionary Ecology*. Monographs in Behavior and Ecology, Princeton University Press, Princeton.

Haccou, P., De Vlas, S.J., van Alphen, J.J.M. and Visser, M.E. (1991) Information processing by foragers: effects of intra-patch experience on the leaving tendency of *Leptopilina heterotoma*. *Journal of Animal Ecology* **60**: 93–106.

Hamelin, F., Bernhard, P., Shaiju, A. and Wajnberg, É. (2007) Foraging under competition: evolutionarily stable patch-leaving strategies with random arrival times. 2. Interference competition. In: Jørgensen, S., Quincampoix, M. and Vincent, T.L. (eds.) *Annals of the International Society of Dynamic Games*, vol. 9. Birkhäuser, Basel, pp. 349–65.

Hardy, I. (1992) Non-binomial sex allocation and brood sex ratio variances in the parasitoid Hymenoptera. *Oikos* **65**: 143–58.

Hassell, M.P. (1978) *The Dynamics of Arthropod Predator–Prey Systems*. Monographs in Population Biology, Princeton University Press, Princeton.

Hassell, M.P. and May, R.M. (1973) Stability in insect host–parasite models. *Journal of Animal Ecology* **42**: 693–736.

Hassell, M.P. and May, R.M. (1974) Aggregation in predators and insect parasites and its effect on stability. *Journal of Animal Ecology* **43**: 567–94.

Heimpel, G.E., Mangel, M. and Rosenheim, J.A. (1998) Effects of time limitation and egg limitation on lifetime reproductive success of a parasitoid in the field. *American Naturalist* **152**: 273–89.

Hemerik, L., van der Hoeven, N. and van Alphen, J.J.M. (2002) Egg distributions and the information a solitary parasitoid has and uses for its oviposition decisions. *Acta Biotheoretica* **50**: 167–88.

Hochberg, M.E. (1997) Hide or fight? The competitive evolution of concealment and encapsulation in host–parasitoid associations. *Oikos* **80**: 342–52.

Hochberg, M.E. and Ives, A.R. (2000) *Parasitoid Population Biology*. Princeton University Press, Princeton.

Ives, A.R. (1989) The optimal clutch size of insects when many females oviposit per patch. *American Naturalist* **133**: 671–87.

Kolss, M., Hoffmeister, T.S. and Hemerik, L. (2006) The theoretical value of encounters with parasitized hosts for parasitoids. *Behavioral Ecology & Sociobiology* **61**: 291–304.

Kraaijeveld, A.R. and van Alphen, J.J.M. (1994) Geographical variation in resistance of the parasitoid *Asobara tabida* against encapsulation by *Drosophila melanogaster* larvae – the mechanism explored. *Physiological Entomology* **19**: 9–14.

Kraaijeveld, A.R. and van Alphen, J.J.M. (1995) Geographical variation in encapsulation ability of *Drosophila* melanogaster larvae and evidence for parasitoid-specific components. *Evolutionary Ecology* **9**: 10–17.

Kraaijeveld, A.R., Hurcheson, K.A., Limentani, E.C. and Godfray, H.C.J. (2001) Costs of counter-defences to host resistance in a parasitoid of *Drosophila*. *Evolution* **55**: 1815–21.

Krebs, J.R. and Davies, N.B. (1997) *Behavioural Ecology: An Evolutionary Approach*, 4th edn. Blackwell Science Ltd, Oxford.

Lei, G.-C. and Camara, M.D. (1999) Behaviour of a specialist parasitoid, *Cotesia melitaearum*: from individual behaviour to metapopulation processes. *Ecological Entomology* **24**: 59–72.

Mangel, M. (2006) *The Theoretical Biologist's Toolbox*. Cambridge University Press, Cambridge.

Mangel, M. and Clark, C.W. (1988) *Dynamic Modeling in Behavioral Ecology*. Princeton University Press, Princeton.

May, R.M. (1978) Host parasitoid systems in patchy environments: a phenomenological model. *Journal of Animal Ecology* **47**: 249–67.

Maynard Smith, J. and Price, G.R. (1973) The logic of animal conflict. *Nature* **246**: 15–18.

Metz, J.A.J., Nisbet, R.M. and Geritz, S.A.H. (1992) How should we define 'fitness' for general ecological scenarios. *Trends in Ecology & Evolution* **7**: 198–202.

Metz, J.A.J., Geritz, S.A.H., Meszéna, G., Jacobs, F.J.A. and van Heerwaarden, J.S. (1996) Adaptive dynamics, a geometrical study of the consequences of nearly faithful reproduction. In: van Strien, S.J. and Verduyn Lunel, S.M. (eds.) *Stochastic and Spatial Structures of Dynamical Systems.* North-Holland, Amsterdam, pp. 183–231.

Mylius, S.D. and Diekmann, O. (1995) On evolutionarily stable life-history strategies, optimization and the need to be specific about density dependence. *Oikos* **74**: 218–24.

Nagelkerke, C.J., van Baalen, M. and Sabelis, M.W. (1996) When should a female avoid adding eggs to the clutch of another female? A simultaneous oviposition and sex allocation game. *Evolutionary Ecology* **10**: 475–97.

Nicholson, A.J. and Bailey, V.A. (1935) The balance of animal populations. Part 1. *Proceedings of the Zoological Society of London* 551–98.

Parker, G.A. and Courtney, S.P. (1984) Models of clutch size in insect oviposition. *Theoretical Population Biology* **26**: 27–48.

Plantegenest, M., Outreman, Y., Goubault, M. and Wajnberg, É. (2004) Parasitoids flip a coin before deciding to superparasitize. *Journal of Animal Ecology* **73**: 802–6.

Rohlfs, M. and Hoffmeister, T.S. (2003) An evolutionary explanation of the aggregation model of species coexistence. *Proceedings of the Royal Society of London Series B Biological Science* **270**: 33–5.

Roitberg, B.D. and Mangel, M. (1989) On the evolutionary ecology of marking pheromones. *Evolutionary Ecology* **2**: 289–315.

Roitberg, B.D. and Prokopy, R.J. (1987) Insects that mark host plants. *BioScience* **37**: 400–6.

Roitberg, B.D., Boivin, G. and Vet, L.E.M. (2001) Fitness, parasitoids, and biological control: an opinion. *Canadian Entomologist* **133**: 429–38.

Sjerps, M. and Haccou, P. (1994) Effects of competition on optimal patch leaving: a war of attrition. *Journal of Theoretical Biology* **46**: 300–18.

Stephens, D.W. and Krebs, J.R. (1986) *Foraging Theory. Monographs in Behavior and Ecology.* Princeton University Press, Princeton.

Streams, F.A. (1968) Factors affecting the susceptibility of *Pseudeucoila bochei* eggs to encapsulation by *Drosophila melanogaster*. *Journal of Invertebrate Pathology* **12**: 379–87.

van Alphen, J.J.M. and Visser, M.E. (1990) Superparasitism as an adaptive strategy for insect parasitoids. *Annual Review of Entomology* **35**: 59–79.

van Baalen, M. (2000) The evolution of parasitoid egg load. In: Hochberg, M.E. and Ives, A.R. (eds.) *Parasitoid Population Biology.* Chapman & Hall, New York, pp. 103–20.

van Baalen, M. and Sabelis, M.W. (1993) Coevolution of patch selection strategies of predators and prey and the consequences for ecological stability. *American Naturalist* **142**: 646–70.

van Baalen, M. and Sabelis, M.W. (1999) Non-equilibrium dynamics of 'ideal and free' prey and predators. *American Naturalist* **154**: 69–88.

van der Hoeven, N. and Hemerik, L. (1990) Superparasitism as an ESS: to reject or not reject, that is the question. *Journal of Theoretical Biology* **146**: 467–82.

Visser, M.E. (1991) Prey selection by predators depleting a patch: an ESS model. *Netherlands Journal of Zoology* **41**: 63–80.

Visser, M.E., van Alphen, J.J.M. and Hemerik, L. (1992a) Adaptive superparasitism and patch time allocation in solitary parasitoids: an ESS model. *Journal of Animal Ecology* **61**: 93–101.

Visser, M.E., Luyckx, B., Nell, H.W. and Boskamp, G.J.F. (1992b) Adaptive superparasitism in solitary parasitoids marking of parasitized hosts in relation to the pay-off from superparasitism. *Ecological Entomology* **17**: 76–82.

Vos, M. and Hemerik, L. (2003) Linking foraging behavior to lifetime reproductive success for an insect parasitoid: adaptation to host distributions. *Behavioral Ecology* **14**: 236–45.

Waage, J.K. (1979) Foraging for patchily distributed hosts by the parasitoid, *Nemeritis canescens*. *Journal of Animal Ecology* **48**: 353–71.

Wajnberg, É. (2006) Time allocation strategies in insect parasitoids: from ultimate predictions to proximate behavioral mechanisms. *Behavioral Ecology & Sociobiology* **60**: 589–611.

Wajnberg, É., Bernhard, P., Hamelin, F. and Boivin, G. (2006) Optimal patch time allocation for time-limited foragers. *Behavioral Ecology & Sociobiology* **60**: 1–10.

Appendix: competition among parasitoids

Assume that the probability that a given parasitized host is also located by exactly p resident parasitoids is given by $\Pr(\underline{p} = p) = z(p)$, then

$$\sigma = \sum_{p=0}^{\infty} z(p) \frac{1}{1+p} \tag{A2.1}$$

This equation is difficult to solve in general but if encounters are random we can take a Poisson distribution (with mean $\bar{p}$) for z,

$$z(p) = \frac{\bar{p}^p}{p!} e^{-\bar{p}} \tag{A2.2}$$

and we obtain

$$\sigma = \sum_{p=0}^{\infty} \frac{\bar{p}^p}{p!} e^{-\bar{p}} \frac{1}{1+p} = \sum_{p=0}^{\infty} \frac{\bar{p}^p}{(p+1)!} e^{-\bar{p}} = \frac{1}{\bar{p}} \sum_{q=1}^{\infty} \frac{\bar{p}^q}{q!} e^{-\bar{p}} = \frac{1}{\bar{p}} (1 - e^{-\bar{p}}) \tag{A2.3}$$

When we substitute this expression for σ into the expression for the parasitoids R_0 in Equation (2.12), while setting $\bar{n} = a_Q N_t$ and $\bar{p} = a_P P_t$, we recover Equation (2.11).

3

Parasitoid foraging and oviposition behavior in the field

George E. Heimpel and Jérôme Casas

Abstract

Since parasitoids can be reared from their hosts and leave evidence of parasitism after they emerge from hosts, some general outlines of their foraging behavior can be inferred from field studies that do not involve direct observation. However, direct observations are often needed to uncover critical aspects of patch use, host finding, and host use. Studies of parasitoid foraging in the field utilizing direct observation were rare until relatively recently. However, a nascent literature on parasitoid foraging in the field can now be identified that is shedding some light on how parasitoids allocate their time in the field and what they do when they encounter hosts. These studies have been performed on a variety of parasitoids in natural as well as agricultural systems. They range from observations of host-species selection by *Drosophila* parasitoids foraging in fermenting fruits and sap fluxes in Dutch woodlands to investigations of the amount of time that parasitoids of stem-boring Lepidoptera will wait for their concealed hosts to emerge from within corn plants in the USA, and a number of other interesting studies. In this chapter we review studies of parasitoid behavior in the field, focusing primarily on the implications of these behaviors for density-dependent parasitism and optimal foraging.

3.1 Introduction

In this chapter we consider the foraging and oviposition behavior of adult parasitoids in the field. While the ease with which parasitoids can be studied in laboratory settings has made them superb subjects for testing some aspects of foraging and evolutionary theory, field studies have lagged behind laboratory studies. This has in some cases made it difficult to interpret the results of laboratory studies or to put them into a realistic context. One problem with relying primarily on laboratory data to test foraging theory is that parasitoid behavior is often context dependent. Indeed, laboratory studies are the primary means by which this context dependence has been demonstrated. A classical example is the body of work from the laboratory showing that female parasitoids allocate increasing fractions of sons to hosts as the foundress number increases under conditions of local mate competition (Hamilton 1967, Hardy 1994). We know very little about sex allocation decisions by females under field circumstances, including the underlying conditions thought to influence behavior.

A second problem with relying on laboratory studies is that such studies are much better at telling us what parasitoids can do rather than telling us what they actually do. For example, hundreds of elegant laboratory studies have demonstrated the extent to which parasitoids use olfactory cues to locate hosts and host habitats (Vet & Dicke 1992), but we are relatively ignorant of the extent to which olfactory or other cues are used by parasitoids foraging freely in the field (Sheehan & Shelton 1989, De Farias & Hopper 1997).

However, recognition of the paucity of field-based studies of parasitoid behavior has been growing (Hardy et al. 1995, Casas 2000) and the situation is improving. Thus, we are able to review a number of studies in this chapter that are improving our understanding of what parasitoids do in the field. Most of the advances that we review have come from an increasing willingness of researchers to conduct observations of parasitoids foraging in the field over the past two decades (Casas 2000). Advances in our understanding of parasitoid dispersal have also come from new developments in insect marking methodologies (Hagler & Jackson 2001, Lavandero et al. 2004, Wanner et al. 2006), but a review of parasitoid dispersal is outside the range of this review. Similarly, the adoption of biochemical tests that can be used to infer the feeding status of field-caught parasitoids has improved our understanding of sugar feeding by parasitoids in the field (Casas et al. 2003, Heimpel et al. 2004, Heimpel & Jervis 2005, Lavandero et al. 2005, Lee et al. 2006), which we also do not cover here (see also Chapter 7 by Bernstein and Jervis). Finally, we do not cover field studies of mating behavior (Crankshaw & Matthews 1981, Tagawa & Kitano 1981, Antolin & Strand 1992, Field & Keller 1993).

Rather, this chapter is organized around two topics that have received attention over the past few decades with respect to investigations of parasitoid oviposition and foraging behavior in field settings. The first of these is spatial density-dependent parasitism. Here we will discuss the role that field behavior studies have played in contributing to the long-standing quest to determine the mechanisms leading to density-dependent parasitism rates. In particular, we will discuss aggregation to patches of high host density and behaviors once patches are found. Second, we will focus on reproductive strategies, concentrating on behaviors that parasitoids engage in once hosts have been found. Much of the progress in this area has been centered on tests of state-dependent foraging models (see also Chapter 2 by van Baalen and Hemerik and also Chapter 15 by Roitberg and Bernhard). In addition though, females of some parasitoid species experience difficulty in identifying whether hosts are dead or alive in the field. We deal with this point in more detail in the chapter as well.

3.2 Density dependence

By regressing the parasitism rate against the host density found at various locations, we can determine whether parasitism is positively density dependent, inversely density dependent, or density independent in a spatial context. While the link between density dependence and population dynamics certainly provides a good reason to document patterns of density dependence for their own sake (Hassell 2000, Murdoch et al. 2003), we would also like to know how particular patterns of spatial density dependence are achieved in the field. For instance, positive density dependence can come about through aggregation of many parasitoids to sites of higher host density or through an increased per capita search rate in areas of higher host density. However, the examination of patterns of parasitism cannot help us to distinguish between these two explanations. Similarly, two explanations are available for inverse density dependence: handling time limitation and/or egg limitation at higher host densities. Examination of parasitism rates cannot distinguish between these two hypotheses either. Thus, while parasitoids offer great advantages for detecting patterns of density dependence, a more detailed understanding of behavior in the field is necessary to understand the determinants of density dependence.

In the following paragraphs we review case studies that employed direct observation of parasitoids foraging in the field to explore the determinants of density dependence. We deal specifically with three questions that are germane to this topic. First, we ask whether parasitoids aggregate to patches of higher host density. Second, we ask whether parasitoids endeavor to use as many hosts as possible in a patch once it is found or whether they tend to spread their reproductive effort over multiple host patches. Finally, we ask whether female parasitoids lay more eggs at higher host densities.

3.2.1 Do parasitoids aggregate to patches of higher host density?

A pioneering study in this area was done by Waage (1983) who used binoculars to census naturally occurring *Diadegma* spp. foraging on Brussels sprouts plants in the field that were artificially infested with various densities of diamondback moth larvae. Waage (1983) did document aggregation of *Diadegma* to plants with higher host densities, but the rate of parasitism was density independent, hovering around 70% for all host densities tested. These results were corroborated by a similar study done at the same site by Legaspi and Legaspi (2005). Waage (1983) posed two hypotheses to explain the lack of positively density-dependent parasitism despite aggregation. The first was higher rates of superparasitism at higher densities, a hypothesis he was able to reject using host dissections and the second was increased devotion to non-searching activities at higher host densities. These activities could include host handling time or mutual interference with conspecifics. Using both field and laboratory observations, Waage (1983) rejected the mutual interference hypothesis, but concluded that handling time could indeed have limited oviposition rates at the highest densities (Wang & Keller 2002). Increased handling time at higher densities typically leads to a type II functional response and so inverse density dependence (Hassell 2000), but the combination of handling time limitation at higher densities and aggregation could, in principle, lead to density-independent parasitism, as observed by Waage (1983).

A few other studies have recorded the density of adult parasitoids in patches of hosts that varied in density. Two behavioral studies of host finding in leafminer parasitoids conducted in the field found that parasitoids disproportionately visited leaves with higher miner densities (Casas 1989, Connor & Cargain 1994). However, in the case of *Closterocerus tricinctus*, the proportion of leafmines visited per leaf decreased with increasing miner density and the per-leaf parasitism rate was inversely density dependent (Connor & Beck 1993, Connor & Cargain 1994). This result was explained by the effects of mutual interference: increasingly more time was spent handling and rejecting previously parasitized hosts as the density increased. On the other hand, Thompson (1986) found that the attack rate of *Agathis* sp., a braconid parasitoid of a seed-mining moth, was not related to host density but, in this case, aggregation to patches of varying host density was not quantified. Umbanhowar et al. (2003) recorded females of the tachinid *Tachinomyia similis* foraging freely on patches of western tussock moth larvae feeding on lupine bushes in coastal California. These researchers found both aggregation to patches of higher host density and patterns of positively density-dependent parasitism. They also performed separate studies showing that the per-parasitoid attack rate decreased with host density, an effect that would, on its own, lead to inverse density dependence. Thus, it appears that, in this case, the aggregative response of the flies was strong enough to overcome a decelerating functional response to maintain an overall pattern of positive density dependence. The precise behavioral mechanism(s) leading to aggregation were not identified, however. For example, it is not known whether the observed aggregation was due to higher arrival rates, lower leaving rates at patches of higher host density, or both.

The question of arrival versus leaving rates was broached by Sheehan and Shelton (1989), who observed the aphid parasitoid *Diaeretiella rapae* on seven artificially prepared patches of cabbages near a cabbage field. These included one patch of 40 plants, one of 20 plants, one of 10 plants, and four patches of five plants. All of the patches with the exception of two of the five-plant patches included one cabbage aphid-infested plant for every four aphid-free plants. The sample size in this study was clearly small, so the results have to be viewed with caution, but the authors found no evidence for aggregation to either high-density patches of plants or to plant patches that included aphid-infested plants or not. In contrast, *D. rapae* tended to stay longer on plants with higher aphid densities and also in patches with more rather than fewer plants. These observations, although best considered preliminary because of the limited sample size, are consistent with random host finding followed by density-dependent host-patch leaving tendencies. Unfortunately, no estimates of density-dependent parasitism were made in this study.

Using direct hourly observations of *Aphytis melinus* in the field over the entire foraging period on 23 citrus leaves containing various densities of California red scale insects, Casas (2000) noted that the number of parasitoids aggregating on leaves was not related to the number of hosts on leaves nor to the number of eggs laid on single leaves. There was also no correlation between the number of hosts and the number of eggs laid. In fact, only five eggs were laid during the observations and they were all laid on the same leaf, which was itself not the most highly-visited leaf. These observations are consistent with previous findings of density-independent parasitism by *A. melinus* at the scale of the leaf, fruit twig, and whole tree (Reeve & Murdoch 1985, Reeve 1987). Field entomologists know all too well that certain spots in the wild will be highly attractive to butterflies and other insects over many years and generations. The mechanisms responsible for such fined-tuned behavioral preference represent a novel field of study beyond parasitoids. In addition to biotic factors such as resource abundance, a complex mix of abiotic factors such as the color of light in the immediate surroundings and the presence of particular architectural features of the physical environment, among others, may be at play as well.

White and Andow (2005) used a creative approach to manipulate host densities. They planted conventional (non-Bt) corn plants into a Bt-corn field in patches of different sizes. Since the Bt corn is toxic to the European corn borer *Ostrinia nubilalis*, while the non-Bt corn is non-toxic, the non-Bt patches could be expected to have higher host densities than the surrounding Bt corn (Orr & Landis 1997). To make sure, White and Andow (2005) also out-planted corn borers onto these plants. Using this method, they created replicated patches of 2, 8, and 32 plants that had much higher per-plant corn borer densities than the surrounding corn plants. Into this field they released about 10,000 laboratory-reared *Macrocentrus grandii* (Hymenoptera: Braconidae), a specialist parasitoid of the European corn borer. Surveys of *M. grandii* foraging on the patches of three sizes were made and these observational data were compared to parasitism rates. In this case, no aggregation to high densities was detected, yet parasitism rates were positively density dependent and the number of *M. grandii* found foraging in a plot was not a good predictor of the parasitism level in that plot. The authors concluded that the survey observations were an insensitive measure of parasitoid aggregation since survey times were relatively short. The determinants of the density-dependent parasitism found in this study therefore remain unknown.

We conclude this section by reviewing a set of studies that investigated the relationship between parasitoid aggregation and density-dependent parasitism involving cleptoparasitoids of solitary ground-nesting sphecid wasps. The parasitoids considered in these studies include chrysidid wasps and sarcophagid flies and the studies were done in the spirit of the classical observations of hunting wasps done by comparative ethologists dating back to the work of J-H. Fabre and N. Tinbergen (for general reviews of this body of work, see Tinbergen 1958, Crompton 1987, Evans & O'Neill 1988). The most complete studies of this kind were those of the chrysidid *Argochrysis armilla*, a cleptoparasitoid of *Ammophila* ground-nesting hunting wasps. Parasitoids foraging freely for nests of *Ammophila* to oviposit into were observed in natural settings by Rosenheim (1987a,b, 1988, 1989, Rosenheim et al. 1989) and manipulative field experiments showed that *Argochrysis* oriented to *Ammophila* nests using visual cues and also that they used landmarks to locate the nests (Rosenheim 1987a). Rosenheim (1989) also found that, while *Argochrysis* consistently aggregated to sites of local *Ammophila* nest density, patterns of parasitism showed density dependence in only one of 2 years where this was investigated. To investigate aggregating foraging more thoroughly, Rosenheim et al. (1989) regressed the abundance of foraging *Argochrysis* females onto *Ammophila* nest abundance at 13 spatial scales and were able to determine that the parasitoids aggregated to patches of higher host density at the spatial scale of 3–50 m². Strohm et al. (2001) obtained a similar result by using field observations of a chrysidid parasitoid of beewolves. They found positive density-dependent foraging at a spatial scale of 16 m², but not at 4 m². In their study, parasitism was positively density dependent when measured at a site different from the site at which wasp behavior was monitored, but evaluations of both factors at the same site were not possible. For parasitoids of ground-nesting wasps, density-dependent foraging can occur in conjunction with positively density-dependent, inversely density-dependent, and density-independent patterns of parasitism (Wcislo 1984, Rosenheim 1990, Strohm et al. 2001). Thus, this microcosm of field behavioral studies on host–parasitoid interactions confirms the theoretical expectation that behavior cannot be inferred from patterns of parasitism (Hassell 1982, Morrison 1986) and that aggregation to patches of high host density does not necessarily lead to positive density dependence (see also Chapter 13 by Bonsall and Bernstein).

3.2.2 Do parasitoids exploit all hosts in a patch?

Host patch use should be related to density dependence (see also Chapter 13 by Bonsall and Bernstein). Increased residence and oviposition times in larger patches have the potential to lead to positive density dependence, while inverse density dependence is a likely outcome when parasitoids provide equal time and/or eggs to host patches regardless of their size. Host patch use as a determinant of density-dependent parasitism has been studied using field observations or similar techniques for a number of host–parasitoid systems, as reviewed in the following paragraphs.

A study of the relationship between the reproduction of the mymarid parasitoid *Anagrus delicatus* (= *Anagrus sophiae*) in the field and the size of host patches revealed an initially puzzling pattern: individual females oviposited into a small fraction of the hosts in each patch encountered, leaving many apparently high-quality hosts unparasitized and spreading their eggs over a series of host patches (Cronin & Strong 1993). Similar patterns were seen for the congener *Anagrus columbi* (Cronin 2003). Per capita field oviposition rates in this study were not estimated based upon direct observation, but by capturing parasitoids as they were entering host patches and comparing these catches to the numbers of parasitoid eggs found per patch to arrive at an estimate of the per capita eggs laid. Cronin and Strong (1993) called this pattern of host use 'substantially submaximal oviposition' and argued that it contributes to the general pattern of density-independent parasitism found by *A. delicatus* in the field (Stiling & Strong 1982, Strong 1989, Cronin & Strong 1990). This observation of submaximal oviposition by *Anagrus* in the field has led to interesting discussion and insights. Cronin and Strong (1993) originally interpreted the pattern as an example of risk spreading (or 'bet hedging'), where the strategy of spreading eggs over multiple patches is favored as a response to a high rate of catastrophic patch failure. They documented that 20–30% of the host-plant leaves senesced in the field, leading to the death of all host eggs and any parasitoids developing in them. The argument then was that the high variability in such patch failure selected for a strategy (risk spreading) that decreased the variance of fitness at the cost of a lower arithmetic mean per-female fitness. Their field and laboratory experiments were able to exclude a number of other potential factors that could explain submaximal oviposition, including egg limitation, long handling times, host refuges, variation in host quality, and parasitoid density.

Cronin and Strong's (1993) risk spreading theory proved to be controversial, however. Godfray (1994) doubted that risk spreading alone could account for *A. delicatus* behavior because the risk in this system would purportedly be spread out spatially and not temporally, as is the case in classical models of risk spreading (Hopper 1999, Hopper et al. 2003). Some other researchers have also proposed possible reasons for submaximal oviposition in this species. Rosenheim and Mangel (1994) noted that *A. delicatus* was reportedly unable to discriminate between parasitized and unparasitized hosts and they constructed an optimality model showing that the risk of self-superparasitism alone could explain the oviposition patterns documented by Cronin and Strong (1993) (see also Chapter 8 by van Alphen and Bernstein for details on the discrimination abilities of parasitoids). In addition, Bouskila et al. (1995) concentrated on the fact that a fair number of hosts were probed and then rejected by *A. delicatus*, leading to the suggestion that differences in host quality did exist and could explain at least some of the submaximal oviposition rates. Other possibilities that have been hypothesized to lead to submaximal patch utilization by parasitoids include avoidance of density-dependent hyperparasitism or predation (Ayal & Green 1993, Mackauer & Völkl 1993) and reducing the risk of inbreeding among offspring.

Further studies by Cronin and Strong (1999) showed that *A. delicatus* parasitize more hosts per patch after bouts of long-distance dispersal than the more typical dispersal bouts of 10 cm or less. These authors found that *A. delicatus* depleted their entire egg load on a single host patch if it was more than 250 m from other hosts. Similarly, *A. columbi* depleted 60% of their egg load on a single patch when arriving at patches located at least 25 m from a source of parasitoids and hosts versus 30% in the source itself (Cronin 2003). Thus, it appears that the parasitoids perceive the lower opportunities for future reproduction on these far-flung patches and adjust their oviposition strategy accordingly. The inevitable outcome will be some level of self-superparasitism given the inability of these parasitoids to discriminate between previously-parasitized hosts but, as Rosenheim and Mangel's (1994) model suggested, this cost should be accepted under conditions of time-limitation.

A second example of what could be termed under-exploitation of host patches comes from aphid parasitoids. Völkl (1994) conducted field observations of *Aphidius rosae* that were manually released onto patches of rose aphids that varied in size from 19 to 42 individuals. He found that *A. rosae* females were more likely to alight on parts of the rose bushes that had aphid colonies on them and spent more time on shoots and bushes with aphids rather than without aphids. However, the parasitoids laid, on average, only 2.8 eggs per colony and the number of eggs laid, residence time in the colony, and the oviposition rate were all unrelated to colony size. This lack of a relationship was found at the spatial scales both of single shoots and whole rose bushes and led to an inversely density-dependent pattern of parasitism at both spatial scales. Völkl (1994) explained this by noting that the inter-patch travel times were very short in this case, leading to an expectation of low patch utilization (Weisser et al. 1994). Similar patterns were observed for another aphid parasitoid attacking the grey pine aphid, *Schizolachnus pineti*, in the field (Völkl & Kraus 1996).

Another interesting case of what could be considered early patch leaving was uncovered by van Nouhuys and Ehrnsten (2004) for the specialist ichneumonid *Hyposoter horticola* attacking eggs of the Glanville fritillary butterfly *Mellitaea cinxia* throughout an archipelago of dozens of small islands in southern Finland. Laboratory studies showed that host eggs are only susceptible to parasitism by *H. horticola* when within a few hours of hatching. In the field, host eggs are laid in large batches of 100–200 and mature slowly and asynchronously so that, at any given moment, only a small fraction of eggs are susceptible to parasitism. van Nouhuys and Ehrnsten (2004) conducted field observations and other experiments suggesting that *H. horticola* females monitored host egg masses as they matured and returned to masses once they had eggs that were susceptible to parasitism. They suspected that *H. horticola* uses spatial learning of the kind documented in hunting wasps and the parasitoid *A. sarmilla* (Rosenheim 1987a) to locate host patches. This behavior, along with generally high mobility by *H. horticola* females (van Nouhuys & Hanski 2002, van Nouhuys & Ehrnsten 2004) is the likely explanation for a relatively constant rate of parasitism of *M. cinxia* by *H. horticola* over both time and space, a pattern that is not typical for a specialist parasitoid.

3.2.3 Are more eggs laid at higher host densities?

For the last topic of this section, we focus on a system in which an attempt was made to determine the relationship between oviposition behavior and host density. Adults of the thelytokous *Aphytis aonidiae* were collected in the field while foraging on almond trees infested with San Jose scale, the primary host of *A. aonidiae*. The parasitoids were put on ice in the field and brought to the laboratory for egg load dissections and the per-tree San Jose scale density was estimated (Heimpel & Rosenheim 1998). Observations of behavior were also made and these will be discussed below, but in this section we focus on the relationship between egg load and host density. Since the oviposition rate is expected to be higher at high host densities, it is predicted that egg loads will decline with host density (Rosenheim 1996). Contrary to this expectation, however, host density had no effect on the egg load of *A. aonidiae*. One hypothesis to explain this result involves a previous observation that these parasitoids (and other *Aphytis* species) are more choosy with respect to hosts that they use for oviposition in the field and laboratory as their egg load (number of mature eggs) declined (Rosenheim & Rosen 1991, 1992, Collier et al. 1994, Rosenheim & Heimpel 1994, Heimpel & Rosenheim 1995, Heimpel et al. 1996, Casas et al. 2004). Thus, it was hypothesized that lower oviposition rates at lower egg loads could weaken or even eliminate the relationship between the egg load and host density. This hypothesis was tested using a dynamic state variable model, which confirmed that egg loads should drop with increased host encounter rates and that increased choosiness at low egg loads (dynamic behavior) could weaken but not eliminate this relationship (Mangel & Heimpel 1998). So it is still not clear why no relationship was found between the egg load and host density in *A. aonidiae*. Remaining hypotheses include the possibility that the host density was measured at an inappropriate scale to detect density dependence of the egg load, mutual interference of parasitoids at high host densities, and the possibility that egg maturation rates are higher at higher host densities (Rivero-Lynch & Godfray 1997, Wu & Heimpel, 2007).

3.3 Reproductive strategies

Individual host-based patterns of parasitism, such as brood size and offspring sex ratio, can be discerned simply by collecting parasitized hosts in the field and either dissecting them or rearing them out in the laboratory. This can give clues to the reproductive strategies that parasitoids are employing, but the per capita behavior may remain unknown since we do not know how many female parasitoids visited the hosts that have been collected. Thus, it will not be clear whether a given number of offspring per host reflects clutch size or superparasitism decisions (even in solitary parasitoids) (Rosenheim & Hongkham 1996) although, under conditions of low parasitism rates, superparasitism can in some cases be considered unlikely (Bezemer & Mills 2003). Similarly, the number of foundresses contributing to the overall sex ratio produced by a collected host is unknown. Exceptions to this are cases where genetic methods can be employed to differentiate between the offspring of different females within a host. For example, Edwards and Hopper (1999) used molecular markers to show that more than a single female of the polyembryonic parasitoid *Macrocentrus cingulum* (= *M. grandii*) routinely contributed to single hosts. The high levels of superparasitism that they uncovered, along with the relatively low overall parasitism rates, supported the hypothesis that the hosts of *M. cingulum* (larvae of the European corn borer) were only available for short portions of their life cycle and invulnerable to parasitism for a large fraction of their larval life. Unfortunately, the use of genetic analyses for assessing parasitoid reproductive strategies in the field is not well developed.

In this section, we will review a number of studies that have used direct observations to uncover parasitoid oviposition strategies. We first discuss host species selection, then the complications that can arise when numerous dead hosts occur alongside living hosts and the influence of physiological state on oviposition behavior in the field and we finish by discussing optimal patch use in the field.

3.3.1 Host species selection in the field

In a pioneering study using direct observation in the field of a relatively small parasitoid, Janssen (1989) followed females of *Asobara tabida* and *Leptopilina heterotoma*, both parasitoids of *Drosophila* larvae, in natural field settings. Janssen (1989) brought a stereomicroscope into the field, suspended it on a tripod, and observed these parasitoids searching for and attacking their hosts on fermenting apples and pears and on sap fluxes of wounded trees. He recorded details of foraging behavior by speaking into a tape recorder or by entering data into a portable data logger. A total of 19.5 hours of observations were logged in this way over the summers of 1984 and 1985, providing novel data on encounter rates and handling times in the field of these parasitoids whose behavior had been well studied in laboratory settings. To help interpret data from these observations, hosts were collected and parasitoids reared from them in the laboratory and handling times for both parasitoid species on nine *Drosophila* spp. were obtained in laboratory studies.

One of Janssen's (1989) key findings was that the encounter rate between both *A. tabida* and *L. heterotoma* and host larvae was rather low, ranging between 0.2 and 5.0 per hour, depending upon the year and substrate. Of the nine *Drosophila* species that were present at the sites that Janssen (1989) studied, three comprised over 97% of the samples reared from field collections of field material in approximately equal proportions (*Drosophila subobscura*, *Drosophila immigrans*, and *Drosophila simulans*). Of these, *D. subobscura* was a highly suitable host for both *A. tabida* and *L. heterotoma*, *D. simulans* was suitable for *L. heterotoma* but a very poor host for *A. tabida*, and *D. immigrans* was a poor host for both parasitoid species. Janssen's (1989) field observations involved host larvae and it was therefore impossible to distinguish between host species. However, given the relative abundance of the various hosts and assuming no differences in detectability of the different host species (as appeared to be the case in the laboratory), it is likely that approximately one-third of *L. heterotoma*'s encounters were with a low-quality host (*D. immigrans*) and approximately two-thirds of *A. tabida*'s encounters were with low quality hosts (*D. immigrans* or *D. simulans*). Despite this, only three of the 33 host encounters observed in the field for both parasitoid species resulted in rejections. Thus, it appears that these parasitoids readily accept very low-quality hosts in the field. Janssen (1989) used this result to conclude that fitness in these parasitoids was limited by time (or the number of hosts that they could encounter during their lifetime) and not by eggs. Indeed, the high acceptance of low-quality hosts that was observed matched the prediction of an optimality model that was based upon an assumption of time limitation (Janssen 1989). The use of low-quality hosts by parasitoids is relatively common (Heimpel et al. 2003) and this is broadly consistent with widespread time limitation.

Janssen's (1989) conclusion of time limitation by *Asobara* and *Leptopilina* has been investigated further. Driessen and Hemerik (1992) used estimates of egg load at emergence, oviposition rates within host patches, patch residence time, travel times between patches, and life expectancy to simulate the time and egg budget of *Leptopilina clavipes* (note that this is not the same species of *Leptopilina* that Janssen (1989) worked with). Their model suggested that a relatively high fraction (13%) of females would deplete their egg supply during their lifetime. Contrary to Janssen (1989), they concluded that at least some of the population was probably under selection to reject low-quality hosts at some times during the season. Females of *A. tabida* were also able to deplete their egg load under field conditions, although seemingly at a lower rate than *L. clavipes*, with most females dying with high egg loads (Ellers et al. 1998, 2001).

Another case in which direct field observations were used to determine host species selection involves the parasitoids *Cotesia glomerata* and *Cotesia rubecula*. Laboratory studies and field data have shown that *C. rubecula* is a specialist, showing a strong preference for larvae of the small cabbage white butterfly *Pieris rapae* over other *Pieris* spp. (Brodeur et al. 1996, 1998, Geervliet et al. 2000). *Cotesia glomerata*, on the other hand, attacks larvae of at least three species of *Pieris*, including *P. rapae* and the large cabbage white butterfly *Pieris brassicae*. Geervliet et al. (2000) set up cabbage plants in the field on which one leaf was provisioned with *P. rapae* larvae and another leaf was provisioned with *P. brassicae* larvae. Observers then sat near the plants and observed naturally foraging *C. rubecula* and *C. glomerata* locate and attack the host larvae. They found, as suspected, that *C. rubecula* attacked exclusively *P. rapae* and that *C. glomerata* attacked both *P. rapae* and *P. brassicae*.

3.3.2 Host quality and parasitism rate

The low hourly parasitism rate of *A. melinus* observed by Casas et al. (2000, 2004) on the basis of egg load dynamics is due to a high rejection rate of unsuitable hosts. In this case, we are not referring to unsuitable host species, but instead to individuals of high-quality host species that are either a suboptimal stage or previously parasitized or dead. So, finding and even encountering members of the correct host species may not be a problem *per se*. Instead, finding high-quality individuals might be a greater problem. There is a large variety of host types that can be considered unsuitable, including already parasitized hosts, hosts that are too small or otherwise physiologically unsuitable, and even dead hosts. This latter case is particularly interesting, as it is difficult to imagine how spending time examining dead hosts can be considered anything but suboptimal. Dead hosts appear to play a key role in two totally different systems: for eulophids attacking leafminers (Casas 1989, Connor & Cargain 1994) and for *Aphytis* attacking scale insects (Heimpel et al. 1996, Casas et al. 2000, 2004). In both cases, the hosts remain present in their habitat and continue to attract parasitoids long after their death.

This simple fact is in itself already surprising, as it is difficult to reconcile with the highly complex array of sensors that parasitoids are equipped with. *Aphytis melinus* has to insert its ovipositor into the host for examination and most scale probed in the field were recorded dead once dissected in the laboratory (Casas et al. 2004). Heimpel et al. (1996) calculated that over 20% of the scale examined long enough to proceed to ovipositor insertion by *A. aonidiae* were in fact dead. *Sympiesis* also spends time on empty host mines (Casas 1989).

The parasitism rate reported in the field of *A. melinus* (between 0.5 and 1.0 eggs per hour) using behavioral observations is lower than that reported from laboratory experiments (1.5 eggs per active hour). These low attack rates are, however, consistent with the estimates from egg load maturation and egg laying rates obtained through ovarian dynamics in the field (Heimpel et al. 1998, Casas et al. 2000). As stated above, the key difference between laboratory and field observations is in the much lower acceptance rate in the field. The discrepancy is therefore not the result solely of the time spent dealing with dead scale insects. Indeed, a simple calculation gives an increase in search time of only 20%, once the time spent on dead scale is added to the actual search time (Casas et al.

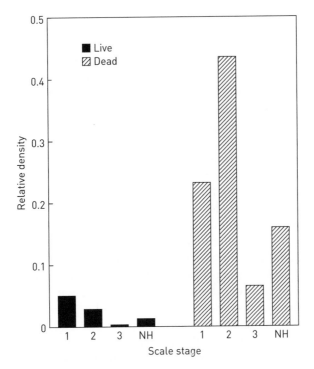

Fig. 3.1 Relative density of live and dead California red scale in a sample from the bark in the interior of a tree. A scale of stage 1 to 3 can be use as hosts (the smallest for host feeding) and NH represents non-host stages. Even though *Aphytis* foraging in such an environment has enough hosts within reach to use up its egg load in a few minutes, it will use the whole day to lay its eggs, rejecting most scale, 90% being dead.

2004). If *Aphytis* in the field had shown the same acceptance rate as in the laboratory, they would have laid their entire egg complement (up to 12 eggs) in only a few minutes, as the females were literally surrounded by live hosts (three hosts per square centimeter (Fig. 3.1)). The results for the leafminer systems are less clear, but dead hosts and empty mines are an impediment to the female's search here as well. More work is needed on the lack of mechanisms to recognize dead hosts and their impact on behavioral decisions, as it is difficult to envision adaptive explanations for the lack of recognition and the time spent dealing with dead hosts.

3.3.3 Physiological state and host use patterns

Further work on the importance of egg versus time limitation and its influence on reproductive decisions in the field was done on *Aphytis* parasitoids, which attack armored scale insects.

Heimpel and colleagues conducted observations of female *A. aonidiae* and *Aphytis vandenboschi* foraging for and attacking the San Jose scale in California almond orchards (Heimpel et al. 1996, 1997b, 1998, Heimpel & Rosenheim 1998). These observations were focused primarily on the decision of the parasitoids to use hosts for oviposition or host feeding. Previous laboratory studies had shown that A. *melinus* were more likely to oviposit when they had high egg loads and host feed when they had low egg loads (Collier et al. 1994, Heimpel & Rosenheim 1995) and the question was whether this trend would be upheld in the field. A sufficient number of observations to address this question was obtained for A. *aonidiae* and, indeed, females with lower egg loads were more likely to use hosts for host feeding than oviposition (Heimpel et al. 1996), supporting predictions from dynamic host use models (Chan & Godfray 1993, Heimpel et al. 1994, Collier 1995, Heimpel & Collier 1996, McGreggor 1997, Heimpel et al. 1998, see also Chapter 7 by Bernstein and Jervis). Using these results, it was possible to derive a simple behavioral rule that takes into account only the parasitoid egg load and host size. In this rule, a parasitoid with a given egg load oviposits on hosts that are larger than a particular size and host feed on hosts that are below this size. In simulation models, this simple (empirically derived) behavioral rule was compared to 'optimal' dynamic behavior as a predictor of field egg loads. The rule was much better at predicting field egg loads than the dynamic model. In particular, the dynamic model predicted that many more females should deplete their egg loads than actually did. One implication of this discrepancy is that parasitoids engaged in more host feeding than was deemed optimal. And while it is possible that the high level of host feeding observed was suboptimal or maladaptive, it is also possible that the estimates of host feeding gain were underestimated in laboratory studies (Heimpel et al. 1997a). An alternative explanation is the lack of a clear relationship between the time spent host feeding and the amount obtained, as well as the small quantities of food often obtained (Rosenheim & Rosen 1992, Giron et al. 2004).

Measuring the oviposition rate per hour in the field is a daunting task. Statistical inference using the dynamics of egg load is much easier, a case where physiology is the best route to quantify behavior rather than measuring behavior directly. By conducting both a series of cage experiments in the field on A. *melinus* and stochastic modeling of the egg maturation and egg laying rates using wild females caught the same day in the same environment, Casas et al. (2000) showed that the oviposition rate is a positive function of the egg load itself. Parasitoids either pause more or get choosier as they deplete their egg load. The stochastic model developed also showed that a major portion, up to 40%, of the A. *melinus* population was experiencing transient time and egg limitation, sometimes in sequence during a single day. A similar result was found for the A. *aonidiae* system discussed above in which an increased tendency to host feed versus oviposit was documented as the egg load declined (Heimpel et al. 1998). The time window for foraging becomes crucial and a female can end up being time limited one day and being egg limited the next, depending on weather conditions.

We know surprisingly little about the daily time window for foraging and its influence (see Casas 2000, for a quantification in the field). In collaboration with J. Casas, P. D'Ettorre measured the egg load of the parasitoid *Pteromalus sequester* attacking a seed weevil (*Apion ulicis*) on *Ulex europeaus* in the field over the entire daily foraging over three days. As shown in Fig. 3.2, the egg load declined over the day for this species, but did not reach zero. Why not lay all of the eggs? Is the female unable to assess her own ovarian state closely enough? Does the entire machinery of hormonal control and egg production maintain a minimal supply of eggs (see also Chapter 7 by Bernstein and Jervis)? Is it optimal to behave in a way that leads to some small leftover at the end of the day, given the stochastic nature of host availability and metabolic needs? This is not some oddity of this biological

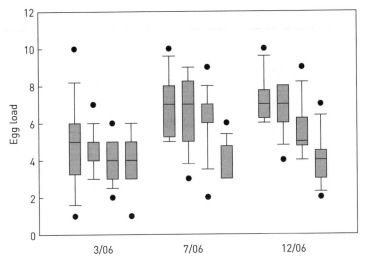

Fig. 3.2 Daily egg load dynamics of *P. sequester* in the wild under high host density conditions over 3 days corresponding to the peak activity of this species. The sampling times were 09:00, 12:00, 15:00, and 19:00 hours. A clear decrease in egg load can be observed, as well as the conspicuous absence of females without any mature eggs. In the vertical direction, the boxes extend from the lower to the upper quartile (25th, 50th, and 75th empirical quartiles). Whiskers are drawn for the largest and lowest observations at a distance of 1.5 the interquartile range. All observations, if any, lying beyond the whiskers are marked by a dot.

system, as it was also observed for *A. melinus* (Casas et al. 2000). The documentation of switching between time and egg limitation implies that the debates about whether time or eggs limit may be misleading. Indeed, models of the dynamics of synovigenic parasitoids predict transient egg (and time) limitation (Heimpel et al. 1998, Rosenheim et al. 2000). This switching is bound to occur often for synovigenic species having only a few eggs at a time, much like a reflecting random walk hovering near the origin. Selection may rather act upon the speed at which the physiological machinery is able to increase and decrease in speed as a function of egg load and host availability.

3.3.4 Optimal patch use in the field

One of the key reproductive strategies that parasitoids face is how long to stay in a patch of hosts (Godfray 1994, see also Chapter 8 by van Alphen and Bernstein). In an earlier section of this chapter, we discussed observations of patch use behavior and the effect of patch leaving on density-dependent parasitism. Here, we focus on tests of optimal patch use theory done on parasitoids in the field. While these studies have implications for density dependence, their main purpose was to test behavioral optimality models.

Much of the work on patch use by parasitoids and other animals has been done in the context of Charnov's (1976) marginal value theorem (MVT), which posits that foragers should leave a patch of resources in search of another only once it has been depleted to the average level of other patches in the habitat. A number of tests of this theory have been conducted on parasitoids in laboratory settings (Godfray 1994, van Alphen et al. 2003, Wajnberg et al. 2003, Wajnberg 2006) and the first explicit field test of predictions associated with the MVT using parasitoids that we are aware of was done by Tenhumberg et al. (2001). These researchers evaluated the patch leaving behavior of *C. rubecula*, a parasitoid of larvae of cabbage white butterflies *P. rapae*, by releasing laboratory-reared parasitoids into matrices of 16 potted cabbage plants that harbored varying densities of host larvae. They monitored the oviposition and flight behavior of the released parasitoids and investigated the effect of per-plant host density and parasitoid oviposition rate on the tendency of parasitoids to leave a plant. As predicted by the MVT, parasitoids were more likely to leave plants with low rather than high host density and a high oviposition rate increased the tendency to leave the patch. Interestingly, these effects had not been previously found in greenhouse studies on the same host–parasitoid system performed by Vos et al. (1998). No effect of per-leaf host density was found in these studies. In addition, rather than decreasing the patch residence time as found by Tenhumberg et al. (2001), contacts with hosts increased patch residence times in the study by Vos et al. (1998). Why these differences? Tenhumberg et al. (2001) speculated that strain differences may be involved since they worked with Australian *C. rubecula* and Vos et al. (1998) worked with a native Dutch strain. Another possibility is that the fact that one study from the field and the other from the laboratory somehow caused the differences, although it is not clear how this would lead to the different results obtained by the two research groups.

In a recent set of studies, White and Andow (2007) explored a different kind of patch use behavior in the parasitoid *M. grandii* (= *M. cingulum*). *Macrocentrus grandii* attacks larvae of the European corn borer, which spend much of their time in a physical refuge from attack (Edwards & Hopper 1999). The parasitoid can only attack corn borer larvae when they are either outside of their tunnels or within a half centimeter or so from the tunnel entrances. Upon encountering a European corn borer tunnel that does not have an exposed larva, a female parasitoid must 'decide' whether to stay at the tunnel entrance and wait for the larva to emerge and thus be available for oviposition (if indeed there is a larva inside) or to leave in search of other tunnels that may have exposed larvae. White and Andow (2007) modified an optimal foraging model first introduced by McNamara and Houston (1985) that explored the influence of the following factors on the decision to wait or leave: the proportion of tunnels that contain either inaccessible or accessible hosts, the emergence rate of hosts (i.e. the transition probability from an inaccessible to an accessible state), and the travel time between patches. The model distinguished between two strategies: waiting (akin to sit-and-wait foraging) and leaving (akin to active foraging). A robust prediction of this model was that a leaving strategy is favored by low inter-host travel times and a waiting strategy is favored by high inter-host travel times.

White and Andow (2007) were able to test this prediction using direct observations in the field, by using laboratory-reared *M. grandii* and gingerly placing them upon host tunnels in the field. They found that, if the parasitoids were primed by exposure to a dollop of host frass on a forceps arm just prior to being moved onto the host 'patch', they would actively investigate and probe the tunnel entrance and associated frass. Once placed on the tunnel opening, parasitoids were not manipulated further and freely moved to patches other than the one onto which they were placed. Wasps were observed for as long as possible and probed on average between two and three sites, most of which turned out to be empty tunnels. Information of data was recorded via a running commentary into a tape recorder. Parasitoid 'travel time' was investigated by comparing behavior on the first patch encountered (which followed days of host deprivation in the laboratory) to behavior on subsequent patch visits. The assumption was that perceived travel time was 'long' for the first patch visit and 'short' for subsequent patch visits. The results from these observations supported the prediction that a waiting strategy is favored by longer inter-host travel times. Tenure time averaged more than 30 minutes on initial patches and less than 15 minutes on subsequent patches. Observations showed that it takes between 6 and 16 minutes to properly evaluate a potential host patch for *M. grandii*, so that most parasitoids were essentially engaging in 'leaving' behavior on the subsequent patch visits. As may be expected, parasitoids spent longer on occupied than unoccupied patches for both initial and subsequent patch visits. As in the work of Tenhumberg et al. (2001) discussed above, these results provide support, from the field, for the major predictions of the MVT, namely that long patch tenure times are expected for higher quality patches and ones that require more travel time to locate.

3.4 Concluding statements

The gap between the ever-increasing toolbox of laboratory techniques and the paucity of field studies that was alluded to by Casas (2000) is widening. The reasons are well known: fieldwork is tedious, nothing happens much most of the time, and the material worth publication may be meagre for a 3–5-year PhD study. On the positive side, the number of different species studied in the field has increased, giving us a larger base for the statement that can be made. The wave of work dealing with adult nutrition has shown that physiological work in the field is indeed feasible and the last few years have seen a major advance in the way we do this bit of science (Heimpel et al. 2004). After a long period of mainly observational studies of behavior in the field, Geervliet et al. (2000) showed how laboratory studies could be used to generate qualitative predictions of behavior in the field. Casas et al. (2004) carried on this step further to quantitative ethological predictions. What is still missing as of today is the reverse approach: setting up highly controlled conditions in the field and testing females caught on the spot, enabling clear-cut conclusions. The 'lack' of optimality in the behavior of parasitoids in the field, compared to their behavior in the laboratory, should tell us a lot about the constraints acting in the field.

The number of combinations of environmental conditions experienced by an animal in the field is infinite and changes over time and space. Laboratory experiments can therefore be considered an endless chase after the right combination, very much like Alice trying to find her way in a kaleidoscopic wonderland. Ideally, field tests of theory would produce a probabilistic envelope of behavioral trajectories from the field and, hence, a dose or realism into an otherwise deterministic world found only in the laboratory.

Acknowledgments

We thank Éric Wajnberg for the invitation to write this chapter. We also thank Éric Wajnberg and an anonymous reviewer for comments on the manuscript and Jennifer White, Saskya van Nouhuys, Jetske De Boer, Simon Hsu, and Jeremy Chacon for fruitful discussion.

References

Antolin, M.F. and Strand, M.R. (1992) Mating system of *Bracon hebetor* (Hymenoptera: braconidae). *Ecological Entomology* **17**: 1–7.

Ayal, Y. and Green, R.F. (1993) Optimal egg distribution among host patches for parasitoids subject to attack by hyperparasitoids. *American Naturalist* **141**: 120–38.

Bezemer, T.M. and Mills, N.J. (2003) Clutch size decisions of a gregarious parasitoid under laboratory and field conditions. *Animal Behavior* **66**: 1119–28.

Bouskila, A., Robertson, I.C., Robinson, M.E. et al. (1995) Submaximal oviposition rates in a mymarid parasitoid: choosiness should not be ignored. *Ecology* **76**: 1990–3.

Brodeur, J., Geervliet, J.B.F. and Vet, L.E.M. (1996) The role of host species, age and defensive behavior on ovipositional decisions in a solitary specialist and gregarious generalist parasitoid (*Cotesia* species). *Entomologia Experimentalis et Applicata* **81**: 125–32.

Brodeur, J., Geervliet, J.B.F. and Vet, L.E.M. (1998) Effects of *Pieris* host species on life history parameters in a solitary specialist and gregarious generalist parasitoid (*Cotesia* species). *Entomologia Experimentalis et Applicata* **86**: 145–52.

Casas, J. (1989) Foraging behaviour of a leaf miner parasitoid in the field. *Ecological Entomology* **14**: 257–65.

Casas, J. (2000) Host location and selection in the field. In: Hochberg, M.E. and Ives, A.R. (eds.) *Parasitoid Population Biology*. Princeton University Press, Princeton, pp. 17–26.

Casas, J., Nisbet, R.M., Swarbrick, S. and Murdoch, W.W. (2000) Eggload dynamics and oviposition rate in a wild population of a parasitic wasp. *Journal of Animal Ecology* **69**: 185–93.

Casas, J., Driessen, G., Mandon, N. et al. (2003) Energy dynamics in a parasitoid foraging in the wild. *Journal of Animal Ecology* **72**: 691–7.

Casas, J., Swarbrick, S. and Murdoch, W.W. (2004) Parasitoid behavior: predicting field from laboratory. *Ecological Entomology* **29**: 657–65.

Chan, M.S. and Godfray, H.C.J. (1993) Host-feeding strategies of parasitoid wasps. *Evolutionary Ecology* **7**: 593–604.

Charnov, E.L. (1976) Optimal foraging, the marginal value theorem. *Theoretical Population Biology* **9**: 129–36.

Collier, T.R. (1995) Adding physiological realism to dynamic state variable models of parasitoid host feeding. *Evolutionary Ecology* **9**: 217–35.

Collier, T.R., Murdoch, W.W. and Nisbet, R.M. (1994) Egg load and the decision to host-feed in the parasitoid *Aphytis melinus*. *Journal of Animal Ecology* **63**: 299–306.

Connor, E.F. and Beck, M.W. (1993) Density-related mortality in *Cameraria hamadryadella* (Lepidoptera, Gracillariidae) at epidemic and endemic densities. *Oikos* **66**: 515–25.

Connor, E.F. and Cargain, M.J. (1994) Density-related foraging behaviour in *Closterocerus tricinctus*, a parasitoid of the leaf-mining moth, *Cameraria hamadryadella*. *Ecological Entomology* **19**: 327–34.

Crankshaw, O.S. and Matthews, R.W. (1981) Sexual behavior among parasitic *Megarhyssa* wasps. *Behavioral Ecology & Sociobiology* **9**: 1–7.

Crompton, J. (1987) *The Hunting Wasp*. Lyons Press, Augusta.

Cronin, J.T. (2003) Patch structure, oviposition behavior, and the distribution of parasitism risk. *Ecological Monographs* **73**: 283–300.

Cronin, J.T. and Strong, D.R. (1990) Density-independent parasitism among host patches by *Anagrus delicatus* (Hymenoptera: Mymaridae): experimental manipulation of hosts. *Journal of Animal Ecology* **59**: 1019–26.

Cronin, J.T. and Strong, D.R. (1993) Substantially submaximal oviposition rates by a mymarid egg parasitoid in the laboratory and field. *Ecology* **74**: 1813–25.

Cronin, J.T. and Strong, D.R. (1999) Dispersal-dependent oviposition and the aggregation of parasitism. *American Naturalist* **154**: 23–36.

De Farias, A.M.I. and Hopper, K.R. (1997) Responses of female *Aphelinus asychis* (Hymenoptera: Aphelinidae) and *Aphidius matricariae* (Hymenoptera: Aphididae) to host and plant-host odors. *Environmental Entomology* **26**: 989–94.

Driessen, G. and Hemerik, L. (1992) The time and egg budget of *Leptopilina clavipes* a parasitoid of larval *Drosophila*. *Ecological Entomology* **17**: 17–27.

Edwards, O.R. and Hopper, K.R. (1999) Using superparasitism by a stem borer parasitoid to infer a host refuge. *Ecological Entomology* **24**: 7–12.

Ellers, J., van Alphen, J.J.M. and Sevenster, J.G. (1998) A filed study of size-fitness relationships in the parasitoid *Asobara tabida*. *Journal of Animal Ecology* **67**: 318–24.

Ellers, J., Bax, M. and van Alphen, J.J.M. (2001) Seasonal changes in female size and its relation to reproduction in the parasitoid *Asobara tabida*. *Oikos* **92**: 309–14.

Evans, H.E. and O'Neill, K.M. (1988) *The Natural History and Behavior of North American Beewolves*. Cornell University Press, Ithaca, New York.

Field, S.A. and Keller, M.A. (1993) Alternative mating tactics and female mimicry as post-copulatory mate-guarding behaviour in the parasitic wasp *Cotesia rubecula*. *Animal Behavior* **46**: 1183–9.

Geervliet, J.B.F., Verdel, M.S.W., Snellen, H., Schaub, J., Dicke, M. and Vet, L.E.M. (2000) Coexistence and niche segregation by field populations of the parasitoids *Cotesia glomerata* and *C. rubecula* in the Netherlands: predicting field performance from laboratory data. *Oecologia* **124**: 55–63.

Giron, D., Pincebourde, S. and Casas, J. (2004) Lifetime gains of host-feeding in a synovigenic parasitic wasp. *Physiological Entomology* **29**: 436–42.

Godfray, H.C.J. (1994) *Parasitoids: Behavioral and Evolutionary Ecology*. Princeton University Press, Princeton.

Hagler, J.R. and Jackson, C.G. (2001) Methods for marking insects: current techniques and future prospects. *Annual Review of Entomology* **46**: 511–44.

Hamilton, W.D. (1967) Extraordinary sex ratios. *Science* **156**: 477–88.

Hardy, I.C.W. (1994) Sex ratio and mating structure in the parasitoid Hymenoptera. *Oikos* **69**: 3–20.

Hardy, I.C.W., van Alphen, J.J.M., Heimpel, G.E. and Ode, P.J. (1995) Entomophagous insects: progress in evolutionary and applied ecology. *Trends in Ecology & Evolution* **10**: 96–7.

Hassell, M.P. (1982) Patterns of parasitism by insect parasitoids in patchy environments. *Ecological Entomology* **7**: 365–77.

Hassell, M.P. (2000) *The Spatial and Temporal Dynamics of Host–Parasitoid Interactions*. Oxford University Press, Oxford.

Heimpel, G.E. and Collier, T.R. (1996) The evolution of host-feeding behavior in insect parasitoids. *Biological Reviews* **71**: 373–400.

Heimpel, G.E. and Rosenheim, J.A. (1995) Dynamic host feeding by the parasitoid *Aphytis melinus*: the balance between current and future reproduction. *Journal of Animal Ecology* **64**: 153–67.

Heimpel, G.E. and Rosenheim, J.A. (1998) Egg limitation in parasitoids: a review of the evidence and a case study. *Biological Control* **11**: 160–8.

Heimpel, G.E. and Jervis, M.A. (2005) Does floral nectar improve biological control by parasitoids? In: Wäckers, F., van Rijn, P. and Bruin, J. (eds.) *Plant-provided Food and Plant–Carnivore Mutualism*. Cambridge University Press, Cambridge, pp. 267–304.

Heimpel, G.E., Rosenheim, J.A. and Adams, J.M. (1994) Behavioral ecology of host feeding in *Aphytis* parasitoids. *Norwegian Journal of Agricultural Sciences*. **Supplement 16**: 101–15.

Heimpel, G.E., Rosenheim, J.A. and Mangel, M. (1996) Egg limitation, host quality, and dynamic behavior by a parasitoid in the field. *Ecology* **77**: 2410–20.

Heimpel, G.E., Rosenheim, J.A. and Kattari, D. (1997a) Adult feeding and lifetime reproductive success in the parasitoid *Aphytis melinus*. *Entomologia Experimentalis et Applicata* **83**: 305–15.

Heimpel, G.E., Rosenheim, J.A. and Mangel, M. (1997b) Predation on adult *Aphytis* parasitoids in the field. *Oecologia* **110**: 346–52.

Heimpel, G.E., Mangel, M. and Rosenheim, J.A. (1998) Effects of egg and time limitation on lifetime reproductive success of a parasitoid in the field. *American Naturalist* **152**: 273–89.

Heimpel, G.E., Neuhauser, C. and Hoogendoorn, M. (2003) Effects of parasitoid fecundity and host resistance on indirect interactions among hosts sharing a parasitoid. *Ecology Letters* **6**: 556–66.

Heimpel, G.E., Lee, J.C., Wu, Z., Weiser, L., Wäckers, F. and Jervis, M.A. (2004) Gut sugar analysis in field-caught parasitoids: adapting methods originally developed for biting flies. *International Journal of Pest Management* **50**: 193–8.

Hopper, K.R. (1999) Risk-spreading and bet-hedging in insect population biology. Annual *Review of Entomology* **44**: 535–60.

Hopper, K.R., Rosenheim, J.A., Prout, T. and Oppenheim, S.J. (2003) Within-generation bet hedging: a seductive explanation. *Oikos* **101**: 219–22.

Janssen, A. (1989) Optimal host selection by *Drosophila* parasitoids in the field. *Functional Ecology* **3**: 469–79.

Lavandero, B., Wratten, S., Hagler, J. and Jervis, M. (2004) The need for effective marking and tracking techniques for monitoring the movements of insect predators and parasitoids. *International Journal of Pest Management* **50**: 147–51.

Lavandero, B., Wratten, S., Shishehbor, P. and Worner, S. (2005) Enhancing the effectiveness of the parasitoid *Diadegma semiclausum* (Helen): movement after use of nectar in the field. *Biological Control* **34**: 152–8.

Lee, J.C., Andow, D.A. and Heimpel, G.E. (2006) Influence of floral resources on sugar feeding and nutrient dynamics of a parasitoid in the field. *Ecological Entomology* **31**: 470–80.

Legaspi, B.C. and Legaspi, J.C. (2005) Foraging behavior of field populations of *Diadegma* spp. (Hymenoptera: Ichneumonidae): testing for density-dependence at two spatial scales. *Journal of Entomological Science* **40**: 295–306.

Mackauer, M. and Völkl, W. (1993) Regulation of aphid populations by aphidiid wasps: does parasitoid foraging behaviour or hyperparasitism limit impact? *Oecologia* **94**: 339–50.

Mangel, M. and Heimpel, G.E. (1998) Reproductive senescence and dynamic oviposition behaviour in insects. *Evolutionary Ecology* **12**: 871–9.

McGreggor, R. (1997) Host-feeding and oviposition by parasitoids on hosts of different fitness value: influences of egg load and encounter rate. *Journal of Insect Behavior* **10**: 451–61.

McNamara, J. and Houston, A. (1985) A simple model of information use in the exploitation of patchily distributed food. *Animal Behavior* **33**: 553–60.

Morrison, G. (1986) 'Searching time aggregation' and density dependent parasitism in a laboratory host–parasitoid interaction. *Oecologia* **68**: 298–303.

Murdoch, W.W., Briggs, C.J. and Nisbet, R. (2003) *Consumer–Resource Dynamics*. Princeton University Press, Princeton.

Orr, D.B. and Landis, D.A. (1997) Oviposition of European corn borer (Lepidoptera: Pyralidae) and impact of natural enemy populations in transgenic versus isogenic corn. *Journal of Economic Entomology* **90**: 905–9.

Reeve, J.D. (1987) Foraging behavior of *Aphytis melinus*: effects of patch density and host size. *Ecology* **68**: 530–8.

Reeve, J.D. and Murdoch, W.W. (1985) Aggregation by parasitoids in the successful control of the California red scale: a test of theory. *Journal of Animal Ecology* **54**: 797–816.

Rivero-Lynch, A.P. and Godfray, H.C.J. (1997) The dynamics of egg production, oviposition and resorption in a parasitoid wasp. *Functional Ecology* **11**: 184–8.

Rosenheim, J.A. (1987a) Host location and exploitation by the cleptoparasitic wasp *Argochrysis armilla*: the role of learning (Hymenoptera: Chrysididae). *Behavioral Ecology & Sociobiology* **21**: 401–6.

Rosenheim, J.A. (1987b) Nesting behavior and bionomics of a solitary ground-nesting wasp, *Ammophila dysmica* (Hymenoptera: Sphecidae): influence of parasite pressure. *Annals of the Entomological Society of America* **80**: 739–49.

Rosenheim, J.A. (1988) Parasite presence acts as a proximate cue in the nest-site selection process of the solitary digger wasp, *Ammophila dysmica* (Hymenoptera: Sphecidae). *Journal of Insect Behavior* 1: 333–42.

Rosenheim, J.A. (1989) Behaviorally mediated spatial and temporal refuges from a cleptoparasite, *Argochrysis armilla* (Hymenoptera: Chrysididae), attacking a ground-nesting wasp, *Ammophila dysmica* (Hymenoptera: Sphecidae). *Behavioral Ecology & Sociobiology* 25: 335–48.

Rosenheim, J.A. (1990) Density-dependent parasitism and the evolution of aggregated nesting in the solitary Hymenoptera. *Annals of the Entomological Society of America* 83: 277–86.

Rosenheim, J.A. (1996) An evolutionary argument for egg limitation. *Evolution* 50: 2089–94.

Rosenheim, J.A. and Heimpel, G.E. (1994) Sources of intraspecific variation in oviposition and host-feeding behavior. In: Rosen, D. (ed.) *Advances in the Study of* Aphytis. Intercept Press, Andover, pp. 41–78.

Rosenheim, J.A. and Hongkham, D. (1996) Clutch size in an obligately siblicidal parasitoid wasp. *Animal Behaviour* 51: 841–52.

Rosenheim, J.A. and Mangel, M. (1994) Patch-leaving rules for parasitoids with imperfect host discrimination. *Ecological Entomology* 19: 374–80.

Rosenheim, J.A. and Rosen, D. (1991) Foraging and oviposition decisions in the parasitoid *Aphytis lingnanensis*: distinguishing the influences of egg load and experience. *Journal of Animal Ecology* 60: 873–93.

Rosenheim, J.A. and Rosen, D. (1992) Influence of egg load and host size on host-feeding behaviour of the parasitoid *Aphytis lingnanensis*. *Ecological Entomology* 17: 263–72.

Rosenheim, J.A., Meade, T., Powch, I.G. and Schoenig, S.E. (1989) Aggregation by foraging insect parasitoids in response to local variations in host density: determining the dimensions of a host patch. *Journal of Animal Ecology* 58: 101–17.

Rosenheim, J.A., Heimpel, G.E. and Mangel, M. (2000) Egg maturation, egg resorption and the costliness of transient egg limitation in insects. *Proceedings of the Royal Society of London Series B Biological Science* 267: 1565–73.

Sheehan, W. and Shelton, A.M. (1989) Parasitoid response to concentration of herbivore food plants: finding and leaving plants. *Ecology* 70: 993–8.

Stiling, P. and Strong, D.R. (1982) Egg density and the intensity of parasitism in *Prokelisia marginata* (Homoptera: Delphacidae). *Ecology* 63: 1630–5.

Strohm, E., Laurien-Kehnen, C. and Bordon, S. (2001) Escape from parasitism: spatial and temporal strategies of sphecid wasp against a specialized cuckoo wasp. *Oecologia* 129: 50–7.

Strong, D.R. (1989) Density independence in space and inconsistent temporal relationships for host mortality caused by a fairyfly parasitoid. *Journal of Animal Ecology* 58: 1065–76.

Tagawa, J. and Kitano, H. (1981) Mating behaviour of the braconid wasp, *Apanteles glomeratus* L. (Hymenoptera: Braconidae) in the field. *Applied Entomology & Zoology* 16: 345–50.

Tenhumberg, B., Keller, M.A., Possingham, H.P. and Tyre, A.J. (2001) Optimal patch-leaving behaviour: a case study using the parasitoid *Cotesia rubecula*. *Journal of Animal Ecology* 70: 683–91.

Thompson, J.N. (1986) Oviposition behaviour and searching efficiency in a natural population of a braconid parasitoid. *Journal of Animal Ecology* 55: 351–60.

Tinbergen, N. (1958) *Curious Naturalists*. Anchor Books, Garden City.

Umbanhowar, J., Maron, J. and Harrison, S. (2003) Density-dependent foraging behaviors in a parasitoid lead to density-dependent parasitism of its host. *Oecologia* 137: 123–30.

van Alphen, J.J.M., Bernstein, C. and Driessen, G. (2003) Information acquisition and time allocation in insect parasitoids. *Trends in Ecology & Evolution* 18: 81–7.

van Nouhuys, S. and Ehrnsten, J. (2004) Wasp behavior leads to uniform parasitism of a host available only a few hours per year. *Behavioral Ecology* 15: 661–5.

van Nouhuys, S. and Hanski, I. (2002) Colonization rates and distances of a host butterfly and two specific parasitoids in a fragmented landscape. *Journal of Animal Ecology* 71: 639–50.

Vet, L.E.M. and Dicke, M. (1992) Ecology of infochemical use by natural enemies in a tritrophic context. *Annual Review of Entomology* **37**: 141–72.

Völkl, W. (1994) Searching at different spatial scales: the foraging behaviour of the aphid parasitoid *Aphidius rosae* in rose bushes. *Oecologia* **100**: 177–83.

Völkl, W. and Kraus, W. (1996) Foraging behaviour and resource utilization of the aphid parasitoid *Pauesia unilachni*: adaptation to host distribution and mortality risks. *Entomologia Experimentalis et Applicata* **79**: 101–9.

Vos, M., Hemerik, L. and Vet, L.E.M. (1998) Patch exploitation by the parasitoids *Cotesia rubecula* and *Cotesia glomerata* in multi-patch environments with different host distributions. *Journal of Animal Ecology* **67**: 774–83.

Waage, J.K. (1983) Aggregation in field parasitoids populations: foraging time allocation by a population of *Diadegma* (Hymenoptera: Ichneumonoidea). *Ecological Entomology* **8**: 447–53.

Wajnberg, É. (2006) Time-allocation strategies in insect parasitoids: from ultimate predictions to proximate behavioural mechanisms. *Behavioral Ecology & Sociobiology* **60**: 589–611.

Wajnberg, É., Gonsard, P.A., Tabone, E., Curty, C., Lezcano, N. and Colazza, S. (2003) A comparative analysis of patch-leaving decision rules in a parasitoid family. *Journal of Animal Ecology* **72**: 618–26.

Wang, X.G. and Keller, M.A. (2002) A comparison of the host-searching efficiency of two larval parasitoids of *Plutella xylostella*. *Ecological Entomology* **27**: 105–14.

Wanner, H., Gu, H., Hattendorf, B., Gunther, D. and Dorn, S. (2006) Using the stable isotope marker ^{44}Ca to study dispersal and host-foraging activity in parasitoids. *Journal of Applied Ecology* **43**: 1031–9.

Wcislo, W.T. (1984) Gregarious nesting of a digger wasp as a 'selfish herd' response to a parasitic fly (Hymenoptera: Sphecidae; Diptera: Sarcophagidae). *Behavioral Ecology & Sociobiology* **15**: 157–60.

Weisser, W.W., Houston, A.I. and Völkl, W. (1994) Foraging strategies in solitary parasitoids – the trade-off between female and offspring mortality risks. *Evolutionary Ecology* **8**: 587–97.

White, J.A. and Andow, D.A. (2005) Host–parasitoid interactions in a transgenic landscape: spatial proximity effects of host density. *Environmental Entomology* **34**: 1493–500.

White, J.A. and Andow, D.A. (2007) Foraging for intermittently refuged prey: theory and fileld observations of a parasitoid. *Journal of Animal Ecology* (in press).

Wu, Z. and Heimpel, G.E. (2007) Dynamic egg maturation strategies in an aphid parasitoid. *Physiological Entomology* **32**: 143–9.

4

Behavior influences whether intra-guild predation disrupts herbivore suppression by parasitoids

William E. Snyder and Anthony R. Ives

Abstract

Intra-guild predation (IGP), the consuming of one natural enemy by another, is a widespread phenomenon that could weaken the top-down control of herbivores. Parasitoids are often among the most important natural enemies of herbivores and are generally the victims rather than perpetrators of IGP. IGP disruption of herbivore control by parasitoids might thus be expected to be the rule. However, numerous empirical studies show that potential intra-guild predators (IGPredators) often act in synergy with parasitoids to suppress herbivore densities. We constructed mathematical models to examine conditions under which disruption of herbivore control is most likely when a parasitoid interacts with an IGPredator. In general, an IGPredator will disrupt herbivore control when the IGP rate on parasitoids is greater than 0.5 the predation rate on herbivores. We reviewed the literature on host–parasitoid systems including IGPredators to illustrate the numerous behaviors shown by predators and parasitoids that probably modulate the outcome of IGP. In examples in which IGPredators directly attack parasitoid adults or preferentially attack parasitized rather than unparasitized hosts, IGP seems to disrupt herbivore suppression. Conversely, the impact of IGP is apparently minimal in examples in which parasitoid females avoid placing eggs into hosts likely to encounter predators and in examples in which parasitoid larvae alter host behavior or utilize physical defenses to avoid IGP. A more detailed understanding of the behavioral details of intra-guild interactions may ameliorate ecologists' concerns about a powerfully disruptive effect of IGP on the control of herbivore pests.

4.1 Introduction

Hairston et al. (1960) proposed an explanation for why the world is green: strong top-down control of herbivores by their natural enemies limits herbivore densities, in turn freeing plants from continuous and intense herbivory. Subsequent modifications of this theory included varying the numbers of trophic levels (Oksanen et al. 1981), but retained the linear structure and strong top-down regulation proposed by Hairston and co-authors. Clear empirical evidence in support of this view has come from aquatic lake systems dominated by one or a few predator species (Carpenter et al. 1985, Carpenter & Kitchell 1993, Lathrop et al. 2002) and from terrestrial communities from northern regions, for example where communities are somewhat species poor (Strong 1992). Successes in classical biological control programs, where natural enemies from an invasive pest's natural range are collected and released into the exotic range, provide further support for the strong top-down control that natural enemies can provide (DeBach & Rosen 1991).

However, within species-diverse communities, the assignment of species to particular trophic levels can sometimes be difficult. For example, intra-guild predation (IGP), the consumption of one natural enemy by another, appears common in nature (Polis et al. 1989). IGP complicates the assignment of predator species to a particular trophic level and if a distinct 'predator' trophic level cannot be delineated, then top-down trophic cascades of the type envisioned by Hairston et al. (1960) could not be common (Polis 1991, Polis & Holt 1992, Strong 1992). Even the classic aquatic and agricultural systems, where much evidence for simple trophic cascades originated, sometimes demonstrate complicated trophic roles for some predator species (Brett & Goldman 1996, Rosenheim 2001).

Disagreement among ecologists about whether predators can exert strong top-down control sparked a lively debate in the literature of the 1980s and 1990s (Strong 1992, Hairston & Hairston 1993, Polis & Strong 1996, Hairston & Hairston 1997). This controversy triggered a great deal of experimental work and these studies have shed light on and somewhat tempered the ferocity of the debate. Several groups have reported meta-analyses that compile data from predator-removal experiments in a variety of aquatic and terrestrial systems (Schmitz et al. 2000, Halaj & Wise 2001, Shurin et al. 2002). The compiled experimental data reveal that trophic cascades are clearly evident, on the whole, across both aquatic and terrestrial systems, although cascading effects are generally somewhat stronger in aquatic systems (Schmitz et al. 2000, Halaj & Wise 2001, Shurin et al. 2002). Most noticeably in terrestrial communities, however, the strength of top-down effects often becomes successively weaker as interactions are transmitted from predators to herbivores and then from herbivores to plants (Halaj & Wise 2001, Shurin et al. 2002). This amelioration of top-down effects might in part reflect disruptive effects of IGP and other trophic complexities.

It appears that, while IGP commonly occurs, only in a limited number of systems does IGP lead to strong disruption of top-down herbivore control. Why? This question represents a challenge for ecologists. We need to reconcile the ubiquity of IGP across many communities with the consistently strong top-down control in these same communities.

Here, we discuss the role of IGP in mediating the strength of top-down control of herbivores by their parasitoids, an important group of natural enemies in many ecological communities. We focus on parasitoid–host systems for two reasons. First, hosts and their specialist parasitoids often have tightly coupled dynamics, with parasitoids capable of exerting strong top-down control. However, parasitoids are also susceptible to IGP from generalist predators that attack either parasitized hosts or the parasitoids directly (Brodeur & Rosenheim 2000). Therefore, host–parasitoid systems might be good candidates to see the role of IGP in disrupting top-down control. Second, the possible disruption of host regulation by parasitoids is of great practical importance, as many damaging agricultural pests are at least potentially regulated by parasitoids (Cornell & Hawkins 1995,

Hawkins et al. 1997). Disruption from IGP in these systems has a direct economic impact (Rosenheim et al. 1995).

To address the impact of IGP on host–parasitoid systems, we first discuss some key attributes of host–parasitoid systems that seem to make them especially susceptible to disruptive effects of IGP. We then use simple mathematical models to ask how high IGP rates on a parasitoid must be for IGP to disrupt host control by the parasitoid. Finally, we compare these results to examples in the literature where IGP has been found to be particularly important or relatively unimportant for host population control.

IGP can be either reciprocal or unidirectional. When reciprocal, both interacting predator species prey upon one another at all stages or larger stages of one predator feed on smaller stages of the other and vice versa (Polis et al. 1989). For parasitoids, IGP will most often be unidirectional (Brodeur & Rosenheim 2000). This will certainly be true when generalist predators are the intra-guild predators (IGPredators) on parasitoids. It will also be true when the IGPredator is a facultative hyperparasitoid – a parasitoid that can attack not only unparasitized hosts, but also parasitized hosts containing the parasitoid (Godfray 1994). In systems with hosts, parasitoids, and facultative hyperparasitoids (Hunter et al. 2002, Borer et al. 2004), there is a unidirectional effect of facultative hyperparasitoid on the parasitoid. The combination of the unidirectional effects of an IGPredator on intra-guild prey (IGPrey) and the often tight coupling between parasitoid and host dynamics sets up a scenario that is particularly likely to see dramatic disruption of herbivore control when IGPredators are present.

Conversely, it is possible to argue that IGP on parasitoids is less likely to disrupt herbivore suppression than IGP on predators (Ehler 1995, Rosenheim et al. 1995). If IGP on parasitoids occurs when parasitoids are still within hosts, then IGP on parasitoids is inextricably linked to predation on hosts. In this case, IGP is 'coincidental' with predation, in contrast to omnivorous IGPredators that feed on IGPrey and herbivores as separate events (Polis et al. 1989). The close link between IGP and predation in coincidental IGP might limit the IGP rate on parasitoids to be no greater than that on the herbivore, thereby guarding against severe effects of IGP to disrupt herbivore control.

In many communities, the behavior of the IGPredator and IGPrey are keys to determining the outcome of IGP. For example, in a community of arthropods attacking a herbivorous mite (*Tetranychus cinnabarinus*) on papaya, Rosenheim and co-workers (Rosenheim & Corbett 2003, Rosenheim et al. 2004) found that IGP was particularly disruptive of herbivore control when IGP was carried out by a sit-and-wait predator attacking an active foraging predator. In this community, the herbivorous mite is sedentary so that the key mite control agent, the ladybird beetle *Stethorus siphonulus*, must search for its prey. But while foraging, the ladybird is susceptible to predation by the sit-and-wait spider IGPredator *Nesticodes rufipes*. In many other cases, behavioral defenses of IGPrey or hosts can act to weaken or strengthen the importance of IGP (Werner & Peacor 2003). Thus, understanding the impact of IGP necessarily involves studying the behavior of IGPredators, IGPrey, and hosts to understand the patterns of attack rates in the system.

The key question is whether any reduction in host control caused by loss of parasitoids from IGP can be compensated by direct predation on hosts by the IGPredator. What does the relative predation rate of an IGPredator on parasitoids relative to hosts need to be for IGP to disrupt host control? To answer this question quantitatively, we developed simple models of host–parasitoid–predator dynamics.

4.2 Models

To develop models of host–parasitoid–predator dynamics, we considered first the case of a generalist predator whose dynamics are not coupled to those of the hosts or parasitoids (the IGPrey). This will occur, for example, if the predator species attacks a large number of prey species, with our focal host and parasitoid making up only a small part of the IGPredator's diet. We then develop a model for a specialist IGPredator that attacks only hosts and parasitoids, and whose dynamics are consequently tightly coupled to the host–parasitoid system. Systems with hosts, parasitoids, and facultative hyperparasitoids might fit this description. These two cases clearly represent extremes between generalist and specialist IGPredators. Nonetheless, we show that the conditions for the IGPredator to disrupt parasitoid control of host populations are similar in both cases.

4.2.1 Generalist IGPredators

For the generalist IGPredator, we model host–parasitoid dynamics using a modified Nicholson–Bailey model. Let $x(t)$ and $y(t)$ denote the densities of susceptible hosts and adult parasitoids, respectively, in generation t, assuming hosts and parasitoids have synchronized, discrete, and non-overlapping generations. Assume that susceptible hosts are juveniles (so parasitism occurs before reproduction) and that juveniles experience density-dependent survival such that a fraction $e^{-sx(t)}$ survive, where s is the instantaneous survival rate. If a denotes the instantaneous attack rate of parasitoids on hosts and if parasitoids are searching for hosts randomly, then a fraction $e^{-ay(t)}$ of hosts escape parasitism and a fraction $1 - e^{-ay(t)}$ are parasitized. Thus, the density of hosts escaping parasitism is $e^{-ay(t)} x(t)$ and the density of parasitized hosts is $(1 - e^{-ay(t)}) x(t)$. Let $z(t)$ denote the density of IGPredators and assume the predation rate on hosts is p_1. Then the fraction of hosts that escape predation is $e^{-p_1 z(t)}$. Similarly, let p_2 denote the predation rate on parasitoids. It makes no difference whether predation occurs on parasitized but still-living hosts or parasitoids directly (e.g. during parasitoid pupation or even on parasitoid adults). Finally, let λ be the number of offspring produced per surviving host. This set of assumptions gives the model

$$\begin{cases} x(t + 1) = \lambda e^{-p_1 z(t)} e^{-ay(t)} e^{-sx(t)} x(t) \\ y(t + 1) = e^{-p_2 z(t)} (1 - e^{-ay(t)}) e^{-sx(t)} x(t) \end{cases} \tag{4.1}$$

Note that these equations do not contain a term for parasitoid survival. Instead, because decreasing survival has the same effect as decreasing the instantaneous attack rate, parasitoid survival is incorporated into the parameter a.

This is an extreme simplification. Nonetheless, it reveals some qualitatively instructive results. Figure 4.1 graphs two cases of the dynamics given by Equation (4.1). In the first (Fig. 4.1a) the predation rate on parasitoids is zero ($p_2 = 0$) and, with the introduction of the predator, the host population eventually reaches a lower level than in the absence of the predator. Thus, the introduction of a predator when there is no IGP leads to a synergistic effect of predator and parasitoid on host population control. If the predation rate on parasitoids is the same as on hosts ($p_2 = p_1$), then initially the host density drops as the IGPredator is introduced. However, this effect is only short term and after the parasitoid

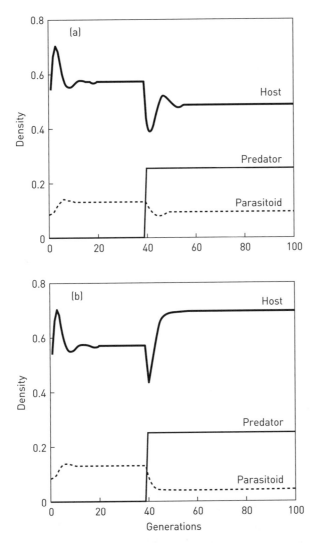

Fig. 4.1 For the case of a generalist predator, population dynamics of hosts and parasitoid governed by Equation (4.1) when the predation rate on parasitoids is (a) $p_2 = 0$ and (b) $p_2 = 1$. IGPredators were introduced at a density of 0.25 at generation 40. The parameter values for Equation (4.1) are $\lambda = 3$, $s = 1$, $a = 4$, and $p_1 = 1$.

population drops due to IGP, the host density increases and ultimately reaches a density above that before the introduction of the IGPredator. Thus, if IGP on the parasitoid has the same magnitude as direct predation on the host, then IGP disrupts host control. But how strong does IGP need to be for disruption to occur?

Figure 4.2a graphs the long-term (equilibrium) host population density as a function of the density of the predator $z(t)$, considering different strengths of IGP p_2. When there is no IGP ($p_2 = 0$), increasing predator density decreases the population density of hosts. When the predator density reaches about 0.75, the parasitoid becomes extinct, because the

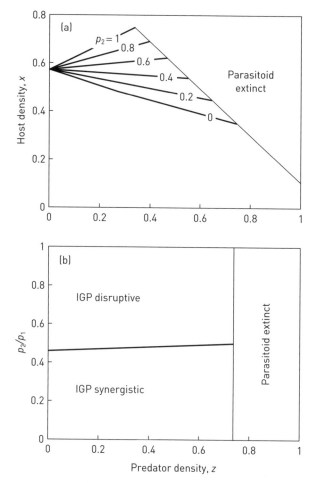

Fig. 4.2 The effect of IGP on the control of a host population by a specialist parasitoid for the case of a generalist predator (Equation 4.1). (a) The equilibrium host density as a function of the density of IGPredators, z, for different values of the predation rate on parasitoids p_2 (labelled). (b) Combinations of the IGPredator density z and the relative predation rate on parasitoids versus hosts p_2/p_1 that give disruptive or synergistic effects of IGP on host control. The parameter values for equation (4.1) are $\lambda = 3$, $s = 1$, $a = 4$, and $p_1 = 1$.

predator has depressed host densities below the level required for parasitoid persistence. As the predator density increases further, the host density drops more rapidly. In contrast, when there is strong IGP ($p_2 = 1$), the initial increase in predator density causes an increase in the equilibrium host density, as IGP suppresses the parasitoid. This continues up to the point at which the parasitoid becomes extinct. For predator densities beyond this, increasing predator density decreases the host density.

Since the cases of $p_2 = 0$ and $p_2 = 1$ give opposite effects of IGP on host control, at some intermediate predation rate $0 < p_2 < 1$, the effects of IGP on host and parasitoid must balance each other out, leading to no net effect on the equilibrium host density. Figure 4.2b shows the threshold between synergistic and disruptive effects of the predator. For low values of p_2/p_1, predation on the host outweighs that on the parasitoid, leading to lower host densities with increasing predator density. Conversely, for high values of p_2/p_1, predation on the parasitoid outweighs that on the host, leading to disruption of host control. The threshold between synergism and disruption is given by the inequality

$$\frac{p_2}{p_1} > \frac{e^{ay^*} - ay^* - 1}{ay^*(e^{ay^*} - 1)} \tag{4.2}$$

where y^* is the equilibrium density of the parasitoids. Any value of p_2/p_1 above the threshold given on the right-hand side of Equation (4.2) implies IGP is disruptive. A particularly attractive feature of this mathematical result is that the threshold is always less than 0.5 for any model parameters, approaching 0.5 as y^* approaches zero (i.e. when the IGPredator becomes sufficiently common to put the parasitoid on the verge of extinction). This property of the threshold is seen in Fig. 4.2b.

Thus, IGP disruption of parasitoid control of host densities can occur readily even when the IGPredator attack rate on the parasitoid is much less than the IGPredator attack rate on hosts. For the model given by Equation (4.1), if the predation rate on parasitoids reaches roughly half the predation rate on hosts, then disruption begins to occur. This threshold of 0.5 is robust to any parameter values in the model. If p_2/p_1 exceeds 0.5, disruption always occurs. Further, the threshold of 0.5 occurs regardless of the functional response, that is the dependence of the parasitoid attack rate a on host density. However, when there is some form of parasitoid interference, such that the per capita success of parasitoids attacking hosts decreases with increasing parasitoid density (Ives 1992), then the p_2/p_1 threshold can be greater than 0.5. Specifically, suppose that parasitism follows a negative binomial distribution, so the proportion of hosts escaping parasitism is $\left(1 + \frac{ay(t)}{k}\right)^{-k}$ where k is the aggregation parameter (Hassell 1978). In this case, the p_2/p_1 threshold equals $\frac{k+1}{2k}$, so the more aggregated parasitism is (the smaller k), the higher the threshold. Nonetheless, the threshold remains less than 1.0 until parasitism becomes strongly aggregated ($k = 1$). Although analyses of more general models are needed to investigate IGP disruption of the control of hosts by parasitoids, these simple models suggest that disruption can occur even when IGP on parasitoids is relatively low.

4.2.2 Specialist IGPredators

The case of a generalist IGPredator is relatively simple, because the density of the IGPredator is determined by prey other than the host and parasitoid. However, when IGPredators are specialists, with dynamics coupled to those of hosts and parasitoids, the situation is more complex. When the IGPredator requires hosts to sustain their population, the IGPredator is a competitor with the parasitoid; if the parasitism rate is high, the parasitoid can out-compete the IGPredator and drive it to extinction. Thus, the first question to ask is whether a specialist IGPredator can persist (Briggs 1993, Holt & Polis 1997, Borer et al. 2004) and only if the IGPredator can persist can we ask whether the IGPredator disrupts the parasitoid's control of the host population. The case of a specialist IGPredator that we consider is similar to original models of IGP (Holt & Polis 1997), and Briggs & Borer (2005) gave an excellent treatment of this issue. Here, we give only an abbreviated discussion to compare with our results for a generalist IGPredator.

To model a system of host, parasitoid, and specialist IGPredator, we use the equations

$$\begin{cases} x(t+1) = \lambda e^{-p_1 z(t)} e^{-ay(t)} e^{-sx(t)} x(t) \\ y(t+1) = e^{-p_2 z(t)} (1 - e^{-ay(t)}) e^{-sx(t)} x(t) \\ z(t+1) = ((1 - e^{-p_1 z(t)}) e^{-ay(t)} + c(1 - e^{-p_2 z(t)})(1 - e^{-ay(t)})) e^{-sx(t)} x(t) \end{cases} \qquad (4.3)$$

Here, the equations for host and parasitoid are identical to those in the model for a generalist IGPredator (Equation 4.1). The equation for the IGPredator dynamics is appropriate for a facultative hyperparasitoid. We modeled the IGPredator as a facultative hyperparasitoid, reasoning that facultative hyperparasitoids are the most likely form of highly specialized IGPredators. The reproduction of the IGPredator depends on both the number of hosts and the number of parasitoids they attack. We have included a parameter c that discounts reproduction of IGPredators from parasitoids. If, for example, $c = 0.5$, then each depredated parasitoid gives rise to on average one-half the number of IGPredators compared to the number produced from hosts. For solitary hyperparasitoids, this difference could be due to differences in larval survival within hosts versus parasitoids. Because the overall survival (including adult survival) is incorporated into the IGPredator attack rates p_1 and p_2, the value of c is not constrained to be less than 1.0. c will be greater than 1.0 whenever survival of an IGPredator within IGPrey is higher than within hosts. The importance of c is that it gives the benefit of predation on parasitoids to the IGPredator relative to the benefit of predation on hosts. Since the IGPredators are competitors with parasitoids for hosts, IGPredators may be able to overcome this competition by getting a direct benefit from consuming parasitoids.

The model described by Equation (4.3) can give dynamics similar to those from the generalist IGP model. Specifically, as can be seen Fig. 4.3, the introduction of IGPredators can either decrease or increase the long-term, equilibrium host density. The case of a specialist IGPredator, however, is more complex than for a generalist IGPredator, because only certain conditions allow the IGPredator to persist and only a persisting IGPredator will disrupt or augment control of the host population.

To address what conditions allow the IGPredator to persist, Fig. 4.4 graphs the possible outcomes of IGP for combinations of the conversion rate of parasitoids into IGPredators c and the predation rate on parasitoids relative to that on hosts p_2/p_1. There are four general outcomes: (i) the parasitoid can exclude the IGPredator; (ii) the IGPredator can exclude the parasitoid; (iii) the IGPredator and parasitoid can coexist; and (iv) and the winner in competition between the IGPredator and parasitoid can be determined by their initial densities. In the last case, high initial densities of the IGPredators relative to parasitoids cause the IGPredator to exclude the parasitoid, whereas high initial densities of the parasitoid relative to the IGPredator cause the parasitoid to exclude the IGPredator (Briggs & Borer 2005).

A particularly interesting point can be made from Fig. 4.4. In the absence of the IGPredator obtaining direct benefit from consuming parasitoids ($c \ll 1$), coexistence of IGPredator and parasitoids is impossible. Therefore, the coexistence of an IGPredator and IGPrey requires either that the IGPredator obtain a direct benefit from attacking the IGPrey or that the IGPredator be a generalist, reproducing on alternative prey as in the generalist IGPredator model described earlier.

When coexistence is not possible and the IGPredator excludes the parasitoid, the condition for IGPredators to disrupt host population control is simple: if the IGPredator has

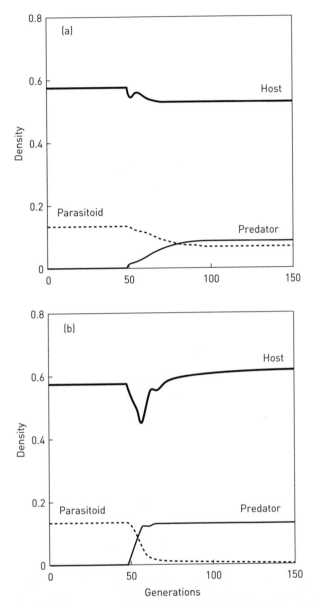

Fig. 4.3 Population dynamics of hosts, parasitoids, and IGPredators for the case of a specialist predator governed by Equation (4.3) when the predation rate on parasitoids is (a) relatively low ($p_2 = 0.3\,p_1$) and (b) relatively high ($p_2 = 0.6\,p_1$). IGPredators were introduced at a density of 0.01 at generation 50. The parameter values for Equation (4.1) are $\lambda = 3$, $s = 1$, $a = 4$, $c = 2$, and $p_1 = 3.75$.

a lower attack rate on the host than the parasitoid ($p_1 < a$), then disruption will occur, with the equilibrium host population higher with the IGPredator than with the parasitoid. Conversely, when coexistence occurs, the condition for disruption of host control is exactly the same as for the case of the generalist IGPredator (Equation 4.2). In this case,

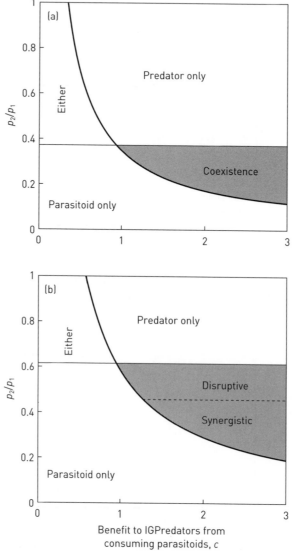

Fig. 4.4 The effect of IGP on host–parasitoid interactions for the case of a specialist predator when (a) parasitoids have a lower attack rate on hosts than IGPredators ($p_1 = 4.25$) and (b) parasitoids have a higher attack rate on hosts than IGPredators ($p_1 = 3.75$). Four distinct dynamics occur for different combinations of values of parasitoids to IGPredator, c, and the relative predation rate on parasitoids versus hosts p_2/p_1. The parasitoid out-competes the IGPredator (parasitoid only), the IGPredator out-competes the parasitoid (predator only), the parasitoid or predator out-competes the other depending on their relative initial densities (either), and the parasitoid and predator coexist (shaded gray). In the last case, when parasitoids have a lower attack rate than IGPredators (b), the IGPredator can either disrupt control of the host by the parasitoid and lead to higher host densities (disruptive) or the combination of an IGPredator and parasitoid can lead to greater host control (synergistic). The other parameter values for Equation are $\lambda = 3$, $s = 1$, and $a = 4$.

if the ratio of the IGPredator attack rates on parasitoids relative to hosts p_2/p_1 is greater than approximately 0.5, then IGP will lead to higher equilibrium host densities (Fig. 4.4b). But it turns out that the requirement for IGPredator and parasitoids to coexist introduces another condition: for p_2/p_1 to be greater than 0.5, the IGPredator must have a lower attack rate on the host than the parasitoid ($p_1 < a$). If instead the IGPredator has a higher attack rate on the host than the parasitoid ($p_1 > a$), then values of p_2/p_1 greater than 0.5 cause the IGPredator to exclude the parasitoid (Fig. 4.4a).

Thus, the conditions for a specialist IGPredator to disrupt host control by parasitoids are more complicated than for the case of a generalist IGPredator, but not much more complicated. First, the IGPredator must have a lower attack rate on the host than the parasitoid ($p_1 < a$). If the IGPredator drives the parasitoid extinct, this condition is sufficient. If the IGPredator and parasitoid coexist, then there is an additional condition that the ratio of IGPredation rates on parasitoids relative to hosts p_2/p_1 be greater than approximately 0.5, the same condition as for a generalist IGPredator.

4.2.3 Summary

Although the simplicity of the models we investigated cautions against any general conclusions, the models nonetheless suggest the following.

1 When an IGPredator is a generalist, the IGPredation rate on parasitoids must be considerable for IGP to disrupt parasitoid control of host populations, but it does not have to be as great as the IGPredation rate on hosts. The model given by Equation (4.1) gives a rule of thumb that IGP disruption occurs whenever the IGPredation rate on parasitoids relative to hosts p_2/p_1 exceeds 0.5, although for other models this threshold can be higher.

2 When an IGPredator is a specialist whose population dynamics are tightly coupled to those of the host and parasitoid, competition with the parasitoid may exclude the IGPredator. For coexistence to occur, the specialist IGPredator must get substantial benefit by reproducing on the parasitoid (c having a value near 1.0). This is likely the case for facultative hyperparasitoids.

3 Specialist IGPredators will only disrupt parasitoid control of the host if the attack rate of IGPredators on hosts is less than that of parasitoids ($p_1 < a$). If IGPredators and parasitoids coexist, disruption further requires roughly that $p_2/p_1 > 0.5$.

4 For the most part we can focus on IGPredators that coexist with parasitoids, since these are the systems of IGP that occur in nature. Assuming coexistence, the conditions for IGP to disrupt host population control are the same for specialist and generalist IGPredators.

5 For coexisting IGPredators and parasitoids, any factor that decreases the IGPredator attack rate on parasitoids relative to hosts p_2/p_1 will tend to mitigate IGP disruption of top-down host control.

4.3 Literature review

To organize our review of the literature, we will begin by asking first how IGPredator behavior may increase attack rates on IGPrey and then how IGPrey behaviors may decrease IGP. We then turn to specialist IGPredators by considering cases in which IGP of parasitoids is carried out by other parasitoids. Finally, we address studies of multiple IGPredator and/or IGPrey species, in which the net effect of IGP depends upon the actions and interactions among several species in the predator trophic guild. Our literature review is not exhaustive. Instead, we have tried to give diverse examples from different systems to illustrate the many ways in which predator and parasitoid behaviors modulate the outcomes of IGP and host suppression. Because the models highlight the importance of the IGPredator attack rate on parasitoids versus hosts, this is our main focus.

4.3.1 Parasitoids more likely to be attacked by IGPredator

The models suggest that predators that target parasitoids while rarely attacking hosts are particularly disruptive. Indeed, there is evidence that some predators act in this way. Rees and Onsager (1982, 1985) studied grasshoppers (*Melanoplus sanguinipes*) attacked by a guild of sarcophagid (Diptera) parasitoids in the genus *Blaesoxipha*. Robber flies (Asilidae) attack both parasitized and unparasitized grasshoppers, apparently without bias, but also attack foraging adult parasitoids. In large field cages asilids sometimes attacked grasshoppers, but were far more effective predators of adult parasitoids: cages including asilids experienced the near complete cleansing of parasitoids from the grasshopper population, whereas cages without asilids experienced an approximately twofold increase in parasitism rates (Rees & Onsager 1982). The net effect of asilids was a dramatic improvement in grasshopper survival, compared to cages including only grasshoppers and parasitoids. In a rare and heroic follow-up experiment, Rees and Onsager (1985) supported their field cage work with what must have been an exhausting open-field experiment. The authors established a series of large open field plots and removed asilids from some of them using sweep nets. Hand removal resulted in a 40% reduction in asilid densities and a doubling of percent parasitism of grasshoppers by fly parasitoids (Rees & Onsager 1985). Adding to this convincing evidence, asilids experienced a natural reduction of more than 90% in one year and, in this year, parasitism of grasshoppers by flies increased as much as fourfold (Rees & Onsager 1985). Clearly, the selective predation of adult parasitoids by robber flies was contributing to the typically low parasitism rate of grasshoppers in the field, which often hovered around 1% at sites and/or during years when asilids were typically common (Rees & Onsager 1985).

There is some other evidence that predation on adult parasitoids, as observed by Rees and Onsager (1985), might be common in nature. For example, Heimpel et al. (1997) observed *Aphytis* parasitoids as they foraged in the open field for their host, the San Jose scale (*Quadraspidiotus perniciosus*), on almond (*Prunus dulcis*) trees. Several hundred wasps were observed over nearly 90 hours of observations, providing the type of data on field behavior that is painful to collect but exceedingly useful. These authors reported that, during the fall, when parasitoids were most abundant, mortality from predators reduced adult wasp survivorship from about 30 days as measured (under albeit benign conditions) in the laboratory, to as little as 2 days. Predators observed feeding upon wasps included spiders of several types, the Argentine ant (*Linepithema humile*), and the assassin bug *Zelus renardii*. Thus, the effects of predators were as damaging to wasp survival as were starvation and extreme temperatures (Heimpel et al. 1997).

Tostawaryk (1971) provided an example of predation directed toward parasitized hosts and away from unparasitized hosts. This author examined the impact of aggregation behavior by the sawfly species *Neodiprion swainei* and *Neodiprion pratti banksiana*, herbivores of jack pine (*Pinus banksiana*), on interactions among the sawflies' natural enemies. These sawfly larvae form tight feeding colonies that result in interior larvae being relatively protected from natural enemies, with only those larvae at the colony's edge exposed to attack. A guild of parasitoids (*Spathimeigenia spingera*, *Lamachus* species, and *Perilampus hyalinus*) successfully attack larvae at the periphery of the colony, but many of these parasitized larvae are later killed by predatory pentatomids (*Podisus modestus*) that also limit their attacks to the larvae at the edge of the colony. Thus, aggregation behavior by the sawflies indirectly heightens IGP of parasitized larvae by pentatomids while protecting most unparasitized larvae (Tostawaryk 1971).

4.3.2 Parasitoid strategies to avoid IGP

Avoiding IGP through selective oviposition behavior
Taylor et al. (1998) first reported a now well-studied example of IGPredator avoidance by foraging female parasitoids. The work was conducted with the pea aphid, *Acyrthosiphon pisum* and its wasp parasitoid *Aphidius ervi*. Like other aphid parasitoids, *A. ervi* is but one member of a diverse community attacking pea aphids, including several species of generalist predator. One common true predator in this community is the ladybird beetle *Coccinella septempunctata*, which acts as an IGPredator of *A. ervi* when ladybird larvae or adults consume parasitized aphids or parasitoid pupae within the aphid cadaver. Taylor et al. (1998) compared the patch residence time of parasitoid females foraging on leaves either currently or previously housing a foraging ladybird larva or adult. They found that wasps spent roughly half the time foraging on plants that either housed or formerly had housed a ladybird, compared to plants never contacted by ladybirds. These data indicated both that *A. ervi* avoids placing larvae into hosts at risk of being consumed by ladybirds and that chemical cues deposited by the beetles alone are sufficient to initiate IGP avoidance behavior by female wasps.

Nakashima and Senoo (2003) and Nakashima et al. (2004) provided additional detail on the avoidance of the ladybird *C. septempunctata* by the parasitoid wasp *A. ervi*. In simple laboratory choice arenas consisting of broad bean (*Vicia faba*) leaves upon which a ladybird adult or larva had walked for 24 hours or control leaves that had not housed a ladybird, they recorded residence time of parasitoids on patches of each type. Parasitoids spent less time on leaves that had housed a *C. septempunctata*, about 30 seconds less over a 10-minute observation period. However, the deterrence effect was of limited duration. Leaves that had housed ladybirds deterred parasitoid foraging for up to 18 hours, but thereafter this effect disappeared (Nakashima & Senoo 2003). The chemical responsible for the deterrent effect was then isolated (Nakashima et al. 2004) and was found to be caused by either of two surface hydrocarbons common to several coccinellid species (Kosaki & Yamaoka 1996) and deposited in the 'footprints' of coccinellids as they foraged. These hydrocarbons are also used by coccinellids to avoid egg laying in patches already visited earlier by other beetles (Hemptinne et al. 2001), perhaps reflecting the ubiquity of such chemicals on coccinellid body surfaces. Spraying plant surfaces with the active hydrocarbons only, in the absence of a coccinellid, elicited patch avoidance behavior similar to that imposed by intact ladybird beetles and a roughly 50% reduction in percent parasitism (Nakashima et al. 2004).

Avoidance of ladybirds by parasitoids appears to have measurable impact on aphid parasitism in the field. Raymond et al. (2000) found that the percent parasitism of bean aphid (*Aphis fabae*) by the parasitoid *Lysiphlebus fabarum* generally increased with aphid density across several aphid host plants. However, the exception was on the plant *Chenopodium album*, which housed the greatest density of aphids but also the greatest number of ladybird beetles. On these plants, parasitism rates did not increase with growing aphid density. It appeared that parasitoids may have been avoiding aphid colonies also housing ladybirds and, indeed, subsequent laboratory experiments provided some suggestion that parasitoids avoid coccinellids when choosing patches to forage within. This avoidance of IGP perhaps explains the relatively low parasitism rates when aphids occurred on *C. album* in the field (Raymond et al. 2000).

In a host–parasitoid system not including aphids, Shiojiri and Takabayashi (2005) examined interactions between ants of several species that feed on nectar of the plant *Rorippa indica*, but also will opportunistically feed on herbivorous diamondback moth larvae (*Plutella xylostella*) once on plants. Diamondback larvae are also attacked by a specialist parasitoid, *Cotesia plutellae*. Like unparasitized larvae, those parasitized by *C. plutellae* suffer high ant predation when on flowering plants, because ant densities are much higher on plants offering nectar. The wasps avoid flowering plants, perhaps as a strategy to avoid IGP by ants, although other possibilities, such as physiological or chemical differences between younger plants without flowers and older plants with flowers, could not be excluded (Shiojiri & Takabayashi 2005).

Avoiding IGP by manipulating host behavior

Larvae of aphid parasitoids of several species manipulate the behavior of their hosts, most dramatically just before the host's death. Brodeur and McNeil (1989, 1992) examined the manipulation of potato aphid (*Macrosiphum euphorbiae*) behavior by its parasitoid *Aphidius nigripes*. Parasitoids about to enter diapause often trigger their hosts to move from the plant to concealed sites shortly before host death, whereas aphids parasitized by non-diapausing parasitoids are more likely to remain on the plant (Brodeur & McNeil 1989). One possible advantage of this behavioral modification is that hyperparasitoids might have a more difficult time locating concealed parasitoid pupae and, indeed, in the laboratory the hyperparasitoid *Asaphes vulgaris* was less likely to find and attack concealed pupae than pupae remaining on the plant (Brodeur & McNeil 1989). Non-diapausing *A. nigripes* also induce a more subtle alteration of potato aphid behavior before host death, causing the aphid to move from typical feeding positions on the underside of the leaf to the top of the leaf. In a series of field observations and experimental manipulations of pupa location, hyperparasitism and predation rates on pupae sometimes were indeed lower when pupae were on the top rather than bottom of leaves (Brodeur & McNeil 1992).

Brodeur and Vet (1994) reported a more unusual manipulation of host behavior to fend off IGPredators. The parasitoid *Cotesia glomerata* exits its host, *Pieris rapae*, to pupate. The doomed, now vacated caterpillar then crawls onto the parasitoid pupa and defends the parasitoid that has killed it from any IGPredators through a flicking behavior directed at approaching predators (Brodeur & Vet 1994). It is unclear whether host behavior is designed to push away IGPredators physically or, alternatively, to act as a decoy by attracting predator attacks and leaving the predator too satiated to then attack the wasp pupa.

Parasitoid physical defenses against IGP

Parasitoids attacking Homoptera often induce a physical hardening of the host cuticle at host death, forming what is called the mummy. No doubt this coating provides some physical defense against harsh environmental conditions during pupation, but an additional advantage is protection from IGP. For example, Hoelmer et al. (1994) examined IGP by the coccinellid beetle *Delphastus pusillus* of the parasitoid *Encarsia transvena*. This parasitoid attacks the sweet potato whitefly *Bemisia tabaci*. Coccinellids attack unparasitized hosts and those containing early instar parasitoids at equal rates, but as parasitoid development progresses predators increasingly avoid attacking parasitized hosts. The avoidance apparently is caused by a hardening of the whitefly cuticle that is triggered by the parasitoid developing inside, which deters predator attack (Hoelmer et al. 1994). Several other authors have reported similar cases of IGP repelled by a tough mummy case (Kindlmann & Ruzicka 1992, Snyder & Ives 2003).

Physical defenses of parasitoid pupae can eliminate disruptive effects of IGP for herbivore regulation. Colfer and Rosenheim (2001) examined IGP by ladybird beetles (*Hippodamia convergens*) on *Lysiphlebus testaceipes* parasitoids attacking cotton aphids *Aphis gossypii*. In field cages, IGP on parasitoids was intense, with coccinellids reducing mummy densities by more than 98%. Still, aphid suppression was significantly greater when both ladybirds and parasitoids were present, compared to replicates where only parasitoids were present. In feeding trials, ladybirds were somewhat more likely to feed on unparasitized aphids than upon mummies, presumably because the mummy case retarded predator attacks (Colfer & Rosenheim 2001). A similar example of the avoidance of mummy predation leading to synergistic effects of predator and parasitoid on aphid control has been reported for a greenhouse rose system with potato aphid (*M. euphoribae*) as the herbivore, *Aphelinus asychis* as the parasitoid, and the coccinellid *Harmonia axyridis* as the IGPredator (Snyder et al. 2004).

4.3.3 Specialist IGPredators

The most common type of specialist IGPredator are facultative hyperparasitoids, defined broadly as specialist parasitoids that can attack both unparasitized hosts and hosts previously parasitized by another species. Some species of parasitoids are unambiguously facultative hyperparasitoids. For autoparasitoids, females develop as primary parasitoids on a host, whereas males develop as obligatory hyperparasitoids either on females of their own species or on other species of primary parasitoids (Godfray 1994). Thus, autoparasitoids clearly fall into the category of facultative hyperparasitoids when they attack hosts previously parasitized by other species. Other cases are less clear. For example, take the case of two parasitoids of pea aphids, *A. ervi* and *Praon pequodorum*. Both species are solitary parasitoids, so even if multiple eggs are oviposited into the same host, only one parasitoid adult emerges. *Praon pequodorum* embryos develop an extra-serosa envelope in the presence of heterospecific competitors, which serves as a defense against older larvae (Danyk & Mackauer 1996). Thus, unless *A. ervi* larvae get a head start, with *A. ervi* females ovipositing at least 24 hours before *P. pequodorum* females, a doubly parasitized pea aphid will invariably give rise to *P. pequodorum* (Danyk & Mackauer 1996). Although *A. ervi* and *P. pequodorum* are generally considered as competitors, from a population dynamics perspective, *P. pequodorum* is indistinguishable from an IGPredator. Taking this broad definition of facultative hyperparasitoids, IGPredation involving two (or more) parasitoids might be common.

Schellhorn et al. (2002) studied competition between *A. ervi* and *P. pequodorum* for pea aphid hosts in lucerne, specifically addressing the following quandary. Although

P. pequodorum is the superior within-host competitor (i.e. effectively a facultative hyper-parasitoid), *A. ervi*, after being introduced into North America for biological control, has apparently displaced *P. pequodorum*. Using detailed studies of foraging behavior, Schellhorn et al. (2002) showed that *A. ervi* had considerably higher searching efficiency than *P. pequodorum*. Pea aphids show large fluctuations in population density driven mainly by harvesting of lucerne. Densities are diminished by orders of magnitude following harvesting, but then often rebound exponentially before the following harvest. When faced with high fluctuations in host abundance, foraging efficiency is at a premium, because it allows parasitoids to take advantage of low host densities. The combination of high *A. ervi* searching efficiency and highly fluctuating pea aphid densities probably explains the ability of *A. ervi* to out-compete *P. pequodorum*, despite *P. pequodorum*'s within-host competitive superiority. In terms of biological control, if *A. ervi* excluded *P. pequodorum* due to its higher searching efficiency, then *A. ervi* is also the better biological control agent, able to suppress pea aphid densities to levels below which *P. pequodorum* can persist in lucerne. Thus, even though *P. pequodorum* is effectively an IGPredator, this does not confer sufficient advantage for it to inhibit control of pea aphids by *A. ervi*.

A suite of examples of IGP involving autoparasitoids acting as facultative hyperparasitoids occurs for *Encarsia* spp. that parasitize whiteflies. Recently, the possible disruptive effects of autoparasitoids have received attention and the wisdom of introducing autoparasitoid species to attack invasive pests has been questioned. Theoretical treatments including detailed accounting of stage-structured development suggest that autoparasitoids will have a net positive impact on whitefly biological control only when the autoparasitoid is superior to the primary parasitoid (IGPrey) at exploiting the host, in which case it excludes the primary parasitoid. Conversely, an autoparasitoid will always increase host density if it can coexist with the primary parasitoid (Briggs & Collier 2001). This is consistent with our theoretical results for specialist IGPredators, because Briggs and Collier (2001) assumed that an autoparasitoid will have the same attack rates on parasitized and unparasitized hosts, leading to $p_2/p_1 = 1$ (>0.5).

Hunter et al. (2002) examined interactions between two parasitoids, the IGPredator *Encarsia sophia* and the IGPrey *Encarsia eremicus* and their shared prey, the sweet potato whitefly *B. tabaci*. *Encarsia sophia* is an autoparasitoid, depositing female eggs in the primary host and male eggs in parasitized hosts. *Encarsia sophia* will hyperparasitize both hetero- and conspecifics, but conspecifics are covered by a protective sheath throughout much of the vulnerable stage. In contrast, *E. eremecus* does not attack *E. sophia*. Therefore, IGP will usually be asymmetric, with the IGPredator having a large impact on the IGPrey and, consequently, a disruptive effect of IGP is possible. Indeed, in a field cage experiment, it was found that *E. eremecus* densities were substantially reduced, sometimes to extirpation, when paired with *E. sophia*. However, displacement of *E. eremecus* did not lead to disrupted whitefly control, because *E. sophia* densities remained unchanged and, apparently, *E. sophia* is superior to *E. eremecus* as a whitefly control agent (Hunter et al. 2002).

Bogran et al. (2002) examined interactions among three parasitoids of silverleaf whitefly (*Bemisia argentifolii*): *Encarsia pergandiella*, *Eretmocerus mundus*, and *Encarsia formosa*. Of these, one species, *E. pergandiella*, a heteronomous hyperparasitoid (meaning that female eggs are laid into unparasitized hosts whereas male eggs are laid into previously parasitized hosts) serves as an IGPredator when placing male eggs into the other two species. The researchers constructed communities consisting of all possible unique combinations of the three species and then looked for synergistic or disruptive effects of the different pairings on resulting whitefly survival. In the majority of cases, parasitoids acted in an additive fashion, with no evidence of interference. In several cases where interference was observed, this appeared to be due to exploitative competition won by E. *Mundus*. *Encarsia mundus* takes smaller host stages than will the other two species. However, when E. *pergandiella* was paired with E. *formosa*, whitefly suppression was also disrupted and this case appeared to be due to hyperparasitism of E. *formosa* by E. *pergandiella* (Bogran et al. 2002).

The final example comes from perhaps the best-understood system of parasitoid biological control, the control of California red scale (Murdoch et al. 2003). The dominant parasitoid controlling red scale is *Aphytis melinus*, an autoparasitoid. Another parasitoid, *Encarsia perniciosi*, is also regularly present but is not a facultative hyperparasitoid of A. *melinus*. Using a gradient in red scale productivity across three citrus cultivars (orange, grapefruit, and lemon), Borer et al. (2003) demonstrated an increase in the density of A. *melinus* associated with increasing red scale productivity. This is consistent with A. *melinus* acting as an IGPredator with lower attack rates than E. *perniciosi*. Low host productivity favors the IGPrey with its higher attack rate, whereas higher host productivity allows the IGPredator to capitalize on its ability to parasitize previously parasitized hosts. Further, the lower attack rate of A. *melinus* than E. *perniciosi* may be explained by its preference for foraging on leaves rather than stems, producing a refuge for red scale on stems (Borer et al. 2004). This contrasts with E. *perniciosi* that forages without preference on leaves and stems, thereby allowing it to attack all hosts present.

4.3.4 Taking a community ecology perspective

In diverse ecological communities, there may be multiple IGPredators and multiple IGPrey. In these systems, disproportionate attack on parasitized and unparasitized hosts by different IGPredator species or the same IGPredator on different primary parasitoid species may both occur simultaneously. For example, Tscharntke (1992) examined IGP by birds (primarily bluetits *Parus caeruleus*) on a guild of parasitoids attacking gall midges (*Giraudiella* sp.) on *Phragmites australis* plants. Different species within the parasitoid guild had different preferences for hosts. Some species preferred smaller gall clusters, while others preferred large gall clusters. Birds were more likely to attack larger gall clusters, with the result that parasitoid species attacking larger clusters were particularly likely to fall victim to IGP by birds (Tscharntke 1992). Presumably, birds would have both positive and negative effects on herbivore regulation by the different parasitoid species.

Snyder and Ives (2001, 2003) provided another example involving multiple possible IGPredators and IGPrey. This work focused on lucerne (*Medicago sativa*), pea aphids, and the parasitoid A. *ervi*. Like some other aphid parasitoids and as we have seen before, A. *ervi* induces its host to climb to upper leaf surfaces higher in the foliage shortly before host death. Lucerne is periodically harvested and the resulting stubble places aphids (both parasitized and unparasitized) and mummies in close proximity to the ground where lurks the carabid beetle *Pterostichus melanarius*. Field cage experiments revealed that, when plants were short after cutting, carabids consumed large numbers of aphids confined to the short stubble and the beetles contributed strongly to aphid suppression. However, as plants regrew, carabids climbing in the foliage became increasingly unable to capture active aphids, but could still effectively capture the immobile mummies. This selective mummy predation led to sharp reductions in parasitoid densities, disrupted aphid control, and the eventual exacerbation of aphid outbreak (Snyder & Ives 2001). In contrast to carabids, however, predators that are more active and adept in the foliage, such as ladybird beetles and nabids, exhibited either no preference for mummies over aphids or actually preferred aphids over mummies. When the entire predator guild, including both carabids and foliar predators, was manipulated in a subsequent field experiment, no net selective predation on parasitoids was observed and the predator guild and parasitoid both contributed to aphid control (Snyder & Ives 2003), at least in the short term (over a single parasitoid generation). However, using models to extrapolate from the short-term experiments to the longer term of multiple parasitoid generations, when the effect of IGP on parasitoid population densities come into play, IGP from the generalist predator guild was predicted to disrupt pea aphid suppression by *A. ervi*.

4.3.5 Summary

Numerous different IGPredator behaviors can lead to greater attack rates on IGPrey and numerous IGPrey behaviors can reduce IGPredation rates. The question for any given set of species is whether, on balance, IGPredation rates on parasitoids are sufficient to cause disruption of long-term host suppression. Our models suggest that, for disruption of host suppression, IGPredator attack rates on IGPrey must be significant, but need not be as high as IGPredator attack rates on hosts. Indeed, for the model given by Equation (4.1), disruption occurs whenever $p_2/p_1 > 0.5$. When IGPredators are generalists, the suites of IGPredator and IGPrey behaviors that set IGPredator attack rates on IGPrey versus hosts are large, making it difficult to make an informed guess about the effects of IGP without detailed knowledge of the species in question. When IGPredators are specialist parasitoids, consistent disruption of host suppression by IGP seems to be more likely, because IGP on parasitized and unparasitized hosts will probably be similar. Nonetheless, there are too few studies of IGP on parasitoids to reach any broad conclusions.

In a recent meta-analysis of IGP, Rosenheim and Harmon (2006) identified 25 experimental studies on IGP and found that roughly half demonstrated a net negative effect of IGPredators on herbivore population densities, while the other half showed a net positive effect. Of the 25 studies, 11 involved parasitoids as the IGPrey and, for this group, the chances of IGP disrupting herbivore control were less than the chances when IGP involved a non-parasitoid IGPrey. Thus, of those systems studied, host–parasitoid systems do not seem more prone to disruption by IGP as might be expected due to the close coupling of host and parasitoid dynamics. Rosenheim and Harmon (2006) added the caution, however, that the timescales of the experiments were short and disruption is more likely to occur over the longer periods of time required for suppression of IGPrey densities. Nonetheless, the results of Rosenheim and Harmon (2006) suggest that the range of behaviors exhibited by parasitoids to avoid predation and the range of behaviors of IGPredators selecting unparasitized hosts, parasitized hosts, and parasitoids themselves, may play major roles in mitigating the negative effect of IGP on herbivore suppression.

4.4 Conclusions

From a theoretical perspective, it is difficult to anticipate whether IGP will cause an increase or decrease in long-term herbivore population densities when the main potential control agent is a parasitoid, because the disruption of host suppression by IGP requires moderately strong but not too strong IGPredator attack rates on parasitized hosts or parasitoids themselves: the outcome of IGP for host suppression teeters on the boundary between disruption and synergism. Studies of real systems seem to bear out this expectation, with examples of both disruption and synergism by IGP. Of course, comparing theoretical models of IGP with empirical studies is always difficult. Our theoretical treatment addressed the long-term consequences of IGP, whereas most empirical studies are restricted to much shorter time periods, often just one or a few parasitoid generations (Snyder & Ives 2003, Briggs & Borer 2005). Nonetheless, the models substantiate an intuitive result: IGPredators, such as facultative hyperparasitoids that do not distinguish between unparasitized and parasitized hosts, will most likely be more disruptive than IGPredators that are more likely to attack unparasitized than parasitized hosts due to either their own behaviors or the defenses of the parasitoid.

References

Bogran, C.E., Heinz, K.M. and Ciomperlik, M.A. (2002) Interspecific competition among insect parasitoids: field experiments with whiteflies as hosts in cotton. *Ecology* **83**: 653–68.

Borer, E.T., Briggs, C.J., Murdoch, W.W. and Swarbrick, S.L. (2003) Testing intraguild predation theory in a field system: does numerical dominance shift along a gradient of productivity? *Ecology Letters* **6**: 929–35.

Borer, E.T., Murdoch, W.W. and Swarbrick, S.L. (2004) Parasitoid coexistence: linking spatial field patterns with mechanism. *Ecology* **85**: 667–78.

Brett, M.T. and Goldman, C.R. (1996) A meta-analysis of the freshwater trophic cascade. *Proceedings of the National Academy of Sciences USA* **93**: 7723–6.

Briggs, C.J. (1993) Competition among parasitoid species on a stage-structured host and its effect on host suppression. *American Naturalist* **141**: 372–97.

Briggs, C.J. and Collier, T. R. (2001) Autoparasitism, interference, and parasitoid–pest population dynamics. *Theoretical Population Biology* **60**: 33–57.

Briggs, C.J. and Borer, E.T. (2005) Why short-term experiments may not allow long-term predictions about intraguild predation. *Ecological Applications* **15**: 1111–17.

Brodeur, J. and McNeil, J.N. (1989) Seasonal microhabitat selection by an endoparasitoid through adaptive modification of host behavior. *Science* **244**: 226–8.

Brodeur, J. and McNeil, J.N. (1992) Host behavior modification by the endoparasitoid *Aphidius nigripes*: a strategy to reduce hyperparasitism. *Ecological Entomology* **17**: 97–104.

Brodeur, J. and Vet, L.E.M. (1994) Usurpation of host behavior by a parasitic wasp. *Animal Behavior* **48**: 187–92.

Brodeur, J. and Rosenheim, J.A. (2000) Intraguild interactions in aphid parasitoids. *Entomologia Experimentalis et Applicata* **97**: 93–108.

Carpenter, S.R. and Kitchell, J.F. (1993) *The Trophic Cascade in Lakes*. Cambridge University Press, Cambridge.

Carpenter, S.R., Kitchell, J.F. and Hodgson, J.R. (1985) Cascading trophic interactions and lake productivity. *Bioscience* **35**: 634–9.

Colfer, R.G. and Rosenheim, J.A. (2001) Predation on immature parasitoids and its impact on aphid suppression. *Oecologia* **126**: 292–304.

Cornell, H.V. and Hawkins, B.A. (1995) Survival patterns and mortality sources of herbivorous insects: some demographic patterns. *American Naturalist* **145**: 563–93.

Danyk, T.P. and Mackauer, M. (1996) An extraserosal envelope in eggs of *Praon pequodorum* (Hymenoptera: Aphidiidae), a parasitoid of pea aphid. *Biological Control* **7**: 67–70.

DeBach, P. and Rosen, D. (1991) *Biological Control by Natural Enemies*. Cambridge University Press, New York.

Ehler, L.E. (1995) Biological control of obscure scale (Homoptera, Diaspididae) in California – an experimental approach. *Environmental Entomology* **24**: 779–95.

Godfray, H.C.J. (1994) *Parasitoids, Behavioural and Evolutionary Ecology*. Princeton University Press, Princeton.

Hairston, Jr, N.G. and Hairston, Sr, N.G. (1993) Cause–effect relationships in energy flow, trophic structure, and interspecies interactions. *American Naturalist* **142**: 379–411.

Hairston, Jr, N.G. and Hairston, Sr, N.G. (1997) Does food-web complexity eliminate trophic-level dynamics? *American Naturalist* **149**: 1001–7.

Hairston, N.G., Smith, F.E. and Slobodkin, L.B. (1960) Community structure, population control and competition. *American Naturalist* **94**: 421–5.

Halaj, J. and Wise, D.H. (2001) Terrestrial trophic cascades: how much do they trickle? *American Naturalist* **157**: 262–81.

Hassell, M.P. (1978) *The Dynamics of Arthropod Predator–Prey Systems*. Princeton University Press, Princeton.

Hawkins, B.A., Cornell, H.V. and Hochberg, M.E. (1997) Predators, parasitoids and pathogens as mortality agents in phytophagous insect populations. *Ecology* **78**: 2145–52.

Heimpel, G.E., Rosenheim, J.A. and Mangel, M. (1997) Predation on adult *Aphytis* parasitoids in the field. *Oecologia* **110**: 346–52.

Hemptinne, J.L., Lognay, G., Doumbia, M. and Dixon, A.F.G. (2001) Chemical nature and persistence of the oviposition deterring pheromone in the tracks of the two spot ladybird, *Adalia bipunctata* (Coleoptera: Coccinellidae). *Chemoecology* **11**: 43–7.

Hoelmer, K.A., Osborne, L.S. and Yokomi, R.K. (1994) Interactions of the whitefly predator *Delphastus pusillus* (Coleoptera: Coccinellidae) with parasitized sweetpotato whitefly (Homoptera: Aleyrodidae). *Environmental Entomology* **23**: 136–9.

Holt, R.D. and Polis, G.A. (1997) A theoretical framework for intraguild predation. *American Naturalist* **149**: 745–64.

Hunter, M.S., Collier, T.R. and Kelly, S.E. (2002) Does an autoparasitoid disrupt host suppression provided by a primary parasitoid? *Ecology* **83**: 1459–69.

Ives, A.R. (1992) Density-dependent and density-independent parasitoid aggregation in model host–parasitoid systems. *American Naturalist* **140**: 912–37.

Kindlmann, P. and Ruzicka, Z. (1992) Possible consequences of a specific interaction between predators and parasites of aphids. *Ecological Modeling* **61**: 253–65.

Kosaki, A. and Yamaoka, R. (1996) Chemical composition of footprints and cuticular lipids of three species of ladybird beetles. *Japanese Journal of Applied Entomology & Zoology* **40**: 47–53.

Lathrop, R.C., Johnson, B.M., Johnson, T.B. et al. (2002) Stocking piscivores to improve fishing and water quality: a synthesis of the Lake Mendota biomanipulation project. *Freshwater Biology* **47**: 2410–24.

Murdoch, W.W., Briggs, C.J. and Nisbet, R.M. (2003) *Consumer–Resource Dynamics*. Princeton University Press, Princeton.

Nakashima, Y. and Senoo, N. (2003) Avoidance of ladybird trails by an aphid parasitoid *Aphidius ervi*: active period and effects of prior oviposition experience. *Entomologia Experimentalis et Applicata* **109**: 163–6.

Nakashima, Y., Birkett, M.A., Pye, B.J., Pickett, J.A. and Powell, W. (2004) The role of semiochemicals in the avoidance of the seven-spot ladybird, *Coccinella septempunctata*, by the aphid parasitoid, *Aphidius ervi*. *Journal of Chemical Ecology* **30**: 1103–16.

Oksanen, L., Fretwell, S.D., Arruda, J. and Niemelä, P. (1981) Exploitation ecosystems in gradients of primary productivity. *American Naturalist* **118**: 240–61.

Polis, G.A. (1991) Complex trophic interactions in deserts: an empirical critique of food-web theory. *American Naturalist* **138**: 123–55.

Polis, G.A. and Holt, R.D. (1992) Intraguild predation: the dynamics of complex trophic interactions. *Trends in Ecology & Evolution* **7**: 151–4.

Polis, G.A. and Strong, D.R. (1996) Food web complexity and community dynamics. *American Naturalist* **147**: 813–46.

Polis, G.A., Myers, C.A. and Holt, R.D. (1989) The ecology and evolution of intraguild predation: potential competitors that eat each other. *Annual Review of Ecology & Systematics* **20**: 297–330.

Raymond, B., Darcy, A.C. and Douglas, A.E. (2000) Intraguild predators and the spatial distribution of a parasitoid. *Oecologia* **124**: 367–72.

Rees, N.E. and Onsager, J.A. (1982) Influence of predators on the efficiency of *Blaesoxipha* spp. parasites of the migratory grasshopper. *Environmental Entomology* **11**: 426–8.

Rees, N.E. and Onsager, J.A. (1985) Parasitism and survival among rangeland grasshoppers in response to suppression of robber fly (Diptera: Asilidae) predators. *Environmental Entomology* **14**: 20–3.

Rosenheim, J.A. (2001) Source–sink dynamics for a generalist insect predator in habitats with strong higher-order predation. *Ecological Monographs* **71**: 93–116.

Rosenheim, J.A. and Corbett, A. (2003) Omnivory and the indeterminacy of predator function: can a knowledge of foraging behavior help? *Ecology* **84**: 2538–48.

Rosenheim, J.A. and Harmon, J.P. (2006) The influence of intraguild predation on the suppression of a shared prey population: an empirical reassessment. In: Brodeur, J. and Boivin, G. (eds.) *Trophic and Guild Interactions in Biological Control*. Springer, New York.

Rosenheim, J.A., Kaya, H.K., Ehler, L.E., Marois, J.J. and Jaffee, B.A. (1995) Intraguild predation among biological-control agents: theory and practice. *Biological Control* **5**: 303–35.

Rosenheim, J.A., Glik, T.E., Goeriz, R.E. and Ramert, B. (2004) Linking a predator's foraging behavior with its effects on herbivore population suppression. *Ecology* **85**: 3362–72.

Schellhorn, N.A., Kuhman, T.R., Olson, A.C. and Ives, A.R. (2002) Competition between native and introduced parasitoids of aphids: nontarget effects and biological control. *Ecology* **83**: 2745–57.

Schmitz, O.J., Hamback, P.A. and Beckerman, A.P. (2000) Trophic cascades in terrestrial systems: a review of the effect of predator removals on plants. *American Naturalist* **155**: 141–53.

Shiojiri, K. and Takabayashi, J. (2005) Parasitoid preference for host-infested plants in affected by the risk of intraguild predation. *Journal of Insect Behavior* **18**: 567–76.

Shurin, J.B., Borer, E.T., Seabloom, E.W. et al. (2002) A cross-ecosystem comparison of the strength of trophic cascades. *Ecology Letters* **5**: 785–91.

Snyder, W.E. and Ives, A.R. (2001) Generalist predators disrupt biological control by a specialist parasitoid. *Ecology* **82**: 705–16.

Snyder, W.E. and Ives, A.R. (2003) Interactions between specialist and generalist natural enemies: parasitoids, predators, and pea aphid biocontrol. *Ecology* **84**: 91–107.

Snyder, W.E., Ballard, S.N., Yang, S. et al. (2004) Complementary biocontrol of aphids by the ladybird beetle *Harmonia axyridis* and the parasitoid *Aphelinus asychis* on glasshouse roses. *Biological Control* **30**: 229–35.

Strong, D.R. (1992) Are trophic cascades all wet? Differentiation and donor-control in speciose ecosystems. *Ecology* **73**: 747–54.

Taylor, A.J., Muller, C.B. and Godfray, H.C.J. (1998) Effect of aphid predators on oviposition behavior of aphid parasitoids. *Journal of Insect Behavior* **11**: 297–302.

Tostawaryk, W. (1971) Relationship between parasitism and predation of diprionid sawflies. *Annals of the Entomological Society of America* **64**: 1424–7.

Tscharntke, T. (1992) Cascade effects among 4 trophic levels: bird predation on galls effects density-dependent parasitism. *Ecology* **73**: 1689–98.

Werner, E.E. and Peacor, S.D. (2003) A review of trait-mediated indirect interactions in ecological communities. *Ecology* **84**: 1083–100.

5

Chemical and behavioral ecology in insect parasitoids: how to behave optimally in a complex odorous environment

Monika Hilker and Jeremy McNeil

Abstract

Volatiles from herbivorous insects and their host plants may serve as reliable cues for parasitoids in search of suitable hosts. However, one major challenge for the parasitoid female is to distinguish these cues from the myriad of others simultaneously present in the environment. In this chapter we will focus on the following questions related to chemically mediated host finding by parasitoids.

1 Which volatiles play a role in determining the reliability of olfactory cues? The usefulness of volatiles to parasitoids when searching for hosts and host habitats will have a quantitative component, based on the concentrations and ratios of compounds present. However, there are also qualitative aspects, as the signals may include compounds that are either ubiquitous or specific in origin.

2 How do parasitoids detect such information among the numerous other volatiles present? The relative importance of these different parameters may vary depending on whether the parasitoids are generalists or specialists. Even when the appropriate cues are present, the detection of these volatiles by foraging parasitoids will be influenced by other compounds present in the environment. Background odors are generally considered as 'noisy' volatiles that impede the perception of appropriate information. However, there is some evidence that this is not always the case. We outline an 'olfactory contrast hypothesis' postulating that a background odor may provide an important and necessary contrast that facilitates the detection of the informative odor.

3 What ecological factors could affect the value of these cues? In nature, both the rate at which volatiles are released from different sources, as well as their spatial and temporal persistence in the environment, will be affected by numerous abiotic and biotic factors. However, the impact of these factors on the reliability of chemical cues, as well as on parasitoid foraging, has so far received little attention. Research investigating

how parasitoids actually cope with an unstable infochemical environment will lead to a better understanding of the evolution of complex multitrophic interactions. In addition, such knowledge can be utilized to improve the efficacy of current biological control programs, including the development of novel approaches using infochemicals to control important agricultural and forest pests.

5.1 Introduction

Adult parasitic wasps must complete several major tasks during their lifetime to optimize their reproductive success, which include finding adequate food sources and a mate if they do not reproduce parthenogenetically. In addition, they must locate and select suitable hosts for their offspring, as the immature stages of parasitoids usually have limited mobility and live in an intimate relationship with their hosts. Thus, the success of the progeny strongly depends on the mother's ability to use a wide range of physical and chemical cues (semiochemicals or infochemicals) to locate a suitable host for their progeny (Vinson 1981, 1991, Godfray 1994, Rutledge 1996, Steidle & van Loon 2003).

Volatile cues can guide parasitoids to the vicinity of their hosts, while contact chemical and physical cues, such as shape and size of the host, or texture of the host plant, are important at close range. In this chapter we will focus on the use of volatile cues by which adult parasitoids orientate when they search for eggs or larvae of their herbivorous host insects. We will address the question of how parasitoids searching for hosts discriminate between reliable cues and the jungle of other volatiles that emanate not only from the herbivorous host or its food plant, but also from other living organisms and dead organic matter in the habitat.

High odor specificity indicating the suitability of hosts can help parasitoids orientate during host searching. Specificity can be provided by specific plant or host chemicals, as well as by a specific quantitative blend of volatiles. A parasitoid female can recognize host-indicating cues innately or may learn them during her life (Vet et al. 1990, 1995). The detection of such cues can also be influenced by the background of the larger web of chemicals present in the habitat, as outlined below.

However, as neither the background odors nor the specific host-indicating cues are fixed in nature, parasitoids require enormous adaptive and integrative abilities to cope with such a high variability. Biotic (plant and host phenotype and genotype) and abiotic factors (e.g. wind speed, temperature and humidity) will markedly influence environmental odor profiles, and similar factors will be modulating the responses of parasitoids.

We will outline the parameters influencing orientation of parasitoids by odor during their search for a suitable host focusing on: (i) the role of specificity of cues; (ii) the detectability of cues; and (iii) the variability of odor information. Our aim is to emphasize the context specificity of orientation by odor, and to show that response to a specific blend of volatiles depends on a wide range of abiotic and biotic factors, many of which are poorly understood for parasitoids.

5.2 Informative value of odor

A positive response to an odor profile only makes sense for a foraging parasitoid when the odor is indicative of the probable presence of a host. Travis (1999) entitled his review about fascinating recent results on olfactory coding in mammals 'Making sense of scents', and we need to ask: Which volatiles make sense for parasitoids when searching for hosts? Both plant and host volatiles are known to provide information for foraging parasitoids. A plant, because of its high biomass, is able to emit volatiles in the mg range, whereas an herbivorous host usually releases volatiles in the ng range (Turlings et al. 1995). Thus, it is not surprising that plant cues play an important role in the orientation of many parasitoid species during host searching (Fig. 5.1).

5.2.1 Informative value of plant odor

Ubiquitous, so-called general green leaf volatiles (GGLV), such as C_6-aldehydes (alcohols and their esters), are released both from intact plants and attacked ones. Usually, feeding

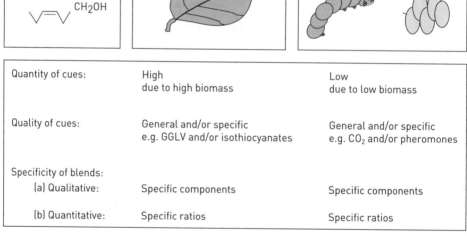

Volatiles CO_2 CH_2OH	Plant	Herbivorous host

Quantity of cues:	High due to high biomass	Low due to low biomass
Quality of cues:	General and/or specific e.g. GGLV and/or isothiocyanates	General and/or specific e.g. CO_2 and/or pheromones
Specificity of blends: (a) Qualitative:	Specific components	Specific components
(b) Quantitative:	Specific ratios	Specific ratios

Fig. 5.1 General characters of volatiles from herbivorous hosts (eggs and larvae) and their food plants that may be used by parasitoids for host location. We differentiate between general volatile cues (e.g. general green leaf volatiles (GGLV) from plants or CO_2 from host insects) and specific volatile cues (e.g. isothiocyanates typical for Brassicacean plants, or pheromones that are generally highly species-specific in herbivorous insects). As outlined in this chapter, both the quantity and quality of a cue, as well as the quantitative and qualitative composition of a volatile blend, may be used by parasitoids for orientation and host location. The specificity of a blend may be due to (a) specific components released from a plant or a host (qualitative composition of blend) and/or (b) specific quantities of the components.

damaged plants release much more GGLV than undamaged ones (Geervliet et al. 1997, Agelopoulos et al. 1999). Parasitoids of polyphagous host insects may use a broad range of food plant odors (Steidle & van Loon 2003), as these cues could indicate at least some probability of host presence. For example, the egg parasitoid *Trichogramma chilonis* (Reddy et al. 2002) and the larval parasitoids *Campoletis sonorensis* and *Microplitis croceipes* attack hosts with a wide range of host plants, and GGLV are amongst the cues they use to locate suitable hosts (Vinson & Williams 1991, Whitman & Eller 1992). However, parasitoids specializing on monophagous or oligophagous host insects are also able to exploit widespread plant volatiles (Geervliet et al. 1994), although orientation by ubiquitous plant volatiles will be only a good guide for specialist parasitoids if a specific blend or combination with other volatiles provides information on the host species present.

Specific odors from a given plant species can indicate the presence of specific herbivorous hosts, if the host specializes on these plants. For example, the aphid parasitoid *Diaeretiella rapae*, that prefers host insects feeding on Brassicaceae, is attracted by isothiocyanates that are typical for these plants (Bradburne & Mithen 2000, Baer et al. 2004). In fact, Vaughn et al. (1996) found indications suggesting that this parasitoid has receptors responding specifically to allyl isothiocyanate.

The presence of specific herbivore-induced plant volatiles may also serve as reliable cues for the presence of a specific host species or host stage. Egg parasitoids may orientate using oviposition-induced plant cues (Hilker & Meiners 2002, Hilker et al. 2002b, Colazza et al. 2004a, 2004b, Hilker & Meiners 2006), while larval parasitoids can exploit plant cues induced by larval feeding (Turlings et al. 1995, Dicke & Vet 1999). Interestingly, some egg parasitoids also respond to plant cues induced by larval feeding (Lou et al. 2005, Moraes et al. 2005). This behavior could be beneficial when eggs and larvae co-occur, as larval feeding damage is expected to induce higher amounts of plant volatiles than induction by egg deposition does. To date, however, there is no evidence that larval parasitoids use oviposition-induced plant cues to orientate. Such a precocious response could be important in intraguild competition (see also Chapter 4 by Snyder and Ives) and might become beneficial if the offspring of the parasitoids that first encountered the newly emerged host larvae gained an advantage over the offspring of other parasitoids exploiting the same developmental stage of the host.

The specificity of herbivore-induced plant blends has been intensively studied with respect to the herbivore and plant species, as well as to the age of herbivores and plants (Dicke 1999). While adults of some larval parasitoids cannot recognize the presence of their host species or suitable host larval stages from the blend of feeding-induced plant volatiles (Gouinguené et al. 2003, and references therein), other species can (Takabayashi et al. 1995, De Moraes et al. 1998, Guerrieri et al. 1999). Furthermore, several egg parasitoids are able to distinguish between oviposition-induced plant volatiles produced by a plant following egg deposition by different herbivore species (Meiners et al. 2000, Mumm et al. 2005). In addition, herbivore-induced plant cues not only provide parasitoids with information about the host species and/or developmental stage, but also on host quality. Fatouros et al. (2005b) found that *Cotesia* species could discriminate plant volatiles induced by parasitized *Pieris* caterpillars from those induced by non-parasitized host larvae. The specificity of herbivore-induced plant volatiles may be due to a mixture of specific components or to specific quantities of components mixed in specific ratios, and thereby providing a specific signal (Fig. 5.1).

The GGLV profiles emitted from a plant may vary when several different species of herbivores simultaneously attack the same individual host plant (Vos et al. 2001, Dicke & Hilker 2003). To what extent such changes arising from multiple attacks will modify the usefulness of the GGLV as cues for foraging parasitoids remains to be determined.

5.2.2 Informative value of host volatiles

Surprisingly, little is known about the role of general host volatiles for searching behavior of insect parasitoids. Carbon dioxide released from respiring host insects could be a highly detectable cue, especially when host insects occur in high abundances. While other parasitic organisms, like trematode cercariae, are well-known for using carbon dioxide to locate their hosts (Haas 1992, Haas et al. 2002), the role of carbon dioxide for insect parasitoids remains unclear. Several insect species have been shown to be able to respond and orientate along carbon dioxide gradients: herbivorous insects (Stange et al. 1995, Langan et al. 2001, 2004, Johnson & Gregory 2006); those living on decaying organic material like *Drosophila* (Faucher et al. 2006); and blood-sucking insects (Gillies 1980, McCall 2002, Barrozo & Lazzari 2004). Furthermore, it is unknown whether the carbon dioxide gradient established by respiration of herbivorous insects on a plant is even fortified by reduced photosynthesis activity of plant tissue close to feeding damage (Welter 1989, Zangerl et al. 2002, Haile & Higley 2003). Thus, even though orientation by carbon dioxide might provide some information for generalist parasitoids on the presence of hosts, to the best of our knowledge, carbon dioxide sensitivity in parasitoids of herbivorous insects has not been studied so far (Fig. 5.1).

General and widespread host cues may be released from faeces or honeydew of herbivorous insects (Steidle & van Loon 2003, Buitenhuis et al. 2004, and references therein), although they may also release specific volatiles or specific ratios of volatiles that provide valid information on the host species for both larval and egg parasitoids. For example, egg parasitoids with a narrow host range, but with hosts feeding on a wide variety of plants, recognize specific patterns of chemicals released from adult host faeces regardless of the plant species used by the herbivore (Hilker & Meiners 1999). Hydrocarbons on the surface of host scales are also widespread chemical cues used by several egg parasitoids (Jones et al. 1973, Lewis et al. 1975, Shu et al. 1990, Paul et al. 2002). However, they probably serve as foraging cues over very short distances or upon contact only.

Several host specific cues, such as pheromones, are used as kairomones by egg and larval parasitoids (Steidle & van Loon 2003). Many egg parasitoids have been shown to use sex, anti-aphrodisiac, or oviposition marking pheromones of their hosts to find microhabitats where host eggs could be expected (Rutledge 1996, Powell 1999, Nufio & Papaj 2001, Anderson 2002, Fatouros et al. 2005a). This use of cues from other stages, other than the one used as an oviposition site by the female parasitoid, has been labeled 'infochemical detour' (Vet & Dicke 1992). Such a detour is worthwhile, as eggs *per se* hardly release any volatiles (Kaiser et al. 1989), except those adsorbed on the egg surface or those released from compounds attaching eggs to the plant (Frenoy et al. 1992, Renou et al. 1992, Bin et al. 1993). The specificity of response to host pheromones can be extremely high, as seen with the eulophid wasp *Chrysonotomyia ruforum* that attacks pine sawfly eggs. Females of this wasp only respond to those stereoisomers that the sawflies use as intraspecific mating signals (Hilker et al. 2000) (Fig. 5.1).

5.3 Detection of informative odor

5.3.1 The importance of 'contrast' cues

Parasitic wasps must be able to distinguish the volatiles signaling the presence of hosts from background noise. The effectiveness of signals, whether they are visual, auditory, or chemical in nature, will be influenced by the degree of contrast with the background. For example, with visual cues, reduced background contrast would be optimal for camouflage, while marked contrast would be most effective in the case of aposematic coloration. Background volatiles may have a positive effect if they provide a 'sharper view' of the specific informative volatile (blend), or may be negative if they mask detection of the crucial, informative scent.

To date, the effects of background odor on herbivorous insects have received the greatest attention. When repellent components from non-host plants are present in the background, they may interfere with the positive cues that herbivores use to locate their host plants (Isaacs et al. 1993, Hori & Komatsu 1997, Yamasaki et al. 1997, Held et al. 2003, Mauchline et al. 2005). In addition, components from non-host plants, with no repellent properties *per se*, may also act as masking volatiles (Visser & Avé 1978, Thiery & Visser 1986, Nottingham et al. 1991, Yamasaki et al. 1997, Costantini et al. 2001). Masking effects of non-host plants have often been mentioned as one possible reason for the lower abundance of specialized herbivores in habitats with high plant diversity (Bukovinszky 2004, and references therein).

The positive influence of background odors on the response to chemical cues has been reported for a number of different insect species. *Drosophila melanogaster* adults responded better to minute changes of carbon dioxide when the background odor included vinegar rather than pure air (Faucher et al. 2006). Similarly, the response of corn rootworm *Diabrotica virgifera virgifera* larvae to CO_2 was enhanced when the gas was presented in combination with corn volatiles (Hibbard & Bjostad 1988). Background dependent changes in olfactory sensitivity have also been seen in the housefly *Musca domestica* (Kelling et al. 2002). There are also several examples showing that both the release of, and response to, sex pheromones in herbivorous insects is modified by the presence of host plant volatiles (McNeil & Delisle 1989, Dickens et al. 1993, Landolt & Phillips 1997, Ochieng et al. 2002, Said et al. 2005).

There is no question that background odors will affect the use of volatiles by foraging parasitoids, but this phenomenon has not received a great deal of attention. The egg parasitoid *Chrysonotomyia ruforum*, a eulophid wasp specializing on pine sawflies, is attracted to oviposition-induced pine odors (Hilker et al. 2002a) that have higher quantities of (*E*)-*β*-farnesene than control trees (Mumm et al. 2003). However, (*E*)-*β*-farnesene presented alone does not attract the parasitoid. It only elicits a positive response when presented in combination with non-induced pine odor (Mumm & Hilker 2005). Thus, the change in (*E*)-*β*-farnesene quantities will only serve as a reliable cue for the parasitoid if this is detected within a specific odor context. Modified responses related to background odors may also be learnt, as seen with *Leptopilina heterotoma*, a parasitoid of *Drosophila* larvae. The response to yeast odor was enhanced in the presence of a green leaf volatile, but only if the parasitoid had previously experienced the green leaf volatile together with host kairomones (Vet & Groenewold 1990).

5.3.2 Odor coding

To unravel how background odor affects the response to a specific volatile (blend), it is necessary to understand the perception of complex volatile mixtures. Several reviews have addressed the question on how chemical information is processed in insects (Visser 1986, Masson & Mustaparta 1990, Smith & Getz 1994, Bruce et al. 2005). A complex odor can be deciphered in a combinatorial way where: (i) different volatiles can activate a single receptor cell expressing a single receptor protein; and (ii) a single specific volatile can activate different receptor cells (Hildebrand & Shepherd 1997, Malnic et al. 1999, Galizia & Menzel 2000, 2001). This coding strategy gains a further level of complexity when considering that olfactory receptor cells have two modes of signaling: excitation and inhibition. Moreover, this 'combinatorial coding' has been suggested to act 'in concert' with another strategy, coding by 'labeled lines' where neurons mediate only the signal of a specific volatile. Such coding strategies might explain the findings that the response to highly species-specific pheromones can be enhanced by general plant volatiles. Thus, combinatorial coding would enable animals to decipher a plethora of different bouquets, and Ache and Young (2005) have suggested that it is an evolutionary conserved strategy used by a wide range of organisms, including insects. In addition to coding of complex volatile blends in the antennal lobe (Joerges et al. 1997, Lei et al. 2002), evidence is growing that interactions of volatiles may also occur at the receptor level in insects (De Jong & Visser 1988, Getz 1999, and references therein, de Bruyne et al. 2001, Said et al. 2005).

The information conveyed by a complex odor is not just determined by the quality of components, but also by quantities and relative ratios of components. The above-mentioned study of the egg parasitoid *C. ruforum* indicates that a small increase of the quantities of (E)-β-farnesene in the headspace of oviposition-induced pine determines whether the odor is attractive or not (Hilker et al. 2002a). Compared to the huge amounts of monoterpenes present in pine odor, the sesquiterpene (E)-β-farnesene is just a minor component within this complex mixture, even in oviposition-induced pine.

How can a minor component of a mixture determine what the odor 'tells' the parasitoid? A small increase of the concentration of a minor component within a blend could raise this component above the threshold concentration, at which point it elicits a direct response or modifies the response to other components in the blend. Thus, the perception of a blend of volatiles is determined by the synergistic or antagonistic interactions between components, the kinetics of the dose-response curves, and the kinetics of ligand-receptor binding (Hildebrand & Shepherd 1997, Getz 1999, Duchamp-Viret et al. 2003). Therefore, the reliability of informative odor conveyed to a parasitoid within a complex mixture of environmental odor may strongly depend on both the quantity and quality of the volatiles perceived (Fig. 5.1).

Vet (1999) asked whether the actual compounds present in odor blends (qualitative pattern) provide more specific information than the actual concentrations and relative ratios of the compounds making up the blend (quantitative pattern). Clearly, in the case of specialist parasitoids, the qualitative aspects of blends that contain specific cues from the host plant and/or herbivore may provide highly reliable information. Also for discriminative learning, quality of components was shown to be more important than quantities of volatiles in *L. heterotoma*, a larval parasitoid of *Drosophila* (Vet et al. 1998). However, for parasitoids that orientate using a wide range of ubiquitous volatiles, it would be difficult, if not impossible, to determine whether qualitative or quantitative characteristics provide more reliable information. An increase in the concentration of a single volatile component released among numerous other volatiles from a plant may provide a new odor 'Gestalt', as suggested by the results on the parasitoid's response to oviposition-induced pine cues mentioned above (Hilker et al. 2002b, Mumm et al. 2003). The increased concentration of a single component within a complex blend might positively or negatively affect the perception of other components and thus, enhance or lower the 'contrast' of informative cues to background volatiles. Therefore, both quality and quantity of components of an infochemical blend seem to be very important in conveying reliable information.

To draw a conclusion from these facts and ideas, we propose the following 'olfactory contrast hypothesis': When considering the total amount of a blend, we would generally expect that low concentrations would have low detectability, while high ones would be more readily detected (Vet & Dicke 1992). However, as noted previously, the degree to which an infochemical (blend) 'contrasts' with the other odors in the habitat may be crucial. Therefore, a low amount of volatiles from an herbivore-induced plant may still have high detectability if there is strong contrast relative to the odor of non-induced plants, while high concentrations that are poorly contrasted with the background could be hard to detect (Fig. 5.2). Which chemicals contrast with each other, and at what concentrations, for any given parasitoid species will be interesting questions for future studies.

The response to odor may depend on previous experience and learning in numerous parasitoids (Vet et al. 1995). Experience with an infochemical (blend) might 'sharpen the view' of a parasitoid for these particular volatiles, so that they may be detected even when the contrast with background odor is low. Some parasitoids are able to learn a wide array of cues, while others only learn very specific signals (Vet et al. 1995, Steidle & van Loon 2003, Mumm & Hilker 2005). For both types of 'learners', it has been shown that the odor context (or background) at which volatiles have been experienced is important (Meiners et al. 2003, Mumm & Hilker 2005).

Learning to respond to an infochemical blend might be particularly adaptive in highly variable environments. This would allow female parasitoids to respond quickly to the quality of the host patch by modifying their searching and exploitation behavior (van Alphen et al. 2003). The ability of parasitoids to learn may be linked with their degree of dietary specialization. A generalist parasitoid with high dietary variation needs to be flexible in its responses to host-related odor because the infochemicals supporting host search may vary considerably. Thus, we would expect them to have the capacity to learn many odors that would help locate a host, rather than exhibiting innate responses to specific odors. In contrast, we would predict that specialists respond innately to specific host-relevant odors (Vet & Dicke 1992). A meta-analysis by Steidle and van Loon (2003) supported the idea that learning to respond to host and plant infochemicals is more widespread among generalist than specialist parasitoid species.

Specialist parasitoids may also be confronted with the problem of high variability of infochemical odor. The eulophid parasitoid *C. ruforum*, as noted above, exploits eggs of sawflies on pine, so it is a specialist with a narrow host range and its hosts feed upon a small number of plant species. However, the oviposition-induced plant volatiles that the parasitoid uses to locate hosts may vary greatly both between and within trees (Mumm & Hilker 2006). In such cases, being able to learn to respond to a specific volatile pattern and adjust foraging behavior would be more adaptive than just responding innately to a fixed pattern of infochemicals. Therefore, the ability to learn host cues might be a strategy

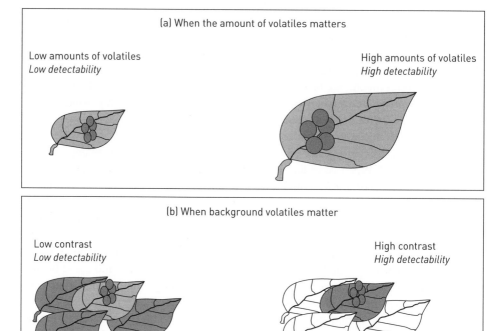

Fig. 5.2 A general outline of the olfactory contrast hypothesis. (a) When the amount of volatiles matters, a low concentration presented in a complex environment would have lower detectability than a high one (here exemplified with oviposition-induced plant volatiles). (b) However, detectability of plant and host volatiles may not only depend on the amounts released but also on the background odor that, in the past, has often been considered as 'noise' hampering detection of signals. However, we argue that background odor may also facilitate detection of signaling odor (see text), even when present in low amounts, because of the 'contrast' that is created (here exemplified with volatiles from oviposition-induced plant volatiles). Left: signaling odor has little 'contrast' to background odor. Right: signaling odor has high 'contrast' to background odor.

to cope with high environmental variability for both specialist and generalist parasitoids (Steidle & van Loon 2003).

5.4 Variation in a complex odorous environment

The release of volatiles from both plants and hosts, as well as the responsiveness of the parasitoids, will strongly depend on the phenotypic and genotypic variation at all trophic levels. Furthermore, the plume released, and the behavior of the parasitoids will directly be affected by abiotic factors such as wind speed, temperature, or air pressure. However, most studies examining the response of parasitoids to volatiles have been carried out under controlled laboratory conditions where the majority of factors mentioned have been standardized, with few actually being conducted under natural conditions (Bernasconi Ockroy et al. 2001, De Moraes et al. 2001).

5.4.1 Variable emission of volatiles by plants

There is evidence of considerable variability in volatile profiles emitted by different members of the same plant species, when held under similar conditions (Takabayashi et al. 1994, Degen et al. 2004) and even within an individual plant (Takabayashi et al. 1994, Mumm & Hilker 2006). However, whether these inter- and intra-individual differences increase, decrease, or remain the same following attack by an herbivore is unknown. The resulting response could be of considerable importance with respect to both the actual intensity of the cue released by a given plant and the resulting contrast that occurs against the background noise. If a parasitoid responds to a specific compound that increases following defoliation, then a plant with lower emission rates would provide less easily detectable signals than a conspecific one emitting higher concentrations of volatiles.

Further, volatile emissions from different plant parts may vary, and volatiles released during the day may differ from those emitted at night (Loughrin et al. 1990, Bertin & Staudt 1996, Dudareva & Pichersky 2000, Gouinguené et al. 2001, Vallat et al. 2005). These day/night changes in plant volatile patterns are probably genetically controlled, with their expression triggered by abiotic factors such as light intensity. However, abiotic factors like temperature (Takabayashi et al. 1994, Llusià & Peñuelas 1999, 2000, Vallat et al. 2005) may also directly affect release and evaporation of plant volatiles. The range of temperature over which effects are observed, as well as the magnitude of the changes, will be affected by the physical properties of each compound in the blend. The degree and duration of water stress, associated with RH and rainfall under field conditions, will also alter the patterns of volatile releases (Takabayashi et al. 1994, Bertin & Staudt 1996).

Further, the temperature/emission rate relationship may be significantly modified by the level of drought stress the plant experiences, which will be determined by relative humidity and/or rainfall (Takabayashi et al. 1994, Bertin & Staudt 1996, Llusià & Peñuelas 1999). For example, in the non-terpene storing holm oak, *Quercus ilex*, the significant positive relationship between the rates of release of alpha-pinene, sabinene and beta-pinene, and ambient temperature, that was evident under low water stress conditions, disappeared when the plants experienced high water stress (Bertin & Staudt 1996). Rates of photosynthesis and transpiration dropped quickly after the onset of water stress, while the decline in terpene emissions was only observed several days later when the CO_2 balance was close to zero, probably when reserves necessary for terpene synthesis had been depleted (Bertin & Staudt 1996). This hypothesis was supported by a subsequent study under controlled conditions that examined the effects of temperature and relative humidity on the emission of terpenes from the non-terpene storing *Q. ilex*, and from *Pinus halepensis*, a terpene storing species (Llusià & Peñuelas 1999).

In *P. halepensis*, the positive relationship of terpene release with temperature was seen at both high (>60%) and low (<60%) RH, as well as in *Q. ilex* under high RH. However, in the non-storing holm oak *Q. ilex*, the emission rates of terpenes declined with increasing temperatures at low RH. Thus, the direct effects of temperature and water/RH, as well as potential interactions on the signals produced will be species-dependent. A significantly greater data set will be required to determine if there are any general trends for plant species with similar biology.

In addition, there may be marked seasonal differences in the release rates of volatiles from plants (Llusià & Peñuelas 2000, Vallat et al. 2005) due to changes in the abiotic factors mentioned above, as well as others such as light intensity (Takabayashi et al. 1994). Thus, a parasitoid that attacks a multivoltine herbivore could be confronted with very different inter-generational volatile profiles even though they are exploiting the same herbivore species on the same host plant.

5.4.2 Variability of perception of plant and host volatiles by insects

It is evident from the literature that, even when the majority of parasitoid adults tested show a positive response to an infochemical (blend), some exhibit no response at all, while others do not discriminate between treatment and control volatiles. The absence of response or the lack of discrimination could have underlying genotypic or phenotypic causes.

Very little work has been done on the genetics of parasitoid responsiveness to volatiles. However, the few cases that have been studied provide strong evidence that the propensity with which female parasitoids perceive odors is heritable (Prévost & Lewis 1990, Gu & Dorn 2000, Wang et al. 2003). For example, in an intensive bidirectional selection study using *Cotesia glomerata*, Wang et al. (2003) were able to select lines that showed significant differences in the perception of volatile cues with respect to both upwind flight and the ability to reach the source that was independent of the ability to fly. The level of responsiveness of both lines increased when females had previous exposure to the volatiles, indicating that the learning component remained in both lines despite the selection on responsiveness.

Thus, the genetic composition of the parasitoid population relative to olfactory responsiveness will affect the efficacy with which individuals can effectively exploit volatile cues from plants. Consequently, we would predict a higher degree of genetic variability in populations of parasitoids that exploit polyphagous herbivores than in those that exploit monophagous ones. Similarly, a higher genetic variability might be expected in populations of polyvoltine species, for as noted above, they have to cope with large temporal changes in volatile emissions. However, a considerably greater database will have to be generated to test these hypotheses.

Phenotypic variation, due to biotic factors such as age, hunger, or the reproductive state may influence how insects, including parasitoids, respond to odor cues (Wäckers 1994, Greiner et al. 2002, Mercer & Hildebrand 2002), and are associated with physiology changes that occur during different times in the insect's lifetime (Takasu & Lewis 1993, Gadenne & Anton 2000). For example, mated females of the spruce budworm preferentially oviposit on sites treated with waxes from the needles of a host plant, while unmated individuals did not discriminate between treated and control sites (Wallace et al. 2004). A similar case was reported in the braconid *Habrobracon hebetor*, where naïve mated females responded to volatile cues from host frass, but naïve virgin females did not (Parra et al. 1996). In these particular cases, the changes observed in ability to process chemical cues may be caused, directly or indirectly, by male-derived substances transferred at the time of mating.

The question on how the actual sensory system of parasitoids may be affected by variations in climatic conditions has received little attention to date but some possible effects may be predicted from the results of studies on herbivorous insects. For example, the responsiveness of European corn borer (ECB), *Ostrinia nubilalis*, males to different concentrations of female sex pheromone is significantly affected by relative humidity (Royer & McNeil 1993), and the concentration/humidity interactions with respect to male flight behavior suggest that the changes are related to the integration of the chemical message at the receptor level. Humidity related changes in the response to pheromones have also been reported for an aphid (Wiener & Capinera 1979) and a tick (Hassanali et al. 1989). The response of ECB males to a fixed concentration of female sex pheromone was also significantly affected by the temperature conditions during pupal development, with those exposed to low temperatures exhibiting significantly reduced upwind flight, even several days following emergence (Royer & McNeil unpublished). Increasing the concentration of the pheromone increased the number of males that oriented to the lure but did not increase the number that successfully reached the source. As there were no differences in the physiological responses of the antennae from males in the different pupal temperature treatments, it would appear that the temperature effect is related to the integration of the signals in the olfactory lobe. Thus, the ability of parasitoids to detect and respond to info-chemicals could be influenced by the effects of abiotic conditions on both the peripheral and central processing systems.

The prevailing weather conditions can have a significant impact on the ability of para-sitoids, which are generally weak flyers, to actually reach a source once it has been detected. While low temperatures may inhibit flight (Fournier & Boivin 2000, and references therein), it is unlikely that a parasitoid's ability to respond to a cue will be significantly restricted as a direct consequence of temperature conditions encountered in the habitat once the temperature threshold for flight is reached. Similarly, it is unlikely that relative humidity itself will impact directly on the parasitoid's ability to forage. However, small insects have a large surface-to-volume ratio and are thus confronted with constraints associated with water loss. Consequently, relative humidity, in concert with ambient temperature, can modify the timing and intensity of certain behavioral activities (Quiring & McNeil 1987, Royer & McNeil 1991), and thus might modify how a parasitoid responds to volatiles. Rainfall, that will influence humidity, may be a significant mortality factor and can significantly limit parasitoid foraging behavior, particularly between different host plants (Fink & Völkl 1995, Weisser et al. 1997, Schwörer et al. 1999, Schwörer & Völkl 2001). It has been shown that raindrops hitting the plant may interfere with the specific vibrational signals female parasitoids use to find host-infested leaves (Casas et al. 1998), and water on the surface of aphids interferes with parasitoids' host location, probably due to the masking of chemical signals (Weinbrenner & Völkl 2002). Thus, it would be of interest to know if water or dew on plants affects the emission of volatile signals and, if so, how this impacts on parasitoid activity.

Wind conditions are not only important in determining different aspects of the odor plume and the distance over which the signal is transmitted (Elkinton & Cardé 1984, Bell 1991, and references therein, Murlis et al. 1992), but also impact on the parasitoid's ability to fly. In most parasitoid species tests, flight is generally inhibited at wind speeds >2m/sec (Juillet 1964, Keller 1990, Fink & Völkl 1995, Messing et al. 1997, Schwörer et al. 1999, Marchand & McNeil 2000, Gu & Dorn 2001). However, a recent study found that *Aphidius ervi* males will walk toward a pheromone source at wind speeds that inhibit flight if a physical bridge connects the source and the release site (McClure & McNeil, unpublished data). A similar use of ambulatory behavior when wind velocity inhibited flight was reported for males of the potato aphid, *Macrosiphum euphorbiae* (Goldansaz & McNeil 2006), suggesting that walking may be an important behavior for weak flying species when following volatile cues. Thus, the highly variable wind conditions in the field will not only impact on the direct responses of foraging parasitoids, but will also affect the integrity of the plume structure within the canopy where walking toward a source would be possible.

There is a growing body of literature showing that parasitoids respond to prevailing atmospheric conditions, or to changes that occurred at some time in the preceding 24 hours. There is evidence that flight behavior (Fournier et al. 2005), pheromone mediated mating (Marchand & McNeil 2000), the acceptance of previously parasitized hosts (Roitberg et al. 1993), and the response to host plant volatiles (Steinberg et al. 1992) may change with changing atmospheric pressure. As noted above, wind and rain may increase the probability of injury or death, so the ability to modify foraging behavior in response to atmospheric pressure would reduce activity during periods of adverse climatic conditions. A more extensive database is required in order to determine exactly how the responsiveness of parasitoids to volatile cues that are important for host location is affected by atmospheric pressure, and how this in turn affects reproductive success.

5.5 Concluding remarks

Parasitoids must navigate in a highly complex odorous environment, and their ability to exploit volatile cues indicating the presence of a host under such conditions is amazing. This is especially true when we consider the variability arising from the effects of abiotic and biotic factors at each trophic level, making successful host location comparable to finding of a needle in a swirling haystack.

One of the major future challenges will be to understand the mechanisms by which parasitoids successfully orientate by odor. As noted, the effects of background odor on signal perception are poorly understood in parasitoids, and we must find answers to questions like 'how does background odor enhance the response to an informative signal?' Is it just synergy between background odor and signal, like the synergistic interaction of essential components in a blend, or is it through an adaptive 'comparison' of background odor with the signal? Similarly, how does background odor lower the response to informative cues? Does the presence of many other components just 'overload' the antennae of parasitoids, or are there inhibitory interactions between the background and informative odors? In addition, we do not know at what level, peripheral or central, background odors positively or negatively influence signal perception in parasitoids. To elucidate the integrative capacities of parasitoids in a highly variable environment, it will also be interesting to see if a parasitoid's ability to filter informative volatiles from background odor follows statistical paradigms like 'lazy learning', a statistical method used in computer spam filters or weather forecasting. This method just memorizes previous events, often using the k-nearest neighbor algorithm (Aha 1997, Marling et al. 2002).

Future studies will also need to take into account the variability of odor information. For example, what factors actually affect the stability of the molecules that make up the chemical signal, as the duration and active space of a reliable scent could vary considerably depending on the climatic conditions (Ideses et al. 1982, Veit et al. 2001, Baldwin et al. 2006). Further, the temporal variability of volatile emissions and its impact on short and long term foraging activity needs more attention.

If we are to really understand the ecological context of such tritrophic interactions we will need to obtain a better comprehension of how the cues are effectively exploited by parasitoids relative to background noise. The answers will undoubtedly vary depending on the systems under consideration, but some general categories of response should exist. Both further knowledge on the effects of background odor as well as of variability of odor and odor perception will also help us to improve biological control of pest insects by parasitoids. As previously argued for the effective use of pheromones in the management of lepidopteran pests (McNeil 1991), the successful use of synthetic volatiles to manipulate parasitoid behaviors (Pickett et al. 1997) will require a much better fundamental understanding of how insects actually exploit odors in complex and highly variable environments.

Acknowledgments

We thank Giovanni Galizia, University Konstanz, Germany, Éric Wajnberg, INRA Sophia Antipolis, France, and three anonymous reviewers for their very valuable comments on the manuscript. Moreover, many thanks are due to Junji Takabayashi, University Kyoto, Japan, who provided a wonderful place to write this chapter during a sabbatical. Last, but not least, we are very grateful to the editors of this book for their kind invitation to write this chapter.

References

Ache, B.W. and Young, J.M. (2005) Olfaction: diverse species, conserved principles. *Neuron* **48**: 417–30.

Agelopoulos, N.G., Hooper, A.M., Maniar, S.P., Picket, J.A. and Wadhams, L.J. (1999) A novel approach for isolation of volatile chemicals released by individual leaves of a plant *in situ*. *Journal of Chemical Ecology* **25**: 1411–25.

Aha, D.W. (1997) *Lazy Learning*. Springer, Heidelberg.

Anderson, P. (2002) Oviposition pheromones in herbivorous and carnivorous insects. In: Hilker, M. and Meiners, T. (eds.) *Chemoecology of Insect Eggs and Egg Deposition*. Blackwell, Oxford, Berlin, pp. 235–63.

Baer, C.F., Tripp, D.W., Bjorksten, A. and Antolin, M.F. (2004) Phylogeography of a parasitoid wasp (*Diaeretiella rapae*): No evidence of host-associated lineages. *Molecular Ecology* **13**: 1859–69.

Baldwin, I.T., Halitschke, R., Paschold, A., Dahl, C.C. and Preston, C.A. (2006) Volatile signaling in plant–plant interactions: 'Talking trees' in the genomics era. *Science* **311**: 812–15.

Barrozo, R.B. and Lazzari, C.R. (2004) The response of the blood-sucking bug *Triatoma infestans* to carbon dioxide and other host odours. *Chemical Senses* **29**: 319–29.

Bell, W.J. (1991) *Searching Behavior, the Behavioral Ecology of Finding Resources*. Chapman & Hall, London.

Bernasconi Ockroy, M.L., Turlings, T.C.J., Edwards, P.J., et al. (2001) Response of natural populations of predators and parasitoids to artificially induced volatile emissions in maize plants (*Zea mays* L.). *Agricultural & Forest Entomology* **3**: 201–9.

Bertin, N. and Staudt, M. (1996) Effect of water stress on monoterpene emissions from young potted holm oak (*Quercus ilex* L.) trees. *Oecologia* **107**: 456–62.

Bin, F., Vinson, S.B., Strand, M.R., Colazza, S. and Jones, W.A. (1993) Source of an egg kairomone for *Trissolcus basalis*, a parasitoid of *Nezara viridula*. *Physiological Entomology* **18**: 7–15.

Bradburne, R.P. and Mithen, R. (2000) Glucosinolate genetics and the attraction of the aphid parasitoid *Diaeretiella rapae* to *Brassica*. *Proceedings of the Royal Society of London Series B Biological Science* **267**: 89–95.

Bruce, T.J.A., Wadhams, L.J. and Woodcock, C.M. (2005) Insect host location: A volatile situation. *Trends in Plant Science* **10**: 269–73.

Buitenhuis, R., McNeil, J.N., Boivin, G. and Brodeur, J. (2004) The role of honeydew in host searching of aphid hyperparasitoids. *Journal of Chemical Ecology* **30**: 273–85.

Bukovinszky, T. (2004) *Tailoring complexity. Multitrophic interactions in simple and diversified habitats*. PhD thesis. Wageningen University, The Netherlands.

Casas, J., Bacher, S., Tautz, J., Meyhofer, R. and Pierre, D. (1998) Leaf vibrations and air movements in a leafminer–parasitoid system. *Biological Control* **11**: 147–53.

Colazza, S., Fucarino, A., Peri, E., Salerno, G., Conti, F. and Bin, F. (2004a) Insect oviposition induces volatile emission in herbaceous plants that attracts egg parasitoids. *Journal of Experimental Biology* **297**: 47–53.

Colazza, S., McElfresh, J.S. and Millar, J.G. (2004b) Identification of volatile synomones, induced by *Nezara viridula* feeding and oviposition on bean spp., that attract the egg parasitoid *Trissolcus basalis. Journal of Chemical Ecology* **30**: 945–64.

Costantini, C., Birkett, M.A., Gibson, G. et al. (2001) Electroantennogram and behavioural responses of the malaria vector *Anopheles gambiae* to human-specific sweat components. *Medical & Veterinary Entomology* **15**: 259–66.

De Bruyne, M., Foster, K. and Carlson, J.R. (2001) Odor coding in *Drosophila* antenna. *Neuron* **30**: 537–52.

Degen, T., Dillmann, C., Marion-Poll, F. and Turlings, T.C.J. (2004) High genetic variability of herbivore-induced volatile emission within a broad range of maize inbred lines. *Plant Physiology* **135**: 1928–38.

De Jong, R. and Visser, J.H. (1988) Specificity-related suppression of responses to binary mixtures in olfactory receptors of the Colorado potato beetle. *Brain Research* **447**: 18–24.

De Moraes, C.M., Lewis, W.J., Paré, P.W., Alborn, H.T. and Tumlinson, J.H. (1998) Herbivore-infested plants selectively attract parasitoids. *Nature* **393**: 570–3.

De Moraes, C.M., Mescher, M.C. and Tumlinson, J.H. (2001) Caterpillar-induced nocturnal plant volatiles repel conspecific females. *Nature* **410**: 577–80.

Dicke, M. (1999) Specificity of herbivore-induced plant defences. In: *Insect–Plant Interactions and Induced Plant Defence*. Novartis Foundation Symposium 223. Wiley, Chichester, pp. 43–55.

Dicke, M. and Vet, L.E.M. (1999) Plant–carnivore interactions: evolutionary and ecological consequences for plant, herbivore and carnivore. In: Olff, H., Brown, V.K. and Drent, R.H. (eds.) *Herbivores: Between Plants and Predators*. Blackwell Science, Oxford, pp. 483–520.

Dicke, M. and Hilker, M. (2003) Induced plant defences: from molecular to evolutionary Ecology. *Basic & Applied Ecology* **4**: 3–14.

Dickens, J.C., Visser, J.H. and van der Pers, J.N.C. (1993) Detection and deactivation of pheromone and plant odor components by the beet armyworn, *Spodoptera exigua* (Hubner) (Lepidoptera, Noctuidae). *Journal of Insect Physiology* **39**: 503–16.

Duchamp-Viret, P., Duchamp, A. and Chaput, M.A. (2003) Single olfactory sensory neurons simultaneously integrate the components of an odour mixture. *European Journal of Neuroscience* **18**: 2690–6.

Dudareva, N. and Pichersky, E. (2000) Biochemical and molecular genetic aspects of floral scents. *Plant Physiology* **122**: 627–33.

Elkinton, J.S. and Cardé, R.T. (1984) Odor dispersion. In: Bell, W.J. and Cardé, R.T. (eds.) *Chemical Ecology of Insects*. Sinauer Associates Inc., Sunderland, pp. 73–91.

Fatouros, N.E., Huigens, M.E., van Loon, J.J.A., Dicke, M. and Hilker, M. (2005a) Butterfly anti-aphrodisiac lures parasitic wasps. *Nature* **433**: 704.

Fatouros, N.E., van Loon, J.J.A., Hordijk, K.A., Smid, H.M. and Dicke, M. (2005b) Herbivore-induced plant volatiles mediate in–flight host discrimination by parasitoids. *Journal of Chemical Ecology* **31**: 2033–47.

Faucher, C., Forstreuter, M., Hilker, M. and de Bruyne, M. (2006) Behavioral responses of *Drosophila* to biogenic levels of carbon dioxide depend on life stage, sex, and olfactory context. *Journal of Experimental Biology* **209**: 2739–48.

Fink, U. and Völkl, W. (1995) The effect of abiotic factors on foraging and oviposition success of the aphid parasitoid, *Aphidius rosae*. *Oecologia* **103**: 371–8.

Fournier, F. and Boivin, G. (2000) Comparative dispersal of *Trichogramma evanescens* and *Trichogramma pretiosum* (Hymenoptera: Trichogrammatidae) in relation to environmental conditions. *Population Ecology* **29**: 55–63.

Fournier, F., Pelletier, D., Vigneault, C., Goyette, B. and Boivin, G. (2005) Effect of barometric pressure on flight initiation by *Trichogramma pretiosum* and *Trichogramma evanescens* (Hymenoptera: Trichogrammatidae). *Environmental Entomology* **34**: 1534–40.

Frenoy, C., Durier, C. and Hawlitzky, N. (1992) Effect of kairomones from egg and female adult stages of *Ostrinia nubilalis* (Hübner) (Lepidoptera, Pyralidae) on *Trichogramma brassicae* Bezdenko (Hymenoptera, Trichogrammatidae) female kinesis. *Journal of Chemical Ecology* **18**: 761–73.

Gadenne, C. and Anton, S. (2000) Central processing of sex pheromones stimuli is differentially regulated by juvenile hormone in a male moth. *Journal of Insect Physiology* **46**: 1195–206.

Galizia, C.G. and Menzel, R. (2000) Odour perception in honeybees: Coding information in glomerular patterns. *Current Opinion in Neurobiology* **10**: 504–10.

Galizia, C.G. and Menzel, R. (2001) The role of glomeruli in the neural representation of odours: Results from optical recording studies. *Journal of Insect Physiology* **47**: 115–30.

Geervliet, J.B.F., Vet, L.E.M. and Dicke, M. (1994) Volatiles from damaged plants as major cues in long range host searching by the specialist parasitoid *Cotesia rubecula*. *Entomologia Experimentalis et Applicata* **73**: 289–97.

Geervliet, J.B.F., Posthumus, M.A., Vet, L.E.M. and Dicke, M. (1997) Comparative analysis of headspace volatiles from different caterpillar-infested or uninfested food plants of *Pieris* species. *Journal of Chemical Ecology* **23**: 2935–54.

Getz, W.M. (1999) A kinetic model of the transient phase in the response of olfactory receptor neurons. *Chemical Senses* **24**: 497–508.

Gillies, M.T. (1980) The role of carbon dioxide in host-finding by mosquitoes (Diptera: Culicidae): A review. *Bulletin of Entomological Research* **70**: 525–32.

Godfray, H.C.J. (1994) *Parasitoids. Behavioural and Evolutionary Ecology*. Princeton University Press, Princeton.

Goldansaz, S.H. and McNeil, J.N. (2006) Effect of wind speed on the behaviour of sexual morphs of the potato aphid, *Macrosiphum euphorbiae* (Thomas) (Homoptera: Aphididae) under laboratory and field conditions. *Journal of Chemical Ecology* **32**: 1719–29.

Gouinguené, S., Degen, T. and Turlings, T.C.J. (2001) Differential attractiveness of induced odour emissions among maize cultivars and their wild ancestors (teosinte). *Chemoecology* **11**: 9–16.

Gouinguené, S., Alborn, H. and Turlings, T.C.J. (2003) Induction of volatile emissions in maize by different larval instars of *Spodoptera littoralis*. *Journal of Chemical Ecology* **29**: 145–62.

Greiner, B., Gadenne, C. and Anton, S. (2002) Central processing of plant volatiles in *Agrotis ipsilon* males is age-independent in contrast to sex pheromone processing. *Chemical Senses* **27**: 45–8.

Gu, H. and Dorn, S. (2000) Genetic variation in behavioral response to herbivore-infested plants in the parasitic wasp, *Cotesia glomerata* (L.) (Hymenoptera: Braconidae). *Journal of Insect Behavior* **13**: 141–56.

Gu, H. and Dorn, S. (2001) How do wind velocity and light intensity influence host-location success in *Cotesia glomerata* (Hym., Braconidae)? *Journal of Applied Entomology* **125**: 115–20.

Guerrieri, E., Poppy, G.M., Powell, W., Tremblay, E. and Pennachio, F. (1999) Induction and systemic release of herbivore-induced plant volatiles mediating in-flight orientation of *Aphidius ervi*. *Journal of Chemical Ecology* **25**: 1247–61.

Haas, W. (1992) Physiological analysis of cercarial behavior. *Journal of Parasitology* **78**: 243–55.

Haas, W., Stiegeler, P., Keating, A. et al. (2002) *Diplostomum spathaceum* cercariae respond to a unique profile of cues during recognition of their fish host. *International Journal of Parasitology* **32**: 1145–54.

Haile, F.J. and Higley, L.G. (2003) Changes in soybean gas-exchange after moisture stress and spider mite injury. *Environmental Entomology* **32**: 433–40.

Hassanali, A., Nyandat, E., Obenchain, F.A., Otieno, D.A. and Galun, R. (1989) Humidity effect of response of *Argas persicus* (Oken) to guanine, an assembly pheromone in ticks. *Journal of Chemical Ecology* **15**: 791–7.

Held, D.W., Gosinska, P. and Potter, D.A. (2003) Evaluating companion planting and non-host masking odors for protecting roses from the Japanese beetle (Coleoptera: Scarabaeidae). *Journal of Economical Entomology* **96**: 81–7.

Hibbard, B.E. and Bjostad, L.B. (1988) Behavioral responses of western corn rootworm larvae to volatile semiochemicals from corn seedlings. *Journal of Chemical Ecology* **14**: 1523–39.

Hildebrand, J.G. and Shepherd, G.M. (1997) Mechanisms of olfactory discrimination: Converging evidence for common principles across phyla. *Annual Review of Neuroscience* **20**: 595–631.

Hilker, M. and Meiners, T. (1999) Chemical cues mediating interactions between chrysomelids and parasitoids. In: Cox, M.L. (ed.) *Advances in Chrysomelidae Biology*. Backhuys Publishers, Leiden, pp. 197–216.

Hilker, M. and Meiners, T. (2002) Induction of plant responses towards oviposition and feeding of herbivorous arthropods: a comparison. *Entomologia Experimentalis et Applicata* **104**: 181–92.

Hilker, M. and Meiners, T. (2006) Early herbivore alert: Insect eggs induce plant defense. *Journal of Chemical Ecology* **32**: 1379–97.

Hilker, M., Bläske, V., Kobs, C. and Dippel, C. (2000) Kairomonal effects of sawfly sex pheromones on egg parasitoids. *Journal of Chemical Ecology* **26**: 221–31.

Hilker, M., Kobs, C., Varama, M. and Schrank, K. (2002a) Insect egg deposition induces *Pinus sylvestris* to attract egg parasitoids. *Journal of Experimental Biology* **205**: 455–61.

Hilker, M., Rohfritsch, O. and Meiners, T. (2002b) The plant's response towards insect egg deposition. In: Hilker, M. and Meiners, T. (eds.) *Chemoecology of Insect Eggs and Egg Deposition*. Blackwell, Oxford, Berlin, pp. 205–34.

Hori, M. and Komatsu, H. (1997) Repellency of rosemary oil and its components against the onion aphid, *Neotoxoptera formosana* (Takahashi) (Homoptera, Aphididae). *Applied Entomology & Zoology* **32**: 303–10.

Ideses, R., Shani, A. and Klug, J.T. (1982) Sex pheromone of the European grapevine moth (*Lobesia botrana*). Its chemical transformations in sunlight and heat. *Journal of Chemical Ecology* **8**: 973–80.

Isaacs, R., Hardie, J., Hick, A.J. et al. (1993) Behavioral responses of *Aphis fabae* to isothiocyanates in the laboratory and field. *Pesticide Science* **39**: 349–55.

Joerges, J., Küttner, A., Galizia, C.G. and Menzel, R. (1997) Representations of odours and odour mixtures visualized in the honeybee brain. *Nature* **387**: 285–8.

Johnson, S.N. and Gregory, P.J. (2006) Chemically-mediated host-plant location and selection by root-feeding insects. *Physiological Entomology* **31**: 1–13.

Jones, R.L., Lewis, W.J., Beroza, M., Bierl, B.A. and Sparks, A.N. (1973) Host-seeking stimulants (kairomones) for the egg parasite, *Trichogramma evanescens*. *Environmental Entomology* **2**: 593–6.

Juillet, J.A. (1964) Influence of weather on flight activity of parasitic Hymenoptera. *Canadian Journal of Zoology* **42**: 1133–41.

Kaiser, L., Pham-Delegue, M.H., Backchine, E. and Masson, C. (1989) Olfactory response of *Trichogramma maidis*: Effect of chemical cues and behavioural plasticity. *Journal of Insect Behavior* **2**: 701–10.

Keller, M.A. (1990) Responses of the parasitoid *Cotesia rubecula* to its host *Pieris rapae* in a flight tunnel. *Entomologia Experimentalis et Applicata* **57**: 243–9.

Kelling, F.J., Ialenti, F. and Den Otter, C.J. (2002) Background odour induces adaptation and sensitization of olfactory receptors in the antennae of houseflies. *Medical & Veterinary Entomology* **16**: 161–9.

Landolt, P.J. and Phillips, T.W. (1997) Host plant influences on sex pheromone behavior of phytophagous insects. *Annual Review of Entomology* **42**: 371–91.

Langan, A.M., Wheater, C.P. and Dunleavy, P.J. (2001) Does the small white butterfly (*Pieris rapae* L.) aggregate eggs on plant gas exchange activity? *Journal of Insect Behavior* **14**: 459–68.

Langan, A.M., Wheater, C.P. and Dunleavy, P.J. (2004) Biogenic gradients of CO_2 and H_2O and oviposition by the small white butterfly (*Pieris rapae* L.) in cages. *Applied Entomology & Zoology* **39**: 55–9.

Lei, H., Christensen, T.A. and Hildebrand, J.G. (2002) Local inhibition modulates odor-evoked synchronization of glomerulus-specific output neurons. *Nature Neuroscience* **5**: 557–65.

Lewis, W.J., Jones, R.L., Nordlund, D.A. and Gross, J.R. (1975) Kairomones and their use for management of entomophagous insects. II. Mechanisms causing increase in rate of parasitization by *Trichogramma* spp. *Journal of Chemical Ecology* **1**: 349–60.

Llusià, J. and Peñuelas, J. (1999) *Pinus halepensis* and *Quercus ilex* terpene emission as affected by temperature and humidity. *Biologia Plantarum* **42**: 317–20.

Llusià, J. and Peñuelas, J. (2000) Seasonal patterns of terpene content and emission from seven Mediterranean woody species in field conditions. *American Journal of Botany* **87**: 133–40.

Lou, Y.G., Ma, B. and Cheng, J.A. (2005) Attraction of the parasitoid *Anagrus nilaparvatae* to rice volatiles induced by the rice brown planthopper *Nilaparvata lugens*. *Journal of Chemical Ecology* **31**: 2357–72.

Loughrin, J.H., Hamilton-Kemp, T.R., Andersen, R.A. and Hildebrand, D.F. (1990) Volatiles from flowers of *Nicotiana sylvestris*, *N. otophora*, and *malus* x *domestica*: Headspace components and day/night changes in their relative concentrations. *Phytochemistry* **29**: 2473–7.

Malnic, B., Hirono, J., Sato, T. and Buck, L.B. (1999) Combinatorial receptor codes for odors. *Cell* **96**: 713–23.

Marchand, D. and McNeil, J.N. (2000) Effects of wind speed and atmospheric pressure on mate searching behaviour in the aphid parasitoid *Aphidius nigripes* (Hymenoptera: Aphidiidae). *Journal of Insect Behavior* **13**: 187–99.

Marling, C., Sgalli, M., Rissland, E., Munoz-Avila, H. and Aha, D. (2002) Case-based reasoning integrations. *Artificial Intelligence Magazine* **23**: 69–86.

Masson, C. and Mustaparta, H. (1990) Chemical information processing in the olfactory system of insects. *Physiological Reviews* **70**: 199–245.

Mauchline, A.L., Osborne, J.L., Martin, A.P., Poppy, G.M. and Powell, W. (2005) The effects of non-host plant essential oil volatiles on the behaviour of the pollen beetle *Meligethes aeneus*. *Entomologia Experimentalis et Applicata* **114**: 181–3.

McCall, P.J. (2002) Chemoecology of oviposition in insects of medical and veterinary importance. In: Hilker, M. and Meiners, T. (eds.) *Chemoecology of Insect Eggs and Egg Deposition*. Blackwell, Oxford, Berlin, pp. 235–64.

McNeil, J.N. (1991) Behavioral ecology of pheromone-mediated communication in moths and its importance in the use of pheromone traps. *Annual Review of Entomology* **36**: 407–30.

McNeil, J.N. and Delisle, J. (1989) Host plant pollen influences calling behaviour and ovarian development of the sunflower moth, *Homoeosoma electellum*. *Oecologia* **80**: 201–5.

Meiners, T., Westerhaus, C. and Hilker, M. (2000) Specificity of chemical cues used by a specialist egg parasitoid during host location. *Entomologia Experimentalis et Applicata* **95**: 151–9.

Meiners, T., Wäckers, F. and Lewis, W.J. (2003) Associative learning of complex odours in parasitoid host location. *Chemical Senses* **28**: 231–6.

Mercer, A.R. and Hildebrand, J.G. (2002) Developmental changes in the electrophysiological properties and response characteristics of *Manduca* antennal lobe neurons. *Journal of Neurophysiology* **87**: 2650–63.

Messing, R.H., Klungness, L.M. and Jang, E.B. (1997) Effects of wind on movement of *Diachasmimorpha longicaudata*, a parasitoid of tephritid fruit flies, in a laboratory flight tunnel. *Entomologia Experimentalis et Applicata* **82**: 147–52.

Moraes, M.C.B., Laumann, R., Sujii, E.R., Pires, C. and Borges, M. (2005) Induced volatiles in soybean and pigeon pea plants artificially infested with the neotropical brown stink bug, *Euschistus heros*, and their effect on the egg parasitoid, *Telenomus podisi*. *Entomologia Experimentalis et Applicata* **115**: 227–37.

Mumm, R. and Hilker, M. (2005) The significance of background odour for an egg parasitoid to detect plants with host eggs. *Chemical Senses* **30**: 337–43.

Mumm, R. and Hilker, M. (2006) Direct and indirect chemical defence of pine against folivorous insects. *Trends in Plant Science* **11**: 351–8.

Mumm, R., Schrank, K., Wegener, R., Schulz, S. and Hilker, M. (2003) Chemical analysis of volatiles emitted by *Pinus sylvestris* after induction by insect oviposition. *Journal of Chemical Ecology* **29**: 1235–52.

Mumm, R., Tiemann, T., Varama, M. and Hilker, M. (2005) Choosy egg parasitoids: Specificity of oviposition-induced pine volatiles exploited by an egg parasitoid of pine sawfly. *Entomologia Experimentalis et Applicata* **115**: 217–25.

Murlis, J., Elkinton, J.S. and Cardé, R.T. (1992) Odor plumes and how insects use them. *Annual Review of Entomology* **37**: 505–32.

Nottingham, S.F., Hardie, J., Dawson, G.W. et al. (1991) Behavioral and electrophysiological responses of aphids to host and nonhost plant volatiles. *Journal of Chemical Ecology* **17**: 1231–42.

Nufio, C.R. and Papaj, D.R. (2001) Host marking behavior in phytophagous insects and parasitoids. *Entomologia Experimentalis et Applicata* **99**: 273–93.

Ochieng, S.A., Park, K.C. and Baker, T.C. (2002) Host plant volatiles synergise responses of sex pheromone-specific olfactory receptor neurons in male *Helicoverpa zea*. *Journal of Comparative Physiology A* **188**: 325–33.

Parra, J.R.P., Vinson, S.B., Gomes, S.M. and Cônsoli, F.L. (1996) Flight response of *Habrobracon hebetor* (Say) (Hymenoptera: Braconidae) in a wind tunnel to volatiles associated with infestations of *Ephestia kuehniella* Zeller (Lepidoptera: Pyralidae). *Biological Control* **6**: 143–50.

Paul, A.V.N., Singh, S. and Singh, A.K. (2002) Kairomonal effect of some saturated hydrocarbons on the egg parasitoids, *Trichogramma brasiliensis* (Ashmead) and *Trichogramma exiguum* (Hym., Trichogrammatidae). *Journal of Applied Entomology* **126**: 409–16.

Pickett, J.A., Wadhams, L.J. and Woodcock, C.M. (1997) Developing sustainable pest control from chemical ecology. *Agriculture Ecosystems & Environment* **64**: 149–56.

Powell, W. (1999) Parasitoid hosts. In: Hardie, J. and Minks, A.K. (eds.) *Pheromones of Non-Lepidopteran Insects Associated with Agricultural Plants*. CABI Publishing, Wallingford, pp. 405–27.

Prévost, G. and Lewis, W.J. (1990) Heritable differences in the response of the braconid wasp *Microplitis croceipes* to volatile allelochemicals. *Journal of Insect Behavior* **3**: 277–87.

Quiring, D.T. and McNeil, J.N. (1987) Daily patterns of abundance and temporal distributions of feeding, mating and oviposition of the dipteran leafminer, *Agromyza frontella*. *Entomologia Experimentalis et Applicata* **45**: 73–9.

Reddy, G.V.P., Holopainen, J.K. and Guerrero, A. (2002) Olfactory responses of *Plutella xylostella* natural enemies to host pheromone, larval frass, and green leaf cabbage volatiles. *Journal of Chemical Ecology* **28**: 131–43.

Renou, M., Nagnan, P., Berthier, A. and Durier, C. (1992) Identification of compounds from the eggs of *Ostrinia nubilalis* and *Mamestra brassicae* having kairomone activity on *Trichogramma brassicae*. *Entomologia Experimentalis et Applicata* **63**: 291–303.

Roitberg, B.C., Sircom, J., Roitberg, C.A., van Alphen, J.J.M. and Mangel, M. (1993) Life expectancy and reproduction. *Nature* **364**: 108.

Royer, L. and McNeil, J.N. (1991) Changes in the calling behaviour and mating success of the European corn borer (*Ostrinia nubilalis*), caused by relative humidity. *Entomologia Experimentalis et Applicata* **61**: 131–8.

Royer, L. and McNeil, J.N. (1993) Effects of relative humidity conditions on the responsiveness of European corn borer (*Ostrinia nubilalis*) males to the female sex pheromone in a wind tunnel. *Journal of Chemical Ecology* **19**: 61–9.

Rutledge, C.E. (1996) A survey of kairomones and synomones used by insect parasitoids to locate and accept their hosts. *Chemoecology* **7**: 121–31.

Said, I., Renou, M., Morin, J.-P., Ferreira, J.M.S. and Rochat, D. (2005) Interactions between acetoin, a plant volatile, and pheromone in *Rynchophrous palmarum*: behavioural and olfactory neuron responses. *Journal of Chemical Ecology* **31**: 1789–805.

Schwörer, W. and Völkl, W. (2001) Foraging behavior of *Aphidius ervi* (Haliday) (Hymenoptera: Braconidae: Aphidiinae) at different spatial scales: Resource utilization and suboptimal weather conditions. *Biological Control* **21**: 111–19.

Schwörer, U., Völkl, W. and Hoffmann, K.H. (1999) Foraging for mates in the hyperparasitic wasp, *Dendrocerus carpenteri*: Impact of unfavourable weather conditions and parasitoid age. *Oecologia* **119**: 73–80.

Shu, S., Swedenborg, P.D. and Jones, R.L. (1990) A kairomone for *Trichogramma nubilale* (Hymenoptera: Trichogrammatidae). Isolation, identification, and synthesis. *Journal of Chemical Ecology* **16**: 521–9.

Smith, B.H. and Getz, W.M. (1994) Nonpheromonal olfactory processing in insects. *Annual Review of Entomology* **39**: 351–75.

Stange, G., Monro, J., Stowe, S. and Osmond, C.B. (1995) The CO_2 sense of the moth *Cactoblastis cactorum* and its probable role in the biological control of the CAM plant *Opuntia stricta*. *Oecologia* **102**: 341–52.

Steidle, J.L.M. and van Loon, J.J.A. (2003) Dietary specialization and infochemical use in carnivorous arthropods: testing a concept. *Entomologia Experimentalis et Applicata* **108**: 133–48.

Steinberg, S., Dicke, M., Vet, L.E.M. and Wanningen, R. (1992) Response of the braconid parasitoid *Cotesia* (= *Apanteles*) *glomerata* to volatile infochemicals: Effects of bioassay set-up, parasitoid age, and experience and barometric flux. *Entomologia Experimentalis et Applicata* **63**: 163–7.

Takabayashi, J., Dicke, M. and Posthumus, M.A. (1994) Volatile herbivore-induced terpenoids in plant–mite interactions: Variation caused by biotic and abiotic factors. *Journal of Chemical Ecology* **20**: 1329–54.

Takabayashi, J., Takahashi, S., Dicke, M. and Posthumus, M.A. (1995) Developmental stage of herbivore *Pseudaletia separata* affects production of herbivore-induced synomone by corn plants. *Journal of Chemical Ecology* **21**: 273–87.

Takasu, K. and Lewis, W.J. (1993) Host foraging and food foraging of the parasitoid *Microplitis croceipes* – learning and physiological state effects. *Biological Control* **3**: 70–4.

Thiery, D. and Visser, J.H. (1986) Masking of host plant odour in olfactory orientation of the Colorado potato beetle. *Entomologia Experimentalis et Applicata* **41**: 165–72.

Travis, J. (1999) Making sense of scents. *Science News* **155**: 236.

Turlings, T.C., Loughrin, J.H., McCall, P.J., Roese, U.S., Lewis, W.J. and Tumlinson, J.H. (1995) How caterpillar-damaged plants protect themselves by attracting parasitic wasps. *Proceedings of the National Academy of Sciences USA* **92**: 4164–8.

Vallat, A., Hainan, G. and Dorn, S. (2005) How rainfall, relative humidity and temperature influence volatile emissions from apple trees *in situ*. *Phytochemistry* **66**: 1540–50.

van Alphen, J.J.M., Bernstein, C. and Driessen, G. (2003) Information acquisition and time allocation in insect parasitoids. *Trends in Ecology & Evolution* **18**: 81–7.

Vaughn, T.T., Antolin, M.F. and Bjostad, L.F. (1996) Behavioural and physiological responses of *Diaeretiella rapae* to semiochemicals. *Entomologia Experimentalis et Applicata* **78**: 187–96.

Veit, U., Frank, R., Klumpp, A. and Fomin, A. (2001) Influence of temperature and relative air humidity on the oxidative destruction of pheromones. *IOBC Bulletin* **24**: 107–13.

Vet, L.E.M. (1999) Evolutionary aspects of plant – carnivore interactions. In: *Insect–Plant Interactions and Induced Plant Defence.* Novartis Foundation Symposium 223, Wiley, Chichester, pp. 3–20.

Vet, L.E.M. and Groenewold, A.W. (1990) Semiochemicals and learning parasitoids. *Journal of Chemical Ecology* **16**: 3119–35.

Vet, L.E.M. and Dicke, M. (1992) Ecology of infochemical use by natural enemies in a tritrophic context. *Annual Review of Entomology* **37**: 141–72.

Vet, L.E.M., Lewis, W.J., Papaj, D.R. and van Lenteren, J.C. (1990) A variable response model for parasitoid foraging behavior. *Journal of Insect Behaviour* **3**: 471–90.

Vet, L.E.M., Lewis, W.J. and Cardé, R.T. (1995) Parasitoid foraging and learning. In: Cardé, R.T. and Bell, W.J. (eds.) *Chemical Ecology of Insects* 2. Chapman & Hall, New York, pp. 65–101.

Vet, L.E.M., De Jong, A.G., Franchi, E. and Papaj, D.R. (1998) The effect of complete versus incomplete information on odour discrimination in a parasitic wasp. *Animal Behaviour* **55**: 1271–9.

Vinson, S.B. (1981) Habitat location. In: Nordlund, D.A., Jones, R.L. and Lewis, W.J. (eds.) *Semiochemicals: Their Role in Pest Control.* Wiley, New York, pp. 51–77.

Vinson, S.B. (1991) Chemical signals used by parasitoids. *Redia* **74**: 15–42.

Vinson, S.B. and Williams, H.J. (1991) Host selection behavior of *Campoletis sonorensis*: A model system. *Biological Control* **1**: 107–17.

Visser, J.H. (1986) Host odor perception in phytophagous insects. *Annual Review of Entomology* **31**: 121–44.

Visser, J.H. and Avé, D.A. (1978) General green leaf volatiles in the olfactory orientation of the Colorado beetle, *Leptinotarsa decemlineata. Entomologia Experimentalis et Applicata* **24**: 538–49.

Vos, M., Berrocal, S.M., Karamaouna, F., Hemerik, L. and Vet, L.E.M. (2001) Plant-mediated indirect effects and the persistence of parasitoid–herbivore communities. *Ecology Letters* **4**: 38–45.

Wäckers, F.L. (1994) The effect of food deprivation on the innate visual and olfactory preferences in the parasitoid *Cotesia rubecula. Journal of Insect Physiology* **40**: 641–9.

Wallace, E.K., Albert, P.J. and McNeil, J.N. (2004) Oviposition behavior of the eastern spruce budworm, *Choristoneura fumiferana* (Clemens) (Lepidoptera: Tortricidae). *Journal of Insect Behavior* **17**: 245–54.

Wang, Q., Gu, H. and Dorn, S. (2003) Selection on olfactory response to semiochemicals from a plant–host complex in a parasitic wasp. *Heredity* **91**: 430–5.

Weinbrenner, M. and Völkl, W. (2002) Oviposition behaviour of the aphid parasitoid, *Aphidius ervi*: Are wet aphids recognized as host? *Entomologia Experimentalis et Applicata* **103**: 51–9.

Weisser, W.W., Völkl, W. and Hassell, M.P. (1997) The importance of adverse weather conditions for behaviour and population ecology of an aphid parasitoid. *Journal of Animal Ecology* **66**: 386–400.

Welter, S.C. (1989) Arthropod impact on plant gas exchange. In: Bernays, E.A. (ed.) *Insect–Plant Interactions.* CRC Press, Boca Raton, pp. 135–51.

Whitman, D.W. and Eller, F.J. (1992) Orientation of *Microplitis croceipes* (Hymenoptera, Braconidae) to green leaf volatiles: dose–response curves. *Journal of Chemical Ecology* **18**: 1743–53.

Wiener, L.F. and Capinera, J.L. (1979) Greenbug response to an alarm pheromone analog – temperature and humidity effects, disruptive potential, and analog releaser efficacy (Homoptera, Aphididae). *Annals of the Entomological Society of America* **72**: 369–71.

Yamasaki, T., Sato, M. and Sakoguchi, H. (1997) (–)-Germacrene D: Masking substance of attractants for the cerambycid beetle, *Monochamus alternatus* (HOPE). *Applied Entomology & Zoology* **32**: 423–9.

Zangerl, A.R., Hamilton, J.G., Miller, T.J. et al. (2002) Impact of folivory on photosynthesis is greater than the sum of its holes. *Proceedings of the National Academy of Sciences USA* **99**: 1088–91.

6

Parasitoid and host nutritional physiology in behavioral ecology

Michael R. Strand and Jérôme Casas

Abstract

Nutrient acquisition and allocation critically impacts the fitness of all organisms. However, distinguishing between competing hypotheses about metabolic strategies requires mechanistic details on the physiological requirements and constraints under which the organism exists. Parasitoids acquire nutrients during both their larval stage and as adults. In this chapter, we first explore differences in the acquisition strategies of koinobionts and idiobionts and the physiological adaptations parasitoids have evolved to manipulate host nutrient stores for their own fitness. We then relate these larval strategies to nutrient acquisition strategies during the adult stage by revisiting the link between reproduction (ovigeny index), host-feeding behavior, and oosorption. We conclude our discussion by examining how nutritional interactions potentially impact on other aspects of parasitoid and host fitness. Overall, we argue that the combination of top-down evolutionary approaches with bottom-up insights into physiology and molecular mechanisms offers new avenues for understanding the complex syndromes of the idiobiont and koinobiont life history dichotomy.

6.1 Introduction

Classic life history models assume that fundamental trade-offs will arise between reproduction and determinants of survival when resources are limiting. Many factors can constrain resource availability including ecological conditions and the structure of an organism's life cycle. Parasitoids acquire nutrients as larvae by feeding on hosts (capital resources). They can also acquire additional nutrients as adults by feeding on non-host resources like nectar and/or by host feeding (income resources) (Jervis & Kidd 1986, Heimpel & Collier 1996). Most behavioral and population dynamics models measure the quality of host resources using a single currency such as size (Charnov & Skinner 1984, Mangel 1989). Parasitoid nutrient reserves are also represented usually by single currencies like the number of eggs available for oviposition (egg load) to predict parasitoid foraging decisions or population-level effects on hosts (Mangel 1989, Chan & Godfray 1993, Briggs et al. 1995, Collier 1995, Křivan 1997, Heimpel et al. 1998, Rosenheim et al. 2000). In reality, nutrient acquisition and allocation is much more complex, because individual nutrients can be limiting for specific functions, such as reproduction, even when energetic resources are not (Raubenheimer & Simpson 1999, O'Brien et al. 2002). Changes in diet or metabolism can also affect how nutrients from different life stages are used (Zera & Zhao 2003, O'Brien et al. 2004, Min et al. 2006).

Hosts face similar complexities in nutrient acquisition and allocation. They also face potentially important fitness trade-offs from investing in defense against attack by parasitoids and pathogens relative to reproduction and other maintenance needs (see Chapter 14 by Kraaijeveld and Godfray). Defense against internal parasitoids depends primarily on the ability of the host's immune system to kill the parasitoid egg or larva after oviposition (Strand & Pech 1995, Lavine & Strand 2002, Hoffmann 2003, see also Chapter 14 by Kraaijeveld and Godfray). Host–parasitoid population dynamics are also affected by variation in host resistance, which can arise as a consequence of genetic differences among individuals or from environmental factors such as nutritional state (Chesson & Murdoch 1986, Hochberg 1997, Sasaki & Godfray 1999, Godfray 2000, Carton et al. 2005).

In this chapter we examine the effects of nutritional physiology on the behavioral ecology of parasitoids and their hosts. We begin by outlining the relationship between life history and nutrient dynamics in parasitoids. We then explore the role of nutritional state in immunity and the potential costs of defense to hosts (see also Chapter 14 by Kraaijeveld and Godfray). We finish by arguing that the combination of top-down ecological, population, and evolutionary studies combined with bottom-up insights from physiology and molecular biology offers new avenues for understanding the behavioral ecology of parasitoids.

6.2 Background nutritional physiology

Carbohydrates, proteins, and lipids are the primary nutrient classes and the fat body is the main site of nutrient storage and metabolism for all insects. The fat body stores carbohydrates as glycogen and lipids as tryglycerides (Clements 1992, Candy et al. 1997). The fat body also synthesizes many key molecules including the lipoprotein vitellogenin, the primary constituent of yolk, and trehalose, which is a key sugar in hemolymph. Insect hemolymph is another site of nutrient storage and usually contains high levels of free amino acids, storage proteins, and sugars required for maintenance, metamorphosis, and reproduction. Regulation of nutrient homeostasis (i.e. nutrient sensing) in insects, as in vertebrates, occurs primarily through the insulin and the target of rapamycin (TOR) pathways (Britton et al. 2002, Scott et al. 2004). Insulin signaling is a hormone-based system that regulates metabolism and organismal growth while the TOR pathway responds to nutrient levels to regulate protein synthesis and cell growth (Oldham et al. 2000, Zhang et al. 2000, Wu & Brown 2006). Reciprocally, inhibition of the TOR pathway during starvation induces autophagy, whereby non-essential proteins and organelles are recycled to generate amino acids for other purposes (Scott et al. 2004).

6.3 Nutrient acquisition and allocation by parasitoids varies with life history

Nutrient acquisition and allocation strategies of parasitoids are strongly linked to two components of life-history: (i) egg production and (ii) mode of parasitism. For the former, Flanders (1950) divided parasitoids into pro-ovigenic species that emerge as adults with a fixed complement of mature eggs and synovigenic species that continue to mature eggs during the adult stage. Pro-ovigenic parasitoids allocate nutrient reserves during the adult stage to maintenance, while synovigenic parasitoids confront the decision of whether to allocate reserves to egg production, maintenance, or both (Jervis & Kidd 1986, Heimpel & Collier 1996, Rivero & Casas 1999, Papaj 2000). Jervis et al. (2001) noted that parasitoids actually exhibit a continuum of ovigeny that can be indexed. Relatively few species are strictly pro-ovigenic (ovigeny index = 1) and synovigeny ranges from species that emerge with most eggs mature to species that emerge with no mature eggs (ovigeny index = 0). Askew and Shaw (1986), in contrast, divided parasitoids into idiobionts, whose hosts cease development after parasitism and koinobionts, whose hosts remain mobile and continue to grow. All idiobionts are either ectoparasitoids that paralyze their hosts or endoparasitoids that parasitize sessile host stages like eggs or pupae. Most koinobionts in contrast are endoparasitoids that parasitize insect larvae.

Egg production and mode of parasitism strategies are also interrelated. Parasitoids that exhibit extreme synovigeny, for example, are all idiobionts that produce yolk-rich (anhydropic) eggs and low ovigenic indices, whereas koinobionts tend to produce yolk-deficient (hydropic) eggs and have high ovigenic indices (Mayhew & Blackburn 1999). Since idiobionts parasitize hosts of static size, selection favors oviposition on larger, late-stage hosts, which suffer lower mortality rates and thus select for larger egg sizes and concomitantly lower fecundities (Price 1974, Mackauer & Sequeira 1993, Godfray 1994). Most idiobionts are also ectoparasitoids whose eggs require a pre-packaged yolk source for development. Reciprocally, the ability of koinobionts to attack hosts of variable size favors parasitism of early stage larval hosts that suffer higher mortality rates and thus, favor smaller eggs and larger fecundities. The production of smaller, yolk-deficient eggs is also probably favored in koinobionts by: (i) endoparasitism and access to host nutrients for embryonic development; and (ii) the inability of koinobionts to feed on the mobile, often aggressive, hosts they attack (Strand 2000, Pennacchio & Strand 2006).

6.4 Parasitoid nutrient dynamics

With these life history correlations in mind, it is not surprising that most idiobionts are able to lay a few eggs after adult emergence, but further increases in longevity and egg production require host feeding and/or access to non-host resources like nectar (Jervis & Kidd 1986, Heimpel & Collier 1996, Rivero & Casas 1999). Using biochemical methods and dietary stable isotope signatures, more recent studies have also begun to unravel the contribution of larval and adult diets to parasitoid nutrient budgets. Nutrients that are acquired during the larval stage and used during the adult stage are referred to as capital reserves. Studies of the ectoparasitic idiobiont *Eupelmus vuilleti* indicate that adult females emerge with high capital reserves of lipid, sugars, and glycogen. Host and sugar feeding by adults provides additional protein and carbohydrates but little or none of these nutrients are converted to lipid suggesting an absence of lipogenesis during the adult stage of *E. vuilleti* (Giron & Casas 2003). Lipid ingested by host feeding is also inadequate to replace capital lipid reserves from the larval stage (Giron et al. 2002, Casas et al. 2005). In contrast, the endoparasitic koinobiont *Venturia canescens* is weakly synovigenic, produces small eggs with little yolk, and does not host feed. This species emerges with limited capital reserves, is stored mainly as lipids, and possesses almost no reserves stored in eggs themselves (Casas et al. 2003). Carbohydrate levels increase rapidly in *V. canescens* from feeding on nectar and/or honeydew, yet, similar to *E. vuilleti*, lipid reserves do not (Casas et al. 2003).

Strong synovigeny has previously been associated with weak capital reserves compared to weak synovigeny or pro-ovigeny (Jervis et al. 2001), yet the preceding data suggest this is not the case. Instead, nutrient reserves must be viewed in terms of individual nutrient classes (Section 6.6). Carbohydrates, for example, obtained by adults from host or non-host sources enhance longevity and survival of both idio- and koinobionts. Carbohydrates also appear to be the only nutrient used for flight by Hymenoptera and this nutrient class is rapidly depleted if adults are starved (Vogt et al. 2000, Harrison & Fewell 2002). Proteins and carbohydrates obtained by idiobionts through host feeding increase the production of yolk-rich eggs by allowing females to invest less capital lipid reserves in maintenance functions. In contrast, neither *E. vuilleti* nor *V. canescens* is able to synthesize new lipids from carbohydrates as adults, making lipids acquired during larval development a non-renewable resource that ultimately constrains egg production (see below). Comparative studies have further suggested an absence of lipogenesis during the adult stage may exist across all Hymenoptera (Ellers 1996, Rivero & Casas 1999, Olson et al. 2000, Rivero et al. 2001, Rivero & West 2002). If so, we would hypothesize that constraints on lipid reserves are a major factor influencing the foraging behavior and oviposition strategies of parasitoid wasps.

Given that koinobionts are unable to host feed, we would also suggest that many species have evolved compensatory strategies for enhancing the acquisition of nutrients during larval development that are lacking in the adult diet. This would include lipids, sterols, and non-essential amino acids that are deficient in nutrient sources available to adults like nectar (O'Brien et al. 2002, 2004). Notably, many larval endoparasitoids induce dramatic reductions in weight gain and inhibit metamorphosis by hosts (Harvey & Strand 2002, Beckage & Gelman 2004, Pennacchio & Strand 2006). These developmental alterations are also associated with qualitative and quantitative alterations in how carbohydrates, proteins, and lipids are allocated to host tissues (Thompson 1993, Thompson & Dahlman 1998, Pennacchio & Strand 2006). Key changes include large increases in host hemolymph carbohydrate levels and alterations in protein composition including a loss of major storage proteins like arylphorin (Dahlman & Vinson 1980, Thompson 1982, Vinson 1990, Shelby & Webb 1994). Reductions in triglyceride and glycogen deposits in fat body and increases in hemolymph protein, amino acid, and acyl-glycerol levels have also been reported in aphids parasitized by aphidiine braconids (Pennacchio et al. 1995, Rahbé et al. 2002). Although idiobionts have traditionally been thought to manipulate host physiology less than koinobionts (Askew & Shaw 1986, Jervis et al. 2001), recent studies have suggested this generality may also be inaccurate (Pennacchio & Strand 2006). For example, the venoms produced by idiobionts are well known for their paralytic activity. However, idiobiont venoms also cause several endocrine and metabolic alterations that increase carbohydrate, protein, and lipid levels in host hemolymph that is the primary resource consumed by developing parasitoid larvae (Strand 1986, Rivers & Denlinger 1994, Weaver et al. 2001).

The molecular mechanism(s) underlying these metabolic alterations remain only superficially understood (see Thompson 1993, Beckage & Gelman 2004, Pennacchio & Strand 2006). Nonetheless, Pennacchio and Strand (2006) concluded that their overall effect is to redirect energetic resources away from the host and toward the developing larval stage parasitoid so as to enhance capital reserves. As discussed above, koinobionts almost always juvenilize their larval stage hosts in a manner that increases nutrient availability in the hemolymph and disrupts nutrient uptake by host tissues. Parasitoids that attack nymphal or adult hosts confront a very different host environment but the strategy of manipulating host nutritional physiology remains similar. This is well illustrated by aphid parasitoids like *Aphidius ervi*. The primary metabolic sink of its host is reproduction and, not surprisingly, *A. ervi* produces gene products that suppress host reproduction (Pennacchio et al. 1995, Digilio et al. 2000). Thus, the primary adaptive significance of altering host endocrine physiology and reproduction is likely metabolic with arrested development or inhibition of metamorphosis being indirect consequences of the parasitoid redirecting host nutritional resources. Redirecting nutrient reserves away from the host and toward the parasitoid also potentially reduces immune defenses, which could further favor parasitoid survival (see below).

6.5 Nutrient dynamics and immune defense by hosts

Host insects similarly confront resource constraints that can affect their life history. In Lepidoptera, for example, sugar feeding by adults increases longevity and fecundity, but essential amino acids are acquired primarily from the larval diet, which places an upper limit on the use of adult dietary resources for reproduction (O'Brien et al. 2002, 2004). Even in host insects with similar larval and adult diets, resource allocation of nutrients acquired during the larval and adult phase can vary with time or by tissue. Dietary sucrose and yeast provides virtually all of the carbon *Drosophila melanogaster* allocates to eggs, but the origin of these sugar carbons shifts from larval sources in the first clutches of eggs laid by a female to almost exclusively adult sources in subsequent clutches. In contrast, more than 30% of sugar carbons allocated to adult somatic tissues derive from larval reserves, suggesting that resources acquired during different life stages are allocated differently to maintenance and reproduction (Min et al. 2006).

Since parasitoids and pathogens are often the most important mortality factors facing insects, immune defense has long been recognized as a critically important maintenance function (see Chapter 14 by Kraaijeveld and Godfray). Investment in defense also has possible fitness costs and trade-offs with reproduction and other needs. Such trade-offs have been viewed as either a plastic response, usually referred to as the cost of using the immune system or as a co-evolved trait viewed as the cost of having an immune system (Schmid-Hempel & Ebert 2003, see also Chapter 14 by Kraaijeveld and Godfray). The innate immune response of insects consists of both cellular and humoral components (Strand & Pech 1995, Lavine & Strand 2002, Hoffmann 2003). Some defenses, like melanization, are non-specific and have activity against a range of parasites, while other defenses are directed to a restricted set of parasite species or types. The primary defense response against parasitoids is encapsulation, which involves the recognition and binding of hemocytes to the parasitoid egg or larva. Binding of additional hemocytes ultimately results in a multicellular capsule that fully envelops the parasitoid (see Chapter 14 by Kraaijeveld and Godfray). The encapsulated parasitoid is then likely to be killed by a combination of asphyxiation and toxic compounds produced by the melanization pathway that hemocytes often release in mature capsules (Wertheim et al. 2005). Several immune pathways regulate encapsulation and associated processes such as hemocyte proliferation and activation of the phenoloxidase cascade, which regulates melanization (Irving et al. 2005, Wertheim et al. 2005, see Chapter 14 by Kraaijeveld and Godfray). These include the Toll and immune deficiency (Imd) pathways that also regulate defense responses to microbial infection and the JAK-STAT (Janus kinase and signal transducers) and GATA signaling pathways that regulate hematopoiesis and antiviral defense (Hoffmann 2003, Agaisse & Perrimon 2004). These pathways and the effector responses they regulate together comprise the insect's inflammatory response to infection.

6.5.1 Is immune defense costly?

If there are life history costs associated with resistance, then the benefit to hosts of investing in immune defense will be determined by the risk of attack, which will vary in both space and time (see also Chapter 14 by Kraaijeveld and Godfray). Resistance would also be expected to vary within and between host populations and could be affected by genetic as well as environmental factors. There is considerable evidence for host variation in immune resistance to parasitoid attack both between (Kraaijeveld et al. 1998, Hufbauer 2002, Carton et al. 2005) and within (Henter & Via 1995, Kraaijeveld & Godfray 1997, Fellowes 1999, Kraaijeveld & Godfray 1999, Stacey & Fellowes 2002, Gwynn et al. 2005) host populations (see Chapter 14 by Kraaijeveld and Godfray). A few recent studies have also suggested that immune resistance to parasitoids has fitness costs for survival and future reproduction. For example, selection for increased immune resistance in *D. melanogaster* to the parasitoids *Asabara tabida* and *Leptoplinina boulardi* results in enhanced encapsulation of parasitoids, but at the cost of a reduction in the ability of resistant hosts to compete for food (Kraaijeveld & Godfray 1997, Fellowes et al. 1999, see also Chapter 14 by Kraaijeveld and Godfray). Aphids usually do not eliminate parasitoids by encapsulation but other immune factors in hemolymph prevent parasitoids eggs from developing. This immune response also appears to have a cost, because pea aphid clones (*Acyrthosiphon pisum*) with higher immune resistance to the parasitoid *Aphidium ervi* have lower fecundity than susceptible clones (Gwynn et al. 2005). Outside of the parasitoid literature, other evidence for the cost of immune defense includes reductions in survival and male mating success (Rolff & Siva-Jothy 2002, Schmid-Hempel & Ebert 2003). While the examples cited here provide evidence of trade-offs between immunity, reproduction, and/or other maintenance functions, additional comparative data and experimentation are also needed to determine whether the costs of immune defenses are broadly significant among different host–parasitoid associations.

6.5.2 Is resistance affected by nutrient dynamics?

Evidence that immune defense is energetically expensive in insects and other organisms derives from the observation, in diverse species, that activation of the immune response increases mortality or has other measurable fitness consequences when resources are limited (Rolff & Siva-Jothy 2003, Schmidt-Hempel & Ebert 2003). The prevalence of induced versus constitutive responses in immunity following parasite infection is also thought to have evolved, in part, to avoid the energetic cost of permanent defense. Support for this latter idea derives primarily from genetic modification of immune traits or selection studies. For example, mutations that constitutively activate systemic acquired resistance in plants increase resistance to certain pathogens but at the cost of reduced size and seed production (Heidel et al. 2004). Persistent activation of the pathways regulating antimicrobial peptide production in *D. melanogaster* also reduces fecundity presumably due to resources invested in defense molecules being unavailable for egg production (Zerofsky et al. 2005). As mentioned above, *D. melanogaster* selected for resistance to parasitoid attack produce more hemocytes but at the cost of reduced competitive ability under conditions of high resource competition (Kraaijeveld & Godfray 1997, Kraaijeveld et al. 1998). Surprisingly though, flies selected for increased competitive ability under crowding do not exhibit reduced resistance to parasitoids (Sanders et al. 2005).

Other evidence for the energetic cost of immune defense stems from the observation that stress, including high temperature, starvation, or persistent infection, increases susceptibility to infection due to a presumed reduction in resources available for immunity (Boulétreau 1986, Feder et al. 1997, Faggioni et al. 2000, Yang & Cox-Foster 2005). Insect resistance to parasitoids can also vary by age with younger larvae or older adults usually being most susceptible to infection (Salt 1970, Washburn et al. 2001, Hillyer et al. 2005, Zerofsky et al. 2005). This increased susceptibility could be due to age-dependent changes in the function of the immune system, lower nutrient reserves for investment in defense, or both.

Outside of insects, several linkages have also been identified between metabolic and immune pathways (Lochmiller & Deerenberg 2000, Rolff & Siva-Jothy 2003, Matarese & La Cava 2004). Pro-inflammatory cytokines and persistent infection by pathogens, such as tuberculosis, triggers hyperglycemia in mammals due to insulin insensitivity and release of glucocorticoids (Andersen et al. 2004). This shift in nutrient allocation affects both immune cell function and the production of acute phase proteins, but at the cost of reduced skeletal muscle mass and impaired glycogen synthesis. As noted above, parasitoids induce a similar response in host insects (Pennacchio & Strand 2006). Starvation in insects is also known to alter both TOR and immune pathway signaling (Gordon et al. 2005).

Lastly, a few studies have documented genetic differences in metabolic enzymes that have distinct pleiotropic effects in different life stages that could result in trade-offs between defense, reproduction, and other maintenance functions. For example, the glycolytic enzyme phosphoglucose isomerase (*Pgi*) plays a key role in glucose metabolism and the resupply of energy (ATP). Studies in several insects indicate that different *Pgi* genotypes exhibit variation in their enzyme kinetic and thermal stability properties that correlate with variation in flight performance, adaptation to particular microhabitats, and fecundity (Watt 1992, Hanski & Saccheri 2006, Giron et al. 2007). Lepidopteran larvae feeding on diets with high concentrations of plant toxins have also been reported to suffer higher rates of parasitism due potentially to the energetic costs of detoxification reducing resources available for immune defense (Gentry & Dyer 2002). In contrast, other studies have found little or no evidence that processing of allelochemicals imposes significant energy demands on insect herbivores, making it unclear whether increased vulnerability to parasitoids is actually due to metabolic trade-offs (Appel & Martin 1992).

6.6 The power of a complete nutrient budget

As shown above, nutrient acquisition and allocation clearly impact on fecundity and survival functions such as immune defense in both parasitoids and hosts. Yet, these linkages
are also founded more on logical deductions than hard data, indicating that this field of
study would strongly benefit from additional experimental data. The deficiencies in available empirical data in specific host–parasitoid study systems is also well illustrated in the
theoretical literature by the types of dynamic models that have been developed to predict
the behavior and host usage patterns of parasitoids. These models are often very detailed
and strive for realism. However, resolving differences in the predictions generated by
different models has also been noted to be 'essentially impossible', because of a lack of
physiological knowledge about nutrient dynamics in parasitoids (Clark & Mangel 2000).
Similarly, while several lines of evidence suggest immune defense is energetically expensive to hosts, the precise trade-offs between defense, reproduction, and other fitness traits
remain largely undefined.

To address these challenges, we conclude that quantified energy and nutrient budgets
for both parasitoids and hosts are the only way to make sense of the myriad of evolutionary scenarios that can arise. The key advantage to this approach is that it enables
us to assess the relative benefits and costs of different nutrient acquisition and allocation
strategies. Building complete budgets is not an easy task but it is doable, even in small
parasitoids. This is illustrated by the comprehensive total energy budget recently developed for the idiobiont *Eupelmus vulleti* that we discussed earlier in this chapter (Casas
et al. 2005, see above). For this analysis, the sugar, glycogen, protein, and lipid reserves of
single females at birth and death were quantified, as was daily maintenance. Each host
feeding and oviposition event, along with the nutrient amounts acquired and invested in
eggs, was recorded. The time of death was also used to compare model predictions in the
presence and absence of hosts. In the absence of host feeding, nutrients derive from
larval (capital) reserves. The availability of host feeding, on the other hand, delivers large
quantities of sugars and proteins but very little lipid (Fig. 6.1). In terms of allocation,
carbohydrates are the main energy source for maintenance functions and the decline in
carbohydrate reserves mirrors the time of death. When adults can host feed but have no
supplemental sugar source, proteins and lipids are heavily used for maintenance, which
in turn allows females to use capital lipid reserves acquired during larval feeding for egg
production (Fig. 6.2). Providing adults both sugar and hosts increases longevity, reduces
host feeding, and produces higher realized fecundities because most capital lipid reserves
are used for egg production (Fig. 6.2).

Overall, these data indicate that *E. vulleti* is best described as a capital breeder for one
nutrient class (lipids) and is an income breeder for another (carbohydrates). The nutrient
budget developed for this species also illustrates the inadequacy of a term such as 'energy'
for describing nutrient reserves and the importance of recognizing that different nutrient
classes are not used equivalently for reproduction and maintenance. Even categories
such as protein, lipid, and carbohydrate are potentially too coarse, as illustrated recently
by Mondy et al. (2006), who found that *E. vulleti* is actually an income breeder for at
least one subclass of lipids (sterols), because capital sterol reserves appear sufficient to produce only 30% of the total number of eggs females normally lay when given access to hosts
and sugar.

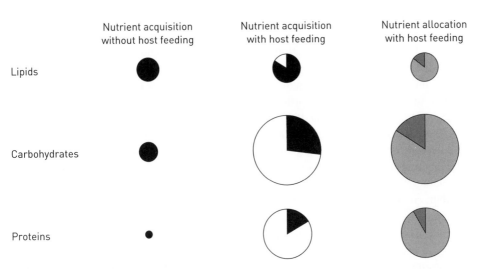

Fig. 6.1 Lifetime acquisition and allocation of nutrients in *Eupelmus vuilleti* when adults are without or with hosts for host feeding. Larval reserves of lipid, carbohydrate, and proteins are indicated in black. Nutrients gained through host feeding are indicated in white. When wasps are able to host feed, most lipid reserves still derive from the larval stage, whereas the majority of carbohydrates and proteins are acquired from host feeding. This indicates that *E. vuilleti* is a capital breeder in terms of lipids, but is an income breeder in terms of carbohydrates and protein. Allocation of lipids, carbohydrates, and proteins to maintenance functions is indicated in light gray, while allocation to eggs is indicated in dark gray.

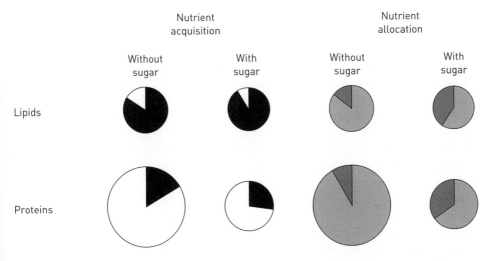

Fig. 6.2 Lifetime acquisition and allocation of nutrients in *Eupelmus vuilleti* when adults can host feed with or without supplemental sugars. Note that acquisition of additional carbohydrates during the adult stage enables females to increase the allocation of lipids and protein to egg production. Sugar acquisition also reduces host-feeding, which results in a decrease in protein reserves.

The development of complete nutrient budgets can also be used to tackle difficult questions about the relative costs and benefits of particular physiological functions and behaviors. For example, egg resorption is a common phenomenon in parasitoids and other insects that is usually explained as a strategy for recovering nutrients under conditions of resource limitation and reallocating them to maintenance (Bell & Bohm 1975). A few studies with parasitoids have reported a positive correlation between the number of eggs resorbed and extended longevity (Collier 1995, Heimpel et al. 1997). Yet, quantitative data with *E. vuilleti* indicate that single eggs contain less than 10% of the protein, lipid, and carbohydrate needed for daily maintenance, suggesting a female cannot extend its longevity even a full day by reabsorption of its entire egg complement. Thus, egg reabsorption is either of value under only the most extreme conditions of resource limitation or it is important for other, currently unrecognized, functions such as oocyte survival. For hosts, a complete nutrient budget could also provide important insights into the actual cost of immune defense against particular parasitoids and trade-offs between processing allelochemicals and defense against parasitoids. The role of different nutrient classes in the metabolic costs of the immune system is only beginning to be characterized in mammals (Lochmiller & Deerenberg 2000) and is almost completely unknown in insects.

6.7 Nutrient dynamics and the future of behavioral and population ecology

The last decade has produced a wave of exciting new data on nutrient dynamics of parasitoids and some hosts, but we are not yet at a stage where a quantitative fitness gain can be ascribed to a single nutrient acquisition or allocation decision. Nonetheless, such data are now possible to generate and are likely to enter the literature in the near future. It is at the population level that the state of the art is more worrisome. Demanding quantitative tools on one hand and a general decrease in the number of researchers studying population dynamics on the other is reducing progress in this area, despite its broadly acknowledged importance to the study of parasitoid–host relationships. Increased emphasis on physiological realism has pervaded host–parasitoid population dynamics models during the last decade but predictions about the stability of these interactions still require details on nutrient acquisition and allocation that are not available in most study systems. For example, recent models predict that allocation of nutrients gained from host feeding to both maintenance and reproduction will result in destabilizing population dynamics (Kidd & Jervis 1991a,b, Briggs et al. 1995, Křivan 1997). Models also predict destabilizing dynamics if eggs are reabsorbed for maintenance purposes (Briggs et al. 1995). In contrast, if the parasitoid death rate is a function of nutrient reserves, then a failure to meet maintenance requirements is predicted to result in stabilizing dynamics. These models, therefore, predict that parasitoid–host systems can shift from population stability to instability depending on how nutrients acquired by host feeding are allocated and the relationship between death rate and reserves. To date, the physiologically structured host–parasitoid population models of Murdoch et al. (2003) provide the most thorough groundwork on the interface between physiology and population dynamics. Unfortunately, parasitoids, like *Aphytis* sp., for which we have a good grasp of population dynamics, are poorly understood from the perspective of nutritional physiology. In other words, inadequate detail about the nutritional physiology of these wasps precludes different model predictions from being thoroughly tested. In addition, while resolving questions of stability and instability are the bred and butter of population dynamicists, these issues are usually of little interest to physiologists and molecular biologists. The challenge then is how to engage these sub-organismal biologists, who can generate quantitative measures of specific nutrient classes and their trafficking, on the importance of producing data that could help distinguish between competing model predictions that lead to opposite dynamic outcomes. This challenge also holds in the study of host immune resistance and parasitoid virulence, where understanding of how variation in non-genetic factors, like nutritional state or the coevolutionary dynamics that maintain genetic variation requires details on the physiological processes involved.

Behavioral and population ecologists would argue that the best route to deciding which physiological processes are most relevant to the foraging and population biology of a parasitoid would be to work in a top-down fashion using dynamic programming models for behavioral decisions or physiologically structured models for population dynamics. Model exploration and sensitivity analysis enable one to identify the relevant processes, i.e. those for which a small change in parameter values or function has a major impact. This can then be followed by experimental studies that generate the necessary physiological data for testing model predictions. Physiologists and molecular biologists, on the other hand, would point to the power and value of bottom-up approaches of characterizing genetic mutants, conducting selection studies, and undertaking functional genomic analyses to identify genes of interest and to understand how variation in specific traits impacts on function. The truth is that both strategies have value in enhancing our understanding of parasitoid–host interactions. This chapter even provides evidence that physiologists/molecular biologists and ecologists/population biologists are capable of working together toward a common goal. We hope that the next decade will experience a move from plausible logical arguments about fitness gains under different ecological scenarios to quantitative predictions matched by experimental data of equivalent precision.

Acknowledgments

We thank Éric Wajnberg and three anonymous referees for valuable suggestions on an earlier draft of this manuscript. Some aspects of the work discussed in this chapter were supported by grants from the National Science Foundation, US Department of Agriculture, and National Institutes of Health to Michael R. Strand.

References

Agaisse, H. and Perrimon, N. (2004) The role of JAK/STAT signalling in the *Drosophila* immune response. *Immunology Reviews* **198**: 72–82.

Andersen, S.K., Gjedsted, J., Christiansen, C. and Tonnesen, E. (2004) The roles of insulin and hyperglycemia in sepsis pathogenesis. *Journal of Leukocyte Biology* **75**: 413–21.

Appel, H.M. and Martin, M.M. (1992) Significance of metabolic load in the evolution of host specificity of *Manduca sexta*. *Ecology* **73**: 216–28.

Askew, R.R. and Shaw, M.R. (1986) Parasitoid communities. Their size, structure and development. In: Waage, J.K. and Greathead, D. (eds.) *Insect Parasitoids*. Academic Press, London, pp. 225–64.

Beckage, N.E. and Gelman, D.B. (2004) Wasp parasitoid disruption of host development: implications for new biologically based strategies for insect control. *Annual Review of Entomology* **49**: 299–330.

Bell, W.J. and Bohm, M.K. (1975) Oosorption in insects. *Biological Review* **50**: 373–96.

Boulétreau, M. (1986) The genetic and coevolutionary interactions between parasitoids and their hosts. In: Waage, J.K. and Greathead, D. (eds.) *Insect Parasitoids*. Academic Press, London, pp. 169–200.

Briggs, C.L., Nisbet, R.M., Murdoch, W.W., Collier, T.R. and Metz, J.A.J. (1995) Dynamical effects of host-feeding in parasitoids. *Journal of Animal Ecology* **64**: 403–16.

Britton, J.S., Lockwood, W.K., Li, L., Cohen, S.M. and Edgar, B.A. (2002) *Drosophila*'s insulin/PI3-kinase pathway coordinates cellular metabolism with nutritional conditions. *Developmental Cell* **2**: 239–49.

Candy, D.J., Becker, A. and Wegener, G. (1997) Coordination and integration of metabolism in insect flight. *Comparative Biochemistry & Physiology B* **117**: 497–512.

Carton, Y., Nappi, A. and Poirié, M. (2005) Genetics of anti-parasite resistance in invertebrates. *Developmental & Comparative Immunology* **29**: 9–32.

Casas, J., Driessen, G., Mandon, N. et al. (2003) Strategies of energy acquisition and use of a parasitoid in the wild. *Journal of Animal Ecology* **69**: 691–7.

Casas, J., Pincebourde, S., Mandon, N., Vannier, F., Poujol, R. and Giron, D. (2005) Lifetime nutrient dynamics reveal simultaneous capital and income breeding in a parasitoid. *Ecology* **86**: 545–54.

Chan, M.S. and Godfray, H.C.J. (1993) Host-feeding strategies of parasitoid wasps. *Evolution* **7**: 593–604.

Charnov, E.L. and Skinner, S.W. (1984) Evolution of host selection and clutch size in parasitoid wasps. *Florida Entomologist* **67**: 5–21.

Chesson, P.L. and Murdoch, W.W. (1986) Aggregation of risk: relationships among host–parasitoid models. *American Naturalist* **127**: 696–715.

Clark, C.W. and Mangel, M. (2000) *Dynamic State Variable Models in Ecology*. Oxford University Press, New York.

Clements, A.N. (1992) *The Biology of Mosquitoes*, vol. I. Chapman & Hall, London.

Collier, T.R. (1995) Adding physiological realism to dynamic state variable models of parasitoid host-feeding. *Evolutionary Ecology* **9**: 217–35.

Dahlman, D.A. and Vinson, S.B. (1980) Glycogen content in *Heliothis virescens* parasitized by *Microplitis croceipes*. *Comparative Biochemistry & Physiology A* **66**: 625–30.

Digilio, M.C., Isidoro, N., Tremblay, E. and Pennacchio, F. (2000) Host castration by *Aphidius ervi* venom proteins. *Journal of Insect Physiology* **46**: 1041–50.

Ellers, J. (1996) Fat and eggs: an alternative method to measure the trade-off between survival and reproduction in insect parasitoids. *Netherlands Journal of Zoology* **46**: 227–35.

Faggioni, R., Moser, A., Feingold, K.R. and Grunfeld, C. (2000) Reduced leptin levels in starvation increase susceptibility to toxic shock. *American Journal of Pathology* **156**: 1781–7.

Feder, D., Mello, C.B., Garcia, E.S. and Azambuja, P. (1997) Immune responses in *Rhodnius prolixus*: influence of nutrition and ecdysone. *Journal of Insect Physiology* **43**: 513–19.

Fellowes, M.D.E. (1999) The relative fitness of *Drosophila melanogaster* (Diptera: Drosophilidae) that have successfully defended themselves against the parasitoid *Asobara tabida* (Hymenoptera: Braconidae). *Journal of Evolutionary Biology* **12**: 123–8.

Fellowes, M.D.E., Kraaijeveld, A.R. and Godfray, H.C.J. (1999) Cross-resistance following artificial selection for increased defence against parasitoids in *Drosophila melanogaster*. *Evolution* **53**: 966–72.

Flanders, S.E. (1950) Regulation of ovulation and egg disposal in the parasitic Hymenoptera. *Canadian Entomologist* **82**: 134–40.

Gentry, G.L. and Dyer, L.A. (2002) On the conditional nature of neotropical caterpillar defenses against their natural enemies. *Ecology* **83**: 3108–19.

Giron, D. and Casas, J. (2003) Lipogenesis in an adult parasitic wasp. *Journal of Insect Physiology* **49**: 141–7.

Giron, D., Rivero, A., Mandon, N., Darrouzet, E. and Casas, J. (2002) The physiology of host-feeding in parasitic wasps: implications for survival. *Functional Ecology* **16**: 750–7.

Giron, D., Ross, K.G. and Strand, M.R. (2007) Presence of soldier larvae determines the outcome of competition in a polyembryonic wasp. *Journal of Evolutionary Biology* **20**: 165–72.

Godfray, H.C.J. (1994) *Parasitoids: Behavioral and Evolutionary Ecology*. Princeton University Press, Princeton.

Godfray, H.C.J. (2000) Host resistance, parasitoid virulence, and population dynamics. In: Hochberg, M.E. and Ives, A.R. (eds.) *Parasitoid Population Biology*. Princeton University Press, Princeton, pp. 121–38.

Gordon, M.D., Dionne, M.S., Schneider, D.S. and Nusse, R. (2005) WntD is a feedback inhibitor of Dorsal/NF-kappaB in *Drosophila* development and immunity. *Nature* **437**: 746–9.

Gwynn, D.M., Callaghan, A., Gorham, J., Walters, K.F.A. and Fellowes, M.D.E. (2005) Resistance is costly: trade-offs between immunity, fecundity and survival in the pea aphid. *Proceedings of the Royal Society of London Series B Biological Science* **272**: 1803–8.

Hanski, I. and Saccheri, I. (2006) Molecular-level variation affects population growth in a butterfly metapopulation. *Plos Biology* **4**: 719–26.

Harrison, J.F. and Fewell, J.H. (2002) Environmental and genetic influences on flight metabolic rate in the honey bee, *Apis mellifera*. *Comparative Biochemistry & Physiology A* **133**: 255–8.

Harvey, J.A. and Strand, M.R. (2002) The developmental strategies of endoparasitoid wasps vary with host feeding ecology. *Ecology* **83**: 2439–51.

Heidel, A.J., Clarke, J.D., Antonovics, J. and Dong, X. (2004) Fitness costs of mutations affecting the systemic acquired resistance pathway in *Arabidopsis thaliana*. *Genetics* **168**: 2197–206.

Heimpel, G.E. and Collier, T.R. (1996) The evolution of host-feeding behaviour in insect parasitoids. *Biological Reviews* **71**: 373–400.

Heimpel, G.E., Rosenheim, J.A. and Kattari, D. (1997) Adult feeding and lifetime reproductive success in the parasitoid *Aphytis melinus*. *Entomologia Experimentalis et Applicata* **83**: 305–15.

Heimpel, G.E., Mangel, M. and Rosenheim, J.A. (1998) Effects of time limitation and egg limitation on lifetime reproductive success of a parasitoid in the field. *American Naturalist* **152**: 273–89.

Henter, H. and Via, S. (1995) The potential for coevolution in a host–parasitoid system: 1. Genetic-variation within an aphid population in susceptibility to a parasitic wasp. *Evolution* **49**: 427–38.

Hillyer, J.F., Schmidt, S.L., Fuchs, J.F., Boyle, J.P. and Christensen, B.M. (2005) Age-associated mortality in immune challenged mosquitoes (*Aedes aegypti*) correlates with a decrease in haemocyte numbers. *Cellular Microbiology* **7**: 39–51.

Hochberg, M.E. (1997) Hide or flight? The competitive evolution of concealment and encapsulation in host–parasitoid associations. *Oikos* **80**: 342–52.

Hoffmann, J.A. (2003) The immune response of *Drosophila*. *Nature* **426**: 33–8.

Hufbauer, R.A. (2002) Evidence for nonadaptive evolution of parasitoid virulence following a biological control introduction. *Ecological Applications* **12**: 66–78.

Irving, P., Ubeda, J.-M., Doucet, D. et al. (2005) New insights into *Drosophila* larval haemocyte function through genome-wide analysis. *Cellular Microbiology* **7**: 335–50.

Jervis, M.A. and Kidd, N.A.C. (1986) Host-feeding strategies in hymenopteran parasitoids. *Biological Reviews* **61**: 395–434.

Jervis, M.A., Heimpel, G.E., Ferns, P.N., Harvey, J.A. and Kidd, N.A.C. (2001) Life-history strategies in parasitoid wasps: a comparative analysis of 'ovigeny'. *Journal of Animal Ecology* **70**: 442–58.

Kidd, N.A.C. and Jervis, M.A. (1991a) Host-feeding and oviposition strategies of parasitoids in relation to host stage. *Researches in Population Ecology* **33**: 13–28.

Kidd, N.A.C. and Jervis, M.A. (1991b) Host-feeding and oviposition by parasitoids in relation to host stage: consequences for parasitoid–host population dynamics. *Researches in Population Ecology* **33**: 87–9.

Kraaijeveld, A.R. and Godfray, H.C.J. (1997) Trade-off between parasitoid resistance and larval competitive ability in *Drosophila melanogaster*. *Nature* **389**: 278–80.

Kraaijeveld, A.R. and Godfray, H.C.J. (1999) Geographical patterns in the evolution of resistance and virulence in *Drosophila* and its parasitoids. *American Naturalist* **153**: 61–74.

Kraaijeveld, A.R., van Alphen, J.J. and Godfray, H.C.J. (1998) The coevolution of host resistance and parasitoid virulence. *Parasitology* **116**: 29–45.

Křivan, V. (1997) Dynamical consequences of optimal host feeding on host–parasitoid population dynamics. *Bulletin of Mathematical Biology* **59**: 809–31.

Lavine, M.D. and Strand, M.R. (2002) Insect hemocytes and their role in cellular immune responses. *Insect Biochemistry & Molecular Biology* **32**: 1237–42.

Lochmiller, R.L. and Deerenberg, C. (2000) Trade-offs in evolutionary immunology: just what is the cost of immunity? *Oikos* **88**: 87–98.

Mackauer, M. and Sequeira, R. (1993) Patterns of development in insect parasites. In: Beckage, N.E., Thompson, S.N. and Federici, B.A. (eds.) *Parasites and Pathogens of Insects*. Academic Press, New York, pp. 1–23.

Mangel, M. (1989) Evolution of host selection in parasitoids: does the state of the parasitoid matter? *American Naturalist* **133**: 688–703.

Matarese, G. and La Cava, A. (2004) The intricate interface between the immune system and metabolism. *Trends in Immunology* **25**: 193–200.

Mayhew, P.J. and Blackburn, T.M. (1999) Does development mode organize life-history traits in the parasitoid Hymenoptera? *Journal of Animal Ecology* **68**: 906–19.

Min, K.-J., Hogan, M.F., Tatar, M. and O'Brian, D.M. (2006) Resource allocation to reproduction and soma in *Drosophila*: a stable isotope analysis of carbon from dietary sugar. *Journal of Insect Physiology* **52**: 763–70.

Mondy, N., Corio-Costet, M.F., Bodin, A., Mandon, N., Vannier, F. and Monge, J.P. (2006) Importance of sterols acquired through host-feeding in synovigenic parasitoid oogenesis. *Journal of Insect Physiology* **52**: 897–904.

Murdoch, W.W., Briggs, C.L. and Nisbet, R.M. (2003) *Consumer–Resource Dynamics*. Princeton University Press, Princeton.

O'Brien, D.M., Fogel, M.L. and Boggs, C.L. (2002) Renewable and nonrenewable resources: amino acid turnover and allocation to reproduction in Lepidoptera. *Proceedings of the National Academy of Sciences USA* **99**: 4413–18.

O'Brien, D.M., Boggs, C.L. and Fogel, M.L. (2004) Making eggs from nectar: the role of life history and dietary carbon turnover in butterfly reproductive resource allocation. *Oikos* **105**: 279–91.

Oldham, S., Montagne, J., Radimerski, T., Thomas, G. and Hafen, E. (2000) Genetic and biochemical characterization of dTOR, the *Drosophila* homolog of the target of rapamycin. *Genes & Development* **18**: 2689–94.

Olson, D.M., Fadamiro, H., Lundgren, J.G. and Heimpel, G.E. (2000) Effects of sugar feeding on carbohydrate and lipid metabolism in a parasitoid wasp. *Physiological Entomology* **25**: 17–26.

Papaj, D.R. (2000) Ovarian dynamics and host use. *Annual Review of Entomology* **45**: 423–48.

Pennacchio, F. and Strand, M.R. (2006) Evolution of developmental strategies in parasitic Hymenoptera. *Annual Review of Entomology* **51**: 233–58.

Pennacchio, F., Digilio, M.C. and Tremblay, E. (1995) Biochemical and metabolic alterations in *Acyrthosiphon pisum* parasitized by *Aphidius ervi*. *Archives of Insect Biochemistry & Physiology* **30**: 351–67.

Price, P.W. (1974) Strategies for egg production. *Evolution* **28**: 76–84.

Rahbé, Y., Digilio, M.C., Febvay, G., Guillaud, J., Fanti, P. and Pennacchio, F. (2002) Metabolic and symbiotic interactions in amino acid pools of the pea aphid, *Acyrthosiphon pisum*, parasitized by the braconid *Aphidius ervi*. *Journal of Insect Physiology* **48**: 507–16.

Raubenheimer, D. and Simpson, S.J. (1999) Integrating nutrition: a geometrical approach. *Entomologia Experimentalis et Applicata* **91**: 67–82.

Rivero, A. and Casas, J. (1999) Incorporating physiology into parasitoid behavioural ecology: the allocation of nutritional resources. *Researches in Population Ecology* **41**: 39–45.

Rivero, A. and West, S.A. (2002) The physiological costs of being small in a parasitic wasp. *Evolutionary Ecology Research* **4**: 407–20.

Rivero, A., Giron, D. and Casas, J. (2001) Lifetime allocation of juvenile and adult nutritional resources to egg production in a holometabolous insect. *Proceedings of the Royal Society Series B Biological Science* **268**: 1231–7.

Rivers, D.B. and Denlinger, D.L. (1994) Redirection of metabolism in the flesh fly, *Sarcophaga bullata*, following envenomation by the ectoparasitoid, *Nasonia vitripennis*, and correlation of metabolic effects with the diapause status of the host. *Journal of Insect Physiology* **40**: 207–15.

Rolff, J. and Siva-Jothy, M.T. (2002) Copulation corrupts immunity: a mechanism for the cost of mating in insects. *Proceedings of the National Academy of Sciences USA* **99**: 9916–18.

Rolff, J. and Siva-Jothy, M.T. (2003) Invertebrate ecological immunity. *Science* **301**: 472–5.

Rosenheim, J.A., Heimpel, G.E. and Mangel, M. (2000) Egg maturation, egg resorption and the costliness of transient egg limitation in insects. *Proceedings of the Royal Society of London Series B Biological Science* **267**: 1565–73.

Sanders, A.E., Scarborough, C., Layen, S.J., Kraaijeveld, A.R. and Godfray, H.C.J. (2005) Evolutionary change in parasitoid resistance under crowded conditions in *Drosophila melanogaster*. *Evolution* **59**: 1292–9.

Salt, G. (1970) *The Cellular Defence Reactions of Insects*. Cambridge University Press, Cambridge.

Sasaki, A. and Godfray, H.C.J. (1999) A model for the coevolution of resistance and virulence in coupled host–parasitoid interactions. *Proceedings of the Royal Society of London Series B Biological Science* **266**: 455–63.

Schmid-Hempel, P. and Ebert, D. (2003) On the evolutionary ecology of specific immune defence. *Trends in Ecology & Evolution* **18**: 27–32.

Scott, R.C., Schuldiner, O. and Neufeld, T.P. (2004) Role and regulation of starvation-induced autophagy in the *Drosophila* fat body. *Developmental Cell* **7**: 167–78.

Shelby, K.S. and Webb, B.A. (1994) Polydnavirus infection inhibits synthesis of an insect plasma protein, arylphorin. *Journal of General Virology* **75**: 2285–94.

Stacey, D.A. and Fellowes, M.D.E. (2002) Influence of temperature on pea aphid *Acyrthosiphon pisum* (Hemiptera: Aphididae) resistance to natural enemy attack. *Bulletin of Entomological Research* **92**: 351–7.

Strand, M.R. (1986) The physiological interactions of parasitoids with their hosts and their influence on reproductive strategies. In: Waage, J. and Greathead, D. (eds.) *Insect Parasitoids*. Academic Press, London, pp. 97–136.

Strand, M.R. (2000) Life history variation and developmental constraints in parasitoids. In: Hochberg, M. and Ives, A.R. (eds.) *Population Biology of Parasitoids*. Princeton University Press, Princeton, pp. 139–62.

Strand, M.R. and Pech, L.L. (1995) Immunological compatibility in parasitoid–host relationships. *Annual Review of Entomology* **40**: 31–56.

Thompson, S.N. (1982) Effects of parasitization by the insect parasite *Hyposoter exiguae* on growth, development, and physiology of its host *Trichoplusia ni*. *Parasitology* **84**: 491–510.

Thompson, S.N. (1993) Redirection of host metabolism and effects on parasite nutrition. In: Beckage, N.E., Thompson, S.N. and Federici, B.A. (eds.) *Parasites and Pathogens of Insects*, vol. 1. Academic Press, New York, pp. 125–44.

Thompson, S.N. and Dahlman, D.L. (1998) Aberrant nutritional regulation of carbohydrate synthesis by parasitized *Manduca sexta*. *Journal of Insect Physiology* **44**: 745–54.

Vogt, J.T., Appel, A.G. and West, M.S. (2000) Flight energetics and dispersal capability of the fire ant, *Solenopsis invicta*. *Journal of Insect Physiology* **46**: 697–707.

Vinson, S.B. (1990) Physiological interactions between the host genus *Heliothis* and its guild of parasitoids. *Archives of Insect Biochemistry & Physiology* **13**: 63–81.

Washburn, J.O., Wong, J.F. and Volkman, L.E. (2001) Comparative pathogenesis of *Helicoverpa zea* S nucleopolyhedrovirus in noctuid larvae. *Journal of General Virology* **82**: 1777–84.

Watt, W.B. (1992) Eggs, enzymes, and evolution: natural genetic variants change insect fecundity. *Proceedings of the National Academy of Sciences USA* **89**: 10608–12.

Weaver, R.J., Marris, G.C., Bell, H.A. and Edwards, J.P. (2001) Identity and mode of action of the host endocrine disrupters from the venom of parasitoid wasps. In: Edwards, J.P. and Weaver, R.J. (eds.) *Endocrine Interactions of Insect Parasites and Pathogens.* BIOS Scientific Publishers, Oxford, pp. 33–58.

Wertheim, B., Kraaijeveld, A.R., Schuster, E. et al. (2005) Genome wide expression in response to parasitoid attack in *Drosophila*. *Genome Biology* **6**: R94.1–20.

Wu, Q. and Brown, M.R. (2006) Signaling and function of insulin-like peptides in insects. *Annual Review of Entomology* **51**: 1–24.

Yang, X. and Cox-Foster, D.L. (2005) Impact of an ectoparasite on the immunity and pathology of an invertebrate: evidence for host immunosuppression and viral amplification. *Proceedings of the National Academy of Sciences USA* **102**: 7470–5.

Zera, A.J. and Zhao, Z. (2003) Life-history evolution and the microevolution of intermediary metabolism: activities of lipid-metabolizing enzymes in history morphs of a wing-dimophic cricket. *Evolution* **57**: 586–96.

Zerofsky, M., Harel, E., Silverman, N. and Tatar, M. (2005) Aging of the innate immune response in *Drosophila melanogaster*. *Aging Cell* **4**: 103–8.

Zhang, H., Stallock, J.P., Ng, J.C., Reinhard, C. and Neufeld, T.P. (2000) Regulation of cellular growth by the *Drosophila* target of rapamycin dTOR. *Genes & Development* **14**: 2112–724.

7

Food-searching in parasitoids: the dilemma of choosing between 'immediate' or future fitness gains

Carlos Bernstein and Mark Jervis

Abstract

Many parasitoid wasps and flies feed habitually on sugar-rich foods, consuming nectar and/or homopteran honeydew in the wild, and substitutes such as diluted honey in the laboratory. Consumption of these 'non-host foods' generally results in increases in female life expectancy, dispersal capacity, and realized fecundity. Typically in the wild, hosts and food sources are spatially separate. A female parasitoid therefore faces a choice as to which of the two types of resource she should forage for. Foraging for hosts will increase the likelihood of her obtaining 'immediate' fitness gains, but it will decrease her life expectancy. Foraging for food will postpone egg-laying to a later stage in adult life (constituting an 'immediate' reproductive opportunity cost), and will also incur energetic and mortality costs. However, it will increase opportunities to locate hosts in the future. From the standpoint of the fitness consequences of parasitoid foraging decisions, the exploitation of non-host foods is thus a highly interesting topic, posing the question of how females should forage so as to optimize trade-offs with respect to the use of both time and metabolic resources, and also mortality risks. In order to understand how such optimization might be achieved, we review current knowledge regarding resource 'income' and 'capital' used by females of parasitoid species that feed solely on non-host materials. Host-foraging versus non-host food-foraging, in such insects, has previously been modeled by means of stochastic dynamic programming. Using this technique, we model the behavioral scenario in which hosts and food patches are concurrently available but spatially separate. We focus initially on the egg maturation strategy of pro-ovigeny (in which all eggs are mature upon female emergence), but then examine the alternative and more common strategy of synovigeny (some eggs are immature). We discuss briefly the implications of variation in food availability for host population dynamics, and relate these to biological control scenarios. Finally, we identify promising avenues of future research.

7.1 Introduction

Feeding by the adult, particularly the female, is common among parasitoid wasps and flies (Jervis & Kidd 1986, Jervis et al. 1993, Heimpel & Collier 1996, Gilbert & Jervis 1998, Jervis 1998, Jervis et al. 2004, Heimpel & Jervis 2005). Generally, parasitoid biologists are well aware of this, and also of the large body of evidence that parasitoid adult survival and lifetime realized fecundity are usually significantly constrained if adults are deprived of food. Therefore, when parasitoids are either reared, mass-cultured, or used in laboratory experiments, their adults are routinely provided with sugar-rich foods, often as diluted honey or a sucrose solution (Waage et al. 1985, Jervis et al. 2005b).

Feeding by adult parasitoids was described by several pioneers of parasitoid biology and biological control (Clark 1901 (cited in Froggatt 1902), Doten 1911, Johnston 1915, d'Emmerez de Charmoy 1917, Illingworth 1921, Box 1927, Fulton 1933, Ahmad 1936), and information regarding the effects of feeding on the parasitoids' key life-history variables, such as fecundity and longevity, have accumulated into a now-massive literature. However, the fitness implications of food-searching, and the decision choices faced by females, were unexplored until relatively recently (Jervis & Kidd 1986, Kidd & Jervis 1991a, Sirot & Bernstein 1996, Křivan & Sirot 1997, Desouhant et al. 2005, Tenhumberg et al. 2006). Insect behavioral ecologists started focusing on them for several reasons: First, for the majority of parasitoid species, the acts of oviposition and feeding do not occur simultaneously (see 'Decisions', Section 7.9), and so feeding constitutes a missed opportunity in terms of 'immediate' fitness gain – if, that is, the female has any eggs to lay (Heimpel & Collier 1996). However, the female, by choosing to feed, stands to gain future oviposition opportunities, due to the potentially positive effects the dietary nutrients have on her life expectancy and also (if she is synovigenic, i.e. matures eggs during her life) on her rate of egg manufacture (e.g. host-feeding parasitoids) (Jervis & Kidd 1986, Heimpel & Collier 1996).

Second, the majority of parasitoids exploit sugar-rich materials such as nectar and honeydew and, for such insects, food and hosts usually occur in separate patches (Jervis et al. 1993, Heimpel & Jervis 2005). Therefore, the female's overall foraging strategy will be partly moulded by the costs (in terms of time and energy) and mortality risks associated with searching for food (Sirot & Bernstein 1996, Tenhumberg et al. 2006). Third, it was realized that a parasitoid's foraging decisions are likely to be state-dependent, the female's particular choice of behavior being contingent upon various physiological variables, among which nutritional state – which itself depends on the parasitoids' previous oviposition and feeding behavior – is likely to figure prominently together with egg load (Iwasa et al. 1984, Jervis & Kidd 1986, Mangel 1987, Mangel & Clark 1988, Jervis & Kidd 1991, Chan & Godfray 1993, Heimpel & Collier 1996, Sirot & Bernstein 1996, Jervis & Kidd 1999).

Sirot and Bernstein (1996) were the first to model optimal time allocation in parasitoid wasps that forage for both hosts and food and, using a stochastic dynamic programming (SDP) approach (see also Chapter 15 by Roitberg and Bernhard), they took account both of the state-dependence of behavior and of the differences that can occur in the availability of food. They sought to calculate the fitness pay-off from searching for hosts or food by assuming there to be a host patch and a food patch. In their model, movement between the patches incurred metabolic costs, as did time spent on either patch. As a first approximation, in this early modeling work, the authors made some simplifying assumptions:

on the host patch, females parasitized hosts at a given rate, and on the food patch they first searched for food, but having found some, they consumed it at a given rate. Sirot and Bernstein (1996) found that, by varying the time spent searching for food, parasitoids modify their expected lifespan, and therefore their lifetime reproductive success. Further, it was predicted that, under conditions of high food availability, parasitoids should delay searching for food until reserves become low. However, when food availability is moderate, parasitoids should not delay, and when food availability is low, parasitoids should never search for food.

By assuming constant rates of feeding or oviposition, Sirot and Bernstein's (1996) model took no account of the more realistic scenario of short-term variation in these rates. Tenhumberg et al. (2006) therefore constructed an SDP model that was similar to Sirot and Bernstein's (1996), but they sought to explicitly incorporate such variability, together with a probabilistic distribution of food rewards and a high rate of energy expenditure of searching for either hosts or food. In contradiction to Sirot and Bernstein's (1996) model, Tenhumberg et al.'s (2006) model predicted that parasitoids should always search for food if energy reserves drop to low levels, even if there is a low probability of finding food and also if the average food reward is low. Tenhumberg et al. (2006) attributed the mismatch between the predictions of the two models to differences in assumptions regarding biological processes.

Both of the aforementioned food-foraging models assumed that parasitoid fitness is determined by food availability and life expectancy, and also that the parasitoids are pro-ovigenic (i.e. all of the female's eggs are mature at adult emergence) as opposed to synovigenic (i.e. some or all of her eggs are matured following emergence). No account was taken of egg load, and so the female parasitoids were considered to be potentially subject only to time-limitation. However, for models of parasitoid food- and host-searching to be realistic, they need to take account of variation in egg load, and to do so with respect to synovigeny as well as pro-ovigeny. The reasons for this are twofold. First, egg-limitation is a significant constraint upon lifetime reproductive success in synovigenic parasitoids (Weisser et al. 1997, Ellers et al. 1998, Heimpel & Rosenheim 1998, Heimpel et al. 1998, Casas et al. 2000, Ellers et al. 2000). In fact, as shown by Ellers et al. (2000), both time- and egg-limitation can constrain the fitness of either reproductive type – synovigenic as well as pro-ovigenic. Second, synovigenic species comprise the vast majority of parasitoid wasps (Jervis et al. 2001). The case could even be made for ignoring pro-ovigenic parasitoids altogether, since SDP has shown the egg maturation strategy of pro-ovigeny to be optimal for a uniform spatial distribution of hosts (Ellers & Jervis 2004) – a rare scenario in nature. Ellers and Jervis (2003) present a different modeling-based prediction regarding the selective advantage of a high ovigeny index (defined below), taking into account another region of parameter space. However, if we assume there to be as many as two million extant parasitoid species (Godfray 1994), pro-ovigenic parasitoids are, in terms of absolute numbers of species, a significant functional group, perhaps amounting to tens of thousands of species, so cannot be regarded as being insignificant.

An assumed corollary of synovigenic insects emerging with some immature eggs is that females are dependent upon nutrient reserves for directly fueling egg manufacture. Several models of the food- and host-foraging behavior of synovigenic parasitoids that host-feed have assumed both egg manufacture and somatic functions to be fueled from a single 'resource pool', which can be added to and/or replenished by feeding (Jervis & Kidd 1986, Chan & Godfray 1993, Heimpel et al. 1998, Rosenheim et al. 2000). Although there

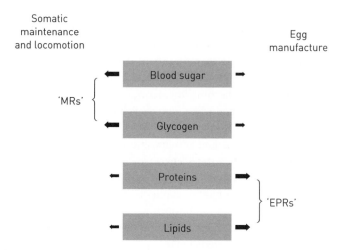

Fig. 7.1 A proposed scenario for the utilization of nutrient resources by non-host-feeding parasitoid wasps. Contributions come from both the adult's diet (i.e. 'income') and from pre-adult life (i.e. there is a carried-over, teneral component – the insect's 'capital'). The width of the arrows indicates the presumed importance of a particular biochemical resource to a particular physiological function. Based on the available evidence, non-host-feeding parasitoids use blood sugar and glycogen mainly for fueling somatic maintenance and locomotion (flight, walking) – and so we refer to these collectively as 'metabolic resources' (MRs), but they use proteins and lipids mainly for fueling egg manufacture, and so we refer to these collectively as 'Egg Production Resources' (EPRs).

is empirical support for the predictions of some of these models regarding the pattern of host density-related changes in the ratio of feeding attacks to oviposition attacks, the critical egg load at which feeding should occur, and also the effect of female nutritional status on the decision to feed (Heimpel & Collier 1996, Jervis & Kidd 1999), the 'resource pool' assumption is likely to be unrealistic, as will become clear from our review of nutrient allocation and utilization, in this chapter. Synovigenic parasitoids draw upon a diverse set of metabolic resources (MRs) that are used to fuel maintenance and ovigenesis, and females utilize these resources differentially. That is, some MRs mainly fuel somatic functions (maintenance and locomotion) whereas others mainly fuel egg manufacture (see Fig. 7.1 for one possible scenario). However, females are expected to alter their pattern of utilization during their lifetime, when fuel for somatic functions becomes limiting, i.e. food availability is low and/or food quality is poor.

Thus, from the standpoint of the fitness consequences of foraging decisions, the exploitation of non-host foods by parasitoids is a highly intriguing subject for investigation, posing the question of how females should forage so as to optimize trade-offs with respect to the use of time and MRs, and also with regard to mortality risks. In order to understand how they achieve this, we first review current knowledge regarding the 'income' and the 'capital' used by females, confining ourselves to species that feed solely on non-host materials, i.e. their females do not practise host-feeding in addition to sugar-feeding (for reviews of the foraging behavior of host-feeding parasitoids, see Jervis & Kidd 1986, Heimpel & Collier 1996, and Jervis & Kidd 1999).

In the second part of this chapter, we model the behavioral scenario in which hosts and food patches are concurrently available but are spatially separate. As with previous models of combined host-searching and non-host food searching by parasitoids (Sirot & Bernstein 1996, Tenhumberg et al. 2006), we employ SDP. Faced with the starkly contrasting predictions generated by the Sirot and Bernstein (1996) and Tenhumberg et al. (2006) models, we felt it necessary to try and discover the reasons for this mismatch – is it some aspect of biology, as was concluded by Tenhumberg et al. (2006), or is it simply a question of the parameter values used? Thus, our first modeling exercise is concerned with foraging by pro-ovigenic parasitoids. One of the key aspects we consider is the influence of the distance between the two resources, hosts and food, on the optimal strategies. We follow this by modeling the foraging behavior of parasitoids that exhibit the more prevalent egg maturation strategy of synovigeny. We then compare the differences in the strategies pro-ovigenic and synovigenic parasitoids are expected to use.

7.2 Types, acquisition, composition, and 'whereabouts' of non-host foods

7.2.1 Food types

The main non-host foods of parasitoids are floral and extrafloral nectar, and the honeydew excreted by some Homoptera. Less commonly, plant materials such as pollen, sap (including the juices of fruits), epidermis, trichomes, and (apparently) leachates are consumed (Jervis et al. 1993, Gilbert & Jervis 1998, Jervis 1998, Sisterton & Averill 2002, Jervis et al. 2005a, Sivinksi et al. 2006). Pollen-feeding is practised by some Scoliidae and Mutillidae, but is apparently extremely rare among most other parasitoid wasp taxa (Gilbert & Jervis 1998, Jervis 1998).

7.2.2 Feeding mechanisms and their adaptive significance

Parasitoids obtain the aforementioned foods by means of either: (i) a short, unspecialized, labiomaxillary complex, which the majority of parasitoid wasps use in obtaining nectar and honeydew (Jervis 1998), or (ii) diverse mouthpart specializations, each type of which is linked to the characteristics of the food most often exploited by the insect. The specializations include:

1 the fleshy proboscis with labella used by many 'short-tongued' parasitoid flies for feeding on either exposed nectar or honeydew;
2 the elongated version of the latter, employed by 'long-tongued' parasitoid flies for feeding on deep-lying (concealed) floral nectar;
3 the 'elongated hairy tongue' (glossa) used by a diverse array of parasitoid wasps for feeding on concealed nectar;
4 the 'drinking straw' mechanism employed by a few Braconidae, also for feeding on such nectar (Gilbert & Jervis 1998, Jervis 1998).

The mouthparts can be exceptionally long in some cases (Gilbert & Jervis 1998, Jervis 1998, Quicke & Jervis 2005).

Jervis (1998) presented a character transformation series for the above and other types of concealed nectar extraction apparatus in parasitoid wasps, showing how they likely evolved, either directly or indirectly, from an unspecialized condition. The fitness advantage of using a 'short-tongue' labellar apparatus appears to be that exposed sugar-rich food sources tend to be more common than concealed sources, and so exploiting them is likely to be inexpensive in terms of time and energy. Honeydew can be especially abundant in the parasitoid's habitat (Gilbert & Jervis 1998). The fitness advantage of concealed nectar extraction apparatus appears to be threefold (Gilbert & Jervis 1998). First, deep-lying nectar is usually relatively dilute and therefore of low viscosity. Consequently it requires little or no dilution with saliva, and can be extracted relatively rapidly, so forming a smaller fraction of the female's foraging time budget compared with the exploitation of exposed nectar (which tends to be highly viscous or crystalline). Second, sources of concealed nectar, by offering a much greater volume of nectar than exposed nectar sources, are usually more sugar-rich in absolute terms (Prys-Jones & Corbet 1983). Third, concealed nectar sources are a potentially rich source of water in arid environments.

Jervis (1998) and Jervis and Vilhelmsen (2000) showed how concealed nectar extraction apparatus has evolved independently many times within the order Hymenoptera, even within the 'basal', largely non-parasitoid 'Symphyta'. Jervis and Vilhelmsen (2000) concluded that, providing there exist both the necessary ecological opportunities together with the necessary selection pressures, such apparatus can evolve unconstrained among Hymenoptera. Parasitoid wasps are clearly not exempt from this process (Jervis 1998). Nevertheless, an elongated proboscis is the exception rather than the rule among parasitoid wasps. This could be because exposed nectar sources and honeydew tend, in nature, to be more abundant, and to be continuously available, compared with concealed nectar sources (Gilbert & Jervis 1998).

7.2.3 Biochemical composition, digestion, and assimilation

Honeydew deposits and exposed nectar, such as that secreted by the nectaries of umbelliferous plants (Apiaceae), eventually crystallize. In contrast, both freshly deposited honeydew and the nectar that is contained in tubular corollas have a high water content (concealed nectar is permanently liquid). The water content of the latter food sources may play an important role in shaping the food-foraging strategy of some parasitoids: it could be that females feed mainly in order to obtain the water required for the maintenance of an appropriate hygrothermal balance (Willmer 1985, 1986). It is noteworthy that many parasitoid wasps possessing elongated mouthparts, and also many 'long-tongued' Bombyliidae and Nemestrinidae, occur in, and in several instances are confined to, arid or semi-arid habitats (Gilbert & Jervis 1998, Quicke & Jervis 2005). In this chapter, however, we assume that the water component of foods has no bearing upon the foraging behavior of wasps.

Based on biochemical information in the literature, honeydew and nectar are rich in carbohydrate but poor in proteinaceous materials (Percival 1961, 1965, Barbier 1970, Wigglesworth 1972, Stanley & Linskens 1974, Maurizio 1975, Faegri & van der Pijl 1979, Baker & Baker 1983, Harborne 1988, Dafni 1992). Nectar is extremely nitrogen-poor, containing only trace amounts of amino acids (0.25–15.5 μmol/ml (Baker & Baker 1973, Dafni 1992), while the total nitrogen content of honeydew is typically less than 0.5% fresh weight and less than 2% dry weight (Lamb 1959). Both nectar and honeydew are lipid-poor (Dafni 1992). Honey, which is often supplied to parasitoids under laboratory conditions as a substitute for nectar or honeydew, closely resembles these foods in being sugar-rich but both nitrogen and lipid-poor (Crane 1975, 1980).

In attempting to derive generalizations concerning the fitness consequences of parasitoid food- and host-foraging, we focus on nectar and honeydew insofar as they are sources of one general type of metabolic (potentially both catabolic and anabolic) substrate – namely carbohydrate. However, nectar and honeydew should not be viewed merely as naturally occurring equivalents of the sucrose solutions that parasitoids are given in the laboratory – they typically contain a diverse array of sugars. Some of the sugars require digestion, pre-orally and/or post-orally, whereas others do not, and some sugars cannot be utilized at all. Although we have found it necessary to gloss over these subtleties in our models, we briefly elaborate on them below, to highlight the complexity of the nutritional ecology of parasitoids.

In nectars, the sugars principally comprise monosaccharides and disaccharides, but some longer-chain sugars are also present (Percival 1961, Wäckers 2001). However. carbohydrate transport across the gut wall is assumed to be primarily restricted to monosaccharides, as has been shown for *Diadegma semiclausum* (Wäckers et al. 2007). *Cotesia glomerata* has the enzymes, either in its saliva or in its gut lumen, for breaking down several complex (i.e. non-monosaccharide) sugars (Wäckers 2001). Such dietary sugars comprise: (i) disaccharides, e.g. sucrose, maltose and melibiose: (ii) trisaccharides, e.g. erlose and melezitose; and (iii) tetrasaccharides, e.g. stachyose.

It appears that *C. glomerata*, like the honeybee, is unable to break down dietary trehalose, a disaccharide occurring in significant quantities in honeydew. This is presumably because its gut lumen α-glucosidase is not of the type required (Wäckers 2001). Such variation among sugars, with regard to assimilation by the parasitoid, has important implications for fitness. For example, in *C. glomerata*, the greatest extension in longevity was obtained by feeding on solutions of either fructose, sucrose, or melezitose, whereas no extension was recorded in wasps that were fed on solutions of either rhamnose or trehalose (Hausmann et al. 2005).

The sugar composition of nectar varies among different plant species (Percival 1961, 1965). Nectar from different flower species has different effects on components of flight activity in *C. glomerata* (Wanner et al. 2006). Longevity and realized fecundity of *Diadegma insulare* vary with nectar source (Idris & Grafius 1995). Given these effects on life-history variables, parasitoids are expected to display floral preferences, as is indeed the case (Shahjahan 1974, Syme 1975, Wäckers 2004, Wanner et al. 2006).

7.2.4 The 'whereabouts' of foods

For some parasitoids, the degree of spatial coincidence between host and non-host foods is high, whereas, for others, it is slight or even zero. For example, some non-host-feeding aphelinids directly imbibe the honeydew that exudes from the anus of scale insects (Cendaña 1937). Whether the same scale individuals are actually oviposited in by a female aphelinid is unclear, but assuming that they are, the female, by choosing also to feed on a host, does not forego an opportunity to oviposit.

Females of field populations of *Venturia canescens* can feed on dried fruits, including carob pods in which hosts can occur. If an egg-carrying female visits a host-containing fruit and also feeds on it, she will not forego the opportunity to oviposit. However, if she visits unoccupied fruits for feeding, she will be trading one or several 'immediate' oviposition opportunities for future ones. Females of *V. canescens* also may be in a situation, in the wild, where host patches do not provide food, as could occur when its hosts infest walnut husks and almonds. Thus, in common with many other parasitoids, females will need to visit food patches that are some distance away from host-containing patches, in which case food patch-visiting behavior will always involve an immediate/future reproduction trade-off.

Some tiphiid wasps, which are large-bodied and also strong fliers, are known to exploit nectar sources situated several kilometers from the nearest host population (Clausen et al. 1933). However, this scenario – one of the few reported cases of commuting behavior – is exceptional. The majority of parasitoids need to travel only relatively short distances between host and food patches, although a 'fluid' type of commuting behavior is likely to be involved, the females foraging across a local host population, and interspersing successive visits to host patches with visits to food patches. Flowers and honeydew deposits (the excreta, either of the parasitoid's hosts or of non-host Homoptera) are an example of such food patches. An intermediate situation is found for thelytokous females of *V. canescens* inhabiting mills and stores that are normally devoid of any food source. These animals are observed leaving the buildings presumably in search of food, but whether they are able to return is not known.

Under field conditions, visiting food patches is likely to incur the risk of mortality not only from starvation (if a patch cannot be found before the insect runs out of energy) but also from extrinsic factors such as weather (Weisser et al. 1997), predation (Jervis 1990, Heimpel et al. 1997, Rosenheim 1998), and desiccation (Jervis et al. 2003).

7.3 Non-host foods: how are they utilized?

As we have already mentioned, parasitoid females utilize an array of different biochemical resources in fueling physiological functions (Fig. 7.1).

7.3.1 Somatic maintenance

Ingested sugars, following eventual absorption (as monosaccharides) into the blood-stream via the gut, can be used immediately for maintenance or they can be converted, for subsequent use, to either of two types of carbohydrate reserve: trehalose or glycogen. Trehalose comprises two linked glucose molecules. It is present in tissues (e.g. musculature) but occurs mainly in the blood, and so is known by many physiologists as the 'insect blood sugar' (Blum 1985, Thompson 2003). However, whereas trehalose may be the principal blood storage sugar for some parasitoids (Olson et al. 2000), it appears not to be so for others. For example, in the braconids *C. glomerata* and *Microplitis mediator*, trehalose was not detected at all using high pressure liquid chromatography (Steppuhn & Wäckers 2004). Trehalose, whether from the blood or elsewhere, is broken down into glucose as and when the latter is required (Suarez et al. 2005). Glycogen is a highly branched polysaccharide that is deposited in the fat body and is present, to a lesser extent, in the cells of flight muscles (Chapman 1998). Glycogenolysis provides the insect with glucose, as and when necessary, for fueling somatic maintenance, locomotion and, to a smaller degree, egg manufacture.

In starved, host-deprived female *Macrocentrus grandii*, the levels both of glycogen and of 'body sugars' (comprising those in the blood and, presumably to a lesser extent, those in tissues), showed a decline as the insects aged (Olson et al. 2000, Fadamiro & Heimpel

2001). In starved, host-deprived *V. canescens*, the level of body sugars likewise declined (glycogen dynamics were not studied) (Casas et al. 2003). Injection of sugar solutions directly into the haemolymph of female *Eupelmus vuilleti*, a host-feeding species, resulted in an increase in longevity (an effect observed with a mixture of sucrose and trehalose but not with either sugar injected on its own) over that observed with injection of water alone (Giron et al. 2002). The results of the aforementioned studies indicate the use, by parasitoids, of sugars in fueling somatic maintenance.

That dietary (i.e. exogenous) sugars fuel somatic maintenance in non-host-feeding parasitoids is evident from the numerous laboratory measurements that have been made of parasitoid longevity (Syme 1977, Foster & Ruesink 1984, Hagley & Barber 1992, Idris & Grafius 1995, Olson & Nechols 1995, Dyer & Landis 1996, Mathews & Stephen 1997, Baggen & Gurr 1998, Olson et al. 2000, Ide & Lanfranco 2001, Siekmann et al. 2001, Wäckers 2001, Sisterton & Averill 2002, Eliopoulos et al. 2003, Azzouz et al. 2004, Lee et al. 2004, Mitsunaga et al. 2004, Spafford Jacob & Evans 2004, Bezemer et al. 2005, Desouhant et al. 2005, Fadamiro & Chen 2005, Fadamiro et al. 2005, Hausmann et al. 2005, Winkler et al. 2005, Sivinski et al. 2006, Vattala et al. 2006). Generally, the life expectancy of starved, non-host-feeding wasps is significantly lower than that of females provided with non-host foods such as nectar or the more commonly used substitutes (e.g. honey or sucrose:water solutions), whether hosts are present or absent in both treatments (van Lenteren et al. 1987, Jervis et al. 1993, Jervis et al. 1996). Siekmann et al. (2001) showed that sugar-feeding by female *Cotesia rubecula* reduces the risk of death by starvation by as much as 73%, the precise degree of reduction depending on sugar concentration. These authors concluded that, under field conditions, females of *C. rubecula* need to feed at least once per day in order to prevent death from starvation. Azzouz et al. (2004) calculated that females of *Aphidius ervi* require, under laboratory conditions, at least two intakes of sugar per day in order to attain maximum longevity.

7.3.2 Locomotion

The fact that honeybee and orchid bee flight is fueled by glucose (Gmeinbauer & Crailsheim 1993, Suarez et al. 2005) suggests that parasitoids likewise can fuel their flight – which is energetically very costly (see Hoferer et al. 2000, on *C. glomerata*) – using this particular sugar). Since glucose can be obtained either directly or indirectly from sugar-rich foods, consumption of the latter is likely to influence the propensity to initiate flight, the duration of flights, and possibly other dispersal-related variables. Flight initiation was slightly increased by honey provision in female *Trichogramma minutum* (Forsse et al. 1992). Flight duration in *Nasonia vitripennis* was not affected by prior exposure to honey (King 1993), but Wanner et al. (2006) showed that pre-flight feeding by female *C. glomerata* on floral and extrafloral nectar increased the insects' flight capacity, defined as either the longest single flight (i.e. duration), the number of flights undertaken, or the total distance flown during a given time period (females were tethered). However, feeding on a crude nectar mimic, comprising a sucrose-only solution, did not increase *C. glomerata*'s flight capacity. This suggests that either this species lacks the enzymes needed to convert dietary sucrose to monosaccharides, or that the action of invertase is very slow in this particular parasitoid species. Contradicting this explanation is the observation that continuously sucrose-supplied *C. glomerata* show, like similarly treated *M. mediator*, maintain a balanced glucose-fructose ratio (measured via 'body sugar' assays) (Steppuhn & Wäckers 2004). Clearly, there is a need for additional research into sugar use by flying *C. glomerata*, but it should involve repeated rather than one-off feeding.

As to evidence for the role of dietary sugars in fueling walking, honey-feeding by female *Trichogramma brassicae* from the time of emergence resulted in a steady increase in activity, over several days, whereas females deprived of food could not sustain an initially high level of activity (which presumably increases the chances of finding food, Pompanon et al. 1999).

7.3.3 Egg manufacture

Regarding the role of non-host foods in directly fueling ovigenesis, extreme caution must be exercised when inferring from the results of laboratory experiments. First, if host-provided parasitoids lay more eggs when fed sugar than when unfed (but water-provided), it may simply be because they live longer. It is thus necessary to establish whether age-specific realized fecundity and/or the rate of ovigenesis are enhanced (see Jervis et al. 2005b, for general methodology). Second, even if there is evidence for a higher rate of ovigenesis in sugar-fed females compared to unfed females (Hagley & Barber 1992, England & Evans 1997, Harvey et al. 2001, Eliapoulos et al. 2003, Bezemer et al. 2005, Sivinski et al. 2006, Winkler et al. 2006), the observed difference could be entirely attributable to the sugars in the food lessening the catabolic drain that somatic maintenance exerts upon carried-over resources that truly are involved in fueling egg manufacture.

Among non-host-feeders there are species in which feeding increases the rate of ovigenesis compared with a diet of water (*V. canescens*, Harvey et al. 2001), but there are also species in which it appears not to do so (e.g. *Phanerotoma franklini*, Sisterton & Averill 2002). In *M. grandii*, the egg maturation rate is higher in starved, water-fed females than in sucrose-fed females (Olson et al. 2000). We discuss this somewhat surprising result below (Section 7.11).

Rigorously determining the direct involvement of exogenous nutrients in egg manufacture requires the use of nutrient tracking techniques, i.e. employing radiotracers/isotopes (Boggs 1997, Rivero & Casas 1999, Rivero et al. 2001, O'Brien et al. 2003, 2004, 2005). In host-feeding parasitoids radioactively labeled dietary amino acids have been shown to become directly incorporated into the eggs (Rivero & Casas 1999, Rivero et al. 2001). Nutrient tracking-based evidence for the direct fueling of ovigenesis by dietary sugars comes from Lepidoptera. In some butterflies, dietary sugars are the primary source of carbon in the *de novo* synthesis of non-essential amino acids in eggs (O'Brien et al. 2002, 2004, 2005).

The extent to which dietary sources contribute to egg manufacture partly depends upon the degree of provisioning of oocytes in teneral adults (O'Brien et al. 2004), and it is likely that it is correlated with the ovigeny index, defined as the proportion of the potential lifetime egg complement that is mature upon emergence (Jervis et al. 2005a). Among parasitoid wasps, the host feeding habit is hypothesized to be linked to the ovigeny index, host feeders being apparently confined to synovigenic species (an index of <1) (Jervis & Kidd 1986, Jervis et al. 2001), although the current evidence for this is equivocal (Jervis et al. 2001). There is also a hypothesized link between host-feeding and the production of so-called 'anhydropic' eggs, as opposed to 'hydropic' eggs (Jervis & Kidd 1986). Anhydropic eggs are yolk-rich, and apparently it is only in host-feeding species that the yolk contains protein bodies that are typical of insects generally (King & Richards 1969, Le Ralec 1995), and host blood meals are inferred to be the source of the inclusions. Hydropic eggs contain a small amount of yolk, which is mainly comprised of lipids (Le Ralec 1995), and it appears that few hydropic egg-producers practise host feeding (Heimpel & Collier 1996). Parasitoids that consume only non-host foods include both hydropic and anhydropic egg producers.

Floral nectars contain mineral salts, vitamins, and pigments in addition to other materials (Gardener & Gillman 2002). The vitamin content of an artificial diet was shown by Bracken (1966) to influence realized fecundity, and it was concluded that the vitamins were directly involved in egg manufacture. Nevertheless, in our models, we assume, for simplicity's sake, that dietary vitamins play no role in egg manufacture.

7.4 Teneral resources

Teneral resources comprise that fraction of their that the females have already acquired via larval feeding, i.e. they are the female's 'capital'. The differential allocation of these resources has important consequences for parasitoid foraging behavior, by determining egg load, the amount of reserves that are available for fueling egg manufacture, and somatic functions (maintenance and locomotion), and the role that dietary 'income' (i.e. exogenous nutrients) will play in nutrient dynamics (Jervis et al. 2001, Jervis & Ferns 2004, Jervis et al. 2005a, 2007).

7.4.1 Allocation patterns

During pupal development in parasitoid wasps, as in other holometabolous insects, there occur trade-offs in the proportionate allocation of different functional kinds of resource that the female carries over from the larval stage. These trade-offs occur at two levels (Fig. 7.2), the first of which is the proportionate allocation to 'soma' versus 'non-soma', the latter being taken to comprise reproductive tissues and oocytes, together with initial nutrient reserves (e.g. teneral quantities of blood sugars, and lipids, glycogen, and protein in the fat body and elsewhere). The form of this trade-off is likely to vary among and within species, and will be linked to ovigeny index, which itself is negatively correlated with life-span both intra- and interspecifically (Jervis et al. 2001, 2003). The soma/non-soma trade-off was first postulated by Boggs (1981), who argued that species whose females are longer-lived should invest relatively more resources during metamorphosis into building a 'sturdy body' (soma, defined as body structures other than internal reproductive tissues, oocytes, and nutrient stores, Fig. 7.2), compared with shorter-lived species. The corollary of this is that shorter-lived species should invest relatively more resources during metamorphosis in reproduction (non-soma, Fig. 7.2) at the expense of building a 'sturdy body', compared with longer-lived species.

Currently, data on soma/non-soma allocation is lacking in parasitoid wasps, but the limited evidence available for caddis flies (Trichoptera) indicates that the interspecific pattern in such insects accords with Bogg's hypothesis, and is linked to the ovigeny index (Jervis et al. 2005a, 2007, Stevens et al. 1999, 2000).

The other hypothesized resource trade-off occurs in the relative allocation between initial eggs and initial reserves, the latter being used to manufacture future eggs and also to fuel somatic functions (Fig. 7.2). Approaching the pro-ovigenic end of the ovigeny index continuum (index = 1), investment in eggs should occur at the expense of investing in

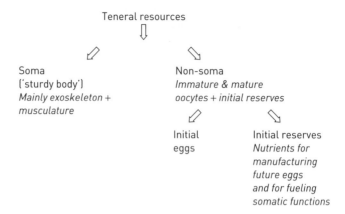

Fig. 7.2 Allocation of carried-over (capital) resources to competing physiological functions in holometabolous insects. Modified from Jervis et al. (2005b). See text for explanation.

initial reserves. Investment in the latter can be minimal or even non-existent: fuel is not required for egg maturation, and expected life-span is short (Jervis et al. 2001, 2003). As ovigeny index decreases from 0.5, investment in initial reserves should occur at the expense of investment in eggs. When the index reaches zero, investment in eggs is either zero or (more likely) consists, at most, of allocation to immature oocytes. Allocation to initial reserves ought to be substantial in non-host-feeding, extremely synovigenic species, because their adult foods are both lipid- and nitrogen-poor (and so cannot fuel egg manufacture to a significant degree), and because, like other Hymenoptera, they are apparently unable to synthesize lipids *de novo* (Giron & Casas 2003).

Currently, there is no empirical information on the pattern of proportionate allocation within non-soma relating directly to variation in ovigeny index at the inter-specific level. However, it is noteworthy that the non-host-feeding parasitoid wasp *Aphaereta pallipes* (Say), when compared size-for-size with its congener *A. genevensis* (Fischer), has a higher initial egg load but smaller fat reserves (Pexton & Mayhew 2002).

A trade-off between initial eggs and initial reserves is expected to occur also at the intraspecific level, and this likewise should be reflected in ovigeny index. There are two lines of evidence indicating that this does indeed occur. First, between populations of the parasitoid wasp *Asobara tabida* (Nees) forming part of a north-south cline in Western Europe, allocation to eggs increased at the expense of allocation to storage as ovigeny index increased. Second, dynamic programming was employed by Ellers and Jervis (2003) to predict the optimal resource allocation strategy of parasitoids with respect to 'compartmentalization' within non-soma (initial egg versus initial reserves) in relation to variation in three variables: body mass, habitat richness (measured as the number of patches encountered multiplied by the mean number of hosts per patch), and habitat stochasticity (measured as the variance relative to the mean of the number of host patches, and of the number of hosts per patch). The Ellers and Jervis (2003) model predicted that as body mass increases, there is increased allocation to both initial eggs and initial reserves (this prediction is supported empirically (Pexton & Mayhew 2002)). However, in a host-rich habitat, allocation to initial eggs is higher, and allocation to initial reserves is lower, than in a poorer habitat – a trade-off that is observed across the range of body sizes but is most marked in larger-bodied individuals. The higher allocation to initial eggs in host-rich habitats is due to the greater oviposition opportunities that such habitats offer. The ovigeny index decreases with increasing body size because the increase in allocation to initial egg load is proportionately smaller than the increase in allocation to initial reserves – which contribute to future eggs, the number of which partly determines the denominator (potential lifetime fecundity) of the ovigeny index.

The predicted negative correlation between body mass and ovigeny index, which has been confirmed empirically in *Aphaereta genevensis* (Thorne et al. 2006), is hypothesized to be selected for under the following scenarios:

1 Small, short-lived individuals experience the environment as being more stochastic, because they would encounter only a few patches as compared to larger, longer-lived individuals (Ellers & Jervis 2003). In more stochastic environments the optimal initial egg load exceeds the expected number of hosts found (Ellers et al. 2000), and so smaller parasitoids need to allocate a larger proportion of their resources to initial egg load than larger individuals, despite small- and large-bodied parasitoids encountering patches from the same host distribution. For large-bodied individuals, the optimal initial egg load also exceeds the expected number of hosts found but, especially in low-quality habitats, it only slightly exceeds that of smaller individuals. This allocation strategy ensures the maximum probability of finding a patch, and maintains reproductive plasticity over the female's life-span.
2 The progeny of small-bodied wasps may, because of the presumably inferior dispersal abilities of the adult females compared to those of larger-bodied insects, experience a higher level of superparasitism and its concomitant mortality costs. This could force allocation to initial eggs far above the level that would match host availability, such that their ovigeny index exceeds that of larger females (Ellers & Jervis 2004).
3 Extrinsic mortality selects for a concentration of egg production early on in adult life in smaller-bodied females (Jervis et al. 2001, 2003). Jervis et al. (2003) posited death due to desiccation as a likely selective factor in the evolution of the negative correlation between the ovigeny index and body mass.

Using the same model as in their previous study, Ellers and Jervis (2004) sought to predict the conditions under which a strategy of pro-ovigeny would be favored. The model in this case predicted pro-ovigeny to be linked to a small body size relative to travel costs (this has some empirical support), large eggs relative to the total amount of allocatable resources (this is not supported empirically), and be adaptive for a uniform host distribution (which, as we have already mentioned, is a most unlikely scenario in the real world). The link to a uniform host distribution is also highly surprising given the earlier finding by Ellers and Jervis (2003) that the ovigeny index-body size relationship can arise because small-bodied wasps experience a higher degree of stochasticity in host availability (see above). However, it is important to note that, while the Ellers and Jervis (2003, 2004) models are identical, they explore different regions of parameter space, and so the relationship between ecological parameters and optimal strategies may not be linear.

The aforementioned trade-off between initial eggs and initial reserves implies there is a cost, imposed by pre-emergence investment in eggs, upon the post-emergence survival of females. In interpreting the adaptive significance of their models' output, Ellers and Jervis (2003) assumed there to be such a cost (the first scenario, above). That assumption is supported empirically at the intraspecific level. First, it has been shown that pre-emergence investment in eggs occurs at the expense of investment in lipids (and vice versa), and that female longevity is positively correlated with lipid levels (Ellers 1996, Ellers & van Alphen 1997, Pexton & Mayhew 2002). Second, as occurs at the interspecific level, ovigeny index and life-span are negatively correlated (Jervis et al. 2001).

7.4.2 Utilization of reserves

What precisely do 'reserves' represent in terms of potential utilization, i.e. as fuel for different metabolic functions? As noted above, the principal components of reserves are sugars – mainly present in the blood (see above) – and glycogen, lipids and proteins – mainly present in the fat body (Chapman 1998), although reserves are, in some species, concentrated in other body regions (e.g. the head in the host-feeding parasitoid *Dinarmus basalis* (Rivero et al. 2001)). We discuss the utilization of each biochemical resource in turn below.

Storage sugars
While being primarily used in somatic maintenance and locomotion (see above), storage sugars, such as trehalose, are also a potential source of glucose for egg manufacture, including yolk synthesis (see above) (Chapman 1998, Suarez et al. 2005). Storage sugars can be replenished and supplemented by feeding. 'Body sugar' levels increased over the life-span of sucrose-fed *M. grandii* females, whereas they declined in females given only water (Olson et al. 2000). Sugar-deprived females of the phorid fly parasitoid *Pseudacteon tricuspis* could maintain the teneral level of body sugars for only one day, whereas honeydew-fed females showed an initial increase over the teneral level, maintaining it throughout life (Fadamiro & Chen 2005). In field-released *V. canescens*, sugar resources increased well above teneral levels, presumably as a result of feeding (Casas et al. 2003).

Glycogen
In *M. grandii*, glycogen levels at adult emergence are below the maximum levels recorded in older, sucrose-fed individuals, indicating that glycogen reserves can be supplemented exogenously (Olson et al. 2000). Both the wasp *D. insulare* and the fly *P. tricuspis* can also supplement their glycogen levels when fed on sugar-rich materials (Fadamiro & Chen 2005, Lee et al. 2006). Glycogen mainly fuels somatic functions in insects (see above) and, according to Chapman (1998), contributes relatively little to egg production in comparison to lipid and protein reserves, but its role as a source of glucose in fueling ovigenesis should not be dismissed (Section 7.11).

In host-deprived *M. grandii*, that were fed on only the first day of life, the level of glycogen declined substantially, its breakdown serving to maintain an appropriate level of body sugars (Fadamiro & Heimpel 2001). By contrast, in host-deprived females continuously provided with dietary sugars, high levels both of sugars and of glycogen were maintained over the females' lifetimes (Olson et al. 2000). Interestingly, host-deprived, water-supplied female *P. tricuspis* were able to maintain the teneral level of glycogen (Fadamiro & Chen

2005). This suggests that the wasps were able to fuel somatic maintenance from some alternative reserve type, possibly lipids (lipid dynamics were not studied).

Lipids
Typically in parasitoid wasps, stored lipids decline steadily in quantity from the time of adult emergence, irrespective of whether food is available (Ellers 1996, Eijs et al. 1998, Olson et al. 2000, Casas et al. 2003, Lee et al. 2004, Casas et al. 2005, Colinet et al. 2006).

Evidence that lipids fuel somatic maintenance in parasitoids is diverse. It includes the following:

1 In host-deprived, fed *A. tabida*, females with a higher total lipid content at eclosion, whether by virtue of genetic strain or larger body size, live longer than females with smaller fat reserves (Ellers 1996).
2 Among *Aphaereta* species, the age-specific survival curve of host-deprived *pallipes* declines more steeply than that of host-deprived *genevensis* when females are starved, whereas the two species do not differ in longevity when females are fed. The species difference in the effect of starvation on survivorship is attributable to *pallipes* being smaller-bodied and having a smaller teneral amount of lipids (Pexton & Mayhew 2002).
3 Within each of the same two species of *Aphaereta*, larger females have higher teneral lipid levels, and live longer under conditions of starvation and host deprivation, compared to smaller females (Pexton & Mayhew 2002).
4 Among host-deprived individuals of the host-feeder *Nasonia vitripennis*, smaller females (smaller teneral lipid reserves), suffer disproportionately the cost, to survival, of not feeding (Rivero & West 2005).
5 Lipid levels decline more rapidly in starved than in fed, host-deprived females of some species (Ellers 1996, Ellers et al. 2006, Lee et al. 2004).

Stored lipids also fuel locomotion. Controlling for variation in longevity, the fat levels of field-released *A. tabida* declined with increasing dispersal distance (dispersal in this species is predominantly, if not exclusively, by flight) (Ellers et al. 1998). It is possible that, in such a parasitoid, the degree of lipid use in locomotion depends upon the availability of dietary sugars. O'Brien (1999) showed that, in the hawk moth *Amphion floridensis*, regularly nectar-fed insects relied primarily, although not exclusively, on carbohydrate to fuel flight, whereas unfed moths relied almost exclusively on fat reserves.

Evidence that lipid reserves are used in egg manufacture in non-host-feeding parasitoids comes from two studies, as follows.

1 A comparison of the sexes in the phorid fly *P. tricuspis* (Fadamiro et al. 2005). In starved females lipid levels declined over time, but in starved males they remained relatively steady. This can be interpreted as reflecting the greater fuel demand of ovigenesis compared to that of spermatogenesis (but see Olson et al. 2000, on a wasp).
2 Laboratory manipulation experiments conducted on *A. tabida* by Ellers and van Alphen (1997): in females allowed to oviposit, replenishment of egg load is accompanied by a significant reduction in fat levels. Also, the age-related depletion rate of fat reserves can only be accounted for if the energetic demand of egg maturation is considered along with that of somatic maintenance.

Proteins

The primary role of a parasitoid wasp's protein reserve is to provide a source of amino acids for use in egg manufacture (Fig. 7.1). However, in sugar-starved, host-deprived females of the host-feeding species *E. vuilleti*, protein levels declined substantially, indicating the catabolism of proteins in somatic maintenance (Casas et al. 2005). Whether this also applies to non-host feeding parasitoids is not known.

Proteins are possibly also used in flight metabolism, providing a source of amino acids such as proline. We consider this more likely to apply to parasitoid flies (Chapman 1998, Gilbert & Jervis 1998), but the use of amino acids in flight by parasitoid wasps cannot be ruled out (Suarez et al. 2005 on other Hymenoptera).

While parasitoid wasps generally can presumably synthesize storage proteins *de novo* from dietary amino acids, only those species that host-feed have a rich enough diet to be able to supplement or replenish their protein reserves from dietary 'income' to a significant degree. Many non-host-feeders are more likely to rely entirely upon teneral protein reserves for future egg production.

Given the above, we considered, for modeling purposes, how our hypothetical non-host-feeding wasps should utilize their diverse biochemical reserves. This was easy in the case of pro-ovigenic parasitoids: females have no need for nutrients to fuel egg production, because all of the egg complement is mature upon emergence and so, irrespective of what the reserves comprise biochemically, the females could be considered as possessing one set of reserves, fueling only somatic functions. Deciding on how to model synovigeny was less straightforward, since several resource allocation/utilization scenarios can be envisaged (Section 7.11). For this, we decided that, due both to the very low nitrogen content of nectar and honeydew and to the apparent inability of most wasps to undertake lipogenesis, neither supplementation nor replenishment of lipid or protein reserves can occur in the model parasitoids. Also, we assumed that ovigenesis is fueled entirely by lipid and protein reserves, with exogenous (i.e. dietary) nutrients being the sole fuel for somatic functions (maintenance and locomotion). Rivero et al. (2001) suggested that, in synovigenic non-host-feeders, larval reserves should play a major and continuous role in egg manufacture over the lifetime of the female. The authors also suggest that this would apply particularly to those species which produce yolk-rich eggs but, in our opinion, egg type might make little difference, since there is a tendency, at the interspecific level, for yolk-richness to be traded for egg numbers.

In our models, we assume, for the purpose of tractability, that sugars and glycogen together comprise one resource, 'metabolic resources' (MRs in both the pro-ovigenic model and the synovigenic models) while, for synovigenic parasitoids, lipids and proteins together comprise another single resource, 'egg production resources' (EPRs in the synovigenic model). Note that, in adopting this nomenclature, we do not imply ovigenesis to be a non-metabolic process (Fig. 7.1). For *V. canescens*, Casas et al. (2003) argued that, because lipid reserves are set at birth, they ought to be used prudently: females should keep them principally as a fuel for egg manufacture and also as a fuel for maintenance during times of low sugar intake. Similar reasoning could be applied to the utilization of protein reserves. Thus, we assume here that non-host-feeding synovigenic parasitoids, in common with pro-ovigenic parasitoids, derive all of the materials needed for egg manufacture from the larval stage, the fundamental difference between the two parasitoid types being that pro-ovigenic parasitoids convert all such resources into mature eggs by the time they emerge as adults whereas the synovigenic parasitoids do not. This contrasts with the assumption, applied to holo-metabolous insects generally (and which we know to be true of some host-feeders) that synovigeny necessarily requires exogenous nutrients to be involved in egg manufacture (Jervis et al. 2005a, 2007).

We further hypothesized that, optimally, lipid and protein reserves will be diverted away from egg manufacture as a 'last resort', when sugars become limiting. To test this hypothesis, we assumed that, in the synovigenic model, EPRs can be drawn upon to fuel somatic functions under any conditions. If optimal, the 'last resort' strategy would be an outcome of the model.

7.5 Egg resorption

Some parasitoid wasps resorb mature eggs during periods of host and/or food scarcity, using the materials in yolk to fuel somatic maintenance, although it is possible that some species also use them to sustain rather than shut down ovigenesis (Jervis & Kidd 1986). Whatever the destination of the egg yolk contents, egg resorption thus serves to re-allocate resources (Rosenheim et al. 2000, Jervis et al. 2005a). It is regarded as a 'last-resort' strategy because, if it is employed, the female will, for several hours or even days, have fewer than the maximum possible number of eggs ready for oviposition. Thus, she can experience 'oosorption-mediated' egg-limitation in host-poor environments, in contrast to the 'oviposition-mediated' egg-limitation that can be experienced in host-rich environments (Rosenheim et al. 2000). Nevertheless, as Rosenheim et al. (2000) pointed out, a female that experiences a brief period of host or food scarcity, and which is subsequently faced with conditions rich in reproductive opportunities, may realize very large increases in lifetime reproductive success, if she is able to survive the lean period by means of egg resorption.

There is a strong inverse relationship, across species, between the ovigeny index, and egg resorption capability, egg resorption being concentrated among synovigenic species (Jervis et al. 2001). This is to be expected since resorption takes several hours to days, and if concurrent in all ovaries, precludes oviposition. It would thus be least costly, in terms of the female's time-budget, to those parasitoid species that have a low ovigeny index, since they are longer-lived compared with high index species (Jervis & Kidd 1986, Jervis et al. 2005a).

While it is theoretically possible for hydropic egg-producing species to practise egg resorption (Jervis et al. 2001), none are known for certain to do so, although *M. grandii* could be an exception (Olson et al. 2000).

7.6 How do parasitoids locate and recognize non-host foods?

For parasitoids that are able to obtain food from patches that can contain both hosts and food, there is no need to decide, pre-visitation, whether food or hosts are the target resource. Females can respond to host patch cues, food cues, or a combination of the two. Parasitoids, such as wild *V. canescens*, whose food patches (dried, fallen fruits) do not necessarily contain hosts, would benefit from deciding, pre-visitation, which patches contain both types of resource. Desouhant et al. (2005) showed that female *V. canescens* can discriminate between food and host odors, and can also distinguish, at a distance, between patches containing both hosts and food and those containing only food.

Parasitoids that need to travel between separate food and host patches must decide which of the two types of resource to forage for. Having decided to feed, females are likely to employ sensory cues in locating the food source, as opposed to locating it by random search. The cues used will be attractant and possibly also arrestant. For some parasitoid species, volatile chemicals are the primary attractant stimuli emanating from food (Shajahan 1974, Lewis & Takasu 1990, Cortesero et al. 2000, Rose et al. 2006), although visual stimuli may also play an important role (Wäckers 1994). Floral nectar-feeding parasitoids may, in common with other flower-visitors, have odor or color preferences in relation to the range of flowering plants they exploit (Wäckers 1994, Begum et al. 2004). Natural honeydew is an arrestant for some aphidiine braconids (Ayal 1987, Budenberg 1990, Du et al. 1997), but it is unclear whether honeydew is a principal food of such wasps in the wild.

One implication of the above is that the parasitoid's cognitive abilities regarding the long-distance or short-distance perception of food cues will influence two of our models' key parameters, namely the probability of finding food, and inter-patch (food patch-host patch) distance.

7.7 State-dependence of parasitoid foraging behavior

We are concerned here with the relationship between physiological state and both reproductive (host-searching, oviposition) and feeding (food-searching, feeding) behavior. Intuitively, we would expect key fitness-related components of foraging behavior, including decisions, to be state-dependent, linked principally to nutritional state and/or egg load (Section 7.9).

Females having low nutrient levels are expected to practise behavior associated with food-searching as opposed to host-searching. Indeed, in *P. franklini*, sugar-deprivation increased the likelihood that females will engage in foraging movements over cranberry foliage since the host's vulnerable stage, the egg, occurs on cranberry fruits. Unfed wasps spent an average of 25% of their available time engaged in 'grazing', apparently on leaf leachates, compared with no time spent grazing by fed wasps (Sisterton & Averill 2002). Nutritional state should also influence patch choice behavior. Wäckers' (1994) experiments on *C. rubecula* provide good empirical evidence for this. Females were presented with different-colored targets, which were attached to cabbage plants. Starved females sought out yellow, as opposed to gray targets, and they searched more actively on yellow targets. Also, starved females, after landing on yellow targets, applied their mouthparts to them. Sugar-fed females, by contrast, concentrated their searching activity on the cabbage foliage. Yellow targets presumably most closely resembled flowers. In another set of experiments, Wäckers (1994) showed that, when given a choice between flower odor and host-damaged leaf odor, starved female *C. rubecula* preferred the former whereas sugar-fed wasps preferred the latter. Siekmann et al. (2004) found, in wind tunnel experiments, that among one-day-old females of *C. rubecula*, only well-fed insects showed a preference for hosts, whereas unfed wasps visited both hosts and flowers (devoid of hosts) in equal proportions.

Desouhant et al. (2005) studied the nutritional state-dependence of odor choice in a 'wild' strain of *V. canescens*. As noted above, in the field the hosts of this species occur in some, but not all, food patches, and it is also the case that some host-containing patches do not provide food. Desouhant et al. (2005) showed that wasps modified their choice according to feeding history. In particular, if they were previously food-deprived and then given a choice, they preferred sites where both food and hosts are present.

Further evidence that nutritional state influences foraging decisions comes from both host-feeding and non-host-feeding parasitoids. In the host-feeder, *Aphytis melinus*, the decision to take a blood meal varies with the food type given beforehand. The propensity to host-feed is lower when females have previously fed on a diet containing yeast than when they have fed on a pure sucrose solution (attributable to the yeast diet containing proteinaceous materials that fuel ovigenesis) (Heimpel & Rosenheim 1995). Eijs et al. (1998) found that *Leptopilina heterotoma*, a non-host-feeder, feeds only when its fat reserves are low.

A female's propensity to feed should also be highest when her gut contents are at a low level. Heimpel and Collier (1996) present a discussion of the empirical evidence.

A priori, we would also expect females with a lower egg load to opt for feeding rather than for oviposition, when they have mature eggs in reserve and are thus faced with choice between the two behaviors (Chan & Godfray 1993, Briggs et al. 1995, Heimpel & Collier 1996, Jervis et al. 1996, McGregor 1997, Jervis & Kidd 1999). This has been confirmed empirically for host feeding parasitoids, both in the laboratory (Heimpel & Rosenheim 1995) and in the field (Heimpel et al. 1998). Note that this expectation assumes that females with a low egg load are in fact egg-limited and that the decision to feed would favor egg production rather than oviposition. This might not be the case, however, since a parasitoid female with oviposition opportunities that match her egg load would not be egg-limited and could prefer to host-search. Also, in the synovigenic parasitoids modeled here (see below) the female's food lacks the nutrients necessary to fuel any amount of egg production, so if egg load is low and egg production reserves happen to be at a very low level, there is not a great deal, fitness-wise, to gained from feeding.

As to egg load influencing patch choice in non-host-feeding parasitoids, there is currently little information. Van Emden (1963) showed that only around a third of females of *Mesochorus* sp. (a synovigenic parasitoid) collected from the center of a wheat field, lacked mature eggs, in contrast to over 90% of females caught in a flowering verge adjacent to the field.

7.8 The effects of experience, and learning

Nutritional state-dependence of parasitoid behavior (Section 7.7, and see also Chapter 6 by Strand and Casas) is also evident with regard to conditioned odor preferences. Lewis and Takasu (1990) showed that the non-host-feeder *Microplitis croceipes* is capable of associative learning of, and subsequent discrimination between, host- and food-associated odors. Conditioned odor preferences have also been reported in other parasitoids (Patt et al. 1999, Rose et al. 2006). Having conditioned wasps to associate these odors with their respective resources, Lewis and Takasu (1990) showed, in a wind tunnel experiment, that sucrose-deprived females preferred to fly toward the food-associated odor, whereas sucrose-fed females preferred to fly toward the host-associated odor. The flight response was strictly a function of nutritional state rather than a function of how well wasps learned the odor associations.

7.9 Decisions

7.9.1 To feed or not feed

A synovigenic parasitoid female carrying only immature eggs is unable to oviposit. Therefore the choice she faces during her reproductive life is whether to seek food or not to do so. This would apply to: (i) extremely synovigenic species, whose females emerge with no mature eggs; and (ii) synovigenic species whose females have exhausted their mature egg load but have immature eggs still to be matured. By contrast, either a pro-ovigenic parasitoid that has exhausted her supply of mature eggs, or a synovigenic parasitoid that has exhausted her supplies of immature as well as mature eggs, has nothing whatsoever to gain from feeding. Note that a non-trivial period of post-reproductive life in female parasitoids appears to be a laboratory artifact (Jervis et al. 1994).

7.9.2 To feed or oviposit

A parasitoid female carrying some mature eggs is faced with trade-off. If she forages for food instead of hosts, the potential exists for her to deposit more eggs, and therefore add to her future fitness. However, by choosing to feed, she sacrifices an immediate fitness gain (Heimpel & Collier 1996).

As noted in Section 7.2, this trade-off is not equally stringent for all parasitoids. In cases where hosts and food are spatially coincident, the time and energy invested in switching between the two activities would be minor. However, in the more common scenario, where hosts and food occur in separate locations, the decision as to which resources (larval or adult) should be sought needs to be made before the search trip is embarked upon, and alternating between the two activities might be costly in terms of energy, time, and mortality risk.

7.10 Modeling the optimal decisions

7.10.1 Previous models and unexplored issues

In our introduction to this chapter, we discussed the pioneering modeling of Sirot and Bernstein (1996, 1997) and the subsequent work of Tenhumberg et al. (2006). Although these studies generated interesting insights and questions, they did not consider certain crucial aspects of the trade-off between food acquisition and parasitism. The major shortcoming of the studies is the lack of an explicit representation of egg dynamics. Also, the models are inapplicable to synovigenic parasitoids insofar as they assume the parasitoid's biochemical resources to form a single pool. This prompted us to develop new models that explicitly incorporate egg dynamics. One of these models takes account of the differential utilization of MRs (synovigeny). The difference between the Sirot and Bernstein (1996, 1997), the Tenhumberg et al. (2006), and the new set of models is summarized in Table 7.1.

The main questions we aim at addressing with our models are:

1 What is the influence of egg load on parasitoid optimal decisions?
2 How do the optimal strategies of pro-ovigenic and synovigenic wasps differ?
3 Under conditions of low food availability, should parasitoids avoid searching for food and devote all of their time to host searching, as predicted by the Sirot and Bernstein (1996) model, or should they concentrate their efforts toward acquiring 'income', as predicted by the Tenhumberg et al. (2006) model?
4 Would synovigenic animals draw energy for maintenance from EPRs (Fig. 7.1) only as a 'last resort' strategy?
5 What is the effect of inter-patch (i.e. host patch-food patch) travel, when travel costs, in terms of both time and energy expenditure, are taken into account?

7.10.2 Model assumptions

Following Sirot and Bernstein (1996, 1997), in our models, a parasitoid is either located on either of two patches, a food patch or a host patch, or is traveling between them. The patch of residence is treated as a state variable ($i = 0$ for the food patch and $i = 1$ for the host patch). The other two state variables are MRs (e) and EPRs (w), and the maximum expected fitness at time t with time horizon T, is $F(w,e,i,t)$ (see also Chapter 15 by Roitberg and Bernhard, for further discussion of state dependent modeling). The insect dies of starvation if the level of MRs drops to a given level (parameter E starvation in Table 7.2).

In our pro-ovigenic model, w represents the current egg load, i.e. all of a female's EPRs have already been converted to mature eggs by the start of adult life. In the synovigenic model, the female's EPRs and eggs are pooled in the same variable w. This means that there is no delay in egg production (a unit of EPRs is instantaneously transformed into an egg at oviposition). We assume that this is not an important shortcoming given that the parasitoid will be able to lay only a single egg per time interval. The assumption of an instantaneous conversion of EPRs into eggs implies that the parasitoid has an egg available whenever an oviposition opportunity arises (and $w > 0$).

The pro-ovigenic model
If the female is in the food patch and decides to remain on it, she will survive a time step with probability $(1 - \mu_f)$ and she will find food and obtain a food reward of er units with probability λ_f. Energy rewards are allocated only to MRs and, when a food rewards is found, e increases to $e + er$. The probability of the wasp not finding food is $(1 - \lambda_f)$. At each time step in the food patch, the parasitoid has an energy expenditure of ee that is drawn from e.

The parasitoid might decide to leave for a host patch. Travel between the food patch and the host patch takes d units of time and costs $ee{\cdot}d$ energy units. Mortality during travel is μ_v per time step and the probability of surviving the voyage is thus $(1 - \mu_v)^d$. From this, the dynamic programming equation for a pro-ovigenic parasitoid in the food patch is

$$F(w,e,0,t) = Max\begin{cases}(1 - \mu_f)[\lambda_f F(w,e + er - ee,0,t + 1) + (1 - \lambda_f)F(w,e - ee,0,t + 1)]; \\ (1 - \mu_v)^d F(w,e - ee{\cdot}d,1,t + d)];\end{cases} \quad (7.1)$$

When the female is in the host patch ($i = 1$), she should decide to remain in it or to move to a food patch. If she stays, she will survive the time-step with probability $(1 - \mu_h)$ and

Table 7.1 Comparison between the models of Tenhumberg et al. (2006) and Sirot and Bernstein (1996) and the models presented in this chapter. The first two columns are similar to those presented by Tenhumberg et al. (2006), while the third column and the second part of the table present the issues addressed by our work.

Tenhumberg et al. (2006)	Sirot and Bernstein (1996)	Models presented in this chapter
Parasitism success specified by a probability distribution*. Implications: it is never guaranteed that parasitoids will find hosts before dying, especially so if energy reserves are low.	**Constant rates of parasitism on host patch.** Implications: at each time-step, parasitoids on host patches receive a guaranteed fitness reward based on the expected number of offspring for one time-period on the host patch.	**Parasitism success specified by a probability distribution***. As for Tenhumberg et al. (2006).
Normally distributed energetic values of food resources that usually allow only partial replenishment of energy reserves. Implications: parasitoids do not simply decide whether they should invest energy and time to search for more food.	**Constant consumption rates on food source.** Implications: females decide how much to consume before leaving.	**Food foraging success is specified by a probability distribution.** As for Tenhumberg et al. (2006).
Energy expenditure per time-step is independent of behavioral choice. The probability of finding food or hosts determines the total energy expenditure required to locate food or hosts. Consequently, the cost of foraging for food or hosts differs.	**Energy expenditure per time-step depends on behavioral choice.** Low energy expenditure while on host or food patches, but high energy expenditure while moving between patches.	**Energy expenditure per time-step is independent of behavioral choice.** As for Tenhumberg et al. (2006).
No energy expenditure is associated with moving between patches.	**Partial representation of energy and time expenditure associated with moving between patches.**	**Full representation of energy and time expenditure associated to moving between patches.**
Mortality risk during food-searching is different from that during host-searching.	**Constant background mortality risk independent of parasitoid's behavior.**	**Mortality risk during food and host searching or when moving between patches are model parameters.**
Model predictions are independent of parasitoid location.	**Model predictions are dependent on parasitoid location (host or food patch)**	**Model predictions are dependent on parasitoid location (host or food patch).**

Table 7.1 (cont'd)

Tenhumberg et al. (2006)	Sirot and Bernstein (1996)	Models presented in this chapter
Exclusively pro-ovigenic parasitoids. Egg dynamics (depletion by oviposition) not represented in the model. This means that the parasitoids are never egg-limited.	**Exclusively pro-ovigenic parasitoids.** Egg dynamics (depletion by oviposition) not represented in the model. This means that the parasitoids are never egg-limited.	**Both pro-ovigenic and synovigenic parasitoids considered.** Explicit representation of egg dynamics (depletion by oviposition) and (in synovigenic females) production. This means that parasitoids might be egg-limited. It is the optimization task of parasitoids to avoid being in this condition or being time-limited.
A single pool of resources. MRs and EPRs not distinguished.	**A single pool of metabolic reserves. MRs and EPRs not distinguished.**	**MRs and EPRs and the trade off in the allocation of fat reserves to either ovigenesis or maintenance explicitly represented.**

*See main text for further discussion.

will find a host with probability λ_h. When the female lays an egg, she gains a fitness reward of f, and w decreases to $w - 1$ (provided that $w > 0$). Note that the assumptions of fixed probabilities of finding hosts or food imply that host and food patches are not depletable. The probability of the parasitoid not finding a host is $(1 - \lambda_h)$. We assume that host searching involves much larger costs than subduing a single host. Consequently, oviposition does not involve specific energy expenditure. A time-step in the host patch costs ee energy units that are drawn from e. If the parasitoid decides to leave, travel time takes d time units and costs $ee \cdot d$ energy units. Mortality during travel is μ_v, per time step and the probability of surviving the voyage is thus $(1 - \mu_v)^d$.

$$F(w,e,1,t) = Max \begin{cases} (1 - \mu_h)[\lambda_h(F(w - 1,e - ee,1,t + 1) + f) + (1 - \lambda_h)F(w,e - ee,1,t + 1)]; \\ (1 - \mu_v)^d F(w,e - ee \cdot d,0,t + d)]; \end{cases} \quad (7.2)$$

The synovigenic model
In the synovigenic model, energy rewards are allocated to MRs alone, i.e. no lipid or protein synthesis is assumed, and the contribution of carbohydrate reserves to ovigenesis is considered to be zero. However, locomotory and somatic maintenance energy expenditures can be drawn from MRs or from EPRs. This means that each alternative in the proovigenic model subdivides into two in the synovigenic version. More accurately, in the food patch the alternatives are:

1 To remain on it, and draw energy from MRs:

$$v_1 = (1 - \mu_f)[\lambda_f F(w,e + er - ee,0,t + 1) + (1 - \lambda_f)F(w,e - ee,0,t + 1)] \quad (7.3)$$

2 To remain on the patch and draw energy from EPRs (energy rewards increase MRs):

$$v_2 = (1 - \mu_f)[\lambda_f F(w - ee,e + er,0,t + 1) + (1 - \lambda_f)F(w - ee,e,0,t + 1)] \quad (7.4)$$

3 To move to the host patch and draw energy from MRs:

$$v_3 = (1 - \mu_v)^d F(w,e - er \cdot d,1,t + d) \quad (7.5)$$

4 To move to the host patch and draw energy from EPRs:

$$v_4 = (1 - \mu_v)^d F(w - er \cdot d,e,1,t + d) \quad (7.6)$$

v_2 and v_4 assume that the parasitoid might always decide to allocate EPRs to somatic functions. This allows us to explore under which conditions the 'last resort' consumption of EPRs is an optimal strategy.

Consequently, $F(w,e,0,t)$, the expected fitness of an pro-ovigenic parasitoid that at time t is in the food patch ($i = 0$) with w EPRs and e MRs, becomes

$$F(w,e,0,t) = Max\{v_1,v_2,v_3,v_4\} \quad (7.7)$$

and $F(w,e,1,t)$, the expected fitness of an pro-ovigenic parasitoid that at time t in the host patch, with w EPRs and e MRs, is calculated in a similar manner.

7.10.3 Analysis of the models

The models were solved working backward in time, commencing at the end of an individual's life (T) and calculating at each time-step, and for all combination of states, the behavior that results in the highest fitness (Clark & Mangel 2000, see also Chapter 15 by Roitberg and Bernhard). T is assumed to coincide with the death of the parasitoid, and so we set the terminal fitness function to 0. As in previous work (Sirot & Bernstein 1996, 1997, Tenhumberg et al. 2006), we are concerned with the stationary solutions of this process, i.e. when t is far from T and the optimal decisions are independent of the parasitoid's age (Clark & Mangel 2000). A more thorough analysis of the models and a detailed sensitivity analysis are to be presented elsewhere (Bernstein & Jervis, in preparation).

Optimal behaviors of parasitoid females, as predicted by the models, are presented as two surfaces of energy reserves (e, i.e. MRs), each of them being a function of the probability of finding food (λ_f) and egg load or EPRs (w). For given values of λ_f and of w, if the parasitoid has an amount of e above the upper surface, it should begin/carry on searching for hosts. If e lies anywhere between the surfaces, it should search for hosts, unless it is already on a food source. In the latter case it should keep on feeding until e reaches the upper surface. Below the lower surface, the parasitoid should begin/carry on searching for food. This is consistent with the aforementioned empirical evidence (Section 7.7) that links an increased probability of food-searching to previous food-deprivation.

7.10.4 Parameter values

In the analysis of the models, we used three sets of parameter values:

1 similar to those used by Sirot and Bernstein (1996);
2 those employed in Tenhumberg et al. (2006) that show parasitoids to search for food when it is scarce, as presented in Fig. 4a of their work; and
3 those employed by Tenhumberg et al. (2006) for *C. rubecula*, their model species. Some of the parameters of the new models have no equivalent in either the Sirot and Bernstein (1996) or the Tenhumberg et al. (2006) models. Somewhat arbitrary, but biologically reasonable, values were therefore used for these parameters (a thorough sensitivity analysis will be presented elsewhere). The parameter values are listed in Table 7.2. Note that, in all cases, parameter values were chosen to ensure that a female makes an expected net energy gain when visiting a food patch.

7.10.5 Results

The pro-ovigenic model
Figure 7.3a presents the optimal decisions of a pro-ovigenic parasitoid under the conditions of the Sirot and Bernstein (1996) model, and for a short travel time (a single time unit). The predictions for a high egg load ($w = 20$) correspond, in general terms, to those obtained by the Sirot and Bernstein (1996) model: when food is scarce (low λ_f) parasitoids should avoid searching for food and should remain in the host patch (both surfaces take minimum e values, the lower surface taking a value of $e = 0$). Food-searching will be more frequent for intermediate probabilities of finding food. For higher λ_f values, parasitoids

Table 7.2 Parameter values and their biological interpretation. These values were selectively incorporated into our pro-ovigenic and synovigenic models so that we could explore the consequences of the conditions set by Sirot and Bernstein (1996) and Tenhumberg et al. (2006).

Parameter	Sirot and Bernstein (1996) conditions	Tenhumberg et al. (2006) conditions	*C. rubecula*
e_{max} Maximum energy reserves	20	48	48
$E_{starvation}$ Minimum energy level at which female dies from starvation.	0	6	6
w_{max}* Maximum w value	20	20	20
λ_h Probability of finding hosts	0.55†	0.8	0.8
λ_f Probability of finding food	Variable‡	Variable‡	Variable‡
μ_f Mortality when searching for food	0.005§	0.001¶	0.1
μ_h Mortality when searching for hosts	0.005§	0.01¶	0.01
μ_v Mortality during dispersal	0.005**	0.001**	0.1**
d Distance between patches	1 or 4	1 or 4	1 or 4
ee Metabolic expenditure per time step	1	1	1
er Metabolic reward of finding food	2	12	12
f Fitness gain per oviposition	1	1	1

*This parameter has no equivalence in the original models (Sirot & Bernstein 1996, Tenhumberg et al. 2006).

†Set to an unrealistic value of 1 in the original model (Sirot & Bernstein 1996).

‡$0.05 \leq \lambda_f \leq 1$.

§A low but more realistic value than in the original model (Sirot & Bernstein 1996).

¶See Fig. 7.1a of the original models (Tenhumberg et al. 2006).

**This parameter has no equivalence in the original models (Sirot & Bernstein 1996, Tenhumberg et al. 2006). It was set to μ_f.

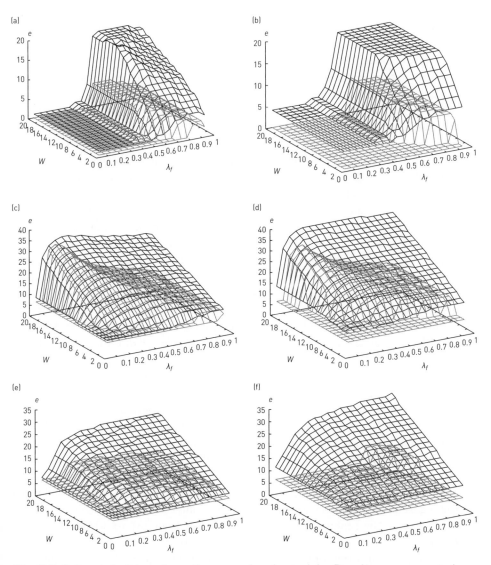

Fig. 7.3 Optimal decisions from the pro-ovigenic models. Results are presented as two surfaces of energy reserves (e), each as a function of the probability of finding food (λ_f) and of egg load or EPRs (w). If the parasitoid has energy reserves above the upper surface, it should begin/carry on searching for hosts. If e lies anywhere between the two surfaces, the parasitoid should continue with its current activity, i.e. searching for food until e reaches the upper surface, or searching for hosts until e reaches the lower surface. Beneath the lower surface, the parasitoid should begin/continue searching for food. (a) Sirot and Bernstein (1996) conditions and travel time equal to unity. (b) Sirot and Bernstein (1996) conditions with travel time equal to four time units. (c) Tenhumberg et al. (2006) conditions and travel time equal to unity. (d) Tenhumberg et al. (2006) conditions and travel time equal to four time units. (e) Parameter values for *C. rubecula* according to Tenhumberg et al. (2006) and travel time equal to unity. (f) Parameter values for *C. rubecula* according to Tenhumberg et al. (2006) and travel time equal to four time units. See main text and Table 7.2 for further details.

should remain longer in the host patch and delay food-searching (as MRs can easily be renewed): the lower surface bends down to lower e values. Figure 7.3 also shows the influence of egg load (w). The upper surface bends to lower e values as w decreases: when egg load is low, animals should not feed to satiation (which would cause them to reach the maximum e value); to continue feeding entails a mortality risk, and moderate resources would suffice to lay the remaining eggs.

Increasing the travel time (d) to four time units (Fig. 7.3b) raises both surfaces, and the surfaces also become less sensitive to the female's physiological state. On the one hand, females become more reluctant to leave their current patch whereas, on the other, they remain longer on the food patch as they have to 'pay' for the metabolic costs of inter-patch travel. For low egg loads, females avoid searching for food even in cases of high food availability, the lower surface taking values of $e = 0$.

Figure 7.3c shows the optimal behavior when Tenhumberg et al.'s (2006) conditions are used in the model. Note that the upper surface in our model's output has no equivalent in the Tenhumberg et al. (2006) model. This is the consequence of dispersal being an instantaneous process in that model: females are in either a food or a host patch but there is no inter-patch movement *per se*. This leads to a single dynamic programming equation and to a single solution. Nevertheless, it can be seen that our model yields similar results: in general terms, when food is scarce (low λ_f) parasitoids should devote most of their time to searching for food (for low food availability, both surfaces reach high e values). The two surfaces reach similar values and this means that, if the parasitoid is in the host patch, it should rapidly abandon it to resume food-searching, and vice versa. Tenhumberg et al. (2006) attribute the difference between their predictions and those of Sirot and Bernstein (1996) to differences in the structure of the two models. The fact that our model output can coincide with either of the predictions suggests that the difference stems more from the parameter values than from the biological scenarios envisaged. Note that in Tenhumberg et al. (2006), a single meal can sustain the energy expenditure for a period six times longer than in Sirot and Bernstein (1996) ($er = 12$ and $er = 2$, respectively).

As in Fig. 7.3a, both surfaces bend down at lower w values. This, again, is because laying a smaller number of eggs entails lower metabolic demands. Remaining on the food patch to gain a surplus of energy incurs unnecessary mortality risks. Note that the starvation threshold assumed by Tenhumberg et al. (2006) exceeded zero ($e = 6$), which is why each of the surfaces, taken as a whole in Fig. 7.3c, are high (note also the differences in scale for the w axes). Longer travel times ($d = 4$, Fig. 7.3d) have similar effects as for the Sirot and Bernstein (1996) conditions, namely:

1 parasitoids stay longer in the food patch (the upper surface takes higher values);
2 females become more reluctant to leave their current patch (the differences between the two surfaces become more marked, particularly for high w and λ_f);
3 at low egg loads, females avoid feeding.

However, in general terms, the optimal strategies appear less sensitive to travel time than in the Sirot and Bernstein (1996) model (the general form of the surfaces remains unchanged).

Figures 7.3e and f show the behavior of pro-ovigenic parasitoids in our model under the conditions used in the Tenhumberg et al. (2006) model for *C. rubecula* for short ($d = 1$) and long ($d = 4$) travel times, respectively. The values of the lower surface for $w = 20$ are those more akin to the conditions of the original paper. Whilst the results for $d = 1$ generally accord with those of Figs 7.3a and c, for longer travel times, females avoid searching for food unless both egg load and food availability are high (for most parameter values, the lower surface is at $e = 6$, the starvation threshold). This result emphasizes the sensitivity of the food searching strategy to travel times.

The synovigenic model
Figures 7.4a and b present the optimal decisions for a synovigenic parasitoid under the conditions of the Sirot and Bernstein (1996) model, for travel times of one or four time units, respectively. The main difference from pro-ovigeny (Figs 7.3a and b) is some displacement of the upper surface toward higher λ_f. For low and intermediate λ_f, the lower surface also remains at lower values: the parasitoids avoid food-searching. Note that the lower values taken by both curves coincide with $e = 1$: parasitoids avoid food-searching unless they have nearly reached the starvation threshold. Under these conditions, pro-ovigenic parasitoids avoid food-searching unless they are already in a food patch (the lower surface take values of $e = 0$, the upper surface values of $e = 1$). The difference stems from the fact that under harsh conditions (low e and λ_f values) synovigenic parasitoids can resort to using EPRs and so sustain a longer food-searching time.

The fact that the general form of the surfaces is unaltered (particularly the upper surface), stems from the optimal solution being to draw energy from EPRs only when MRs are at the lowest values ('last resort' strategy). This is revealed by the examination of the output of the model showing the optimal strategies the insects should adopt. The curves are displaced to the right because, when both food availability is relatively low (low and intermediate λ_f values) and MRs are low (low e), females can resort to EPRs. This is even more marked when EPRs are plentiful (high w values). It explains why, for intermediate and relatively high λ_f, and high w, the lower surface remains at its lower e values. The same conclusions are drawn when comparing the pro-ovigenic model in Tenhumberg et al. (2006) with its synovigenic counterpart, both under the conditions of Figs 7.3c and d (results not shown) and under the conditions used by Tenhumberg et al. (2006) for *C. rubecula* (Figs 7.4b and c, $d = 1$ and $d = 4$, respectively). Note that for longer travel times ($d = 4$) under the conditions for *C. rubecula*, the synovigenic model predicts that females will totally avoid food searching unless their MRs are just above the starvation threshold (i.e. $e = 7$).

In summary, considering both pro-ovigeny and synovigeny, optimal behavior has been shown to be influenced by two environmental factors:

1 inter-patch distance;
2 the probability of finding food

and by three physiological factors:

1 energy levels;
2 the number of available eggs (w) – egg load in the case of pro-ovigeny, and EPRs in the case of synovigeny; and
3 ovigeny index.

With both pro-ovigeny and synovigeny, parameter w influences the likelihood of feeding, but only when food is abundant (high λ_f). Also, with each of the egg maturation strategies,

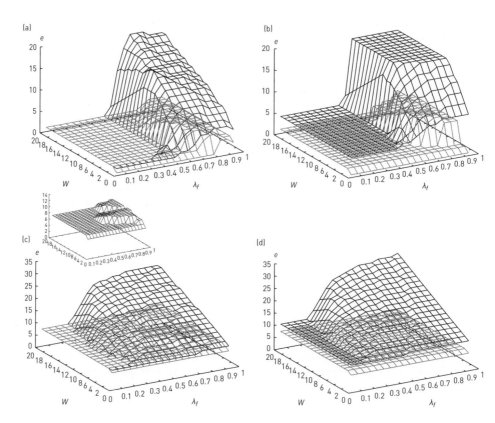

Fig. 7.4 Optimal decisions from the synovigenic models. Results are presented as two surfaces of energy reserves (e), as function of the probability of finding food (λ_f) and egg load or EPRs (w). When both surfaces overlap, only the lower one is represented. (a) Sirot and Bernstein (1996) conditions and travel time equal to unity. (b) Sirot and Bernstein (1996) conditions with travel time equal to four time units. (c) Parameter values for *C. rubecula* according to Tenhumberg et al. (2006) and travel time equal to unity (insert: a plot of the lower surface with a different range of w values). (d) Parameter values for *C. rubecula* according to Tenhumberg et al. (2006) and travel time equal to four time units. See Fig. 7.3, main text, and Table 7.2 for further details.

feeding is less likely at low w. This result is surprising in the case of synovigeny if we take w to be egg load (EPRs are instantaneously converted to eggs), given van Emden's (1963) field observations on *Mesochorus* sp. (see above) and the empirical and theoretical information on host-feeders linking an increased likelihood of feeding to a low egg load. It should be noted, however, that the results of the model correspond to an ideal, optimal parasitoid that would never be subject to either egg- or time-limitation. It is real, not caricatured animals that are egg- or time-limited. This subject warrants further exploration, both empirically and theoretically.

The main differences between pro-ovigeny and synovigeny are first, that with synovigeny, food-searching is less likely not only at low egg loads (as with pro-ovigeny) but also at high egg loads and second, that in terms of the expected frequency of food-searching, pro-ovigenic parasitoids are more likely to feed compared with synovigenic parasitoids (compare the overall amounts of raised, i.e. above-zero e, lower surface in Figs 7.3a and 7.4a). The latter finding is to be expected, given that in our synovigenic model females eventually draw upon EPRs to fuel maintenance, which thus reduces the need to obtain exogenous nutrients. Also, synovigenic females, at high egg loads and submaximal food availability (Fig. 7.4a), offset the mortality risk of delaying oviposition by avoiding feeding altogether. This effect is most evident in the synovigenic model under the conditions for *C. rubecula*, in which parasitoids would seldom search for food.

Our results suggest that the differences in behavior between the Sirot and Bernstein (1996) and Tenhumberg et al. (2006) models stem more from their particular parameter values rather than from the different scenarios envisaged. The two models differ in at least three parameters, e_{max}, the maximum energy reserves, μ_f, the mortality of food-searching, and the starvation level. Parameters such as maximum energy reserves will inevitably be highly variable amongst parasitoid species.

The earliest modeling studies on food-foraging by parasitoids focused on host feeders (Jervis & Kidd 1999). How do our conclusions compare with these? First, it needs to be emphasized that the foraging behavior of host-feeders and non-host-feeders is not strictly comparable, because:

1 most non-host-feeders need to visit food patches, whereas most host-feeders can obtain food (host blood, tissues) from the host patch; and

2 nutrients in blood meals fuel egg production in several species (Heimpel & Collier 1996, Casas et al. 2005)

With models that assume host-feeding to fuel only ovigenesis in synovigenic parasitoids (an effect not encompassed by our models, see Chan and Godfray 1993, Collier et al. 1994, McGregor 1997), a low egg load – specifically, an egg load of zero except in the case of the McGregor (1997) model – increases the likelihood of feeding, in contrast to non-host-feeders in the model presented in this chapter. This reduces future egg-limitation. With models that assume host-feeding to increase only life expectancy (pro-ovigenic host-feeder models, see Houston et al. 1992, Chan and Godfray 1993), females will prefer to feed if e is below some critical threshold. In our model and in Tenhumberg et al. (2006), females similarly prefer to feed when e is low, and they prefer to oviposit if e is high. Other host-feeding models assume host-feeding to contribute both to egg manufacture and to maintenance (Chan & Godfray 1993, Heimpel et al. 1994, Collier 1995, Heimpel et al. 1998). These models predict an increased likelihood of host feeding with decreasing egg load or decreasing nutrient reserves. An added complication when considering host-feeding parasitoids is that many consume non-host-foods in addition to host blood, and some may need to do so in order to realize the benefit to survival, of host-feeding (Heimpel & Collier 1996).

7.11 Future work

7.11.1 Empirical studies

A fundamental empirical need is for nutrient tracking studies to be undertaken. Preferably, these should be performed on a phylogenetically diverse array of non-host-feeding species, in order to determine precisely the pattern of utilization of different biochemical resources and to allow us to test the validity of the chosen grouping/compartmentalization of biochemical resources, in our synovigenic model (i.e. sugars and glycogen into MRs, and proteins and lipids into EPRs). We have assumed the carbohydrate requirement of egg manufacture to be so small as to be ignored for modeling purposes. However, Giron et al. (2004) have shown the eggs of the host-feeding, yolk-rich egg-producing wasp *E. vuilleti* to contain an amount of carbohydrate equivalent to that of either protein or lipid. The finding that host-feeding decreases longevity in *Trichogramma turkestanica* similarly points to a high carbohydrate requirement of egg manufacture (Ferracini et al. 2006). Also, carbohydrates may be involved in the *de novo* synthesis of non-essential amino acids contained in yolk (Section 7.3). Thus, nutrient tracking studies ought perhaps to focus, at least to begin with, on those non-host-feeding parasitoid species that produce anhydropic eggs.

Another important factor to consider is the nitrogen contained in nectar and honeydew. Nitrogen may occur in typically minuscule amounts, but it could nevertheless make a significant contribution to egg manufacture either when: (i) carry-over of EPRs is constrained due to poor larval feeding (Mevi-Schutz & Erhardt 2005, Jervis & Boggs 2005); or (ii) EPRs have been depleted post-emergence. However, in the host-feeder *E. vuilleti*, protein intake via blood meals has no effect on fecundity (Giron et al. 2004). Thus, there may be non-host-feeding parasitoids in which dietary nutrients are involved, to a significant degree, in egg manufacture, and so the resource compartmentalization scheme we adopted would not apply. Instead, at the most basic level, there would need to be one compartment involved in fueling both somatic functions and egg manufacture, while the other would be entirely devoted to egg manufacture (this might comprise only lipids). There may even be parasitoid species to which our compartmentalization scheme applies, but whose parameter values are such that the females' behavior would differ from that of our hypothetical parasitoids in two important respects when MRs become limiting: (i) females should always avoid drawing upon EPRs; and (ii) the rate of egg manufacture increases. The study by Olson et al. (2000) points to *M. grandii* being such a species. As we already noted, starved females of *M. grandii* have a higher egg production rate than sucrose-fed ones. Olson et al. (2000) suggest that this is a viable strategy to increase lifetime reproductive success if the wasp species is either:

1 incapable of resorbing eggs, or able to resorb eggs but unable to achieve a significant increase in life expectancy; and
2 if the metabolic costs of increased egg production are slight or not incurred early on in adult life.

From the behavioral point of view, the cognitive abilities of parasitoids to detect and learn cues related to food-foraging are of major importance as they would influence the inter-patch distances perceived by the parasitoids. Our models, and previous models (Sirot & Bernstein 1996, 1997, Tenhumberg et al. 2006), assume that parasitoids are able simultaneously to perceive stimuli arising from a variety of sources (both environmental and endogenous) and to integrate them in making decisions. Much remains to be known about the actual cognitive abilities of parasitoids in relation to food-foraging.

7.11.2 Modeling

Given their predominance, synovigenic parasitoids ought to be the main focus of future modeling. One obvious way of incorporating greater physiological realism would be to employ a delay in the production of mature eggs. The instantaneous conversion of EPRs to mature eggs, as assumed here, is a gross oversimplification. Modeling of host-feeders has shown that a delay affects the critical egg load at which feeding is predicted to occur (Collier et al. 1994, Heimpel et al. 1998). The length of the delay also could vary with female age (Shirota et al. 1983, Hoffmann et al. 1995) and with experience (Morales-Ramos et al. 1996).

A physiologically realistic representation of egg resorption is also required. Our synovigenic model does, in fact, allow for instantaneous egg resorption insofar as EPRs (w) are drawn upon to fuel maintenance, but more realistically, resorption should take several hours (Jervis & Kidd 1986, Heimpel et al. 1998). There is also a strong case for partitioning of EPRs into mature eggs and the EPRs that have yet to be converted to eggs, and having different transformation rates for each.

A key question is whether different degrees of synovigeny make a difference to model output. Extreme and moderately synovigenic species could differ with respect to the proportionate allocation to initial reserves (see above) and, if lifetime potential fecundity is assumed to be constant, a stronger degree of synovigeny would involve a smaller initial egg load (Jervis & Ferns 2004).

Another important question to address is whether fitness (measured as the lifetime number of eggs laid) varies with ovigeny index. This would have to be explored through forward iteration. Synovigeny in parasitoids does not simply result from a nutritional 'bottleneck' occurring between the larval and adult stages, but is an adaptive strategy, conferring plasticity in matching egg availability to variation in oviposition opportunities (Rosenheim 1999, Ellers et al. 2000, Rosenheim et al. 2000). Previous modeling indicates that in fitness terms, synovigeny is, on balance, superior to pro-ovigeny (Ellers et al. 2000), a view supported by the empirical finding that fewer than 2% of 638 parasitoid wasp species surveyed are strictly pro-ovigenic, whereas there is a much greater percentage of species exhibiting strong to extreme synovigeny (Jervis et al. 2001). However, comparisons of pro-ovigenic and synovigenic strategies have so far not been made with food-foraging incorporated into the models.

There is also a need to study parasitoid behavior under sub-optimal conditions, for example to see how often synovigenic females find themselves in 'last resort' conditions with regard to the use of EPRs, and to shed light on the occurrence of time- and egg-limitation. Insights into both of these phenomena would be achieved through forward iteration of models. Also, our model considers only a single forager that is not influenced by the decisions of conspecifics. Further work should examine the influence of frequency dependency on the feeding decisions in members of a population.

7.12 Implications for parasitoid-host population dynamics and biological control

There has long been an awareness that the effectiveness of parasitoids in suppressing pest populations might depend on the availability of sugar-rich foods such as nectar and honeydew (Clark 1901 (cited in Froggatt 1902), d'Emmerez de Charmoy 1917, Illingworth 1921, Box 1927, Wolcott 1942, Simmonds 1949). However, to date, few biological control practitioners have embraced the idea. Nowadays, the need for sustainability in agriculture has provided fresh impetus to empirical and theoretical studies on the manipulation of parasitoid populations, particularly in a conservation biocontrol context (Gurr et al. 2004, Kean et al. 2003).

Despite the intensity of research undertaken in the 1970s and 1980s into the role of female foraging behavior in parasitoid-host population dynamics (Hassell 1978, Waage & Hassell 1982, Hassell & Waage 1984), no effort was made to explore the dynamical consequences of food-foraging (and its physiological correlates) until the late 1980s (Yamamura & Yano 1988, Kidd & Jervis 1989, Jervis & Kidd 1991, Kidd & Jervis 1991a,b, Jervis & Kidd 1995). However, there have previously been pioneering, intuitive speculations by biocontrol practitioners on the dynamic significance of host-feeding in biological control (Johnston 1915, DeBach 1943). Currently, population modeling remains concentrated on host feeding parasitoids (Kidd & Jervis 1989, Briggs et al. 1995, Křivan 1997, Murdoch et al. 1997, 2005). The role that food-foraging plays in the population dynamics of non-host feeding parasitoids and of their hosts has hardly been considered (Jervis et al. 1996, Křivan & Sirot 1997, Jervis & Kidd 1999).

Despite the considerable amount of research effort now being directed at the manipulation of parasitoids by supplemental food provision (so-called 'food subsidies') in conservation biological control programs (Lewis et al. 1998, Landis et al. 2000, Gurr et al. 2004, Jervis et al. 2004, Heimpel & Jervis 2005, Landis et al. 2005, Lavandero et al. 2006), little of this work is underpinned by population modeling, although Kean et al. 2003) have made a useful start.

What our models tell us, regarding the population dynamic effects of food supplementation, is that the effect of providing floral strips or spraying crops with artificial honeydew is to make the probability of finding food (λ_f, in our models) generally high. Assuming the food to be nutritionally suitable, the result would be similar to confining parasitoids to the 'upland' region of the upper surface in the graphs. This effect, and the decrease in travel costs, means that females will switch to feeding much less readily, resulting in an increase in percentage parasitism. This prediction has empirical support. For example, the prevalence of sugar- or nectar-feeding by *D. insulare* was shown by Lee et al. (2006) to be positively correlated with parasitism within floral plots. However, as Lee et al. (2006) and Heimpel and Jervis (2005) pointed out, several other studies found no beneficial effect of sugar-rich food provision on parasitism levels.

Thus, increased parasitism should result in increased suppression of the host population. However, as Heimpel and Jervis (2005) pointed out, provision of 'food subsidies' may, for various reasons, not result in increased pest suppression, including instances where percentage parasitism is improved. Further, food supplementation may even raise pest numbers above the parasitoid-free level, i.e. exacerbate the pest problem, if mortality from parasitism is followed by overcompensating density dependence in the host population due to factors other than the parasitism in question (van Hamburg & Hassell 1984). The stabilizing effects of optimal food- and host-foraging by parasitoids on the parasitoid-host population interaction were demonstrated by Křivan and Sirot (1997). However, this topic needs to be re-explored, given the fresh insights that have subsequently been gained into optimal food-foraging behavior.

7.13 Concluding remarks

Our literature review and modeling exercises show that a deep understanding of parasitoid foraging behavior requires an integrated approach, taking account not only of behavior but also of parasitoid life-history and nutritional physiology, the latter encompassing nutrient acquisition, allocation, and utilization. A similar view has been expressed by Giron et al. (2002) and Casas et al. (2005). This chapter shows that an important body of empirical knowledge already exists, and that theoretical work can take a more realistic path, i.e. the modeling of synovigeny. However, from an epistemological standpoint, we lack the most important element in this endeavor, i.e. the experimental data that are necessary for rigorously testing the predictions of the models. It is imperative that models be tested using such data obtained from field, not just from the laboratory.

Acknowledgments

We thank Dick Dickinson, Peter Ferns, Hianan Gu, George Heimpel, Julia Jervis, Neil Kidd, and David Lloyd for providing useful answers to various queries, and both Brigitte Tenhumberg and Éric Wajnberg for helpful criticisms of the chapter. We also thank Lionel Humblot for his technical assistance in drawing Figs 7.3 and 7.4.

References

Ahmad, T. (1936) The influence of ecological factors on the Mediterranean flour moth *Ephestia kuehniella* and its parasite, *Nemeritis canescens. Journal of Animal Ecology* **5**: 67–93.

Ayal, Y. (1987) The foraging strategy of *Diaeretiella rapae*. 1. The concept of the elementary unit of foraging. *Journal of Animal Ecology* **56**: 1057–68.

Azzouz, H., Giordanango, P., Wäckers, F. and Kaiser, L. (2004) Effects of feeding frequency and sugar concentration on behavior and longevity of the adult parasitoid: *Aphidius ervi* (Haliday) (Hymenoptera: Braconidae). *Biological Control* **31**: 445–52.

Baggen, L. and Gurr, G.M. (1998) The influence of food on *Copidosoma koehleri* (Hymenoptera: Encyrtidae), and the use of flowering plants as a habitat management tool to enhance biological control of Potato Moth, *Pthorimaea operculella* (Lepidoptera: Gelichiidae). *Biological Control* **11**: 9–17.

Baker, H.G. and Baker, I. (1973) Amino-acids in nectar and their evolutionary significance. *Nature* **241**: 543–5.

Baker, H.G. and Baker, I. (1983) A brief historical review of the chemistry of floral nectar. In: Bentley, B. and Elias, T. (eds.) *The Biology of Nectaries.* Columbia University Press, New York, pp. 126–52.

Barbier, M. (1970) Chemistry and biochemistry of pollens. *Progress in Phytochemistry* **2**: 1–34.

Begum, M., Gurr, G.M., Wratten, S.D. and Nicol, H.I. (2004) Flower color affects tri-trophic biocontrol interactions. *Biological Control* **30**: 584–90.

Bezemer, T.M., Harvey, J.A. and Mills, N.J. (2005) Influence of adult nutrition on the relationship between body size and reproductive parameters in a parasitoid wasp. *Ecological Entomology* **30**: 571–80.

Blum, M.S. (1985) *Fundamentals of Insect Physiology.* Wiley-Interscience, New York.

Boggs, C.L. (1981) Nutritional and life-history determinants of resource allocation in holometabolous insects. *American Naturalist* **117**: 692–709.

Boggs, C.L. (1997) Reproductive allocation from reserves and income in butterfly species with differing adult diets. *Ecology* **78**: 181–91.

Box, H.E. (1927) The introduction of braconid parasites of *Diatraea saccharalis* Fabr., into certain of the West Indian islands. *Bulletin of Entomological Research* **28**: 365–70.

Bracken, G.K. (1966) The role of ten dietary vitamins on fecundity of the parasitoid *Exeristes comstockii* (Cress.) (Hymenoptera: Ichneumonidae). *Canadian Entomologist* **98**: 918–22.

Briggs, C.J., Nisbet, R.M., Murdoch, W.W., Collier, T.R. and Metz, J.A.J. (1995) Dynamical effects of host-feeding in parasitoids. *Journal of Animal Ecology* **64**: 403–16.

Budenberg, W.J. (1990) Honeydew as a contact kairomone for aphid parasitoids. *Entomologia Experimentalis et Applicata* **55**: 139–48.

Casas, J., Nisbet, R.M., Swarbrick, S. and Murdoch, W.W. (2000) Egg load dynamics and oviposition rate in a wild population of a parasitic wasp. *Journal of Animal Ecology* **69**: 185–93.

Casas, J., Driessen, G., Mandon, N. et al. (2003) Strategies of energy acquisition and use of a parasitoid in the wild. *Journal of Animal Ecology* **72**: 691–7.

Casas, J., Pincebourde, S., Mandon, N., Vannier, F., Poujol, R. and Giron, D. (2005) Lifetime nutrient dynamics reveal simultaneous capital and income breeding in a parasitoid. *Ecology* **86**: 545–54.

Cendaña, S.M. (1937) Studies on the biology of *Coccophagus* (Hymenoptera), a genus parasitic on nondiaspidine Coccidae. *University of California Publications in Entomology* **6**: 337–400.

Chan, M.S. and Godfray, H.C.J. (1993) Host-feeding strategies of parasitic wasps. *Evolutionary Ecology* **7**: 593–604.

Chapman, R.F. (1998) *The Insects: Structure and Function*. Cambridge University Press, Cambridge.

Clark, C.W. and Mangel, M. (2000) *Dynamic State Variable Models in Ecology: Methods and Applications*. Oxford University Press, New York.

Clausen, C.P., Jaynes, H.A. and Gardner, T.R. (1933) Further investigations of the parasites of *Popillia japonica* in the Far East. *United States Department of Agriculture Technical Bulletin* **366**: 1–58.

Colinet, H., Hance, T. and Vernon, P. (2006) Water relations, fat reserves, survival, and longevity of a cold-exposed parasitic wasp *Aphidius colemani* (Hymenoptera: Aphidiinae). *Environmental Entomology* **35**: 228–36.

Collier, T.R. (1995) Adding physiological realism to dynamic state-variable models of host-feeding behavior. *Evolutionary Ecology* **9**: 217–35.

Collier, T.R., Murdoch, W.W. and Nisbet, R.M. (1994) Egg load and the decision to host-feed in the parasitoid, *Aphytis melinus*. *Journal of Animal Ecology* **63**: 299–306.

Cortesero, A.M., Stapel, J.O. and Lewis, W.J. (2000) Understanding and manipulating plant attributes to enhance biological control. *Biological Control* **17**: 35–49.

Crane, E. (ed.) (1975) *Honey: A Comprehensive Survey*. Heinemann, London.

Crane, E. (1980) *A Book of Honey*. Oxford University Press, Oxford.

Dafni, A. (1992) *Pollination Ecology: A Practical Approach*. IRL Press, Oxford.

DeBach, P. (1943) The importance of host-feeding by the adult in the reduction of host populations. *Journal of Economic Entomology* **36**: 647–58.

d'Emmerez de Charmoy, D. (1917) Notes relative to the importation of *Tiphia parallela* Smith, from Barbados to Mauritius for the control of *Phytalus smithi* Arrow. *Bulletin of Entomological Research* **8**: 93–102.

Desouhant, E., Driessen, G., Amat, I. and Bernstein, C. (2005) Host and food searching in a parasitoid wasp *Venturia canescens*: a trade-off between current and future reproduction. *Animal Behaviour* **70**: 145–52.

Doten, S.B. (1911) Concerning the relation of food to reproductive activity and longevity in certain hymenopterous parasites. *Technical Bulletin of the Agricultural Experiment Station, the University of Nevada* **78**: 7–30.

Du, Y.J., Poppy, G.M., Powell, W. and Wadhams, L.J. (1997) Chemically mediated associative learning and in the host foraging behaviour of the aphid parasitoid *Aphidius ervi* (Hymenoptera: Barconidae). *Journal of Insect Behavior* **10**: 509–22.

Dyer, L.E. and Landis, D.A. (1996) Effects of habitat, temperature, and sugar availability on longevity of *Eriborus terebrans* (Hymenoptera: Ichneumonidae). *Environmental Entomology* **25**: 1192–201.

Eijs, I., Ellers, J. and van Duinen, G.J. (1998) Feeding strategies in drosophilid parasitoids: The impact of natural food resources on energy reserves in females. *Ecological Entomology* **23**: 133–8.

Eliopoulos, P.A., Harvey, J.A., Athanassiou, C.G. and Stathas, G.J. (2003) Effect of biotic and abiotic factors on reproductive parameters of the synovigenic endoparasitoid *Venturia canescens*. *Physiological Entomology* **28**: 268–75.

Ellers, J. (1996) Fat and eggs: an alternative method to measure the trade-off between survival and reproduction in insect parasitoids. *Netherlands Journal of Zoology* **46**: 227–35.

Ellers, J. and van Alphen, J.J.M. (1997) Life history evolution in *Asobara tabida*: Plasticity in allocation of fat reserves to survival and reproduction. *Journal of Evolutionary Biology* **10**: 771–85.

Ellers, J. and Jervis, M. (2003) Body size and the timing of egg production in parasitoid wasps. *Oikos* **102**: 164–72.

Ellers, J. and Jervis, M.A. (2004) Why are so few parasitoid wasp species pro-ovigenic? *Evolutionary Ecology Research* **6**: 993–1002.

Ellers, J., van Alphen, J.J.M. and Sevenster, J. (1998) A field study of size-fitness relationships in the parasitoid *Asobara tabida*. *Journal of Animal Ecology* **67**: 318–24.

Ellers, J., Sevenster, J.G. and Driessen, G. (2000) Egg load evolution in parasitoids. *American Naturalist* **156**: 650–65.

England, S. and Evans, E.W. (1997) Effects of pea aphid (Homoptera: Aphididae) honeydew on longevity and fecundity of the alfalfa weevil (Coleoptera: Curculionidae) parasitoid *Bathyplectes curculionis* (Hymenoptera: Ichneumonidae). *Environmental Entomology* **26**: 1437–41.

Fadamiro, H.Y. and Heimpel, G.E. (2001) Effects of partial sugar deprivation on lifespan and carbohydrate mobilization in the parasitoid *Macrocentrus grandii* (Hymenoptera: Braconidae). *Annals of the Entomological Society of America* **94**: 909–16.

Fadamiro, H.Y. and Chen, L. (2005) Utilization of aphid honeydew and floral nectar by *Pseudacteon tricuspis* (Diptera: Phoridae), a parasitoid of imported fire ants, *Solenopsis* spp. (Hymenoptera: Formicidae). *Biological Control* **34**: 73–82.

Fadamiro, H.Y., Chen, L., Onagbola, E.O. and Graham, L.F. (2005) Lifespan and patterns of accumulation and mobilization of nutrients in the sugar-fed phorid fly, *Pseudacteon tricuspis*. *Physiological Entomology* **30**: 212–24.

Faegri, K. and van der Pijl, L. (1979) *The Principles of Pollination Ecology*. Pergamon Press, Oxford.

Ferracini, C., Boivin, G. and Alma, A. (2006) Costs and benefits of host feeding in the parasitoid wasp *Trichogramma turkestanica*. *Entomologia Experimentalis et Applicata* **121**: 229–34.

Forsse, E., Smith, S.M. and Bourchier, R.S. (1992) Flight initiation in the egg parasitoid *Trichogramma minutum*: Effects of ambient temperature, mates, food, and host eggs. *Entomologia Experimentalis et Applicata* **62**: 147–54.

Foster, M.A. and Ruesink, W.G. (1984) Influence of flowering weeds associated with reduced tillage in corn on a black cutworm (Lepidoptera: Noctuidae) parasitoid, *Meteorus rubens* (Nees von Esenbeck). *Environmental Entomology* **13**: 664–8.

Froggatt, W.W. (1902) A natural enemy of the sugar-cane beetle in Queensland. *New South Wales Agricultural Gazette* **13**: 63.

Fulton, B.B. (1933) Notes on *Habrocytus cerealellae*, a parasite of the Angoumois Grain Moth. *Annals of the Entomological Society of America* **26**: 536–53.

Gardener, M.C. and Gillman, M.P. (2002) The taste of nectar – a neglected area of pollination biology. *Oikos* **98**: 552–7.

Gilbert, F.S. and Jervis, M.A. (1998) Functional, evolutionary and ecological aspects of feeding-related mouthpart specializations in parasitoid flies. *Biological Journal of the Linnean Society* **63**: 495–535.

Giron, D. and Casas, J. (2003) Lipogenesis in an adult parasitoid wasp. *Journal of Insect Physiology* **49**: 141–7.

Giron, D., Rivero, A., Mandon, N., Darrouzet, E. and Casas, J. (2002) The physiology of host feeding in parasitic wasps: implications for survival. *Functional Ecology* **16**: 750–7.

Giron, D., Pincebourde, S. and Casas, J. (2004) Lifetime gains of host-feeding in a synovigenic parasitoid wasp. *Physiological Entomology* **29**: 436–42.

Gmeinbauer, R. and Crailsheim, K. (1993) Glucose utilization during flight of honeybee (*Apis mellifera*) workers, drones and queens. *Journal of Insect Physiology* **39**: 959–67.

Godfray, H.C.J. (1994) *Parasitoids: Behavioural and Evolutionary Ecology.* Princeton University Press, Princeton.

Gurr, G.M., Scarratt, S.L., Wratten, S.D., Berndt, L. and Irvin, N. (2004) Ecological engineering, habitat manipulation and pest management. In: Gurr, G.M., Wratten S.D. and Altieri, M.A. (eds.) *Ecological Engineering: Advances in Habitat Manipulation for Arthropods.* CSIRO Publishing, Melbourne, pp. 1–12.

Hagley, E.A.C. and Barber, D.R. (1992) Effect of food sources on the longevity and fecundity of *Pholetesor ornigis* (Weed) (Hymenoptera: Braconidae). *Canadian Entomologist* **124**: 341–6.

Hamburg, H. and Hassell, M.P. (1984) Density dependence and the augmentative release of egg parasitoids against graminaceous stalkborers. *Ecological Entomology* **9**: 101–8.

Harborne, J.B. (1988) *Introduction to Ecological Biochemistry.* Academic Press, London.

Harvey, J.A., Harvey, I.F. and Thompson, D.J. (2001) Lifetime reproductive success in the solitary endoparasitoid, *Venturia canescens. Journal of Insect Behavior* **14**: 573–593.

Hassell, M.P. (1978) *The Dynamics of Arthropod Predator-prey Systems.* Monographs in Population Biology. Princeton University Press, Princeton.

Hassell, M.P. and Waage, J.K. (1984) Host-parasitoid population interactions. *Annual Review of Entomology* **29**: 89–114.

Hausmann, C., Wäckers, F.L. and Dorn, S. (2005) Sugar convertibility in the parasitoid *Cotesia glomerata* (Hymenoptera: Braconidae). *Archives of Insect Biochemistry & Physiology* **60**: 223–9.

Heimpel, G.E. and Collier, T.R. (1996) The evolution of host-feeding behaviour in insect parasitoids. *Biological Reviews* **71**: 373–400.

Heimpel, G.E. and Rosenheim, J.A. (1995) Dynamic host feeding by the parasitoid *Aphytis melinus*: The balance between current and future reproduction. *Journal of Animal Ecology* **64**: 153–67.

Heimpel, G.E. and Rosenheim, J.A. (1998) Egg limitation in parasitoids: A review of the evidence and a case study. *Biological Control* **11**: 160–8.

Heimpel, G.E. and Jervis, M.A. (2005) Does floral nectar improve biological control by parasitoids? In: Wäckers, F., van Rijn, P. and Bruin, J. (eds.) *Plant-provided Food for Carnivorous Insects: Protective Mutualism and its Applications.* Cambridge University Press, Cambridge, pp. 267–304.

Heimpel, G.E., Rosenheim, J.A. and Adams, J.M. (1994) Behavioural ecology of host feeding in *Aphytis* parasitoids. *Norwegian Journal of Agricultural Science Supplement* **16**: 101–15.

Heimpel, G.E., Rosenheim, J.A. and Mangel, M. (1997) Predation on adult *Aphytis* in the field. *Oecologia* **110**: 346–52.

Heimpel, G.E., Mangel, M. and Rosenheim, J.A. (1998) Effects of time limitation and egg limitation on lifetime reproductive success of a parasitoid in the field. *American Naturalist* **152**: 273–89.

Hoferer, S., Wäckers, F. and Dorn, S. (2000) Measuring CO_2 respiration rates in the parasitoid *Cotesia glomerata. Mitteilungen der Deutschen Gesellschaft für Allgemeine und Angewandte Entomologie* **12**: 555–8.

Hoffmann, M.P., Walker, D.L. and Shelton, A.M. (1995) Biology of *Trichogramma ostriniae* (Hym.: Trichogramatidae) reared on *Ostrinia nubilalis* (Lep.: Pyralidae) and a survey for additional hosts. *Entomophaga* **40**: 387–402.

Houston, A.I., McNamara, J.M. and Godfray, H.C.J. (1992) The effect of variability on host feeding and reproductive success in parasitoids. *Bulletin of Mathematical Research* **54**: 465–76.

Ide, S. and Lanfranco, D. (2001) Longevity of *Orgilus obscurator* Ness (Hymenoptera: Braconidae) under the influence of different food sources. *Revista Chilena de Historia Natural* **74**: 469–75.

Idris, A.B. and Grafius, E. (1995) Wildflowers as nectar sources for *Diadegma insulare* (Hymenoptera: Ichneumonidae), a parasitoid of diamondback moth (Lepidoptera: Yponomeutidae). *Environmental Entomology* **24**: 1726–35.

Illingworth, J.F. (1921) Natural enemies of sugar-cane beetles in Queensland. *Queensland Bureau Sugar Experimental Station Division of Entomology Bulletin* **13**: 941–4.

Iwasa, Y., Suzuki, Y. and Matsuda, H. (1984) Theory of oviposition strategy of parasitoids: I Effect of mortality and limited egg number. *Theoretical Population Biology* **26**: 205–27.

Jervis, M.A. (1990) Predation of *Lissonota coracinus* (Gmelin) (Hymenoptera: Ichneumonidae) by *Dolichonabis limbatus* (Dahlbom) (Hemiptera: Nabidae). *Entomologist's Gazette* **41**: 231–3.

Jervis, M.A. (1998) Functional and evolutionary aspects of mouthpart structure in parasitoid wasps. *Biological Journal of the Linnean Society* **63**: 461–93.

Jervis, M.A. and Kidd, N.A.C. (1986) Host-feeding strategies in hymenopteran parasitoids. *Biological Reviews* **61**: 395–434.

Jervis, M.A. and Kidd, N.A.C. (1991) The dynamic significance of host-feeding by insect parasitoids – what modellers ought to consider. *Oikos* **62**: 97–9.

Jervis, M.A. and Kidd, N.A.C. (1995) Incorporating physiological realism into models of parasitoid feeding behaviour. *Trends in Ecology & Evolution* **10**: 434–6.

Jervis, M.A. and Kidd, N.A.C. (1999) Parasitoid nutritional ecology. In: Hawkins, B.A. and Cornell, H.V. (eds.) *Theoretical Approaches to Biological Control*. Cambridge University Press, Cambridge, pp. 131–51.

Jervis, M.A. and Vilhelmsen, L.B. (2000) Mouthpart evolution in adults of the 'basal', 'symphytan' hymenopteran lineages. *Biological Journal of the Linnean Society* **70**: 121–46.

Jervis, M.A. and Ferns, P.N. (2004) The timing of egg maturation in insects: Ovigeny index and initial egg load as measures of fitness and of resource allocation. *Oikos* **107**: 449–60.

Jervis, M.A. and Boggs, C.L. (2005) Linking nectar amino acids to fitness in female butterflies. *Trends in Ecology & Evolution* **20**: 585–6.

Jervis, M.A., Kidd, N.A.C., Fitton, M.G., Huddleston, T. and Dawah, H.A. (1993) Flower-visiting by hymenopteran parasitoids. *Journal of Natural History* **27**: 67–105.

Jervis, M.A., Kidd, N.A.C. and Almey, H.A. (1994) Post-reproductive life in the parasitoid *Bracon hebetor* (Say) (Hym., Braconidae). *Journal of Applied Entomology* **117**: 72–7.

Jervis M.A., Kidd, N.A.C. and Heimpel, G.E.H. (1996) Parasitoid adult feeding ecology and biocontrol: A review. *Biocontrol News & Information* **16**: 11–26.

Jervis, M.A., Heimpel, G.E., Ferns, P.N., Harvey, J.A. and Kidd, N.A.C. (2001) Life-history strategies in parasitoid wasps: A comparative analysis of 'ovigeny'. *Journal of Animal Ecology* **70**: 442–58.

Jervis, M.A., Ferns, P.N. and Heimpel, G.E. (2003) Body size and the timing of egg production: A comparative analysis. *Functional Ecology* **17**: 375–83.

Jervis, M.A., Lee, J.C. and Heimpel, G.E. (2004) Use of behavioural and life-history studies to understand the effects of habitat manipulation. In: Gurr, G., Wratten S.D. and Altieri, M. (eds.) *Ecological Engineering for Pest Management: Advances in Habitat Manipulation of Arthropods*. CSIRO Press, Melbourne, pp. 65–100.

Jervis, M.A., Boggs, C.L. and Ferns, P.N. (2005a) Egg maturation strategy and its associated trade-offs in Lepidoptera: A review and synthesis. *Ecological Entomology* **30**: 1–17.

Jervis, M.A., Copland, M.J.W. and Harvey, J.A. (2005b) The life cycle. In: Jervis, M.A. (ed.) *Insects as Natural Enemies: A Practical Perspective*. Springer, Dordrecht, pp. 73–165.

Jervis, M.A., Boggs, C.L. and Ferns, P.N. (2007) Egg maturation strategy and survival trade-offs in holometabolous insects: A comparative approach. *Biological Journal of the Linnean Society* **90**: 293–302.

Johnston, F.A. (1915) Asparagus-beetle egg parasite. *Journal of Agricultural Research* **4**: 303–14.

Kean, J., Wratten, S.D., Tylianakis, J. and Barlow, N. (2003) The population consequences of natural enemy enhancement, and implications for conservation biological control. *Ecology Letters* **6**: 604–12.

Kidd, N.A.C. and Jervis, M.A. (1989) The effects of host-feeding behaviour on the dynamics of parasitoid-host interactions, and the implications for biological control. *Researches on Population Ecology* **31**: 235–74.

Kidd, N.A.C. and Jervis, M.A. (1991a) Host-feeding and oviposition by parasitoids in relation to host stage. *Researches on Population Ecology* **33**: 13–28.

Kidd, N.A.C. and Jervis, M.A. (1991b) Host-feeding and oviposition by parasitoids in relation to host stage: Consequences for parasitoid-host population dynamics. *Researches on Population Ecology* **33**: 87–99.

King, B. (1993) Flight activity in the parasitoid wasp *Nasonia vitripennis* (Hymenoptera: Pteromalidae). *Journal of Insect Behaviour* **6**: 313–21.

King, P.E. and Richards, J.G. (1969) Oogenesis in *Nasonia vitripennis* (Walker) (Hymenoptera: Pteromalidae). *Proceedings of the Royal Entomological Society London Series A* **44**: 143–57.

Křivan, V. (1997) Dynamical consequences of optimal host feeding on host-parasitoid population dynamics. *Bulletin of Mathematical Biology* **59**: 809–31.

Křivan, V. and Sirot, E. (1997) Searching for food or hosts: The influence of parasitoids behavior on host-parasitoid dynamics. *Theoretical Population Biology* **51**: 201–9.

Lamb, K.P. (1959) Composition of the honeydew of the aphid *Brevicoryne brassicae* (L.) feeding on swedes (*Brassica napobrassica* DC). *Journal of Insect Physiology* **3**: 1–13.

Landis, D.A., Wratten, S.D. and Gurr, G.M. (2000) Habitat management to conserve natural enemies of arthropod pests in agriculture. *Annual Review of Entomology* **45**: 175–201.

Landis, D.A., Menalled, F.D., Costamanga, A.C. and Wilkinson, T.K. (2005) Manipulating plant resources to enhance beneficial arthropods in agricultural landscapes. *Weed Science* **53**: 902–5.

Lavandero, B., Wratten, S.D., Didham, R.K. and Gurr, G. (2006) Increasing floral diversity for selective enhancement of biological control agents: A double-edged sward? *Basic & Applied Ecology* **7**: 236–43.

Lee, J.C., Heimpel, G.E. and Leibee, G.L. (2004) Comparing floral nectar and aphid honeydew diets on the longevity and nutrient levels of a parasitoid wasp. *Entomologia Experimentalis et Applicata* **111**: 189–99.

Lee, J., Andow, D.A. and Heimpel, G.E. (2006) Influence of floral resources on sugar feeding and nutrient dynamics of a parasitoid in the field. *Ecological Entomology* **31**: 470–80.

Le Ralec, A. (1995) Egg contents in relation to host-feeding in some parasitic Hymenoptera. *Entomophaga* **40**: 87–93.

Lewis, W.J. and Takasu, K. (1990) Use of learned odours by a parasitic wasp in accordance with host and food needs. *Nature* **348**: 635–6.

Lewis, W.J., Stapel, J.O., Cortesero, A.M. and Takasu, K. (1998) Understanding how parasitoids balance food and host needs: importance to biological control. *Biological Control* **11**: 175–83.

Mangel, M. (1987) Oviposition site selection and clutch size in insects. *Journal of Mathematical Biology* **25**: 1–22.

Mangel, M. and Clark, C.W. (1988) *Dynamic Modelling in Behavioural Ecology*. Princeton University Press, Princeton.

Mathews, P.L. and Stephen, F.M. (1997) Effect of artificial diet on longevity of adult parasitoids of *Dendroctonus frontalis* (Coleoptera: Scolytidae). *Biological Control* **26**: 961–5.

Maurizio, A. (1975) How bees make honey. In: Crane, E. (ed.) *Honey: A Comprehensive Survey*. Heinemann, London, pp. 77–105.

McGregor, R. (1997) Host-feeding and oviposition by parasitoids on hosts of different fitness value: Influences of egg load and encounter rate. *Journal of Insect Behavior* **10**: 451–62.

Mevi-Schutz, J. and Erhardt, A. (2005) Amino acids in nectar enhance butterfly fecundity: A long-awaited link. *American Naturalist* **165**: 412–19.

Mitsunaga, T., Shimoda, T. and Yano, E. (2004) Influence of food supply on longevity and parasitization ability of a larval parasitoid, *Cotesia plutellae* (Hymenoptera: Braconidae). *Applied Entomology & Zoology* **39**: 691–7.

Morales-Ramos, J.A., Rojas, M.G. and King, E.G. (1996) Significance of adult nutrition and oviposition experience on longevity and attainment of full fecundity of *Catolaccus grandis* (Hymenoptera: Pteromalidae). *Annals of the Entomological Society of America* **89**: 555–63.

Murdoch, W.W., Briggs, C.J. and Nisbet, R.M. (1997) Dynamical effects of host-size and parasitoid state-dependent attacks by parasitoids. *Journal of Animal Ecology* **66**: 542–56.

Murdoch, W.W., Briggs, C.J. and Swarbrick, S. (2005) Host suppression and stability in a parasitoid-host system: Experimental demonstration. *Science* **309**: 610–13.

O'Brien, D.M. (1999) Fuel use in flight and its dependence on nectar feeding in the hawkmoth *Amphion floridensis*. *Journal of Experimental Biology* **202**: 441–51.

O'Brien, D.M., Fogel, M.L. and Boggs, C.L. (2002) Renewable and non-renewable resources: Amino acid turnover and allocation to reproduction in Lepidoptera. *Proceedings of the National Academy of Sciences USA* **99**: 4413–18.

O'Brien, D.M., Boggs, C.L. and Fogel, M.L. (2003) Pollen feeding in the butterfly *Heliconius charitonia*: Isotopic evidence for essential amino acid transfer from pollen to eggs. *Proceedings of the Royal Society of London Series B Biological Science* **270**: 2631–6.

O'Brien, D.M., Boggs, C.L. and Fogel, M.L. (2004) Making eggs from nectar: Connections between butterfly life history and the importance of nectar carbon in reproduction. *Oikos* **105**: 279–91.

O'Brien, D.M., Boggs, C.L. and Fogel, M.L. (2005) The dietary sources of amino acids used in reproduction by butterflies: A comparative study using compound specific stable isotope analysis. *Physiological & Biochemical Zoology* **78**: 819–27.

Olson, D.L. and Nechols, J.R. (1995) Effects of squash leaf trichome exudates and honey on adult feeding, survival, and fecundity of the squash bug (Heteroptera: Coreidae) egg parasitoid *Gryon pennsylvaticum* (Hymenoptera: Scelionidae). *Biological Control* **24**: 454–8.

Olson, D.M., Fadamero, H., Lundgren, J.G. and Heimpel, G.E. (2000) Effects of sugar-feeding on carbohydrate and lipid metabolism in a parasitoid wasp. *Physiological Entomology* **25**: 17–26.

Patt, J.M., Hamilton, G.C. and Lashomb, J.H. (1999) Responses of two parasitoid wasps to nectar odors as a function of experience. *Entomologia Experimentalis et Applicata* **90**: 1–8.

Percival, M. (1961) Types of nectar in angiosperms. *New Phytologist* **60**: 235–81.

Percival, M.S. (1965) *Floral Biology*. Pergamon Press, Oxford.

Pexton, J.J. and Mayhew, P.J. (2002) Siblicide and life history evolution in parasitoids. *Behavioral Ecology* **13**: 690–5.

Pompanon, F., Fouillet, P. and Boulétreau, M. (1999) Physiological and genetic factors as sources of variation in locomotion and activity rhythm in a parasitoid wasp (*Trichogramma brassicae*). *Physiological Entomology* **24**: 346–57.

Prys-Jones, O. and Corbet, S.A. (1983) *Bumblebees*. Cambridge University Press, Cambridge.

Quicke, D.J.L. and Jervis, M.A. (2005) A new species of '*Bracon*' (Hymenoptera: Braconidae) from South Africa with exceptionally long mouthparts. *African Entomology* **13**: 367–71.

Rivero, A. and Casas, J. (1999) Rate of nutrient allocation to egg production in a parasitic wasp. *Proceedings of the Royal Society of London Series B Biological Science* **266**: 1169–74.

Rivero, A. and West, S.A. (2005) The costs and benefits of host feeding in parasitoids. *Animal Behaviour* **69**: 1293–301.

Rivero, A., Giron, D. and Casas, J. (2001) Lifetime allocation of juvenile and adult nutritional resources to egg production in a holometabolous insect. *Proceedings of the Royal Society of London Series B Biological Science* **268**: 1231–7.

Rose, U.S.R., Lewis, J. and Tumlinson, J.H. (2006) Extrafloral nectar from cotton (*Gossypium hirsutum*) as a food source for parasitic wasps. *Functional Ecology* **20**: 67–74.

Rosenheim, J.A. (1998) Higher-order predators and the regulation of insect herbivore populations. *Annual Review of Entomology* **43**: 421–47.

Rosenheim, J.A. (1999) The relative contributions of time and eggs to the cost of reproduction. *Evolution* **53**: 376–85.

Rosenheim, J.A., Heimpel, G.E. and Mangel, M. (2000) Egg maturation, egg resorption, and the costliness of transient egg limitation in insects. *Proceedings of the Royal Society of London Series B Biological Science* **267**: 1565–73.

Shahjahan, M. (1974) *Erigeron* flowers as a food and attractive odour source for *Peristenus pseudopallipes*, a braconid parasitoid of the tarnished plant bug. *Environmental Entomology* **3**: 69–72.

Shirota, Y., Carter, N., Rabbinge, R. and Ankersmit, G.W. (1983) The biology of *Aphidius rhopalosiphi*, a parasitoid of cereal aphids. *Entomologia Experimentalis et Applicata* **34**: 27–34.

Siekmann, G., Tenhumberg, B. and Keller, M.A. (2001) Feeding and survival in parasitic wasps: Sugar concentration and timing matter. *Oikos* **95**: 425–30.

Siekmann, G., Tenhumberg, B. and Keller, M.A. (2004) The sweet tooth of adult parasitoid *Cotesia rubecula*: Ignoring hosts for nectar? *Journal of Insect Behavior* **17**: 459–76.

Simmonds, H.W. (1949) On the introduction of *Scolia ruficornis* F., into Western Samoa for the control of *Oryctes rhinoceros* L. *Bulletin of Entomological Research* **40**: 445–6.

Sirot, E. and Bernstein, C. (1996) Time-sharing between host searching and food searching in solitary parasitoids: state-dependent optimal strategies. *Behavioral Ecology* **7**: 189–94.

Sirot, E. and Bernstein, C. (1997) Food searching and superparasitism in solitary parasitoids. *Acta Oecologia* **18**: 63–72.

Sisterton, M.S. and Averill, A.L. (2002) Costs and benefits of food foraging for a braconid parasitoid. *Journal of Insect Behavior* **15**: 571–88.

Sivinksi, J., Aluja, M. and Holler, T. (2006) Food sources for adult *Diachasmimorpha longicaudata*, a parasitoid of tephritid fruit flies: Effects on longevity and fecundity. *Entomologia Expirimentalis et Applicata* **118**: 193–202.

Spafford Jacob, H. and Evans, E.W. (2004) Influence of different sugars on the longevity of *Bathyplectes curculionis* (Hymenoptera: Ichneumonidae). *Journal of Applied Entomology* **128**: 316–20.

Stanley, G. and Linskens, H.F. (1974) *Pollen: Biology, Biochemistry and Management*. Springer-Verlag, Berlin.

Steppuhn, A. and Wäckers, F.L. (2004) HPLC sugar analysis reveals the nutritional state and the feeding history of parasitoids. *Functional Ecology* **18**: 812–19.

Stevens, D.J., Hansell, M.H., Freel, J.A. and Monaghan, P. (1999) Developmental trade-offs in caddis flies: Increased investment in larval defence alters adult resource allocation. *Proceedings of the Royal Society of London Series B Biological Science* **266**: 1049–54.

Stevens, D.J., Hansell, M.H. and Monaghan, P. (2000) Developmental trade-offs and life histories: strategic allocation of resources in caddis flies. *Proceedings of the Royal Society of London Series B Biological Science* **267**: 1511–15.

Suarez, R.K., Darveau, C.-A., Welch Jr., K.C., O'Brien, D.M., Roubik, D.W. and Hochachka, P.W. (2005) Energy metabolism in orchid bee flight muscles: Carbohydrate fuels all. *Journal of Experimental Biology* **208**: 3573–9.

Syme, P.D. (1975) Effects on the longevity and fecundity of two native parasites of the European Pine Shoot Moth in Ontario. *Environmental Entomology* **4**: 337–46.

Syme, P.D. (1977) Observations on the longevity and fecundity of *Orgilus obscurator* (Hymenoptera: Braconidae) and the effects of certain foods on longevity. *Canadian Entomologist* **109**: 995–1000.

Tenhumberg, B., Siekmann, G. and Keller, M.A. (2006) Optimal time allocation in parasitic wasps searching for hosts and food. *Oikos* **113**: 121–31.

Thompson, S.N. (2003) Trehalose – the insect 'blood' sugar. *Advances in Insect Physiology* **31**: 205–85.

Thorne, A.D., Pexton, J.J., Dytham, C. and Mayhew, P.J. (2006) Small body size shifts development towards early reproduction in an insect. *Proceedings of the Royal Society Series B Biological Science* **273**: 1099–103.

van Emden, H.F. (1963) Observations on the effect of flowers on the activity of parasitic Hymenoptera. *Entomologist's Monthly Magazine* **98**: 265–70.

van Lenteren, J.C., van Vianen, A., Gast, H.F. and Kortenhoff, A. (1987) The parasite-host relationship between *Encarsia formosa* Gahan (Hym., Aphelinidae) and *Trialeurodes vaporariorum* (Westwood) (Hom., Aleyrodidae). XVI. Food effects on oogenesis, life span and fecundity of *Encarsia formosa* and other hymenopterous parasites. *Zeitschrift für Angewandte Entomologie* **103**: 69–84.

Vattala, H.D., Wratten, S.D., Phillips, C.B. and Wäckers, F.L. (2006) The influence of flower morphology and nectar quality on the longevity of a parasitoid biological control agent. *Biological Control* **39**: 179–85.

Waage, J.K. and Hassell, M.P. (1982) Parasitoids as biological control agents – a fundamental approach. *Parasitology* **84**: 241–68.

Waage, J.K., Carl, K.P., Mills, N.J. and Greathead, D.J. (1985) Rearing entomophagous insects. In: Singh, P. and Moore, R.F. (eds.) *Handbook of Insect Rearing*, Vol. 1. Elsevier, Amsterdam, pp. 45–66.

Wäckers, F.L. (1994) The effect of food deprivation on the innate visual and olfactory preferences in the parasitoid *Cotesia rubecula*. *Journal of Insect Physiology* **40**: 641–9.

Wäckers, F.L. (2001) A comparison of nectar- and honeydew sugars with respect to their utilization by the hymenopteran parasitoid *Cotesia glomerata*. *Journal of Insect Physiology* **47**: 1077–84.

Wäckers, F.L. (2004) Assessing the suitability of flowering herbs as parasitoid food sources: Flower attractiveness and nectar accessibility. *Biological Control* **29**: 307–14.

Wäckers, F.L., Lee, J.C., Heimpel, G.E., Winkler, K. and Wagenaar, R. (2007) Hymenopteran parasitoids synthesize 'honeydew-specific' oligosaccharides. *Functional Ecology* **20**: 790–8.

Wanner, H., Gu, H. and Dorn, S. (2006) Nutritional value of floral nectar sources for flight in the parasitoid wasp, *Cotesia glomerata*. *Physiological Entomology* **31**: 127–33.

Weisser, W.W., Völkl, W. and Hassell, M.P. (1997) The importance of adverse weather conditions for behaviour and population ecology of an aphid parasitoid. *Journal of Animal Ecology* **65**: 631–9.

Wigglesworth, V.B. (1972) *The Principles of Insect Physiology*, 7th edn. Chapman & Hall, London.

Willmer, P.G. (1985) Size effects on the hygrothermal balance and foraging patterns of a sphecid wasp, *Cerceris arenaria*. *Ecological Entomology* **10**: 469–79.

Willmer, P. (1986) Foraging patterns and water balance: Problems of optimization for a xerophilic bee, *Chalicodoma sicula*. *Journal of Animal Ecology* **55**: 941–62.

Winkler, K., Wäckers, F.L., Stingli, A. and van Lenteren, J.C. (2005) *Plutella xylostella* (diamond-back moth) and its parasitoid *Diadegma semiclausum* show different gustatory and longevity responses to a range of nectar and honeydew sugars. *Entomologia Experimentalis et Applicata* **115**: 187–92.

Winkler, K., Wäckers, F.L. and Bukovinszkine-Kiss, G. (2006) Sugar resources are vital for *Diadegma semiclausum* fecundity under field conditions. *Basic & Applied Ecology* **7**: 133–40.

Wolcott, G.N. (1942) The requirements of parasites for more than hosts. *Science* **96**: 317–18.

Yamamura, N. and Yano, E. (1988) A simple model of host-parasitoid interaction with host-feeding. *Researches on Population Ecology* **30**: 353–69.

8

Information acquisition, information processing, and patch time allocation in insect parasitoids

Jacques J.M. van Alphen and Carlos Bernstein

Abstract

Resources that animals depend on for growth, maintenance, and reproduction are usually heterogeneously distributed. Animals foraging in a heterogeneous environment face complex decisions. This is particularly the case for insect parasitoids foraging for hosts, for whom one of the key questions is how to allocate search time over patches in the habitat. To cope with this problem and to make adaptive decisions, parasitoids must acquire and integrate the information provided by a large variety of cues. We review the wealth of information parasitoids integrate, such as time spent in the patch, ovipositions performed, the chemicals deposited by hosts, contacts with parasitized hosts, etc. Recently, it has also been shown that contacts with conspecifics, the internal state of the animal, and environmental conditions also influence patch-leaving decisions. The demonstration of the existence of genetic variation in patch leaving mechanisms indicates that the way parasitoids respond to information is potentially under the influence of natural selection. Indeed, the patch leaving mechanisms observed in environments with different spatial distributions of hosts and other resources result in patch time allocation patterns that are more often in agreement with the predictions of deductive functional models. We show how most of our detailed knowledge about the responses of parasitoids to different stimuli stems from the use of powerful statistical tools (see also Chapter 18 by Wajnberg and Haccou). However, we also show the danger of ignoring critical influences. If the parasitoid responds to one or more cues not represented in the statistical model, erroneous conclusions on the effect of cues included in the model may be the result. This problem can be solved by a better understanding of parasitoid cognition. We need realistic models on how parasitoids integrate information from the environment with that of their internal state, to design more realistic statistical models. As parasitoids often compete with conspecifics, we review the theory on group foraging and stress the need for more empirical work on this topic. Understanding of the dynamics of group foraging and of dispersal decisions of individuals allows us to make a realistic link between individual behavior and population level processes.

Finally, we will discuss the most promising avenues for future research. The most important message we aim to convey is that the full understanding of a behavioral process depends on the close integration of study of final causes and of the mechanisms involved. When this logic is taken to its final conclusions, the boundary between the two approaches fades.

8.1 Introduction

In this chapter, we consider the foraging behavior of parasitoids in a patchy environment. The hosts of parasitoids occur in discrete patches, separated by areas of unsuitable host habitat. For efficient foraging, a parasitoid must apportion the foraging time which it does not spend in inter-patch travel over patches of different quality. How this foraging time is apportioned affects not only the reproductive success of individual parasitoids, but also the population dynamics of parasitoid host interactions (see also Chapter 13 by Bonsall and Bernstein). The study of patch time allocation has therefore attracted the attention of both theoreticians and empirical students of parasitoid behavior.

Early models on adaptive patch time allocation considered a single foraging female exploiting the habitat and ignored the complicating factors of conspecific competitors exploiting the same host population. In this respect, the first and most influential model was Charnov's Marginal Value Theorem (MVT) (Charnov 1976). It is a rate maximization model predicting the optimal time a parasitoid female should remain in a host patch. This model assumes that, upon entering a patch, the rate of fitness gained by the female progressively drops as a function of patch residence time, as available hosts are progressively depleted. It predicts that female parasitoids should leave the patch when their instantaneous rate of oviposition (dR/dt) falls below the average rate that can be achieved in the environment. This result can be obtained in the following way: Suppose that the net gain of staying a time t in a patch is $g(t)$ and that the average travel time between patches is T. The rate R of gain is then

$$R = \frac{g(t)}{T + t} \tag{8.1}$$

Now assume that the habitat consists of i different patch types, that encounter rate with patches of type i is p_i and that the forager leaves a patch of type i after time t_i. Then the rate of gain becomes

$$R = \frac{\sum_i p_i g_i(t_i)}{T + \sum_i p_i t_i} \tag{8.2}$$

Setting the derivative of g_i with respect to t_i to zero leads to the solution:

$$R = \frac{dg_i}{dt_i} \tag{8.3}$$

Thus, the forager should leave each patch when the rate of gain has dropped to the overall gain of the habitat. From this expression the optimal residence time t^* can be easily deducted.

This model takes into account the spatial distribution of host availability and predicts the time allocation over patches of a single individual. The theorem predicts that all patches in the habitat should be depleted until the rate of gain has dropped to the average gain rate expected for the habitat. It generates three testable qualitative predictions:

1 patch time should increase with host density in the patch;
2 patch times should increase with increasing average travel time in the habitat and should decrease with increasing average host density in the patches; and
3 all patches should be reduced to the same level of profitability.

The model is purely functional (i.e. it addresses the fitness consequences of time allocation) and assumes omniscient parasitoids (i.e. with a full knowledge of host and patch distribution). As a consequence, it does not take into account the behavioral proximate mechanisms used by foragers to control patch time or the mechanisms they use to obtain information about host distribution. However, this functional model prompted questions about the mechanism involved. As parasitoids are unlikely to be omniscient, initially some simple stopping rules (i.e. so-called 'rule of thumb') were proposed, but available evidence suggested that parasitoids used more complex rules than the classical 'fixed giving up time' (i.e. leave after not having found an unparasitized host during a fixed period of time) or 'fixed number expectation' (i.e. leave after a fixed number of hosts has been found), and researchers started to pay more attention to the mechanisms involved (Waage 1979). A further step was taken when behavioral ecologists started to view mechanisms and constraints as adaption problems (Iwasa et al. 1981). Over the years, researchers asking mechanistic questions have shown that a wide variety of environmental stimuli affect parasitoids' patch residence time (Godfray 1994, van Alphen et al. 2003, Wajnberg 2006). In recent years, research has focused on how the information is integrated by the foraging animal. Although promising advances have been made, there are still large gaps in our knowledge, which hamper our progress in understanding adaptation and constraints in foraging behavior of parasitoids. In this chapter, we will briefly review previous work and identify some of the unsolved problems and promising areas for future research.

8.2 Foraging by a single female

In order to forage efficiently in a heterogeneous environment, animals have to solve essential questions: Where to forage? For how long? Should they join or avoid foraging conspecifics? From its early beginnings, behavioral ecology suggested that, in order to forage efficiently in a heterogeneous environment, animals should acquire and process information. This is essential because the environment changes continuously and each individual experiences a different environment as it samples a different subset of the available patches of resources. How foragers acquire and process information is one of the paramount questions in the study of animal behavior. Animals, such as mammals and birds are long-lived, and may learn the geography of their home range and the profitability of different areas during the juvenile stage, before reproduction starts. Subsequently they must update this information either by their own experience (i.e. private information) or by observing the gains obtained by conspecifics (i.e. public information). As newly-born female parasitoids start reproduction shortly after emergence, they lack such information and face the task of acquiring it, mostly on their own, while searching for hosts. Alternatively, females can use an innate, fixed expectation of the gains in the environment (Boivin et al. 2004).

Hence, parasitoid foraging is likely to be constrained by information during at least part of the adult life. On the other hand, unlike birds, parasitoids use a wide variety of chemical and other stimuli to obtain information on their environment, and this allows them to restrict their foraging behavior to that part of the habitat where hosts are most likely to be found (Wajnberg 2006). Yet it is often observed in the field that parasitoids aggregate their attacks in some locations but independently of the distribution of their hosts (Pacala & Hassell 1991). The lack of response to host aggregation indicates that information constraints are important in patch time allocation.

8.2.1 Incremental effects of host encounters

The key question in understanding how parasitoids allocate search time over patches in a habitat is how they integrate the different types of information. The first mechanistic model integrating the effect of several stimuli (Waage 1979) was based on observations of *Venturia canescens*'s behavior and suggested a decision mechanism for the duration of a patch visit by a parasitoid (Fig. 8.1). This model assumes that the motivation of a parasitoid to remain and search on a particular patch decreases with time. The initial level of motivation would be linearly correlated with the concentration of kairomone (i.e. a substance produced by the hosts, providing the parasitoid with information on host species and availability) in the patch, an indication of the host density. As long as this responsiveness is above a given threshold, the parasitoids turn inward each time they reach the

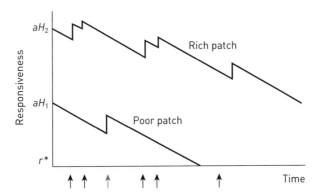

Fig. 8.1 Graphical representation of the Waage (1979) model. The initial responsiveness to the patch edge (aH_i) is proportional to host density (H_i) and decreases as time on the patch proceeds. As a result of ovipositions, the responsiveness increases, the increment being dependent on the time elapsed since the last oviposition. When the level of responsiveness has reached the critical threshold $r*$, the parasitoid leaves the patch. The black arrows along the time axis denote ovipositions in a relatively rich patch, the gray arrow an oviposition in a relatively poor patch.

edge of the patch, otherwise the patch is abandoned. Thus, in the absence of hosts (for instance under experimental conditions), the motivation to search would decrease until a threshold was reached and the parasitoid would leave the patch. This would happen because the parasitoid became accustomed to the stimulus of the kairomone, and no longer perceived the patch edge. An oviposition alters the waning of this response by adding an increment (I) to the current level of responsiveness. The value of I is assumed to be proportional to the time elapsed since the last oviposition, up to a maximum value, I_{max}. T, and the duration of a patch visit is given by

$$T = \frac{aH + \sum_i I_i - r^*}{b} \tag{8.4}$$

where H is the number of hosts in the patch, r^* is the threshold, a is a constant relating initial responsiveness to the number of hosts, and b is the rate of responsiveness decline. Hence, in this model, an oviposition provides an incremental stimulus that would reset the motivation to search to a higher level (the animal is said to use a so-called incremental mechanism). This mechanism results in longer patch times on patches with more hosts. Although this model gives a good qualitative description of patch time allocation in many species of parasitoids (van Alphen & Galis 1983, van Alphen et al. 2003, Wajnberg 2006), it is based on only two stimuli affecting patch time. A test of the model is hampered by the problem that the parameters cannot be estimated from experimental data (Wajnberg 2006, Driessen & Bernstein 1999). It should be noted that the Waage (1979) model assumes that parasitoids gain an accurate estimation of initial host availability through kairomone concentration, and yet increase this estimation through ovipositions (i.e. while the patch is getting poorer). Driessen and Bernstein (1999) suggested that incremental mechanisms would be more appropriate when parasitoids can detect the presence of hosts in a patch but not their number (as it is the case of V. canescens when host numbers are high). In this case, the mechanism will lead to a longer patch residence time and to a more thorough exploitation of the richest patches.

8.2.2 Decremental effects of host encounters

In contradiction to what the Waage (1979) model led us to expect, many parasitoids decrease patch residence time in response to ovipositions (they are said to use a decremental rule, see Fig. 8.2). In this case, a certain value is subtracted from the current responsiveness at each oviposition. This assumes that host availability does not change rapidly as a consequence of conspecific foraging, or any other external causes of host mortality (Spataro 2001).

One of the examples is V. canescens (Driessen et al. 1995), the same parasitoid studied by Waage (1979). Other examples come from Wajnberg et al. (1999), Outreman et al. (2001), and Tenhumberg et al. (2001a,b). Both rules are common in nature, although the number of parasitoid species that increase their patch time as a consequence of ovipositions (i.e. that use incremental rules) seems somewhat larger (although the sample size is still modest (Driessen & Bernstein 1999, Wajnberg et al. 2000, Wajnberg 2006). How could it be that in a process under strong selective pressure, an important event would have opposite consequences? Pre-dating almost all the experimental work, Iwasa et al. (1981) offered the first explanation. Their model suggests that the rules that parasitoids should use depend

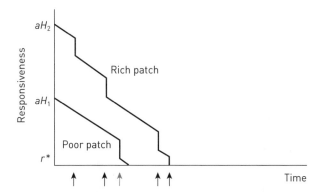

Fig. 8.2 Graphical representation of a decremental patch leaving decision rule. The initial responsiveness to the patch edge (aH_i) is proportional to host density (H_i) and decreases as time on the patch proceeds (decay is assumed linear for simplicity). As a result of ovipositions, the responsiveness decreases. In this example, decrements are taken as constant. For further details, see Fig. 8.1.

on the distribution of the hosts. Parasitoids foraging for hosts following an aggregated distribution should use incremental rules, those foraging for uniformly distributed hosts should use decremental rules, and parasitoids foraging for Poisson distributed hosts should be indifferent to the number of hosts encountered. This model is based on the assumption that the type of host distribution is constant over generations and that the right mechanism is acquired through natural selection. Nevertheless, short-term changes in the environment occur during the life of the individuals and they have evolved mechanisms to gather information to update their knowledge. When the number of hosts in a patch is known, as when kairomone concentration is a reliable indicator of the number of hosts in the patch, each oviposition implies a decrease in host availability. The patch becomes progressively less valuable compared with the rest of the environment and parasitoids should use a decremental mechanism. On the other hand, when information on host availability is lacking (some patches are rich, others poor, but no telltale cue is available), the timing of each additional oviposition reveals the quality of the patch and determines if a parasitoid should continue to search that patch. In this case, an incremental rule would be expected.

Driessen and Bernstein (1999) pursued this idea both theoretically and experimentally. They first showed that *Venturia*'s hosts tend to have a uniform distribution in the field (most fruits contain a single host, see Fig. 8.3). Next, they showed experimentally that, on arrival in a patch, the parasitoid obtains an estimation of the host density by detecting the kairomone concentration (Waage 1978). Using an optimization model, the authors show that, for *V. canescens* searching individually, a decremental mechanism will be out-competed by other rules only when travel times are much larger than those that are likely to occur in the field.

Apart from incremental or decremental effects of encounters with unparasitized hosts, decremental effects of encounters with already parasitized hosts have also been documented (Hemerik et al. 1993, Wajnberg et al. 1999, Outreman 2000, Wajnberg et al. 2000, Wajnberg 2006). Encounters with parasitized hosts can also provide information on the current level of patch exploitation and hence about when to leave a patch. This may be

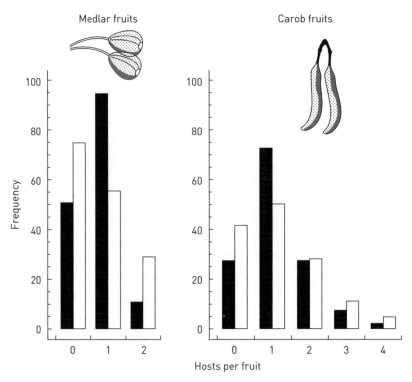

Fig. 8.3 Observed frequency distributions of *Ectomyelois ceratonaie* larvae over carob and the medlar fruits in southern France (dark bars). *E. ceratoniae* is a natural host of *V. canescens* and carob and the medlar fruits are two of the most common trees harboring the larvae. The light bars show the expected frequencies according to a Poisson distribution. Most fruits contain a single larva, and the distribution is more uniform than a Poisson distribution.

especially important for parasitoids searching on patches together with competitors or patches previously visited by one or more competitors. Different responses to parasitized and healthy hosts are only possible if a parasitoid can discriminate between the two categories. A majority of species has this ability, but some species seem to lack it and always accept parasitized hosts. One possible explanation is that these species indeed lack the ability to discriminate, but an alternative explanation is that they superparasitize despite recognizing the hosts as already parasitized. Rosenheim and Mangel (1994) have considered the lack of host discrimination in a parasitoid as an information constraint, and predicted optimal patch time under this condition. Their model predicts that the constraint should result in shorter patch times, as patches become unprofitable sooner because self-superparasitism cannot be detected and avoided.

8.3 Statistical models to determine the effect of different stimuli on the tendency to leave the patch

Most of the knowledge on the responses of parasitoids to host encounters was obtained through statistical methods that allow determining quantitatively the contribution of the different types of information and events on the probability of patch leaving by a parasitoid. Haccou and Hemerik (1985) introduced the use of Cox's proportional hazard model (Cox 1972, Cox & Oakes 1984) into behavioral ecology. Briefly, this statistical model works the following way: If N foragers leave a patch at a constant rate, λ, in deterministic terms, the process can be represented as $dN/dt = -\lambda N$. In probabilistic terms, the number of individuals disappearing in a given time interval would follow a Poisson distribution with mean λNt (t is the observation time). λ need not be constant but most often it would be the outcome of other intervening processes and, as a consequence, a function of time. The proportional hazards model (or Cox's model) allows estimating how λ is affected by different so-called covariates (i.e. events) and their timing, for instance encounters with hosts or conspecifics. Now λ is written as $\lambda(t,z)$, where z is a vector coding the covariates that might be continuous or discrete variables. Basically, this technique estimates $\lambda(t,z)$ as

$$\lambda(t,z) = \lambda_0(t)exp\left(\sum_i \beta_i z_i(t)\right) \tag{8.5}$$

where $\lambda_0(t)$ is an (unspecified) base rate assumed to depend only on time and common to all possible outcomes. $\lambda_0(t)$, the baseline hazard function, corresponds to the situation in which all covariates are set to zero, and β_i are parameters. The exponential function assures that $\lambda(t,z)$ will always take positive values. For the study of patch time allocation of a parasitoid, t could be the time since arrival at the patch or since the last egg was laid, and z could code for the number of eggs already laid, the number of conspecifics in the patch, etc. Non-significant β_i values correspond to processes not influencing the outcome (patch leaving, for instance). Positive (respectively negative) β_i values will reveal processes (number of encounters with parasitized hosts, for instance) that increase (respectively decrease) the leaving rate, so parasitoids stay shorter (respectively longer) times. This technique allows appraising the joint influence of different processes and of the interaction between them with a minimum of *a priori* decisions.

This method allowed progress to a quantitative analysis of the influence of isolated stimuli, the results of which can be integrated into mechanistic or adaptive patch models (Driessen & Bernstein 1999, Wajnberg et al. 2000). The proportional hazards model has now been applied to a number of different data sets representing a variety of parasitoid species (Haccou et al. 1991, Hemerik et al. 1993, Driessen et al. 1995, Driessen & Bernstein 1999, Wajnberg et al. 1999, 2000, Outreman 2000, Goubault et al. 2005, Outreman et al. 2005, Varaldi et al. 2005).

Applied to parasitoid behavior, the Cox model implies untested assumptions about behavior and its results might depend on these assumptions. Therefore, we should carefully study if these assumptions are realistic, and if other assumptions would lead to the same conclusions. The proportional hazards model is basically a multiple regression tool, sharing strength and weaknesses with other similar methods. For instance, the fitted equation (a simple combination of sums and products) lacks a real representation of the biological process implicated. The strength of this is an increased flexibility, because of a relatively low number of *a priori* decisions imposed on the statistical model. However, it also results in the weakness of a less stringent test of theoretical models. More importantly, as with any regression method, Cox's proportional hazard model can identify relationships between variables but not causal effects. Yet the model is applied to identify the causal mechanisms of patch time allocation.

A seldom explored problem in the study of animal behavior that might have important consequences, is the misspecification of the regression model, for example, by ignoring a critical influence. Another difficulty, common to regression methods and stemming from the lack of representation of actual mechanisms, is the interpretation of complex models including the statistical interaction of several variables (Sokal & Rohlf 1994) or when studies of similar conditions lead to divergent results.

One way to explore how robust the Cox model is to the misspecification of the model would be by numerical experiments, i.e. by generating artificial data, simulating the patch leaving decisions of an animal, and then fitting the results to models ignoring some of the key processes. Of course, a way to decrease the risk of misspecification of the model by ignoring a critical influence is to formulate possible alternative hypotheses before collecting the data, on the basis of previous knowledge of patch leaving decisions by parasitoids. In the following, we discuss several examples to show that alternative hypotheses would have resulted in different conclusions, both for the mechanism of time allocation and for functional models predicting optimal time allocation.

8.3.1 The lack of discrimination between parasitized and unparasitized hosts and the response to ovipositions

Varaldi et al. (2003) discovered that a large proportion of the females of *Leptopilina boulardi*, in southern Europe, are infected with a virus, which induces them to accept already parasitized hosts. They hypothesized that the virus incapacitates the ability of parasitoids to discriminate between unparasitized and parasitized hosts. They used virus-infected *L. boulardi* females to study patch time allocation and compared the results with those of uninfected *Leptopilina heterotoma* females in similar experiments. Using a proportional hazards model with, as covariates, ovipositions during current patch visit, rejections during current patch visit, ovipositions during previous patch visits, number of visits, and host species, they did not find any effect of ovipositions during the current patch visit on the tendency to leave in the infected *L. boulardi* females, while ovipositions in *L. heterotoma* resulted in a decreased leaving tendency (Varaldi et al. 2005). Their conclusion that virus-infected *L. boulardi* females do not respond to ovipositions by a change in leaving tendency is based on all observed ovipositions, without discriminating those performed in healthy hosts from those in hosts that had already been parasitized. They assumed that infected *L. boulardi* females do not recognize already parasitized hosts and hence, that each oviposition would have had the same effect on leaving tendency, irrespective of the host's quality. However, this seems unlikely, as the experiments also provided indications that infected females still discriminate between parasitized and unparasitized hosts (see below) and we know that other related parasitoids respond differently to encounters with parasitized and unparasitized hosts.

An alternative explanation for Varaldi et al.'s (2005) findings could be that decreases in leaving tendency by ovipositions in unparasitized hosts are counteracted by increases in leaving tendency due to ovipositions in parasitized hosts, resulting in a value not different from zero. This alternative hypothesis explicitly assumes that the cue for a change in leaving tendency is not the oviposition *per se*, but a host cue perceived upon encounter.

It seems unlikely that virus-infected females of *L. boulardi* do not respond to cues related to ovipositions, because infected females still reject some parasitized hosts and because, in the same experiments, a significant effect on leaving tendency was found from ovipositions during previous patch visits. An alternative explanation could be that the discrimination ability is still functional, but that the virus only lowers the threshold for superparasitism. The distinction between an inability to discriminate and a low threshold for superparasitism is important, because in both functional models for patch time allocation (Rosenheim & Mangel 1994) and in a statistical analysis of the cues affecting leaving tendency, the outcome of the models will differ depending on if we assume *a priori* an inability to perceive the information, or assume the ability is present and the information is available for the parasitoid.

8.3.2 The response to pheromone marking trails

Varaldi et al. (2005) suggest that, for their patch time allocation, virus-infected *L. boulardi* females use cues not related to host encounters. This suggestion stems from the fact that they found no effect of ovipositions and yet females had patch times similar to those of *L. heterotoma*. This is indeed possible, as several species of parasitoids have been found to leave pheromone trails on patches while searching (Price 1970, Galis & van Alphen 1981, Sugimoto et al. 1987, Bernstein & Driessen 1996). Sugimoto et al. (1987) have explored a patch time model on the basis of a pheromone trail that is left by the parasitoid while searching the patch. The concentration of the pheromone increases until it reaches a threshold upon which the parasitoid leaves the patch. This scheme provides an alternative for Waage's (1979) model with habituation to the patch edge. If such a response to trail pheromones occurs, in addition to the response to the patch edge, and is not represented as a covariate in the Cox model, the outcome may be different from that of a model that includes this response. This might be so, even if the effect of time-dependent trail mark accumulation on patch leaving might be partially accounted for by the baseline hazard function. Note that if the effect of trail marks is fully accounted for by the baseline hazard, an important biological process would be left undetected. Waage's (1979) and Sugimoto's et al. (1987) mechanisms might work in concert, but functionally they are not the same, as we will explain in the next section.

Omitting the effects of trail pheromones and odors from the Cox model, when studying the mechanisms of patch leaving, could result in spurious conclusions on the effects of ovipositions on leaving tendency. For example, let us assume that we have studied a parasitoid that responds to ovipositions in unparasitized hosts by a constant decremental effect on its patch leaving tendency (i.e. constant incremental effect on patch time). Early during patch exploitation, encounter rates with unparasitized hosts should be higher while the concentration of trail pheromones will be low. When depletion of the patch has proceeded, the rate of encounter with unparasitized hosts will become lower and the trail pheromone concentration will have built up. If the latter is not taken into account as a covariate, one possible outcome of the model is that early ovipositions will have a stronger effect on patch time, with later ovipositions having a lower effect, while in fact the effect of each oviposition was constant. Driessen and Bernstein (1999) found a decreasing effect of ovipositions with the rank number of the oviposition. An alternative hypothesis is that the increased leaving tendency is caused by a response to trail marks or odors. Indeed, Bernstein and Driessen (1996) found such a trial mark effect, but it was never introduced into models. Without further investigation, we cannot be certain that Driessen and Bernstein's (1999) result are not spurious. Omitting the effects of trail pheromones and odors also has consequences for functional models. Spataro (2001) has shown, in a theoretical study, that when parasitoid density is high, kairomone concentration no longer provides reliable information on the number of healthy hosts to be expected: patches might have already been exploited by others and several hosts might be already parasitized. In this case, parasitoids should use an incremental patch leaving rule. Hence, Spataro (2001) predicts that parasitoids should switch from a decremental rule when foraging alone, to an incremental one when patches are also exploited by competitors. However, this conclusion only holds under the constraint that the parasitoids have no other information than host kairomone concentration about the expected profitability of the patch. In fact, if this constraint can be circumvented by using other sources of information, the use of a decremental rule would possibly be preferable. In the real world, parasitoids might use other cues to assess patch quality in the presence of competitors. Examples include the presence of marking pheromones, contacts with parasitized hosts, or odors produced by the competitors (Höller et al. 1994, Janssen et al. 1995). Pierre et al. (2003) have suggested that parasitoids could use the changing ratio between unparasitized and parasitized hosts. Such factors would provide information about the current patch quality.

Other events, such as encounters with other parasitoids, could possibly also result in either incremental or decremental effects on patch time. Visser et al. (1992) showed that females of *L. heterotoma* stayed longer on patches when they had spent the previous day with competitors, than females that spent the previous day alone. Hence, previous exposure to competitors also affects patch time.

So far, patch leaving tendencies have neither been analyzed in models including both the effects of ovipositions and those of an increasing trail pheromone nor in models including the effects of current or previous exposure to parasitoid odor concentration.

Switching between incremental and decremental mechanisms, as predicted by Spataro (2001), has been found in several studies using Cox's proportional hazards model (Outreman et al. 2001, Lucchetta et al. 2007). Although the outcome of the Cox model in these cases could represent real behavioral processes, the same outcome could also be interpreted as a statistical artifact resulting from the misspecification of the model. For the first time, a switch from an incremental to a decremental rule was found by Outreman et al. (2001) for the parasitoid *Aphidius rhopalosiphi*. This wasp begins by using an incremental rule when exploiting an aggregation of wheat aphids, but switches to a decremental rule when the patch has been partly exploited. This switch in mechanism results in the parasitoid leaving the patch long before all hosts have been parasitized. As aphids become increasingly excited during the visit of the parasitoid, the probability that the parasitoid would become immobilized with a sticky substance ejected by aphids increases. Hence, the accelerated departure progressively decreases the mortality risk of the parasitoid (Outreman et al. 1999).

However, these findings still need to be confirmed in experiments in which the possible effects of trail pheromones, or other potential variables that change during patch exploitation, would be manipulated and included as covariates in the model.

8.3.3 Other cues

The danger of erroneous interpretation also occurs when other cues, used by parasitoids, are not represented as covariates in the model. Other examples of such cues, not mentioned in the previous section, could be the production of alarm pheromones or defense reactions by the host (Outreman et al. 2001), ovipositions in or rejection of parasitized hosts, physical encounters with other parasitoids (Wajnberg et al. 2004), the odor of hyperparasitoids, and induced plant odors (Tentelier et al. 2005, Tentelier & Fauvergue 2007, Wajnberg 2006).

Hence, for a proper understanding of the mechanisms underlying patch time allocation, we should have a detailed knowledge of the possible cues that affect leaving tendency, as suggested by previous work or by theoretical considerations. This means that instead of using proxies that are easily observed and quantified, like ovipositions, we have to analyze which cues related to oviposition are important (e.g. information about the quality of the host: already parasitized or not, size, or stage) and how these affect leaving tendency, and possible alternative hypotheses must be considered. This calls for combining the powerful tool provided by the Cox model with other, neuro-physiological methods.

Examples of such alternative hypotheses are: Is the effect of an oviposition in a healthy host on leaving tendency constant or dependent on its timing (Waage 1979) and/or dependent on its place in the sequence of ovipositions during the exploitation of the patch? Is the fading response to the patch edge (Waage 1979), indeed due to habituation to host kairomone or due to a response to an increasing concentration of trail pheromone, or a combination of both?

If conditioning to the patch edge is the only mechanism, two parasitoids searching the same patch, from which the hosts have been removed, would stay as long as a single female would stay on that patch but, on average, only half of the time if the trail pheromone concentration matters and if females are not able to discriminate between their own traces from those of conspecifics. As shown by Bernstein and Driessen (1996) for *V. canescens*, some parasitoids are able to discriminate their own traces from those of conspecifics. Tests of such simple hypothesis have never been done, as far as we know. An alternative study is, of course, the analysis of two consecutive visits to the same patch. The effect of a preceding visit to the patch by a conspecific was shown to decrease patch residence time in several parasitoids (Wajnberg 2006).

Previous patch visits and other previous experience
In *Asobara tabida*, both the quality of the previous patch and the time elapsed since the previous patch visit affect patch time (Thiel & Hoffmeister 2001). A similar result was obtained by Tentelier et al. (2006) for *Lysiphebus testaceipes*. Vos et al. (1998), Wajnberg et al. (1999), and Tenhumberg et al. (2001a) have shown that patch leaving tendency increases as the number of successive patches already visited increases. In contrast, Outreman (2000) has shown that females which have already visited a high-quality patch spend more time in the next patch. The function of this change in behavior remains unclear.

Roitberg et al. (1992, 1993) showed that previous exposure to circumstances that indicate unfavorable environmental conditions in the near future (i.e. dropping air pressure heralding a thunderstorm, and low temperatures and short day length heralding the onset of winter) result in longer patch times. The effect of a drop in temperature was also investigated by Amat et al. (2006). Extensive reviews are given by van Alphen et al. (2003) and Wajnberg (2006).

Initial response to kairomone concentration should vary with previous experience before arriving in the patch, in particular by travel time. As the MVT predicts longer patch times with increasing inter-patch travel times, initial responsiveness to the patch edge should be higher when travel times are longer.

Internal state
Internal state (e.g. egg load and lipid and carbohydrate reserves, see also Chapter 6 by Strand and Casas and Chapter 7 by Bernstein and Jervis) may provide information about future

lifespan and should also affect patch time. A state dependent approach has proved to be most fruitful in the study of different aspects of animal behavior, both from theoretical and experimental points of view (Sirot et al. 1997, Houston & McNamara 1999, Clark & Mangel 2000, Wajnberg et al. 2006, see also Chapter 15 by Roitberg and Bernhard). In the realm of parasitoid behavior, the study of superparasitism is a clear indication of the power of such an approach.

Conversely, a state dependent approach is, apart from the above-cited papers by Roitberg et al. (1992, 1993) and Wajnberg et al. (2006), most often absent from the study of parasitoid patch leaving decisions. In spite of this, some of the state dependent aspects taken into account in studies include age, mating status, and some types of previous experience (Wajnberg et al. 2006). For instance, the influence of the metabolic state of parasitoids in their patch leaving decisions is an unexplored field, with the exception of Lucchetta et al. (2007). Sirot and Bernstein (1996, 1997) and Tenhumberg et al. (2006) analyzed, from a theoretical point of view, how a parasitoid should allocate its foraging time between host patches and the search for metabolic resources, but experimental tests are painfully lacking (see also Chapter 7 by Bernstein and Jervis). The available evidence suggests that, in parasitoids, no lipogenesis occurs in the adults and that energy reserves during adult life can only be supplemented in the form of carbohydrates (Ellers 1996, Olson et al. 2000, Rivero & West 2007, see also Chapter 7 by Jervis and Bernstein). This adds a new twist to the problem by suggesting that, in state dependent models, metabolic reserves should not be represented as a single component.

Hence, the relevant question here is how parasitoids process information, and the answer can only come from the study of the cognitive processes. Knowing the way in which parasitoids process information will provide insight into the constraints that, together with natural selection, have moulded the rules for patch leaving. Once we know the mechanisms and the constraints, we can start to integrate mechanistic and functional models by asking whether the hierarchy and the relative contribution of the different cues result in optimal patch time allocation in a given habitat, and if parasitoids living in different habitats use a different hierarchy or have other relative contributions of different cues. Neuro-physiological models will be of great help in this endeavor and besides clarifying how information is processed, they will provide information to suggest how the relevant covariates should be introduced into statistical models. Statistical models will carry on playing a major role as they will allow the extraction of information from experimental data on the relative influence of process, as they have done so successfully over the last decade.

8.4 Group foraging

As we have already seen, some of the cues used by parasitoids in determining their patch residence time are related to the presence of competitors and provide a parasitoid with information on the level of exploitation of a patch. However, functional models discussed above, except that of Spataro (2001), do not take into account the presence of competing parasitoids in the same habitat. Theories that do take into account interactions with competitors and that are relevant to patch time allocation are game theories like the Ideal Free Distribution (Fretwell & Lucas 1970) and the war of attrition (Bishop & Cannings 1978).

8.4.1 The Ideal Free Distribution

Suppose that there are several parasitoid females simultaneously foraging in a habitat where several host patches can be exploited. The Ideal Free Distribution (IFD) is a simple theoretical model that predicts the stable distribution of the parasitoids over all available patches (Fretwell & Lucas 1970). Just like for the MVT, foragers in the IFD are assumed to have full information about the quality of all patches in the habitat, and this is why individuals are referred to as 'ideal'. In contrast to the MVT, in its simplest form, the IFD does not include depletion of patches and there is no time cost for traveling between patches. This last assumption, and the fact that no animal can prevent a conspecific from entering the patch of its choice, is why the foraging animals are referred to as 'free'. Under such restrictive hypotheses, parasitoids are predicted to distribute themselves among host patches in such a way that the encounter rate with hosts is equal for all of them. Hence, host patches of better quality will contain more parasitoids than poorer patches. A parasitoid foraging in a patch in which its host encounter rate is lower than the average encounter rate in the environment will move, at no cost, to a patch where its rate of fitness will be higher. The displacement of consumers will continue until equal gains are obtained by all competitors. If equal fitness for all foragers has been reached, fitness cannot be increased by moving to a different patch. Therefore the IDF is a Nash equilibrium and, like the MVT, is a conceptual model that only provides the baseline for biologically more realistic models.

Using a series of simulation models, Bernstein et al. (1988) relaxed the assumption that parasitoids should 'know' the quality of all patches in the habitat. Rather, these authors considered that animals, by sampling their environment, progressively learn the quality of the host patches to be exploited. In a non- or slowly-depleting environment, results show that the distribution of foragers rapidly approaches the IFD. Thus, having an innate knowledge of the quality of all host patches in the habitat is not a prerequisite for female parasitoids to achieve an IFD. However, when an increasing cost of traveling between patches is also included, the distribution of foraging animals follows the predictions of the IFD less and less (Bernstein et al. 1991). Hence, aggregative distributions of parasitism, often observed in the field, can be understood as the result of information and time constraints of individual parasitoids. Bernstein et al. (1999) suggested that perceptual constraints that preclude animals from fully converging to the IFD would make a major contribution to the stability of the system. Experimental tests of the ability of parasitoids to converge to the IFD were performed by Tregenza et al. (1996) under laboratory conditions, and by Fauvergue et al. (2006) under more realistic, glasshouse conditions, showing in general a partial convergence to theoretical predictions. Introducing patch depletion and interference between foraging parasitoids can add further realism to the IFD models.

Such IFD models still assume that all foragers are equal. However, parasitoids may differ in competitive ability (Wajnberg et al. 2004), or in arrival time on a patch. Such differences between foragers can be added to an IFD model (Sutherland 1996, van der Meer & Ens 1997), but can be better analyzed in other game theoretical models that explicitly formulate how parasitoids compete (van Alphen & Visser 1990, Sjerps & Haccou 1994, Haccou et al. 2003).

8.4.2 The war of attrition

Competition between adult females on a patch is either by patch guarding or by super-parasitism. In a superparasitized host, the larvae of solitary parasitoids will compete for possession of the host. The winning larva kills the other(s). In gregarious parasitoids, superparasitism results, in most of the cases, in smaller, less fit emerging parasitoid adults and a more male biased sex ratio. In spite of this, in both types of parasitoids, a female ovipositing in a host already parasitized by a competing female may still obtain offspring. Hence, superparasitism can be advantageous in circumstances where better alternatives, i.e. unparasitized hosts, are scarce. A special situation occurs when competing parasitoids are exploiting a patch simultaneously. A female leaving before the others would leave the hosts that she parasitized available for superparasitism by the others. This would result in a loss of offspring unless she would also stay longer and compensate for the loss of off-spring by superparasitizing herself. This results in all the females staying longer. How long should females stay?

The war of attrition theory provides the conceptual framework to predict how long females should stay in the patch (Maynard Smith, 1974, see also Chapter 9 by Haccou and van Alphen). The solution to the problem is not to provide the opponent with information on when the animal intends to leave, otherwise the opponent could choose a longer persistence time. Drawing a persistence time from an exponential distribution is the evolutionarily stable solution. In the real world, intrinsic variation between individuals in internal state (e.g. in age, feeding condition, egg load, and energy reserves), previous experience, and heritable differences in responses to cues (Wajnberg et al. 1999, 2004), could generate the unpredictability of leaving times. Indeed, Goubault et al. (2005) have shown that competitive strategies between females may vary due to differences in age and/or experience.

This model and the consequences of differences in arrival times between females are discussed in detail in Chapter 9 by Haccou and van Alphen. This is one of the topics in which recent theoretical developments (Haccou et al. 2003) should now be tested experimentally (Giraldeau & Caraco 2000) and is one of the most promising subjects of future research.

Experimental evidence, that females when exploiting a patch together stay longer, comes from Visser et al. (1990) and van Alphen et al. (1992). It is unknown which mechanism leads the females to stay longer. There are clear indications that odors or physical contacts could play a role (Visser et al. 1990, Wajnberg et al. 2004). Alternatively, the information on the presence of conspecific parasitoids can come from the hosts themselves. The ability to recognize hosts parasitized by conspecifics from hosts parasitized by the parasitoid itself has been reported in several species (van Dijken et al. 1992, Marris et al. 1996). Similar capacity in the recognition of trail mark was observed by Bernstein and Driessen (1996). Parasitoids could respond to these types of cues by changing the strength or direction of their response to host encounters.

8.4.3 Other modes of competition and patch time allocation

Several parasitoid species have been found to compete with conspecifics not only by superparasitism, but also by defending a patch in which they have parasitized hosts against intruding conspecifics. This behavior has additional time costs because the time spent in guarding the patch, in patroling the patch edge, and in fighting and chasing intruders cannot be spent in searching and parasitizing hosts. Patch defense is not a widespread behavior. We hypothesize that it can only evolve either when patches are of a defendable size (Waage 1982), or when travel times between patches are either short, such that intruders will easily give up and move to a nearby unoccupied patch, or when travel times are long, such that the probability of a resident wasp finding new patches during the remaining period of her adult life is extremely low.

An example of a parasitoid that defends patches with small distances between them is the African *Asobara citri*, which attacks *Drosophila* larvae in fallen fruit. Examples of parasitoids defending patches, which are far apart, are several scelionid egg parasitoids (Waage 1982, Field et al. 1997, 1998a,b). The theory predicting under which conditions patch defense will evolve is a burgeoning subject (Sirot 2000, Dubois et al. 2003, Dubois & Giraldeau 2003), but we lack the knowledge of how this behavior affects the spatial distribution of parasitism. Yet, this is a central problem in the study of patch time allocation in insect parasitoids.

8.5 Conclusion

Patch time allocation is one of the classical problems of behavioral ecology at the heart of evolutionary biology. We begin now to understand how natural selection acts on mechanisms for patch time allocation in differently structured environments, and how information constraints determine patch time decisions. Insight into the behavioral mechanisms of patch time decisions provides a basis to study the genetics underlying the behaviors involved, as a complementary route to study adaptation in foraging behavior, and no longer limits the problem to the realm of phenotypic modeling (Wajnberg et al. 1999, 2004).

Foraging in a heterogeneous environment necessitates that parasitoids acquire and process information, because the environment changes continuously. To take adaptive decisions, animals have to compare alternatives that present themselves in a sequential manner or to compare actual values of what they have experienced with the mean values for the environment as a whole. We know that parasitoids acquire information from a large variety of sources and exquisitely integrate it. We started by asking how patch time allocation affects fitness of individual parasitoids. This leads us to ask which mechanisms are involved in information acquisition and processing.

Our quest has led us now to ask about the evolutionary consequence of the use of different mechanisms. This is a clear reminder that there are no sharp boundaries between proximate and ultimate causes of behavior.

We have pointed out some of the most promising subjects for future research, such as state dependent patch leaving decisions, group foraging, and the study of optimal values of incremental and decremental responses to different stimuli. We would like to add other important ones: the influence of experiences gained during previous patch visits and the question to what extent parasitoids in nature are constrained by lack of information about the spatial distribution of hosts and competitors.

It will often be difficult to measure inter-patch travel times in the field, or to collect information on the natural distribution of hosts and competitors among patches. Yet such information is desperately needed to judge in a more quantitative way if patch time allocation in parasitoids is indeed adaptive and to understand the differences in patch time determining mechanisms in different species and how parasitoid foraging behavior is constrained.

Many aspects of patch time allocation are still poorly understood. In particular, group foraging, state dependent patch leaving decisions, and behavioral genetics of foraging decisions require further study to understand to what extent patch time decisions are constrained. Hence, patch time allocation will remain a fascinating area of research. Theoretical studies have suggested that the distribution of host mortality risks caused by parasitoids is of crucial importance for the stability properties of parasitoid–host population dynamics (Hassell et al. 1991, Taylor 1994, Bernstein et al. 1999). Obviously, the distribution of host mortality risks is tightly linked to the distribution of parasitoid patch time allocation. All considered, the study of patch time allocation links processes at the population level to the one and only fundamental theory in biology: the theory of evolution by natural selection.

Acknowledgments

We are most thankful to the participants of the ESF/BEPAR program for stimulating discussion.

References

Amat, I., Castelo, M., Desouhant, E. and Bernstein, C. (2006) The influence of temperature and host availability on the host exploitation strategies of sexual and asexual parasitic wasps of the same species. *Oecologia* **148**: 153–61.

Bernstein, C. and Driessen, G. (1996) Patch marking and optimal search patterns in the parasitoid *Venturia canescens*. *Journal of Animal Ecology* **65**: 211–19.

Bernstein, C., Kacelnik, A. and Krebs, J.R. (1988) Individual decisions and the distribution of predators in a patchy environment. *Journal of Animal Ecology* **57**: 1007–26.

Bernstein, C., Kacelnik, A. and Krebs, J.R. (1991) Individual decisions and the distribution of predators in a patchy environment. 2. The influence of travel costs and structure of the environment. *Journal of Animal Ecology* **60**: 205–25.

Bernstein, C., Auger, P. and Poggiale, J.C. (1999) Predator Migration decisions, the Ideal Free distribution and predator-prey dynamics. *American Naturalist* **153**: 267–81.

Bishop, D.T. and Cannings, C. (1978) A generalised war of attrition. *Journal of Theoretical Biology* **70**: 85–124.

Boivin, G., Fauvergue, X. and Wajnberg, É. (2004) Optimal patch residence time in egg parasitoids: Innate versus learned estimate of patch quality. *Oecologia* **138**: 640–7.

Charnov, E.L. (1976) Optimal foraging: The Marginal Value Theorem. *Theoretical Population Biology* **9**: 129–36.

Clark, C.W. and Mangel, M. (2000) *Dynamic State Variable Models in Ecology*. Oxford University Press, Oxford.

Cox, D.R. (1972) Regression models and life tables. *Biometrics* **38**: 67–77.

Cox, D.R. and Oakes, D. (1984) *Analysis of Survival Data*. Chapman & Hall, London.

Driessen, G. and Bernstein, C. (1999) Patch departure mechanisms and optimal exploitation in an insect parasitoid. *Journal of Animal Ecology* **68**: 445–59.

Driessen, G., Bernstein, C., van Alphen, J.J.M. and Kacelnik, A. (1995) A count-down mechanism for host search in the parasitoid *Venturia canescens*. *Journal of Animal Ecology* **64**: 117–25.

Dubois, F. and Giraldeau, L.A. (2003) The forager's dilemma: Food sharing and food defense as risk-sensitive foraging options. *The American Naturalist* 162: 768–79.

Dubois, F., Giraldeau, L.A. and Grant, J.W.A. (2003) Resource defense in a group-foraging context. *Behavioral Ecology* 14: 2–9.

Ellers, J. (1996) Fat and eggs: An alternative method to measure the trade-off between survival and reproduction in insect parasitoids. *Netherlands Journal of Zoology* 46: 227–35.

Fauvergue, X., Boll, R., Rochat, J., Wajnberg, É., Bernstein, C. and Lapchin, L. (2006) Habitat assessment by parasitoids: Consequences for population distribution. *Behavioral Ecology* 17: 522–31.

Field, S.A., Keller, M.A. and Calbert, G. (1997) The pay-off from superparasitism in the egg parasitoid *Trissolcus basalis*, in relation to patch defence. *Ecological Entomology* 22: 142–9.

Field, S.A., Calbert, G. and Keller, M.A. (1998a) Patch defence in the parasitoid wasp *Trissolcus basalis* (Insecta: Scelionidae): The time structure of pairwise contests, and the 'waiting game'. *Ethology* 104: 821–40.

Field, S.A., Keller, M.A. and Austin, A.D. (1998b) Field ecology and behaviour of the egg parasitoid *Trissolcus basalis* (Wollaston) (Hymenoptera: Scelionidae). *Transactions of the Royal Society of South Australia* 122: 65–71.

Fretwell, S.D. and Lucas, H.L. (1970) On territorial behaviour and other factors in habitat distribution in birds. *Acta Biotheoretica* 19: 16–36.

Galis, F. and van Alphen, J.J.M. (1981) Patch time allocation and search intensity of *Asobara tabida* Nees (Braconidea), a larval parasitoid of *Drosophila*. *Netherlands Journal of Zoology* 31: 596–611.

Giraldeau, L.A. and Caraco, T. (2000) *Social Foraging Theory*. Princeton University Press, Princeton.

Godfray, H.C.J. (1994) *Parasitoids, Behavioral and Evolutionary Ecology*. Princeton University Press, Princeton.

Goubault, M., Outreman, Y., Poinsot, D. and Cortesero, A.M. (2005) Patch exploitation strategies of parasitic wasps under intraspecific competition. *Behavioral Ecology* 16: 693–701.

Haccou, P. and Hemerik, L. (1985) The influence of larval dispersal in the cinnabar moth (*Tyria jacobaea*) on predation by the red wood ant (*Formica polyctena*). An analysis based on the proportional hazards model. *Journal of Animal Ecology* 54: 755–69.

Haccou, P., De Vlas, S.J., van Alphen, J.J.M. and Visser, M.E. (1991) Information-processing by foragers – Effects of intra-patch experience on the leaving tendency of *Leptopilina heterotoma*. *Journal of Animal Ecology* 60: 93–106.

Haccou, P., Glaizot, O. and Cannings, C. (2003) Patch leaving strategies and superparasitism: An asymmetric generalized war of attrition. *Journal of Theoretical Biology* 225: 77–89.

Hassell, M.P., May, R.M., Pacala, S.W. and Chesson, P.L. (1991) The persistence of host-parasitoid associations in patchy environments I A general criterion. *American Naturalist* 138: 568–83.

Hemerik, L., Driessen, G. and Haccou, P. (1993) Effects of intra-patch experiences on patch time, search time and searching efficiency of the parasitoid *Leptopilina clavipes*. *Journal of Animal Ecology* 62: 33–44.

Höller, C., Micha, S., Schulz, G.S., Francke, W. and Pickett, J.A. (1994) Enemy induced dispersal in a parasitic wasp. *Experientia* 50: 182–5.

Houston, A.I. and McNamara, J.M. (1999) *Models of adaptive behaviour, an approach based on state*. Cambridge University Press, Cambridge.

Iwasa, Y., Higashi, M. and Yamamura, N. (1981) Prey distribution as a factor determining the choice of optimal foraging strategy. *American Naturalist* 117: 710–23.

Janssen, A., van Alphen, J.J.M., Sabelis, M.W. and Bakker, K. (1995) Odour-mediated avoidance of competition in *Drosophila* parasitoids: The ghost of competition. *Oikos* 73: 356–66.

Lucchetta, P., Desouhant, E., Wajnberg, É. and Bernstein, C. (2007) Small but smart: The interaction between environmental cues and internal state modulates host-patch exploitation in a parasitic wasp. *Behavioral Ecology & Sociobiology* 61: 1409–18.

Marris, G.C., Hubbard, S.F. and Scrimgeour, C. (1996) The perception of genetic similarity by the solitary parthenogenetic parasitoid *Venturia canescens*, and its effects on the occurrence of superparasitism. *Entomologia Experimentalis et Applicata* **78**: 167–74.

Maynard Smith, J. (1974) The theory of games and the evolution of animal conflicts. *Journal of Theoretical Biology* **47**: 209–21.

Olson, D.M., Fadamiro, H., Lundgren, J.O.N.G. and Heimpel, G.E. (2000) Effects of sugar feeding on carbohydrate and lipid metabolism in a parasitoid wasp. *Physiological Entomology* **25**: 17–26.

Outreman, Y. (2000) *Capacité de discrimination et exploitation des colonies d'hôtes chez un parasitoïde solitaire*. Ecologie comportementale et modélisation. PhD thesis, University of Rennes, Rennes, France.

Outreman, Y., Le Ralec, A. and Pierre, J.-S. (1999) Le comportement d'exploitation des patchs d'hotes chez le parasitoïde *Aphidius rhopalosiphi* (Hymenoptera: Braconidae). *Annales de la Société Entomologique de France* **35**: 404–9.

Outreman, Y., Le Ralec, A., Wajnberg, É. and Pierre, J.-S. (2001) Can imperfect host discrimination explain partial patch exploitation in parasitoids? *Ecological Entomology* **26**: 277–80.

Outreman, Y., Le Ralec, A., Wajnberg, É. and Pierre J.-S. (2005) Effects of within- and among-patch experiences on the patch-leaving decision rules in an insect parasitoid. *Behavioral Ecology & Sociobiology* **58**: 208–17.

Pacala, S.W. and Hassell, M.P. (1991) The persistence of host parasitoid associations in patchy environments. II. Evaluation of field data. *American Naturalist* **138**: 584–605.

Pierre, J.S., van Baaren, J. and Boivin, G. (2003) Patch leaving decision rules in parasitoids: Do they use sequential decisional sampling? *Behavioral Ecology & Sociobiology* **54**: 147–55.

Price, P.W. (1970) Trail odors: Recognition by insects parasitic on cocoons. *Science* **170**: 546–7.

Rivero, A. and West, S.A. (2007) The physiological costs of being small in a parasitic wasp. *Evolutionary Ecology Research* (in press).

Roitberg, B.D., Mangel, M., Lalonde, R.G., Roitberg, C.A., van Alphen, J.J.M. and Vet, L. (1992) Seasonal dynamic shifts in patch exploitation by parasitic wasps. *Behavioral Ecology* **3**: 156–65.

Roitberg, B.D., Sircom, J., Roitberg, C.A., van Alphen, J.J.M. and Mangel, M. (1993) Life expectancy and reproduction. *Nature* **36**: 108.

Rosenheim, J.A. and Mangel, M. (1994) Patch-leaving rules for parasitoids with imperfect host discrimination. *Ecological Entomology* **19**: 374–80.

Sirot, E. (2000) An evolutionarily stable strategy for aggressiveness in feeding groups. *Behavioral Ecology* **11**: 351–6.

Sirot, E. and Bernstein, C. (1996) Time sharing between host searching and food searching in solitary parasitoids: State dependent optimal strategies. *Behavioral Ecology* **7**: 189–94.

Sirot, E. and Bernstein, C. (1997) Food searching and superparasitism in solitary parasitoids. *Acta Oecologia* **18**: 63–72.

Sirot, E., Ploye, H. and Bernstein, C. (1997) State dependent superparasitism in a solitary parasitoid: Egg load and survival. *Behavioral Ecology* **8**: 226–32.

Sjerps, M. and Haccou, P. (1994) Effects of competition on optimal patch leaving: A war of attrition. *Theoretical Population Biology* **46**: 300–18.

Sokal, R.R. and Rohlf, F.J. (1994) *Biometry*. Freeman, New York.

Spataro, T. (2001) *De l'individu à la population: Etude théorique de l'influence, au niveau de la population, de traits d'histoire de vie et de comportements individuels dans les systèmes hôte-parasitoïde*. PhD thesis, University of Lyon, Lyon, France.

Sugimoto, T., Murakami, H. and Yamazaki, R. (1987) Foraging for patchily-distributed leaf-miners by the parasitoid *Dapsilartha rufiventris* (Hymenoptera: Braconidae). II. Stopping rule for host search. *Journal of Ethology* **5**: 95–103.

Sutherland, W.J. (1996) *From Individual Behaviour to Population Ecology*. Oxford Series in Ecology and Evolution. Oxford University Press, Oxford.

Taylor, A.D. (1994) Heterogeneity in host parasitoid interactions – Aggregation of risk and the CV^2 greater than one rule. *Trends in Ecology & Evolution* **8**: 400–5.

Tenhumberg, B., Keller, M.A. and Possingham, H.P. (2001a) Using Cox's proportional hazard models to implement optimal strategies: An example from behavioural ecology. *Mathematical & Computing Modelling* **33**: 597–607.

Tenhumberg, B., Keller, M.A., Possingham, H.P. and Tyre, A.J. (2001b) Optimal patch-leaving behaviour: A case study using the parasitoid *Cotesia rubecula*. *Journal of Animal Ecology* **70**: 683–91.

Tenhumberg, B., Siekmann, G. and Keller, M.A. (2006) Optimal time allocation in parasitic wasps searching for hosts and food. *Oikos* **113**: 121–31.

Tentelier, C. and Fauvergue, X. (2007) Herbivore-induced plant volatiles as cues for habitat assessment by a foraging parasitoid. *Journal of Animal Ecology* **76**: 1–8.

Tentelier, C., Wajnberg, É. and Fauvergue, X. (2005) Parasitoids use herbivore-induced information to adapt patch exploitation behaviour. *Ecological Entomology* **30**: 739–44.

Tentelier, C., Desouhant, E. and Fauvergue, X. (2006) Habitat assessment by parasitoids: Mechanisms for patch use behaviour. *Behavioral Ecology* **17**: 515–21.

Thiel, A. and Hoffmeister, T.S. (2001) Wirtssuchverhalten der Brackwespe *Asobara tabida* in einem komplexen Habitat. *Mitteilungen der Deutschen Gesellschaft für Allgemeine und Angewandte Entomologie* **13**: 73–6.

Tregenza, T., Thompson, D.J. and Parker, G.A. (1996) Interference and the ideal free distribution: Oviposition in a parasitoid wasp. *Behavioral Ecology* **7**: 387–94.

van Alphen, J.J.M. and Galis, F. (1983) Patch time allocation and parasitization efficiency of *Asobara tabida*, a larval parasitoid of *Drosophila*. *Journal of Animal Ecology* **52**: 937–52.

van Alphen, J.J.M. and Visser, M.E. (1990) Superparasitism as an adaptive strategy for insect parasitoids. *Annual Review of Entomology* **35**: 59–79.

van Alphen, J.J.M., Visser, M.E. and Nell, H.W. (1992) Adaptive superparasitism and patch time allocation in solitary parasitoids: Searching in groups versus sequential patch visits. *Functional Ecology* **6**: 528–35.

van Alphen, J.J.M., Bernstein, C. and Driessen, G. (2003) Information acquisition and patch time allocation in insect parasitoids. *Trends in Ecology & Evolution* **18**: 81–7.

van Dijken, M.J., van Stratum, P. and van Alphen, J.J.M. (1992) Recognition of individual-specific marked parasitized hosts by the solitary parasitoid *Epidinocarsis lopezi*. *Behavioral Ecology & Sociobiology* **30**: 77–82.

van der Meer, J. and Ens, B.J. (1997) Models of interference and their consequences for the spatial distribution of ideal and free predators. *Journal of Animal Ecology* **66**: 846–58.

Varaldi, J., Fouillet, P., Ravallec, M., Lopez-Ferber, M., Boulétreau, M. and Fleury, F. (2003) Infectious behavior in a parasitoid. *Science* **302**: 1930.

Varaldi, J., Fouillet, P., Boulétreau, M. and Fleury, F. (2005) Superparasitism acceptance and patch-leaving mechanisms in parasitoids: A comparison between two sympatric wasps. *Animal Behaviour* **69**: 1227–34.

Visser, M.E., van Alphen, J.J.M. and Hemerik, L. (1990) Adaptive superparasitism and patch time allocation in solitary parasitoid: The influence of the number of parasitoids depleting the patch. *Behaviour* **114**: 21–36.

Visser, M.E., van Alphen, J.J.M. and Hemerik, L. (1992) Adaptive superparasitism and patch time allocation in solitary parasitoids: An ESS model. *Journal of Animal Ecology* **61**: 93–101.

Vos, M., Hemerik, L. and Vet, L.E.M. (1998) Patch exploitation by the parasitoids *Cotesia rubecula* and *Cotesia glomerata* in multi-patch environments with different host distributions. *Journal of Animal Ecology* **67**: 774–83.

Waage, J.K. (1978) Arrestment response of the parasitoid *Nemeritis canescens* to a chemical produced by its host, *Plodia interpunctella*. *Physiological Entomology* **3**: 135–46.

Waage, J.K. (1979) Foraging for patchily-distributed hosts by the parasitoid *Nemeritis canescens*. *Journal of Animal Ecology* **48**: 353–71.

Waage, J.K. (1982) Sib-mating and sex-ratio strategies in Scelionid wasps. *Ecological Entomology* **7**: 103–12.

Wajnberg, É. (2006) Time allocation strategies in insect parasitoids: From ultimate predictions to proximate behavioral mechanisms. *Behavioral Ecology & Sociobiology* **60**: 589–611.

Wajnberg, É., Rosi, M.C. and Colazza, S. (1999) Genetic variation in patch time allocation in a parasitic wasp. *Journal of Animal Ecology* **68**: 121–33.

Wajnberg, É., Fauvergue, X. and Pons, O. (2000) Patch leaving decision rules and the Marginal Value Theorem: An experimental analysis and a simulation model. *Behavioral Ecology* **11**: 577–86.

Wajnberg, É., Curty, C. and Colazza, S. (2004) Genetic variation in the mechanisms of direct mutual interference in a parasitic wasp: Consequences in terms of patch-time allocation. *Journal of Animal Ecology* **73**: 1179–89.

Wajnberg, É., Bernhard, P., Hamelin, F. and Boivin, G. (2006) Optimal patch time allocation for time-limited foragers. *Behavioral Ecology & Sociobiology* **60**: 1–10.

9

Competition and asymmetric wars of attrition in insect parasitoids

Patsy Haccou and Jacques J.M. van Alphen

Abstract

Competition is widespread in parasitoids. Simultaneous patch depletion involves a dilemma: individuals that leave the patch early get less than the ones which remain, who get more than the expected gain in the environment. However, all payoffs decrease in time. Thus, the longer an individual remains in a patch the lower its payoff will be but, on the other hand, if it manages to remain longer than the rest, it might win a lot. About a decade ago, it was shown that this situation may lead to a generalized war of attrition. This model predicts that leaving times should be random, and that some competitors may leave relatively early, leaving plenty of resources behind. Further, competition can lead to much longer persistence times than predicted by the Charnov's (1976) marginal value theorem (MVT) (see also Chapter 8 by van Alphen and Bernstein).

Recently, the model was extended to include superparasitism. In solitary species, at most one offspring can emerge from a host. Females of such species avoid self-superparasitism, but often lay additional eggs in hosts parasitized by conspecific competitors. Sometimes the offspring of the superparasitizing female emerges, rather than that of the female who parasitized it first. Superparasitism introduces several complications. Since each additional egg diminishes the expected fitness payoff of the female(s) that parasitized the host previously, the payoff of a female that has left a patch is affected by the behavior of the remaining competitors. Further, differences in arrival times cause asymmetries: a female that arrived early has already parasitized more hosts at the moment when superparasitism starts than one that arrived later. Therefore, their expected payoffs from a patch are affected in different ways by superparasitism, leading to an asymmetric situation. In this case, patch leaving strategies should depend on perceived differences in arrival times. Females whose perceived arrival time is early should be prepared to stay the longest. Mistakes in perceived differences also affect persistence times. If the possibility of such mistakes is large, the variance in persistence times decreases.

The theory presented here has a twofold relevance for biological control. It helps to understand mechanisms of aggregation and interference, which are both key factors in determining the population dynamics of parasitoids and their hosts. Further, it can be used to predict effects of mass rearing programs on competitive behavior, and thus help to optimize the efficiency of such programs for biocontrol purposes.

9.1 Introduction

One of the main goals in ecology is to determine the mechanisms underlying observed spatial distributions of interacting species, such as parasitoids and their hosts. The most popular way to study how competition affects such patterns is to compare them with the ideal free distribution (IFD, Fretwell & Lucas 1970), derived as the evolutionarily stable spatial distribution of individuals over resources, given certain basic assumptions (Cressman et al. 2004).

This approach has several disadvantages. Since different behavioral mechanisms can lead to the same IFD, agreement of observations with predictions does not provide direct information about such mechanisms. Partly for the same reason, discrepancies between the predicted IFD and observations can have several causes. Further, the IFD has only been derived for models with unrealistic assumptions, such as non-depleting patches, no time costs of traveling, and instantaneous effects of the number of competitors on their distribution. Although some studies have considered relaxation of the first two assumptions (Bernstein et al. 1988, 1991), it is as yet unclear what the effects of non-instantaneous competitive dynamics, due to the reaction to patch depletion, are.

Patch leaving strategies of parasitoids have been studied extensively (Wajnberg 2006, see also Chapter 8 by van Alphen and Bernstein). In most cases, the effects of competition on patch leaving have not been studied (Wajnberg et al. 2004) and studies to date concern either patch depletion by a single parasitoid, or consider competition as a separate factor (Janssen 1989, Mangel 1992, Sirot & Křivan 1997, Sirot et al. 1997). Parasitoids often compete in field situations (Godfray 1994) and there is ample evidence that the presence of competitors results in behavioral changes (Field & Calbert, 1998a,b).

Reviews of the searching behavior of insect parasitoids (Vinson 1981, 1984, Vet & van Alphen 1985, Vinson 1985, Godfray 1994, van Alphen et al. 2003, Wajnberg 2006) show that parasitoids use a variety of visual, olfactory, and other chemical information to assess the quality of both hosts and host patches, and to obtain information about conspecific as well as heterospecific competitors. Marking pheromones left by the searching female may be used to avoid areas that she has already searched (Price 1980, Galis & van Alphen 1981). Competing parasitoids may respond to these marks by avoiding the area or adjusting their behavior correspondingly (Roitberg & Mangel 1988). Marking pheromones placed on or in the host by the parasitoid after oviposition indicate that it has already been parasitized. It has been shown for several species that females are able to discriminate between their own marks and those of conspecific females. The timing of encounters with already parasitized hosts might also provide a parasitoid with information about the profitability of the patch and its degree of depletion (Haccou et al. 1991).

Competing parasitoids may respond to an encounter with a host already parasitized by a conspecific female either by rejecting it or by laying additional eggs, an act called superparasitism. Superparasitism is a common means of competition in parasitoids and has been recorded for many species of hymenopteran and dipteran parasitoids (Godfray 1994).

In this chapter, we generalize Charnov's (1976) MVT model (see also Chapter 8 by van Alphen and Bernstein) to situations with competition. We show how competition without superparasitism can lead to a generalized war of attrition, with the consequence that randomized patch leaving strategies can evolve, and competitors may stay in a patch long after the point at which they should leave according to the MVT. We then consider situations with superparasitism and show that this may lead to asymmetric wars of attrition, where evolutionarily stable patch leaving strategies depend on perceived differences in arrival times. For mathematical convenience, we consider situations with at most two competitors. Generalization to multiple competitors is straightforward but tedious (Sjerps & Haccou 1994b, for the generalized model without superparasitism).

9.2 Patch depletion by a single female: the marginal value theorem (MVT)

In this section, we consider patch depletion by a single parasitoid female. We list the assumptions that are especially relevant for this situation. However, they will also be assumed to hold in the competitive cases considered in the subsequent sections:

A1 Parasitoid behavior has evolved to optimize the expected number of future offspring that is produced per time unit.

A2 The expected gain rate of offspring in the environment is constant.

A3 Hosts containing only eggs of one female will always yield one offspring to her in the future.

A4 Foragers have complete information about patch state and environmental gain rate.

A1 implies the absence of effects of egg limitation. A2 implies that the environment is stable. A3 is made mainly for convenience sake. Generalizations are easily included in the models, i.e. hosts may be able to yield more than one offspring, or the probability of successful hatching of an offspring may be impaired in multiply parasitized hosts. This will not affect the main results. Some generalizations of A4 will be considered later.

We will first reformulate the MVT in a way that is suitable for generalization to situations with competition. This formulation was introduced by McNamara and Houston (1987) and first applied in a competition framework by Sjerps and Haccou (1994b).

Consider a female parasitoid, denoted by F_v, who forages in a patch alone. Let n represent the host density in a patch upon her arrival, and $v(t)$ the proportion of hosts that she already parasitized by time t (where $t = 0$ corresponds to the moment of her arrival). The environmental gain rate is denoted by γ. The payoff of F_v at time t, denoted by $L_v(t)$, is the difference between her net gain in the patch and the gain that she could have obtained by foraging in the environment for the same amount of time:

$$L_v(t) = nv(t) - \gamma t \tag{9.1}$$

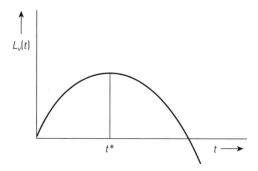

Fig. 9.1 General shape of the payoff of a single female as a function of the time t she is foraging on a host patch. The intra-patch gain rate drops below the environmental gain rate at time t^*. According to the MVT, this is the time when the female should leave the patch.

It is reasonable to assume that $v(t)$ is a monotonically increasing function, and thus $L_v(t)$ has one extremum, which is a maximum (Fig. 9.1). Thus, maximization of the payoff gives the following implicit equation for the optimum patch leaving time:

$$\frac{dL_v}{dt} = 0 \Rightarrow n\frac{dv}{dt} = \gamma \tag{9.2}$$

This equation states the MVT: to maximize her gain rate, a female should leave the patch when the intra-patch gain rate equals the environmental rate.

Although it is difficult to measure environmental gain rate and to obtain direct evidence that patch leaving decisions of parasitoids searching alone can be predicted by the MVT, there is good evidence for a number of species that they are rate maximizers and that their behavior is in accordance with several predictions following from the MVT (Hubbard & Cook 1978, Waage 1979, Galis & van Alphen 1981, van Alphen & Galis 1983, van Alphen et al. 2003, Wajnberg 2006).

9.3 Competitive patch depletion without superparasitism

In addition to assumptions A1 to A3, we assume for competitive situations that:

A5 Females do not differ in patch exploitation behavior, such as searching efficiency.
A6 There are at most two competitors per patch.
A7 Females do not re-enter the patch once they have left.

In competitive situations, each individual has two possible payoffs: (i) the payoff of being the first to leave, and (ii) the payoff if the competitor leaves first. We examine competition between two individuals, F_v and F_w, and denote the time that they have been depleting the patch together by t.

First consider the payoff of F_v if she is the first to leave, at time t. This is again the function given in the equation, although $v(t)$ may be different from the single female case, since

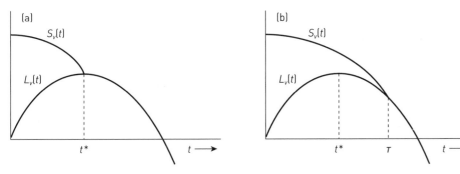

Fig. 9.2 The two different configurations of combinations of the payoff functions in the situation with competition without superparasitism. $L_v(t)$ is the payoff obtained by the individual that leaves first, at time t. $S_v(t)$ is the payoff obtained by the remaining female, if her competitor leaves at time t. (a) Shows the situation without interference, and (b) with interference. If both females stay in the patch, the intra patch gain rate drops below the environmental gain rate at time t^*. According to the MVT, this is the time when both females should leave the patch. From time τ onward, the payoff of leaving first equals that of leaving second. Both females should leave the patch at or before this time.

now two individuals are depleting the patch. It is, however, still a monotonically increasing function (Fig. 9.1) and the place of its maximum is determined by Equation (9.2). Note that the time at which the maximum is reached is generally different from the optimal patch leaving time for a single female, since the rate at which hosts are parasitized will be different when there is competition.

Now consider the payoff of F_v if her competitor F_w is the first one to leave at time t. This function is denoted by $S_v(t)$. Once F_w has left, F_v should adopt the patch leaving strategy that is optimal for a single forager, i.e. she leaves as soon as her gain rate equals γ. This may imply that she also leaves immediately but, especially when t is small, the optimal residual staying time will be positive. Details of this function will be discussed later. For now, we focus on its general shape.

It is clear that the longer competitors are in the patch together, the lower the additional payoff to the remaining female will be. Therefore, $S_v(t)$ decreases in time. Further, this payoff is always larger than or equal to $L_v(t)$, since the remaining female can choose between leaving too at time t or remaining, whichever gives the highest payoff. The optimal option is to stay only if it gives an extra payoff, otherwise F_v will leave at time t, and then $S_v(t)$ equals $L_v(t)$. Figure 9.2 shows the different possible configurations of combinations of the payoff functions.

Below it is shown that without superparasitism, the payoff functions of F_w equal those of F_v. Therefore only one pair of functions has to be considered in this case.

The evolutionarily stable leaving strategy (ESS) is where leaving times are chosen in such a way that no mutant can invade a population that adopts that strategy. Sjerps and Haccou (1994b) pointed out that the ESS can be derived by using the analogy with a war of attrition, which models a contest over a resource (Maynard Smith 1974). Each individual chooses the maximum time it is willing to persist in the contest. The one that chose the longest time wins the resource, while each contestant pays a cost proportional to the time it actually persisted. In terms of payoff functions this implies that

$$L_v(t) = -ct, \quad S_v(t) = R - ct \tag{9.3}$$

where c is the cost per unit of time of persisting in the contest. Thus, the payoff functions are two parallel lines. In the war of attrition, there is no unbeatable fixed leaving time. Thus, the ESS is to choose times according to an exponential distribution with parameter R/c. In other words, each individual should have a constant patch leaving tendency that is proportional to the cost per time unit.

In the generalized war of attrition, however, the payoff functions may be nonlinear and they can, at some point, become equal (Fig. 9.2). Bishop and Cannings (1978), Bishop et al. (1978), and Sjerps and Haccou (1994a,b) derived ESSs for generalized wars of attrition. From their results, it follows that, as in Fig. 9.2a, both competitors should leave at point t^*, when the gain rate in the patch equals γ. In the situation of Fig. 9.2b, a random leaving time is chosen in the period from t^* to the moment when the payoffs become equal, τ. This is the maximum persistence time. The ESS patch leaving tendency equals

$$h(t) = \frac{S_v(t) - L_v(t)}{-L_v'(t)}, \quad t \in (t^*, \tau) \tag{9.4}$$

and the probability that both competitors stay until time τ is

$$\exp\left[-\int_{t^*}^{\tau} h(s)\,\mathrm{d}s \right] \tag{9.5}$$

If both females are still present at this time, they leave simultaneously. Sjerps and Haccou (1994b) showed that the situation given in Fig. 9.2b can only occur when competitors suppress each other's intra-patch gain rate by interference. One example of such behavior is fighting. If they just deplete a patch together without interference, the configuration of the payoff functions is as illustrated in Fig. 9.2a.

General conclusions are that, without interference, the MVT criterion of Equation (9.2) correctly identifies the patch leaving times of both competitors. They should both leave at the same time, as soon as the intra-patch gain rate drops below that of the environment. However, in situations with interference (Fig. 9.2b), no one leaves before this point is reached. The MVT criterion then indicates the minimum leaving time, and average leaving times lay beyond this moment. The results imply that competitors are expected to stay in a patch at lower instantaneous gain rates than γ. However, note that the actual patch residence time depends on the rate of patch depletion, which is higher when there is competition than during patch exploitation by a single female. Therefore, comparison of patch residence times of single foragers with those in competitive situations is not informative: it is the instantaneous gain rate that should be considered. The conditional leaving strategy of the remaining female, after her competitor has left, can be inferred from the MVT criterion.

It is illuminating to have a graphical representation of the patch state in competitive situations. There are three types of hosts in a patch: type U is unparasitized, type V will yield offspring to F_v, and type W will yield offspring to F_w. The proportions of the different

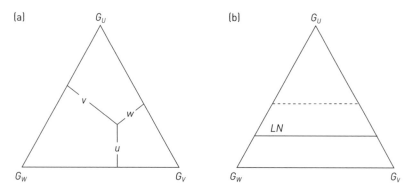

Fig. 9.3 (a) Graphical representation of the patch state in competitive situations, with three types of hosts: type U is unparasitized, type V yields offspring to F_v, and type W yields offspring to F_w. Each point in the triangle corresponds to a different patch state. The lengths of the perpendiculars from the point to the edges correspond to the proportions (u, v, w) of the three different host types. (b) Dotted line: divides the patch state space into the areas where the payoff of leaving first is increasing (top) and decreasing (bottom). Line LN divides the state space into the areas where a singly foraging female should stay in the patch (top) or leave (bottom). Figure adjusted from Haccou et al. (2003).

types are denoted by respectively u, v, and w. Since these numbers sum up to one, the state space can be represented by means of an equilateral triangle. Each point in the triangle corresponds to a different patch state (Fig. 9.3a).

Assumption A7 implies that there is a one-to-one relationship between the state of the patch and the time spent in the patch by the two females. The solid line, denoted by LN in Fig. 9.3b, indicates the patch states where the intra-patch gain rate reaches γ when there is a single forager in the patch. For patch states in the area of the triangle above this line, the gain rate of the forager exceeds γ, and it should stay in the patch. As soon as states below this line are reached she should leave. The broken line denotes the states where the intra-patch gain rate equals γ when both females are present. When these lines are different (Fig. 9.3b), the configuration of payoff functions is as given in Fig. 9.2b. In the situation of Fig. 9.2a, however, the two lines are superposed.

Figure 9.4 gives an example of the trajectory in the state space when F_v arrives in the patch first. At the time of her arrival, there are only unparasitized hosts and the patch is in state G_U. Due to parasitism, the state changes along the edge of the triangle in the direction of G_V. After a certain time, when the patch state is at point A, F_w arrives. As long as both females are in the patch, the trajectory goes straight down, since both females parasitize hosts at the same rate.

If F_w leaves when the patch state is at point B, then F_v stays until her intra-patch gain rate has dropped to γ. In this example, point B lies above the line LN, and therefore F_v stays for a while after her competitor has left. The patch trajectory goes further downward, parallel to the edge $G_U - G_V$, since the proportion of hosts in state W does not change after F_w has left (there is no superparasitism here). F_v leaves at the moment of intersection with LN, indicated by point C. If it is the other female, F_v, that leaves at point B, a similar story holds for F_w, and then the patch state ends up at point D.

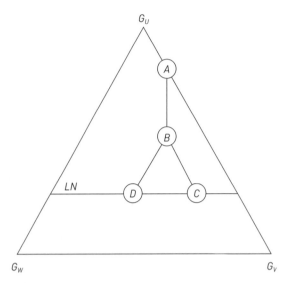

Fig. 9.4 Example of a trajectory in the state space when F_v arrives in the patch first and there is no superparasitism. F_w arrives at point A. At point B, one of the females leaves. If F_v leaves first, the trajectory goes to point D, and then F_w leaves too. If F_w leaves first, the trajectory goes to point C, where F_v leaves.

The payoff functions are calculated as follows. Let a denote the time that F_v spent in the patch before the arrival of F_w. The time t that both are in the patch is the time it took to get from point A to point B. Denote the residual staying time of F_v by y_v^* and that of F_w by y_w^*, then

$$L_v(t) = n(v_B - v_A) - \gamma t$$
$$S_v(t) = n(v_C - v_A) - \gamma(t + y_v^*) \tag{9.6}$$

and

$$L_w(t) = nw_B - \gamma t$$
$$S_w(t) = nw_D - (\gamma + y_w^*)t. \tag{9.7}$$

By means of geometrical methods, it can easily be shown that $v_B - v_A = w_B$ and $v_C - v_A = w_D$ (Fig. 9.5). From this, it also follows that $v_C - v_B = w_D - w_B$ and, since the females parasitize hosts at the same rate, y_v^* equals y_w^*. It can be concluded that the payoffs of F_v and F_w are equal. Therefore, in competitive situations without superparasitism, there is a symmetric generalized war of attrition. The symmetry implies that when there is no superparasitism, differences between arrival times of females do not affect their patch leaving tendencies. Hamelin et al. (2007a,b) recently gave an alternative derivation of this result.

9.4 Competitive patch depletion with superparasitism

Regardless of model specifics, superparasitism has some general implications that can be seen intuitively.

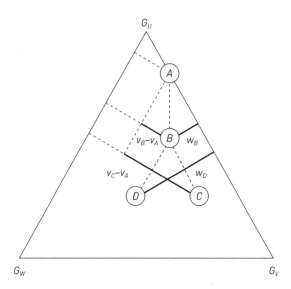

Fig. 9.5 Graphical illustration of the fact that $v_B - v_A = w_B$ and $v_C - v_A = w_D$.

- When a superparasitizing competitor leaves, the intra-patch gain rate of the remaining female will almost always instantly increase. This means that we get a situation as in Fig. 9.2b. Therefore, there is nearly always interference and, as a consequence, there will be a period with a generalized war of attrition. However, it appears that in some situations, this instantaneous increase does not occur (Hamelin et al. in prep.).
- Differences in arrival time create differences in opportunity for, and vulnerability to superparasitism, and thus cause asymmetries between competitors.
- The payoff of a female that has left a patch can still be affected by the patch exploitation behavior of the remaining female.

Three general predictions ensue, as follows.

1 Patch leaving strategies are random when superparasitism occurs.
2 Individuals are expected to stay in the patch even though their gain rate is smaller than γ. This may also happen when there are other types of interference between competitors, but superparasitism will nearly always lead to this situation.
3 Patch leaving strategies of competitors depend on their relative arrival times.

The latter effect implies that we get an asymmetric situation, and this was examined by Haccou and Glaizot (2002) and Haccou et al. (2003) by generalizing the results of Hammerstein and Parker (1982) on the asymmetric war of attrition. To illustrate these effects of superparasitism, we take a look at a specific model. In addition to A1 to A7, the following assumptions are made:

A8 Parasitoids can discriminate between hosts that are unparasitized, parasitized only by themselves, parasitized only by others, or parasitized by themselves and others.

A9 They always attack unparasitized hosts, do not attack hosts that have only been parasitized by themselves and, depending on the circumstances, may superparasitize hosts already attacked at least once by another female.

A10 Once a parasitoid female decides to superparasitize, she does not discriminate between hosts that have only been parasitized by another female and hosts that have been parasitized multiple times (by both herself and another female).

A11 Each time a host is superparasitized, there is, in the absence of further superparasitism, a fixed chance that the future offspring will be from the superparasitizing female.

Figure 9.6 shows diagrams of the transitions between host states for this model, when both females are in the patch, and for three possible situations, according to whether or not

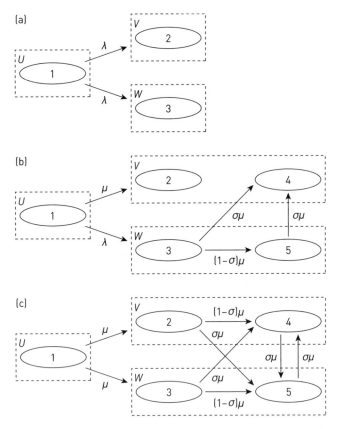

Fig. 9.6 Diagrams of transitions between host states when both females are present in the patch, for three situations: (a) neither superparasitizes, (b) only F_v superparasitizes, (c) both superparasitize. Host states: **1** unparasitized, **2** parasitized once by F_v, **3** parasitized once by F_w, **4** parasitized multiple times, and yielding offspring to F_v, **5** parasitized multiple times and yielding offspring to F_w. The three-state representation is also indicated: $U =$ unparasitized, $V =$ yielding offspring to F_v, and $W =$ yielding offspring to F_w. See the text for further explanation of the parameters. Figure taken from Haccou et al. (2003).

they are superparasitizing. The chance per time unit that an unparasitized host is attacked by a non-superparasitizing female is denoted by λ. For mathematical tractability, handling time is not included explicitly in this model, but to take into account the reduction in available search time due to superparasitism, it is assumed that the chance per time unit of attacking a host is reduced from λ to μ once a female starts superparasitizing. σ denotes the probability that the future offspring will be from the superparasitizing female, in absence of further superparasitism.

Haccou et al. (2003) showed that, for this model, the patch state can still be represented as a point within a triangle (Fig. 9.3a). However, contrary to the situation without super-parasitism, the values of v and w cannot be observed directly, since chance governs the type of offspring produced by superparasitized hosts. However, as before, since there is a one-to-one relationship between the time spent in the patch and the patch state, patch states can be inferred from arrival times of competitors and the time spent in the patch by each of them.

To derive the competitive strategy, we first consider the optimal behavior of a singly foraging female in relation to the patch state. There are three different behavioral options: (i) stay in the patch without superparasitizing; (ii) stay in the patch and superparasitize; and (iii) leave the patch. Haccou et al. (2003) showed that the boundaries of these areas are determined by three straight lines in the state space. Figure 9.7 shows these boundaries for female F_v.

Note that the line LN indicates the optimal leaving strategy of a singly foraging female only in the area without superparasitism, where the intra-patch gain rate is solely determined by u. When superparasitism occurs, the boundary between staying and leaving is given by $LS1$, since then the intra-patch gain rate of F_v is determined by a (linear) combination of u and w.

It is an arbitrary choice which of the females is F_v and which is F_w. Therefore, the threshold line LN is the same for F_w, and the other lines (denoted by $S2$ and $LS2$) are the

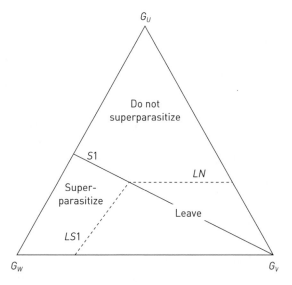

Fig. 9.7 Boundaries of optimal behavioral options for female F_v when she is alone in the patch, and may superparasitize. Figure adjusted from Haccou et al. (2003).

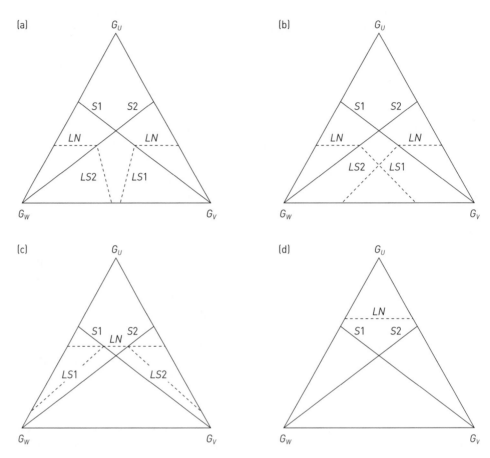

Fig. 9.8 Different possible configurations of the boundaries of optimal behavioral options for the two females if they are alone in the patch and may superparasitize. There are three possible types: (a) type 1: *LN* lies below *S*1 and *S*2, and lines *LS*1 and *LS*2 do not cross, (b,c) type 2: *LN* lies below *S*1 and *S*2, and lines *LS*1 and *LS*2 cross, and (d) type 3: *LN* lies above *S*1 and *S*2. The three types of configurations each imply a different type of ESS. Figure adjusted from Haccou et al. (2003).

mirror images of those for F_v. Figure 9.8 shows the different possible combinations of the sets of lines. The configuration in Fig. 9.8a (type 1) corresponds to a situation with low environmental gain rate and/or high superparasitism success. Figures 9.8b and c (both of type 2) correspond to cases with intermediate values of these parameters, and Fig. 9.8d (type 3) depicts situations with high environmental gain rate and/or low superparasitism success.

Haccou et al. (2003) showed that the lines *S*1 and *S*2 do not change when both females are present in the patch. In type 3 situations, the line *LN* lies entirely above these two lines. Therefore, both females leave without superparasitizing, when the patch trajectory hits the line *LN*, and their intra-patch gain rate equals γ. In this situation, there is simultaneous patch depletion without interference. As a consequence, all pairs of payoff functions are equal and look like those represented in Fig. 9.2a. In the two other types of situations,

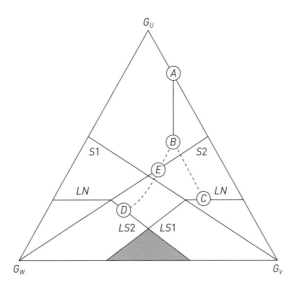

Fig. 9.9 Example of a trajectory in the state space when superparasitism may occur. F_v arrives in the patch first. F_w arrives at point A. At point B, one of the females leaves. If F_v leaves first, the trajectory goes to point E, and then F_w starts superparasitizing. When the trajectory reaches point D, she leaves too. If F_w leaves first, the trajectory goes to point C, where F_v leaves. The shaded area indicates patch states where neither of the females should stay in the patch. Figure adjusted from Haccou et al. (2003).

parasitoids may start to superparasitize before the patch trajectory hits *LN*. This implies that there is competition with interference, and therefore the lines *LN*, *S*1, and *S*2 do not predict competitive patch leaving times (only patch leaving of single foragers). The payoff functions can be derived in a similar way as before (Fig. 9.4). An example is given in Fig. 9.9.

Up to point B the situation is the same as in Fig. 9.4. If F_w leaves at point B, F_v follows her single optimal strategy, so she does not superparasitize and leaves at point C. If F_v leaves at point B, F_w follows her optimal single strategy, which then has a different form: initially she does not superparasitize, until the patch trajectory hits line *S*2 at point E. Then she starts superparasitizing. As a consequence, the patch trajectory becomes curved, since not only w and u, but also v changes from this point onwards. F_w leaves the patch when the trajectory hits the line *S*2, which happens at point D. The payoff functions of the two females are still as given in Equations (9.6) and (9.7). However, in this situation $v_C - v_A \neq w_D$ and it is no longer true that the residual staying times y_v^* and y_w^* are equal. Thus, the payoff functions of F_v and F_w differ. When $y_v^* = y_w^* = 0$, the payoffs of being the first or second to leave become equal, i.e. $L_v(t) = S_v(t)$ and $L_w(t) = S_w(t)$. This happens at the time τ when a trajectory with both females present enters the shaded area in Fig. 9.9. As before, τ corresponds to the maximum persistence time (Haccou & Glaizot 2002, Haccou et al. 2003). Figure 9.10a illustrates how the forms of the payoff functions of the two females in type 2 situations may look. In type 1 situations, there are no patch states where $y_v^* = y_w^* = 0$ (Fig. 9.7a). In this case, all payoff functions become parallel in the long

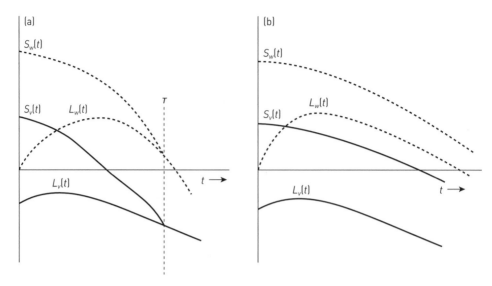

Fig. 9.10 Examples of payoff functions of competing females that may superparasitize. (a) For a type 2 situation (corresponding to Figs 9.8b or c). (b) For a type 1 situation (corresponding to Fig. 9.8a).

run, like in the standard war of attrition, and there is no maximum persistence time. Figure 9.10b gives an example.

Haccou and Glaizot (2002) and Haccou et al. (2003) derived the main properties of the ESS with mistakes in perception of arrival times. Figure 9.11 summarizes their results for a type 2 situation. The figure shows an example of three pairs of payoff functions, each pair corresponding to a payoff of leaving first or second, for different perceptions of arrival times relative to the opponent. Characteristics of the ESS strategy are as follows.

- Do not leave as long as all the perceived payoffs of leaving are increasing, i.e. before time F.
- Do not stay after the point where all the payoffs of leaving first are equal to the payoffs of leaving second, indicated by H.
- Choose a random leaving time in the interval from G to H. The support of the probability distribution that specifies the ESS includes all the points in this interval, where all the payoffs of leaving are decreasing.
- In a type 3 situation, where the payoff functions do not become equal in the long run, there is no upper bound on the ESS distribution of leaving times.

These general characteristics can be derived analytically and do not depend on details of a model, such as the dynamics of host encounters. Haccou et al. (2003) examined the ESS numerically and found that a higher uncertainty in the parasitoids' estimate of relative arrival times decreases the variance in leaving times, which makes the ESS more similar to a fixed patch leaving time strategy. However, they considered a specific model and it remains to be examined how general this result is.

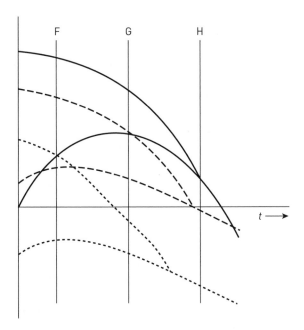

Fig. 9.11 Illustration of main properties of the ESS for type 2 situations. Example of three pairs of payoff functions. Each pair corresponds to a payoff of leaving first or second, for different perceptions of arrival times relative to the opponent. In the interval up to F, all the perceived payoffs of leaving are increasing. For times larger than H, all the payoffs of leaving first are equal to the payoffs of leaving second. In the interval from G to H, all the payoffs of leaving are decreasing.

9.5 Discussion

We have shown that when parasitoids superparasitize, the MVT does not correctly predict their ESS residence time on a host patch. Patch leaving strategies should be random when superparasitism occurs, and individuals are expected to stay in the patch even though their gain rate is smaller than γ. Further, their leaving strategies should depend on their relative arrival times. To date, there are no empirical tests of these predictions, but such tests are feasible, although technically difficult. The difficulties arise from measuring gain, because the values of v and w cannot be observed directly, since chance governs the type of offspring produced by superparasitized hosts. However, with non-moving hosts, the timing of ovipositions in unparasitized hosts and superparasitism can be determined by direct observation and molecular markers can be used to determine the mother of each offspring reared from superparasitized hosts. This, at least, should allow measurements of average gain rates over time intervals. By manipulating estimates of environmental gain rate, for example, through travel times or previous experience with host patches, and by manipulating the arrival time of the second female, different strategies should be observed and it would be possible to decide if these are in accordance with the predictions of the theory.

We have made several assumptions in the theoretical developments presented in this chapter. Assumption A1 implies that we assume there is no egg-limitation. Competition with egg-limitation will give different predictions, and has been considered, for example, by Price 1973 and Iwasa et al. 1984. Assumption A2 implies that the environment is assumed to be stable, or to change much slower than foragers switch between patches. Assumption A3 implies that we consider solitary foragers, and that every parasitized host will yield a parasitoid offspring. The latter assumption can be easily generalized. To consider gregarious parasitoids, differences between clutch sizes also have to be taken into account. This generalization is straightforward, but its effects remain to be examined.

Several studies have shown that parasitoids obtain information about profitability of a patch during foraging (Haccou et al. 1991, Hemerik et al. 1993) and so assumption A4, that parasitoids know the patch state, is not unrealistic. However, when there is superparasitism, patch state cannot be observed, since chance determines which of the competitors will acquire offspring from superparasitized hosts. Since there is a one-to-one relationship between patch state and patch residence times (assumption A7), patch state can be inferred from the arrival times. This implies that females 'know' the initial patch state, time spent on the patch by each, and the behavior of both since then. The second female thus needs to estimate the patch state upon her arrival. This might be done by observing the ratio of encounters with healthy hosts to those with parasitized hosts. Both females may obtain information on their competitor's behavior during their patch visit. Apart from this, if their behavior is shaped by the ESS, it is evolutionarily fixed and can be assumed 'known' from this perspective. Mistakes in perception of relative arrival times were included in our model. Further, Haccou and Glaizot (2002) argued that mistakes in estimation of patch state, given relative arrival time, and as long as they are unbiased, do not affect the expected payoff functions. However, they did not take into account the fact that the errors in the perception of the initial state produces errors in the time of switching to superparasitism, whose effect can be biased. This topic remains to be examined further, and the results of Haccou et al. (2003) must, in this light, be considered as a first-order approximation for the situation with incomplete information.

Effects of asymmetries of females other than arrival times (i.e. generalization of assumption A5), can be incorporated by allowing the payoff functions to depend on other characteristics. Sjerps and Haccou (1994b) give a generalization to situations with more than two competitors (assumption A6). Their methodology can also be applied to the case with superparasitism, but the mathematics will become complicated, and analytical results may be difficult to obtain. If assumption A7 is dropped, patch state and patch residence time are no longer directly related. This will introduce major complications, and it is not clear how model results will be affected. Effects of short absences from the patch, however, can probably be ignored for practical purposes. Assumptions A8 to A10 are reasonable, since females of many species place unique marks on their parasitized hosts (Hubbard et al. 1987, van Dijken et al. 1992, Visser 1993). The probability of superparasitism success is usually known to depend on the time since previous parasitism, as well as the number of eggs already laid in the host and thus, A11 may have to be adjusted. When the expected patch residence time is relatively short, however, these factors may be ignored and our model can still be used.

Superparasitism is not the only way in which parasitoids compete. In several species, patch defense by the first arriving female has been observed (Field & Calbert 1998a,b, Field et al. 1998). Patch defense involves patroling of the patch edge, pursuit, and fighting with intruding competitors. Like superparasitism, this form of competition induces interference and, therefore, situations similar to a generalized war of attrition are expected to occur. However, to precisely examine effects of such behavior, new models have to be developed.

The relevance of the theory presented here for the practice of biological control is twofold. First, understanding the processes that govern dispersal by individuals from exploited patches allows making a realistic link between individual behavior and population level processes, as these processes determine the distribution of parasitoids over patches and the level of mutual interference. As aggregation and interference are key-parameters affecting population dynamics, understanding them is important to understand the population dynamics of a parasitoid and that of the herbivorous host against which it is used as a control agent. Second, these models can help to optimize mass rearing programs of parasitoids for biological control. Mass rearing of parasitoids occurs by its very nature at high densities. This is because high rates of parasitism are required to maximize output, cultures are synchronized to minimize labor, and the costs of climatized breeding space must be minimized. All these circumstances contribute to competition between the individuals of the parental generation, but often also between the individuals of the offspring generation. This may affect the rates of superparasitism and/or rejection of hosts already parasitized by others, and thus may affect *per capita* offspring gain and/or search rate. Thus, competition may decrease the efficiency of parasitoid mass rearing. Currently, such effects are poorly understood. The models presented here form a first step to understand these complicated interactions.

Acknowledgments

The manuscript benefited a great deal from the constructive criticism of Éric Wajnberg and one anonymous referee.

References

Bernstein, C., Kacelnik, A. and Krebs, J.R. (1988) Individual decisions and the distribution of predators in a patchy environment. *Journal of Animal Ecology* **57**: 1107–26.

Bernstein, C., Kacelnik, A. and Krebs, J.R. (1991) Individual decisions and the distribution of predators in a patchy environment II: The influence of travel costs and the structure of the environment. *Journal of Animal Ecology* **60**: 205–26.

Bishop, D.T. and Cannings, C. (1978) A generalised war of attrition. *Journal of Theoretical Biology* **70**: 85–124.

Bishop, D.T., Cannings, C. and Maynard Smith, J. (1978) The war of attrition with random rewards. *Journal of Theoretical Biology* **74**: 377–88.

Charnov, E.L. (1976) Optimal foraging: The marginal value theorem. *Theoretical Population Biology* **9**: 126–36.

Cressman, R., Křivan, V. and Garay, J. (2004) Ideal free distributions, evolutionary games and population dynamics in multiple species environments. *American Naturalist* **164**: 473–89.

Field, S.A. and Calbert, G. (1998a) Patch defence in the parasitoid wasp *Trissolcus basalis*: When to begin fighting? *Behaviour* **135**: 629–42.

Field, S.A. and Calbert, G. (1998b) Don't count your eggs before they're parasitized: Contest resolution and the trade-offs during patch defense in a parasitoid wasp. *Behavioral Ecology* **10**: 122–7.

Field, S.A., Calbert, G. and Keller, M.A. (1998) Patch defense in the parasitoid wasp *Trissolcus basalis* (Insecta: Scelionidae): The time structure of pairwise contests, and the 'waiting game'. *Ethology* **105**: 821–40.

Fretwell, S.D. and Lucas, H.J. (1970) On territorial behavior and other factors influencing habitat distribution in birds. *Acta Biotheroretica* **19**: 16–36.

Galis, F. and van Alphen, J.J.M. (1981) Patch time allocation and search intensity of *Asobara tabida* Nees (Hym.: Braconidae). *Netherlands Journal of Zoology* **31**: 701–12.

Godfray, H.G. (1994) *Parasitoids: Behavioral and Evolutionary Ecology*. Princeton University Press, Princeton.

Haccou, P. and Glaizot, O. (2002) The ESS in an asymmetric generalized war of attrition with mistakes in role perception. *Journal of Theoretical Biology* **214**: 329–49.

Haccou, P., De Vlas, S.J., van Alphen, J.J.M. and Visser, M.E. (1991) Information processing by foragers: Effects of intra-patch experience on the leaving tendency of *Leptopilina heterotoma*. *Journal of Animal Ecology* **60**: 93–106.

Haccou, P., Glaizot, O. and Cannings, C. (2003) Patch leaving strategies and superparasitism: An asymmetric generalized war of attrition. *Journal of Theoretical Biology* **225**: 77–89.

Hamelin, F., Bernard, P., Nain, P. and Wajnberg, É. (2007a) Foraging under competition: Evolutionarily stable patch-leaving strategies with random arrival times. 1. Scramble competition. In: Quincampoix, M., Vincent, T.L. and Jørgensen, S. (eds.) *Advances in Dynamic Game Theory: Numerical methods, algorithms, and applications to ecology and economics*, Vol. 9: Annals of the International Society of Dynamic Games, Birkhaüser.

Hamelin, F., Bernard, P., Shaiju, A.J. and Wajnberg, É. (2007b) Foraging under competition: Evolutionarily stable-patch leaving strategies with random arrival times. 2. Interference competition. In: Quincampoix, M., Vincent, T.L. and Jørgensen, S. (eds.) *Advances in Dynamic Game Theory: Numerical methods, algorithms, and applications to ecology and economics*, Vol. 9: Annals of the International Society of Dynamic Games, Birkhaüser.

Hammerstein, P. and Parker, G.A. (1982) The asymmetric war of attrition. *Journal of Theoretical Biology* **96**: 647–82.

Hemerik, L., Driessen, G. and Haccou, P. (1993) The effects of intra-patch experiences on patch leaving tendency, search time and search efficiency of parasitoids of the species *Leptopilina clavipes*. *Journal of Animal Ecology* **62**: 33–44.

Hubbard, S.F. and Cook, M. (1978) Optimal foraging by parasitoid wasps. *Journal of Animal Ecology* **47**: 593–04.

Hubbard, S.F., Marris, G., Reynolds, A. and Rowe, G.W. (1987) Adaptive patterns in the avoidance of superparasitism by solitary parasitoid wasps. *Journal of Animal Ecology* **47**: 593–4.

Iwasa, Y., Suzuki, Y. and Matsuda, H. (1984) Theory of oviposition strategy of parasitoids. I. Effect of mortality and limited egg number. *Theoretical Population Biology* **26**: 205–27.

Janssen, A. (1989) Optimal host selection by *Drosophila* parasitoids in the field. *Functional Ecology* **3**: 469–79.

Mangel, M. (1992) Descriptions of superparasitism by optimal foraging theory, evolutionarily stable strategies and quantitative genetics. *Evolutionary Ecology* **6**: 152–69.

Maynard Smith, J. (1974) The theory of games and the evolution of animal conflicts. *Journal of Theoretical Biology* **47**: 209–21.

McNamara, J.M. and Houston, A.I. (1987) A general framework for understanding the effects of variability and interruptions on foraging behaviour. *Acta Biotheoretica* **36**: 3–22.

Price, P.W. (1973) Reproductive strategies in parasitoid wasps. *American Naturalist* **107**: 684–93.

Price, P.W. (1980) *Evolutionary Biology of Parasites*. Princeton University Press, Princeton.

Roitberg, B.D. and Mangel, M. (1988) On the evolutionary ecology of marking pheromones. *Evolutionary Ecology* **44**: 623–38.

Sirot, E. and Křivan, V. (1997) Adaptive superparasitism and host-parasitoid dynamics. *Bulletin of Mathematical Biology* **59**: 23–41.

Sirot, E., Ploye, H. and Bernstein, C. (1997) State dependent superparasitism in a solitary parasitoid: Egg load and survival. *Behavioural Ecology* **8**: 226–32.

Sjerps, M. and Haccou, P. (1994a) A war of attrition between larvae on the same host plant: Stay and starve or leave and be eaten? *Evolutionary Ecology* **8**: 269–87.

Sjerps, M. and Haccou, P. (1994b) Effects of competition on optimal patch leaving: A war of attrition. *Theoretical Population Biology* **3**: 300–18.

van Alphen, J.J.M. and Galis, F. (1983) Patch time allocation and parasitization efficiency of *Asobara tabida*, a larval parasitoid of *Drosophila*. *Journal of Animal Ecology* **52**: 937–52.

van Alphen, J.J.M., Bernstein, C. and Driessen, G. (2003) Information acquisition and time allocation in insect parasitoids. *Trends in Ecology & Evolution* **18**: 81–7.

van Dijken, M.J., van Stratum, P. and van Alphen, J.J.M. (1992) Recognition of individual-specific marked hosts by the solitary parasitoid *Epidinocarsis lopezi*. *Behavioral Ecology & Sociobiology* **30**: 77–82.

Vet, L.E.M. and van Alphen, J.J.M. (1985) A comparative functional-approach to the host detection behavior of parasitic wasps 1. A qualitative study on eucoilidae and alysiinae. *Oikos* **44**: 478–86.

Vinson, S.B. (1981) Habitat location. In: Nordlund, D.A., Jones, R.L. and Lewis, W.J. (eds.) *Semiochemicals, Their Role in Pest Control*. John Wiley, New York, pp. 51–78.

Vinson, S.B. (1984) How parasitoids locate their hosts: A case of insect espionage. In: Lewis, T. (ed.) *Insect Communication*. Academic Press, London, pp. 325–48.

Vinson, S.B. (1985) The behaviour of parasitoids. In: Kerkut, G.A. and Gilbert, L.I. (eds.) *Comprehensive Insect Physiology, Biochemistry and Pharmacology*. Pergamon Press, New York, pp. 417–69.

Visser, M.E. (1993) Adaptive self- and conspecific superparasitism in the solitary parasitoid *Leptopilina heterotoma*. *Behavioral Ecology* **4**: 22–8.

Waage, J.K. (1979) Foraging for patchily-distributed hosts by the parasitoid, *Nemeritis canescens*. *Journal of Animal Ecology* **48**: 353–71.

Wajnberg, É. (2006) Time allocation strategies in insect parasitoids: From ultimate predictions to proximate behavioral mechanisms. *Behavioral Ecology & Sociobiology* **60**: 589–611.

Wajnberg, É., Curty, C. and Colazza, S. (2004) Genetic variation in the mechanisms of direct mutual interference in a parasitic wasp: Consequences in terms of patch-time allocation. *Journal of Animal Ecology* **73**: 1179–89.

10

Risk assessment and host exploitation strategies in insect parasitoids

Luc-Alain Giraldeau and Guy Boivin

Abstract

We use the term 'foraging' broadly to mean the search and exploitation of both food and hosts. We approach the problem of food and host selection from a cost-beneficial, behavioral, and ecological point of view. Within behavioral ecology, foraging theory (Stephens & Krebs 1986, Giraldeau & Caraco 2000) proposes a number of economic models among which risk-sensitive models deal with the effect that uncertainty has on a consumer's preference. Risk-sensitivity applies when utility (fitness) functions are curvilinear. For instance, when an animal exploits food, the relationship between intake and fitness tends to be sigmoid; at first, increases in intake may have little effect on fitness when gains remain below the intake required to survive. Beyond this requirement level, gains probably accrue rapidly, but not indefinitely. Once survival is achieved, very little fitness gain will be associated with further increases in food intake. When utility functions are generated for host parasitization and only the number of eggs laid is considered, a linear relationship is expected, indicating that female parasitoids should be risk-insensitive. However, because the fitness return is obtained only if the immature survive to reproduction, using the realized value of the eggs rather than their actual value might be more appropriate. The realized value takes into account the dangers faced by the developing immature (predation, hyperparasitism, superparasitism) and the resulting utility function can be curvilinear if the duration of patch exploitation by the female parasitoid influences the danger level experienced by its developing progeny. Such curvilinear utility functions could then generate risk-sensitive host choices. We end by exploring the consequences of these ideas for applications.

10.1 Introduction

In this chapter we use the logic of risk-sensitive foraging behavior to explore whether it can be applied to the search and exploitation of reproductive resources by insect parasitoids. A similar approach of applying foraging models to reproductive decisions, namely host selection and patch residence time decisions by females, is often used in parasitoid behavioral ecology (Godfray 1994, van Alphen et al. 2003). The approach assumes that the behavior we currently observe is the evolutionary product of a compromise that maximizes the animal's fitness under a set of constraints. A large number of studies using parasitoids have involved the use of such behavioral ecological methods and questions. Most notably, many studies have explored host selection as a diet choice problem, where hosts vary based on their species (Pak et al. 1990, Brodeur et al. 1996), age (Miura & Kobayashi 1998, Monge et al. 1999, Godin & Boivin 2000), or parasitization status (van Baaren & Boivin 1998, Visser & Rosenheim 1998, Royer et al. 1999, Varaldi et al. 2005). Other studies have analyzed the number of hosts exploited in a clump as a patch residence problem (van Alphen et al. 2003, Boivin et al. 2004, Thiel & Hoffmeister 2004, Wajnberg 2006). There has been an interesting interchange between developments of foraging models and their application to parasitoid behavior. However, not all developments of foraging theory have found applications in the parasitoid system and risk-sensitivity in particular has yet to be explored in the context of parasitoid egg-laying behavior. Our goal in this chapter, therefore, is to provide a preliminary exploration of the potential for the logic of risk-sensitivity to spawn new and exciting research questions.

10.2 The optimal foraging approach

All optimality models are made up of a decision, a set of constraints, and a hypothetical currency of fitness (Stephens & Krebs 1986). The general idea is that animals have been selected to adopt decisions that maximize some currency of fitness and so can be seen as optimal decision makers (MacArthur & Pianka 1966). The most common optimal foraging models, some of which figure prominently in many chapters of this book (see also Chapter 8 by van Alphen and Bernstein and Chapter 9 by Haccou and van Alphen), address host choice and patch residence time in a deterministic way. Deterministic models assume that a given decision always corresponds to some determined value of a currency of fitness. Patch residence models, for instance, assume that each patch residence time for a given travel time yields with certainty a given currency of fitness. The success of such deterministic models is indisputable. They have predicted, at least qualitatively, the foraging behavior of a large number of organisms (Stephens & Krebs 1986, Stephens & Dunbar 1993, Nonacs 2001, Sih & Christensen 2001) as well as the oviposition decision of parasitoids (van Alphen et al. 2003, Wajnberg 2006, Wajnberg et al. 2006).

However, the world is rarely deterministic; the same decision does not always yield the same outcome with certainty. It is common to experience variance in the costs and benefits associated with any given decision. For instance, when predators experience variation in the profitability of a given prey item, it implies that the choice between alternatives must be based on some mean expected reward whose certainty depends on the variance in the profitability that characterizes each type of item. To ignore the effect of this variance is equal to assuming that it has no effect on the survival value associated with the choice of

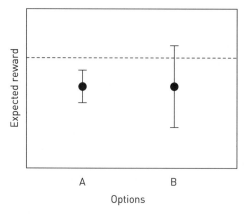

Fig. 10.1 An animal faces two options, A and B, which offer the same mean (filled circle) expected reward. The horizontal dashed line gives a threshold reward level below which the animal suffers some fitness cost. Each available option has a different variance represented by the lines above and below the means. In this case, option B offers the greatest chances of falling above the threshold and animals should be sensitive to an option's uncertainty because it has an influence on its expected fitness.

either alternative. However, in some situations the uncertainty associated with the variance in anticipated reward will probably influence the fitness consequences of a given course of action. Consider two options, A and B, that offer the same mean expected reward but for which B offers the greatest uncertainty (Fig. 10.1). When there is a threshold reward value beyond the mean expected reward of either alternative and below which animals suffer some cost, it becomes clear that the most uncertain option provides the only hope of avoiding the cost (Fig. 10.1). The area of foraging theory that has explored the effect of uncertainty of rewards on animal decisions is known as risk-sensitive foraging.

10.3 The logic of risk sensitivity in a foraging context

Optimality models assume that there exists some positive fitness function describing the relationship between values of a hypothesized currency of fitness and its corresponding utility expressed as fitness. Most first-generation foraging models implicitly assumed linear fitness functions. For a currency such as net rate of energy intake it seemed entirely plausible that the greater the intake obtained from a foraging option the greater the fitness returns associated with it. However, a fitness function cannot be linear over the whole range of values of the currency of fitness. For instance, imagine a small bird in the winter: once it has acquired enough energy to survive the night it is likely that any further increase in energy intake beyond this requirement will not lead to important increases in fitness. Moreover, if it is suffering from an important energetic deficit during the day such that it will probably starve during the night if some minimum requirement is not obtained, any increase in intake that is insufficient to reach this requirement will fail to provide any increase in fitness (Fig. 10.1). It follows that the conditions under which an animal behaves may well be such that fitness functions have either concave or convex shapes (Fig. 10.2).

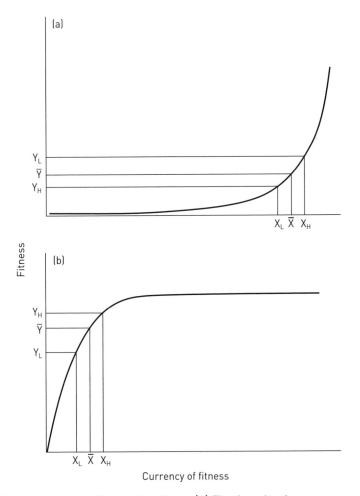

Fig. 10.2 Two hypothetical fitness functions. (a) The function increases exponentially, a situation that corresponds to cases where small increments in reward do not produce equivalent increases in fitness initially. The graph depicts a hypothetical situation where the animal faces a choice between a certain and a variable option, both offering the same mean reward X. However, the variable option is equally likely to provide X_L and X_H reward values. The fitness Y corresponding to the certain option X is smaller than the mean of Y_L and Y_H. An animal in this situation should, all else being equal, prefer the riskier option: it is risk-prone. (b) The function increases at a decelerated pace, a situation that corresponds to options that exceed some threshold value to avoid cost. The graph depicts the same choice between a fixed and variable option. However, this time the shape of the fitness function is such that the mean fitness from the fixed reward Y is larger than the mean of Y_L and Y_H. In this case, risk is detrimental to an animal's fitness and it should be risk-averse.

Such curvilinear fitness functions will have an important consequence on the effect that variability or uncertainty in reward will have on mean expected fitness returns and an animal that changes its behavior according to the uncertainty in fitness gain is thus considered risk-sensitive.

10.4 Explaining risk-proneness: not enough to survive

Consider first an exponentially increasing fitness function (Fig. 10.2a) that, in a foraging context, corresponds to a situation where available intake rate options are below the level required to meet a threshold requirement to avoid some physiological cost. In a deterministic world, a given foraging decision would be characterized by its yield of X units corresponding deterministically to Y units of fitness. In a stochastic world, the same foraging decision will correspond probabilistically to a number of different yields that can be characterized by an expected value associated with some variance that specifies the uncertainty surrounding the expectation. For simplicity we explore the case where the stochastic option offers a 50% chance of obtaining X_H units and a 50% of obtaining X_L units of currency of fitness. Note that the mean expected reward of the stochastic option is the same as the deterministic one, X. However, the fitness corresponding to the deterministic option Y is now lower than the mean fitness expected from the stochastic option. The mean of the uncertain option is larger because of the variance in X and the shape of the curvilinear function, a mathematical reality known as Jensen's inequality (Smallwood 1996). This increase in fitness resulting from the variance implies that, for these types of fitness functions, the uncertainty associated with a given option improves the expected fitness returns from that option. Then, all else being equal, a risk sensitive animal should prefer options that offer a higher level of uncertainty, that is, it should be risk-prone.

10.5 Explaining risk-aversion: more than enough to survive

Consider now a situation where the fitness function is decelerated (Fig. 10.2b), which corresponds to a situation where the available options provide more than enough to avoid some fitness cost. In this case, Jensen's inequality implies the reverse result. Uncertainty in X has a depressing effect on the expected mean fitness compared to a deterministic alternative with the same mean reward. Uncertainty associated with an option now reduces its expected fitness. In this case, all else being equal, the outcome with the lowest uncertainty provides the greatest fitness returns. Once again we expect animals to be risk-sensitive in this situation and to adopt a risk-averse attitude. Thus, all else being equal, the least uncertain alternative should be preferred.

10.6 The energy budget rule

The reasoning associated with risk-sensitive models has given rise to a number of foraging models, such as the z-score model (Stephens & Charnov 1982). Stephens (1981) has analyzed the logic behind risk-sensitivity in a foraging context and summarized it into a simple elegant rule known as the 'energy budget rule'. When an animal is experiencing a negative energy budget such that it is likely to suffer some fitness cost, perhaps even death in extreme cases, given the current conditions it should prefer options that provide more uncertain outcomes. It should become risk-prone. On the other hand, when an animal is on a positive energy budget, it should avoid taking the chances of suffering such costs and hence prefer the more certain options. It should be risk-averse. We are reminded here that 'risk' does not refer to any form of danger such as starvation or predation but to the chance associated with options of greater uncertainty. It is important to distinguish our technical use of the word 'risk' from its vernacular use in the sense of hazard.

10.7 Are insects risk-sensitive?

Risk-sensitivity, describing the tendency of animals for either risk-aversion or risk-proneness, has been found in a number of mammal, bird, and some insect species (Real & Caraco 1986, Stephens & Krebs 1986). Our goal here is not to review all examples of risk-sensitive behavior in insects but simply to illustrate the conditions under which insect risk-sensitive behavior has been observed.

10.7.1 Risk-sensitive food choice

The most common support for risk-sensitive behavior involves providing an individual with a choice between two resource types offering equal mean rewards but differing in variance and hence uncertainty. While many cases involve vertebrates, several use insects. Insect pollinators represent a model system well adapted to the study of risk sensitivity in insects. In many plants, both the concentration and the volume of nectar vary between flowers. Such variability could arise as a mechanism used by plants to decrease the risk of within-plant selfing by promoting fewer flower visits per plant by pollinators while keeping a high attractiveness (Biernaskie et al. 2002). Pollinator insects encounter a variable resource on such plants and, in risk-averse species, reduce the duration of their foraging.

Risk sensitivity has been demonstrated in bumble bees (Real 1981, Waddington et al. 1981, Real et al. 1982, Biernaskie et al. 2002), honey bees (Shafir et al. 1999, Ferguson et al. 2001), and paper wasps (Real 1981), most species being risk-averse. Cartar (1991) provides an elegant experiment demonstrating such risk-sensitive behavior in three species of bumble bees, *Bombus melanopygus, B. mixtus,* and *B. sitkensis,* foraging on two species of flowers, seablush (*Plectritis congesta*) and dwarf huckleberry (*Vaccinium caespitosum*). Before 18h00 the bumble bees deplete the nectar reserves of both plant species to a common mean level such that after this time only the variance in reward differs between plants, dwarf huckleberry offering more variable rewards than seablush. Cartar (1991) experimentally manipulated the energy reserves in some of the colonies by either removing or adding nectar to the colonies' communal honey pots. Depleting these reserves made the bumble bee workers risk-prone and, as predicted, they preferred to exploit dwarf huckleberry after 18h00. Similarly, when adding nectar to the colonies' communal honey pots he observed that the bees became risk-averse and preferred foraging on seablush. Those results not only confirmed risk-sensitivity in field conditions but also showed that bumble bees could alter their preference for risk in the expected way.

10.7.2 Risk-sensitive patch residence

In situations where rewards are cryptic, a forager can determine the variability of a resource clump only by investing some time in sampling it (Biernaskie et al. 2002). Such a situation occurs when the forager cannot identify the quality of the resource through external characteristics, such as visual cues. In that case, risk-sensitive foraging choices will require some minimum amount of sampling time and patch choice may in fact require that we compare apparent patch residence. Patch residence as an indicator of choice occurs especially in pollinators that exploit flowers whose appearance does not change with the amount or quality of nectar or with parasitoids who, having found a host patch, must for example contact hosts to determine if they have already been parasitized.

Mostly applied to situations where an organism must choose among alternative resources, risk-sensitive foraging has more rarely addressed patch residence decisions. Demonstrations of risk-sensitive patch-departure decision are also rare. A theoretical exploration of the effect of stochastic rewards on patch residence has established that risk-averse foragers should stay longer, perhaps up to patch depletion, while risk-prone foragers should leave the patch much sooner (Rita & Ranta 1998). This prediction has been confirmed once in the common shrew (*Sorex araneus*) (Barnard & Brown 1987), and explored for bumble bees foraging on two-flowered inflorescences (Cartar & Abrahams 1996). The predictions, in this case, were generated from computer simulations but are qualitatively similar to those of Rita and Ranta (1998): a reduction in colony reserves that increases the chance of incurring some physiological cost is expected to induce risk-prone workers to adopt lower patch-departure thresholds and hence leave patches (inflorescences) sooner. Experimental tests using artificial flowers presented to *B. occidentalis* workers provided at best equivocal support (Rita & Ranta 1998).

10.8 Applying the logic of risk-sensitivity to host patch choice in parasitoids

The idea of risk-sensitivity that we have presented up to now is static. For a given state, the animal must choose one or the other option. Naturally, an animal's state changes over time, whether time of day or its lifetime, such that the problem of choosing the best alternative may also change in the course of time. When this is the case, dynamic models are required and stochastic dynamic state-dependent models (McNamara et al. 1995, see also Chapter 15 by Roitberg and Bernhard) would most likely apply to analysis of risk-sensitive behavior. However, given that the idea of risk-sensitivity has not been used much in the context of reproduction, we feel it best to start off by developing a static model as a first step, keeping in mind that dynamic versions may come later.

Few studies have considered risk-sensitivity in a reproductive context. One study explores the impact of risk-sensitive foraging decisions on reproductive output (Bednekoff 1996) but no study to our knowledge addresses risk-sensitive host selection decisions. This is not too surprising given that, in most cases, it is probably safe to assume that the fitness function relating egg number to realized fitness is entirely linear and that, therefore, there is no reason to expect risk-sensitivity to play any role in egg-laying. If each egg laid in a host by a parasitoid has the same expected probability of survival, then clearly the more eggs that are laid per patch then the more fitness is obtained. If risk-sensitive models are to be relevant to host exploitation by parasitoids, at least one of the two following situations must be met. Either there is a minimum number of eggs that must be laid if any one of them is to survive to emergence or, beyond some upper threshold the number of eggs leading to emergence ceases to increase linearly with additional eggs laid. We feel there are a number of factors that can affect the linearity of the fitness function for parasitoid host exploitation. We explore these in turn.

10.8.1 Smaller fitness gain per egg in small host patches

Small host patches are likely to provide a smaller fitness gain per egg deposited if the number of hosts is insufficient to allow gregarious or quasi-gregarious parasitoids to allocate an optimal sex-ratio of offspring to allow for efficient sib–sib mating of emerging adults.

Most Hymenoptera parasitoids reproduce by arrhenotokous parthenogenesis, where female gametes develop into females when fertilized and into males when not. The female parasitoids can thus control the sex ratio of the progeny they deposit in a host patch by controlling the access of sperm to eggs during oviposition (see also Chapter 12 by Ode and Hardy). A female parasitoid that allocates the sex of its progeny in a binomial fashion would face a high probability of depositing a non-optimal sex ratio, especially so in small host patches. For example, in a host patch of five hosts, a female using binomial sex allocation and aiming at an optimal sex ratio of 0.2 (proportion of males) would have a 0.328 probability of depositing only females in that patch. These females would thus find no mate on the patch, leading to two consequences. First, if hosts are still available on the emergence patch, the females, being virgin, could only deposit males. Second, these females will eventually disperse as virgins and will look for out-patch mating. The out-patch mates will certainly be unrelated males, therefore diluting the genes of their mother. In both cases the fitness gain obtained through these eggs is smaller than what it could have been in a larger host patch.

Female parasitoids have evolved a mechanism to reduce the costs associated with binomial sex allocation. Females of numerous parasitoid species use precise sex allocation to reach an optimal sex ratio in a host patch (Green et al. 1982). Generally males are deposited early in an oviposition sequence, therefore ensuring that the following females will have at least one male to mate (Mills & Kuhlmann 2004).

Therefore, when a female parasitoid exploits a small host patch, both binomial and precise sex ratio allocation incur a cost, although smaller with the latter. The fitness gain per egg deposited is smaller when the host patch is below a size that allows for an optimal sex ratio. Hence there may exist for some species of parasitoids a lower threshold below which differences in host number do not correspond to directly proportional differences in realized fitness. If this is so, then the fitness function would be of the type promoting risk-proneness.

10.8.2 Decelerated increase in fitness in large host patches

Although realized fitness may be positively related to the number of eggs laid by a female for patches containing some intermediate range of host numbers, it is unlikely that the fitness will continue to increase linearly indefinitely. There are at least three factors that can reduce the expected fitness associated with large numbers of eggs laid in a patch:

hyperparasitoids, predators, and superparasitism that can all reduce the emergence success of parasitoid eggs and are all related to the patch residence time of females exploiting a host patch.

Predation of both the foraging female and its developing progeny, as well as hyperparasitism of the progeny, could lead to a decelerated increase in realized fitness with the number of eggs deposited in a patch. The probability of survival of parasitoid eggs decreases non-linearly as their numbers increase because predators and hyperparasitoids respond to visual and chemical cues emanating from primary parasitoids (Dicke & Grostal 2001) and the longer a female parasitoid exploits a host patch the higher the probability that these natural enemies will detect the patch and exploit her offspring. Hence, as the number of eggs laid increases beyond some detection threshold, the realized fitness ceases to increase in step with the number of eggs laid.

Another cost that changes with the level of exploitation of a host patch is superparasitism. Parasitoids leave their host in place once parasitized and these parasitized hosts can be re-encountered and parasitized again either by the same female (self-superparasitism) or by a different female (conspecific-superparasitism) (van Baaren et al. 1994). Most parasitoid species can discriminate parasitized hosts and reject them until a proportion of the patch has been exploited (van Alphen & Visser 1990). However, in some species, females do not discriminate and rather adjust their patch residence time to decrease the risk of superparasitism (Outreman et al. 2001a,b). In species that do not discriminate parasitized hosts, the risk of self-superparasitism obviously increases with the proportion of hosts parasitized in a patch, leading also to a change in the rate of fitness gain per egg deposited.

10.9 The relative fitness rule

In the context of foraging for resources, the type of risk-sensitivity that should be expressed by the organisms is summarized by the energy budget rule. The rule states that the best policy for a forager depends on its expected state relative to some requirement. What equivalent logic would support risk-sensitivity in the context of reproduction? One possible way of portraying the problem in a reproductive system is to say that the genetic survival requirement for any organisms is to match the expected reproductive success of the population in which it lives in order to avoid loosing genetic representation in the next generation. Animals that, for any reason, do not expect to match the average expected fitness of the population will suffer a cost because they will loose representation in the next generation, perhaps even disappear (genetic death). Consequently, the chances of avoiding the genetic cost for such an organism are maximized by adopting risk-prone reproduction options. However, organisms whose reproductive output is superior to the expected reproductive output of the population they are in will reduce the risk of suffering a genetic cost by avoiding uncertain reproductive options and being risk-averse.

10.10 Predictions

If we are correct in applying the logic of the relative fitness rule to host patch selection and exploitation, then factors that reduce a female's reproductive success relative to the expectation should tend to make her risk-prone in her host selection decisions. For instance, females with fewer eggs to lay or that have lower dispersal ability or oviposition efficiency should tend to adopt risk-prone host selection policies. Naturally, the opposite prediction is also true: any variation that places a female above the expectation for the population should tend to make her risk-averse. These predictions could be tested using female parasitoids that are phenotypically different. The size of parasitoid adults varies with the quality of the host in or on which they developed (Greenberg et al. 1998). Both females and males that have developed in small or low-quality hosts show a decrease in several fitness proxies including size, longevity, potential, and realized fecundity and vagility (Kazmer & Luck 1995, Roitberg et al. 2001). Thus, smaller females that have a lower fecundity and, most importantly, a smaller capacity to find new host patches are then expected to be more risk-prone than larger females.

In the context of patch selection, imagine a female parasitoid confronted with two host patches, one small and one large. If she lays eggs in all hosts of the small patch she could obtain a given realized fitness with greater certainty. The larger host patch could allow her to lay many more eggs but because laying more eggs increases the chance of attracting predators, hyperparasitoids, and the potential for superparasitism, the mean expected realized fitness may be the same as for the smaller patch but the associated uncertainty much greater. If no predators or hyperparasitoids discover the patch she will have obtained a much greater reproductive output. However, she could also lose it all. We predict that, faced with such a choice, female parasitoids will prefer to lay eggs in smaller host patches when they are risk-averse but to lay a large number of eggs in large host patches when they are risk-prone.

However, female parasitoids encounter host patches sequentially and, in addition, will require some time to exploit the patch in order to gain information on its size and quality (see Section 10.7.2. above on risk-sensitive patch residence). Risk-sensitive behavior in this context will express itself in terms of apparent patch residence time. A risk-prone female should spend a long time laying eggs on a large host clump because it is only by laying many more eggs in this clump that she can hope to obtain a large gain if natural enemies should happen not to discover the patch. She should also be observed to abandon small patches quickly once she recognizes them as such. On the other hand, a risk-averse female should leave large patches much sooner in order to avoid the uncertainty that is associated with attraction of natural enemies.

10.11 Evidence for pre-requisites of risk-sensitivity in parasitoid egg-laying

Not surprisingly, no study to date has investigated risk-sensitive egg-laying decisions in parasitoids. However, our goal here is to see whether the literature describes behaviors that appear superficially consistent with such risk-sensitivity in the context of egg-laying. For instance, if natural enemies attack the immatures of a parasitoid, and if the probabilities of attack increase with host patch size, the female could face the following choice. It can exploit a small patch with the associated lower reward but also lower probability of losing that investment, or exploit a large patch with a potentially higher reward but also with a higher probability of losing its progeny. The two first prerequisites for parasitoid females to be risk-sensitive are that they can evaluate the size of a host patch and that they can detect the presence of natural enemies that could cause the fitness gain function to be curvilinear.

10.11.1 The prerequisites

The capacity of female parasitoids to evaluate the size of a host patch and to adjust their behavior accordingly, has been demonstrated in several species (Godfray 1994). As predicted by the marginal value theorem (MVT) (Charnov 1976), the patch residence time of females increases with patch quality in *Aphidius rhopalosiphi* (Outreman et al. 2005), *A. colemani* (van Steenis et al. 1996), or *A. nigripes* (Cloutier & Bauduin 1990).

Several species of primary parasitoids can detect predators through chemical, visual, or mechanical cues (Dicke & Grostal 2001). Predation of adult parasitoids does occur but the mortality due to predators acts mostly on parasitized hosts, killing the developing immature. For example, parasitized pea aphids, *Acyrthosiphon pisum*, are attacked by a guild of 11 species of predators (Wheeler et al. 1968). This mortality can be important: almost 50% of mummies of *Aphis fabae* parasitized by *Lysiphlebus fabarum* were predated in a sugar beet field over a 10-day period (Meyhöfer & Hindayana 2000). As expected, female parasitoids can react to indications of the presence of predators and modify their foraging strategy. Females of *A. colemani* reduced the number of aphids parasitized in a colony when a coccinellid larva was, or had been present in the colony (Takizawa et al. 2000).

Although hyperparasitoids do not attack adult primary parasitoids, they can be an important mortality factor for parasitoid immatures, especially for aphid parasitoids (Brodeur 2000). The primary parasitoid can be attacked as a larva within the aphid or as a prepupa or a pupa inside the aphid mummy (Buitenhuis et al. 2005). Because the capacity of assessing the presence of hyperparasitoids is not without cost, we would expect this capacity to be found mostly in species where the cost for the immature or the adult is higher than the cost of maintaining this capacity (Kamil & Roitblat 1985). In the case of aphid parasitoids, mortality rates reaching 80% are reported (Horn 1988, Höller et al. 1993) and appear sufficient to offset the cost of being able to detect hyperparasitoids and maintain an adapted behavioral response.

10.11.2 Impact of hyperparasitoids on the fitness function

There is also empirical support for an assumption we made earlier concerning the curvilinear nature of the egg-laying fitness function of parasitoids exploiting large host patches. Data suggest that the fitness return during patch exploitation by aphid parasitoids is dome-shaped rather than linear. Hyperparasitoids of aphid parasitoids have been shown to aggregate in patches with higher parasitoid larval density (Horn 1988). This implies that the longer a female aphid parasitoid stays in an aphid colony, the higher the mortality hyperparasitoids will cause to its progeny. Thus, the reproductive success of female aphid parasitoids is not proportional to the size of the aphid colony. In the presence of hyperparasitoids, it becomes advantageous for the parasitoid female to ignore unparasitized aphids in a colony and leave early toward another colony in order to reduce the probabilities for its offspring of being killed by hyperparasitoids.

While the MVT (Charnov 1976) hypothesizes that female parasitoids maximize their long-term rate of oviposition, models have been developed where the patch residence time of female parasitoids is based on a trade-off between female and offspring mortality. In their model, Ayal and Green (1993) showed that the number of adults produced per aphid parasitized is linear in absence of hyperparasitoids but becomes dome-shaped when these hyperparasitoids are added to the system. Based on these results, it appears that the optimum number of hosts to be attacked in an aphid colony depends on: (i) the number of hyperparasitoids expected to visit the colony; (ii) the cost of traveling between colonies; and (iii) the expected number of hosts already parasitized in the colony (Ayal & Green 1993). A foraging model that includes the instantaneous rate of oviposition, female mortality in a host patch, the probability that an egg deposited will develop successfully, and travel costs between patches, arrives at similar results (Weisser 1993). In this model, the optimal patch residence time of female parasitoids that ignored the probability of mortality of their progeny was always superior to those of females that took this mortality into account. The higher mortality incurred by the immatures of females that ignored hyperparasitism led to a loss in total reproductive success.

10.11.3 The aphid-parasitoid-hyperparasitoid system

The predictions of the models of Ayal and Green (1993) and Weisser (1993) have been supported by a number of studies on aphid-parasitoid-hyperparasitoid systems. Females of the primary parasitoid *Aphidius uzbekistanicus* have been shown to react to the presence of the hyperparasitoid *Alloxysta victrix* by leaving the area (Höller et al. 1994). These female parasitoids react to volatiles emitted by female *A. victrix*, and that acts on males as a sexual pheromone and on females as a spacing pheromone (Micha et al. 1993).

When females *A. uzbekistanicus* detect the presence of *A. victrix* they decrease their patch residence time. The number of attacks and the number of aphids parasitized per colony were also decreased but not as a result of a decrease in the rate of attack but simply because the female parasitoid stayed in the aphid colony for a shorter period of time (Petersen et al. 2000). It seems that the effect of the hyperparasitoid on the behavior of the primary parasitoid is limited to a change in the perception of patch quality rather than a change in host evaluation.

These laboratory results are supported by field data, showing that aphid parasitoid females leave areas with high hyperparasitoid densities (Höller et al. 1993). These authors observed that the level of aphid primary parasitoids in cereal fields remained low throughout the season in contrast to hyperparasitism that often reached 100% late in the season. However, neither numerical response nor hyperparasitism were sufficient to explain the low rate of parasitism and analysis indicated that the observed decline in aphid primary parasitism was quantitatively related to hyperparasitoid density (Höller et al. 1993).

These studies clearly indicate that hyperparasitoids influence the foraging behavior of primary parasitoids through chemical cues emitted by the female hyperparasitoids. In response to spatially density-dependant hyperparasitism, primary parasitoids abandon incompletely exploited aphid colonies or, on a larger scale, emigrate from fields that harbor a high density of hyperparasitoids (Rosenheim 1998). By leaving aphid colonies early, parasitoid females decrease the cost to developing immatures. It remains to be seen if parasitoid females change their response to the presence of hyperparasitoids based on their expected reproductive output.

10.12 Conclusions

Because insect parasitoids are used worldwide for biological control programs, it is interesting to predict what could be the impact of risk-sensitivity on the capacity of these insects to control their hosts. The data presented in this chapter point to the fact that female parasitoids change their host acceptance behavior by modifying their patch residence time according to patch size. While both the number of attacks and the number of parasitized aphids by A. *uzbekistanicus* decreased when the hyperparasitoid A. *victrix* was present (Petersen et al. 2000), this was due to the shorter patch exploitation by the parasitoid female, a behavior consistent generally with risk-aversion. When these numbers are expressed in rate of host exploitation, no difference is found. This suggests that parasitoid females were as efficient at finding, evaluating, and accepting hosts. However, they did so for a shorter period of time before leaving the aphid colony. This would have the effect of decreasing the proportion of hosts parasitized in a given patch but increasing the proportion of patches discovered by parasitoids in a habitat. It is tempting to speculate that the use of parasitoids of poorer overall condition, that would adopt risk-prone reproductive options, could have led to a different response altogether, with females exploiting more fully large colonies but investing less time in patch searching.

References

Ayal, Y. and Green, R.F. (1993) Optimal egg distribution among host patches for parasitoids subject to attack by hyperparasitoids. *American Naturalist* **141**: 120–38.

Barnard, C.J. and Brown, C.A.J. (1987) Risk-sensitive foraging and patch residence time in common shrews, *Sorex araneus* L. *Animal Behaviour* **35**: 1255–7.

Bednekoff, P.A. (1996) Risk-sensitive foraging, fitness, and life histories: Where does reproduction fit into the big picture? *American Zoologist* **36**: 471–83.

Biernaskie, J.M., Cartar, R.V. and Hurly, T.A. (2002) Risk-averse inflorescence departure in hummingbirds and bumble bees: could plants benefit from variable nectar volumes? *Oikos* **98**: 98–104.

Boivin, G., Fauvergue, X. and Wajnberg, É. (2004) Optimal patch residence time in egg parasitoids: innate versus learned estimate of patch quality. *Oecologia* **138**: 640–7.

Brodeur, J. (2000) Host specificity and trophic relationships of hyperparasitoids. In: Hochberg, M.E. and Ives, A.R. (eds.) *Parasitoid Population Biology*. Princeton University Press, Princeton, pp. 163–83.

Brodeur, J., Geervliet, J.B.F. and Vet, L.E.M. (1996) The role of host species, age and defensive behaviour on ovipositional decisions in a solitary specialist and gregarious generalist parasitoid (*Cotesia* species). *Entomologia Experimentalis et Applicata* **81**: 125–32.

Buitenhuis, R., Vet, L.E.M., Boivin, G. and Brodeur, J. (2005) Foraging behaviour at the fourth trophic level: a comparative study of host location in aphid hyperparasitoids. *Entomologia Experimentalis et Applicata* **114**: 107–17.

Cartar, R.V. (1991) A test of risk-sensitive foraging in wild bumble bees. *Ecology* **72**: 888–95.

Cartar, R.V. and Abrahams, M.V. (1996) Risk-sensitive foraging in a patch departure context: a test with worker bumble bees. *American Zoologist* **36**: 447–58.

Charnov, E.L. (1976) Optimal foraging, the marginal value theorem. *Theoretical Population Biology* **9**: 129–36.

Cloutier, C. and Bauduin, F. (1990) Searching behavior of the aphid parasitoid *Aphidius nigripes* (Hymenoptera: Aphidiidae) foraging on potato plants. *Environmental Entomology* **19**: 222–8.

Dicke, M. and Grostal, P. (2001) Chemical detection of natural enemies by arthropods: an ecological perspective. *Annual Review of Ecology & Systematics* **32**: 1–23.

Ferguson, H.J., Cobey, S. and Smith, B.H. (2001) Sensitivity to a change in reward is heritable in the honeybee, *Apis mellifera. Animal Behaviour* **61**: 527–34.

Giraldeau, L.-A.G. and Caraco, T. (2000) *Social Foraging Theory.* Princeton University Press, Princeton.

Godfray, H.C.J. (1994) *Parasitoids. Behavioral and Evolutionary Ecology.* Princeton University Press, Princeton.

Godin, C. and Boivin, G. (2000) Effects of host age on parasitism and progeny allocation in Trichogrammatidae. *Entomologia Experimentalis et Applicata* **97**: 149–60.

Green, R.F., Gordh, G. and Hawkins, B.A. (1982) Precise sex ratios in highly inbred parasitic wasps. *American Naturalist* **120**: 653–65.

Greenberg, S.M., Nordlund, D.A. and Wu, Z. (1998) Influence of rearing host on adult size and oviposition behavior of mass produced female *Trichogramma minutum* Riley and *Trichogramma pretiosum* Riley (Hymenoptera: Trichogrammatidae). *Biological Control* **11**: 43–8.

Höller, C., Borgemeister, C., Haardt, H. and Powell, W. (1993) The relationship between primary parasitoids and hyperparasitoids of cereal aphids: an analysis of field data. *Journal of Animal Ecology* **62**: 12–21.

Höller, C., Micha, S.G., Schulz, S., Francke, W. and Pickett, J.A. (1994) Enemy-induced dispersal in a parasitic wasp. *Experiencia* **50**: 182–5.

Horn, D.J. (1988) Secondary parasitism and population dynamics of green peach and cabbage aphids (*Myzus persicae, Brevicoryne brassicae*). In: Niemczk, E. and Dixon, A.F.G. (eds.) *Ecology and Effectiveness of Aphidophaga.* SPB Academic Publishing, the Hague, pp. 305–9.

Kamil, A.C. and Roitblat, H.L. (1985) The ecology of foraging behavior: implications for animal learning and memory. *Annual Review of Psychology* **36**: 141–69.

Kazmer, D.J. and Luck, R.F. (1995) Field tests of the size-fitness hypothesis in the egg parasitoid *Trichogramma pretiosum. Ecology* **76**: 412–25.

MacArthur, R.H. and Pianka, E.R. (1966) On optimal use of a patchy environment. *American Naturalist* **100**: 603–9.

McNamara, J.M., Webb, J.N. and Collins, E.J. (1995) Dynamic optimization in fluctuating environments. *Proceedings of the Royal Society of London Series B Biological Science* **261**: 279–84.

Meyhöfer, R. and Hindayana, D. (2000) Effects of intraguild predation on aphid parasitoid survival. *Entomologia Experimentalis et Applicata* **97**: 115–22.

Micha, S.G., Stammel, J. and Höller, C. (1993) 6-methyl-5-heptene-2-one, a putative sex and spacing pheromone of the aphid hyperparasitoid, *Alloxysta victrix* (Hymenoptera: Alloxystidae). *European Journal of Entomology* **90**: 439–42.

Mills, N.J. and Kuhlmann, U. (2004) Oviposition behavior of *Trichogramma platneri* Nagarkatti and *Trichogramma pretiosum* Riley (Hymenoptera: Trichogrammatidae) in patches of single and clustered host eggs. *Biological Control* **30**: 42–51.

Miura, K. and Kobayashi, M. (1998) Effects of host-egg age on the parasitism by *Trichogramma chilonis* Ishii (Hymenoptera: Trichogrammatidae), an egg parasitoid of the diamondback moth. *Applied Entomology & Zoology* **33**: 219–22.

Monje, J.C., Zebitz, C.P.W. and Ohnesorge, B. (1999) Host and host age preference of *Trichogramma galloi* and *T. pretiosum* (Hymenoptera: Trichogrammatidae) reared on different hosts. *Journal of Economic Entomology* **92**: 97–103.

Nonacs, P. (2001) State dependent behavior and the Marginal Value Theorem. *Behavioral Ecology* **12**: 71–83.

Outreman, Y., Le Ralec, A., Plantegenest, M. and Pierre, J.S. (2001a) Superparasitism limitation in an aphid parasitoid: cornicle secretions avoidance and host discrimination. *Journal of Insect Physiology* **47**: 339–348.

Outreman, Y., Le Ralec, A., Wajnberg, É. and Pierre, J.S. (2001b) Can imperfect host discrimination explain partial patch exploitation in parasitoids? *Ecological Entomology* **26**: 271–80.

Outreman, Y., Le Ralec, A., Wajnberg, É. and Pierre, J.S. (2005) Effects of within- and among-patch experiences in the patch-leaving decision rules in an insect parasitoid. *Behavioral Ecology & Sociobiology* **58**: 208–17.

Pak, G.A., Kaskens, J.W.M. and de Jong, E.J. (1990) Behavioural variation among strains of *Trichogramma* spp.: host-species selection. *Entomologia Experimentalis et Applicata* **56**: 91–102.

Petersen, G., Matthiesen, C., Francke, W. and Wyss, U. (2000) Hyperparasitoid volatiles as possible foraging behaviour determinants in the aphid parasitoid *Aphidius uzbekistanicus* (Hymenoptera: Aphidiidae). *European Journal of Entomology* **97**: 545–50.

Real, L.A. (1981) Uncertainty and pollinator-plant interactions: the foraging behavior of bees and wasps on artificial flowers. *Ecology* **62**: 20–6.

Real, L. and Caraco, T. (1986) Risk and foraging in stochastic environments. *Annual Review of Ecology & Systematics* **17**: 371–90.

Real, L., Ott, J. and Silverfine, E. (1982) On the tradeoff between the mean and the variance in foraging: effect of spatial distribution and color preference. *Ecology* **63**: 1617–23.

Rita, H. and Ranta, E. (1998) Stochastic patch exploitation model. *Proceedings of the Royal Society of London Series B Biological Science* **265**: 309–15.

Roitberg, B.D., Boivin, G. and Vet, L.E.M. (2001) Fitness, parasitoids, and biological control: an opinion. *Canadian Entomologist* **133**: 429–38.

Rosenheim, J.A. (1998) Higher-order predators and the regulation of insect herbivore populations. *Annual Review of Entomology* **43**: 421–47.

Royer, L., Fournet, S., Brunel, E. and Boivin, G. (1999) Intra- and interspecific host discrimination by host-seeking larvae of coleopteran parasitoids. *Oecologia* **118**: 59–68.

Shafir, S., Wiegmann, D.G., Smith, B.H. and Real, L.A. (1999) Risk-sensitive foraging: choice behaviour of honeybees in response to variability in volume of reward. *Animal Behaviour* **57**: 1055–61.

Sih, A. and Christensen, B. (2001) Optimal diet theory: when does it work, and when and why does it fail? *Animal Behaviour* **61**: 379–90.

Smallwood, P.D. (1996) An introduction to risk sensitivity: The use of Jensen's Inequality to clarify evolutionary arguments of adaptation and constraint. *American Zoologist* **36**: 392–401.

Stephens, D.W. (1981) The logic of risk-sensitive foraging preferences. *Animal Behaviour* **29**: 628–9.

Stephens, D.W. and Charnov, E.L. (1982) Optimal foraging: some simple stochastic models. *Behavioural Ecology & Sociobiology* **10**: 251–63.

Stephens, D.W. and Krebs, J.R. (1986) *Foraging Theory*. Princeton University Press, Princeton.

Stephens, D.W. and Dunbar, S.R. (1993) Dimensional analysis in behavioral ecology. *Behavioral Ecology* **4**: 172–83.

Takizawa, T., Yasuda, H. and Agarwala, B.K. (2000) Effect of three species of predatory ladybirds on oviposition of aphid parasitoids. *Entomological Science* **3**: 465–9.

Thiel, A. and Hoffmeister, T.S. (2004) Knowing your habitat: linking patch-encounter rate and patch exploitation in parasitoids. *Behavioral Ecology* **15**: 419–25.

van Alphen, J.J.M. and Visser, M.E. (1990) Superparasitism as an adaptive strategy for insect parasitoids. *Annual Review of Entomology* **35**: 59–79.

van Alphen, J.J.M., Bernstein, C. and Driessen, G. (2003) Information acquisition and time allocation in insect parasitoids. *Trends in Ecology & Evolution* **18**: 81–7.

van Baaren, J. and Boivin, G. (1998) Genotypic and kin discrimination in a solitary Hymenopterous parasitoid: implications for speciation. *Evolutionary Ecology* **12**: 523–34.

van Baaren, J., Boivin, G. and Nénon, J.P. (1994) Intra- and interspecific host discrimination in two closely related egg parasitoids. *Oecologia* **100**: 325–30.

van Steenis, M.J., El-Khawass, M.H., Hemerik, L. and van Lenteren, J.C. (1996) Time allocation of the parasitoid *Aphidius colemani* (Hymenoptera: Aphidiidae) foraging for *Aphis gossypii* (Homoptera: Aphidae) on cucumber leaves. *Journal of Insect Behavior* **9**: 283–95.

Varaldi, J., Fouillet, P., Boulétreau, M. and Fleury, F. (2005) Superparasitism acceptance and patch-leaving mechanisms in parasitoids: a comparison between two sympatric wasps. *Animal Behaviour* **69**: 1227–34.

Visser, M.E. and Rosenheim, J.A. (1998) The influence of competition between foragers on clutch size decisions in insect parasitoids. *Biological Control* **11**: 169–74.

Waddington, K.D., Allen, T. and Heinrich, B. (1981) Floral preferences of bumblebees (*Bombus edwardsii*) in relation to intermittent versus continuous rewards. *Animal Behaviour* **29**: 779–84.

Wajnberg, É. (2006) Time allocation strategies in insect parasitoids: from ultimate predictions to proximate behavioral mechanisms. *Behavioral Ecology & Sociobiology* **60**: 589–611.

Wajnberg, É., Bernhard, P., Hamelin, F. and Boivin, G. (2006) Optimal patch time allocation for time-limited foragers. *Behavioral Ecology & Sociobiology* **60**: 1–10.

Weisser, W.W. (1993) A general approach to oviposition strategies in solitary parasitoids. *European Journal of Entomology* **90**: 429–34.

Wheeler, A.G., Hayes, J.T. and Stephens, J.L. (1968) Insect predators of mummified pea aphids. *Canadian Entomologist* **100**: 221–2.

Part 2

Extension of behavioral ecology of insect parasitoids to other fields

11

Multitrophic interactions and parasitoid behavioral ecology

Louise E.M. Vet and H. Charles J. Godfray

Abstract

Students of parasitoid behavioral ecology have sought to understand how natural selection moulds behavior, using both theoretical and experimental approaches. In these studies, the arena in which the behavior is observed is often a simple and artificial environment. However, parasitoids have evolved to function in a highly complex and spatially and temporally variable multitrophic world, which we ignore at our peril. Research on the interactions between plants, herbivores, and their natural enemies, the field of multitrophic interactions (MTI), is a fast-developing research area that is tackling major new challenges. We will describe some of these challenges and explore their implications for parasitoid behavioral ecology.

Studies on MTI aim to identify the forces that influence the interaction between plants, herbivores, and natural enemies, and which thus affect species population density and community structure. The results of these studies may, for example, be used to enhance crop protection, or to predict how species and communities respond to human-induced global changes. The study of plant defense is central to multitrophic theory. Plants can defend themselves directly against herbivores, but also indirectly by emitting volatiles that attract parasitoids and other natural enemies. Knowledge of the mechanisms underlying the induction of these herbivore-induced-plant volatiles, and of the responses of the parasitoids, is progressing rapidly. MTI research programs are not only broadening in scope to search for the plant genes and plant metabolites involved in plant defense, but also aim at understanding the parasitoid's behavioral and neurobiological responses to variation in plant information.

Herbivore (host and non-host) diversity can affect the reliability of plant information and these plant-mediated indirect effects can influence parasitoid community persistence and stability. Plant defense can also affect higher trophic levels (including hyperparasitoids) and thus community structure. In addition to effects due to primary plant compounds, plant defense links above- and belowground MTIs, so that interactions occurring in the rhizosphere have ramifications for the fitness of insect parasitoids and their hosts. To understand the functioning of multitrophic systems, we need to know how parasitoids deal with such complexity and identify the mechanisms, rules, constraints, trade-offs, and selection pressures involved. We also highlight the importance of making connections between different levels of biological organization (for example, from gene to parasitoid behavior and from parasitoid behavior to community processes), and the emerging importance of applied MTI, which we illustrate with examples from habitat fragmentation. The last decade has seen a marked widening in the scope of behavioral ecology and we conclude by looking forward to the time when parasitoid behavioral ecology will increasingly be part of a broader research agenda spanning the full range of biological complexity from molecules to ecosystems.

11.1 Introduction

Parasitoids provide wonderful subjects for behavioral ecological study. Their variety in form and function creates endless possibilities for exploring adaptation at both a macro- and micro-evolutionary level. Behavioral ecologists ask how natural selection causes animals to do what they do. Their main working method is to assume that natural selection has optimized animal behavior to maximize Darwinian fitness in a particular environment, and to use this assumption to generate hypotheses that can then be tested in the field or laboratory (Stephens & Krebs 1986). Rejecting a hypothesis can be as interesting as the converse, as it suggests that either the model of the animal's life history is wrong, or that the central assumption is not valid in these circumstances. Complementary to these ultimate 'why' questions are a host of proximate 'how and what' questions concerning underlying mechanisms. How do animals respond to the environment in appropriate ways? What are the rules and cues they use? Both approaches are needed to understand fully the ecology and evolution of behavior.

Insect parasitoids have proved to be excellent model systems for asking many of these questions (Godfray 1994, Vet et al. 2002, and several chapters in this volume). Many species are easy to rear and to manipulate experimentally, and it is easy to observe their behavior, especially under laboratory conditions. For studies of sex ratios and related questions, the haplodiploid genetic system of parasitoid Hymenoptera, which allows the ovipositing female proximate control of progeny sex ratio, has meant that work on parasitoids has become hugely influentially (see also Chapter 12 by Ode and Hardy). A plethora of theoretical models predicting parasitoid behavior and life histories have been developed and, though to a far lesser extent, experimentally tested (see also Chapter 2 by van Baalen and Hemerik). These have helped to unravel several key factors that are involved in shaping parasitoid 'decision making', such as the density, distribution, and quality of hosts and the availability of eggs and/or time for foraging (Wajnberg et al. 2003, Wajnberg 2006, and several chapters in this volume).

A primary focus of parasitoid behavioral ecology is to understand how the environment has shaped the evolution of parasitoid behavior. Classical parasitoid behavioral ecology has generally followed a 'bitrophic' approach with the major emphasis on the direct parasitoid-host interaction, and often with the behavior of the parasitoid investigated in particular and simplified environments. However, since the early 1990s, it has become increasingly recognized that parasitoids function and have evolved in a complex multitrophic environment, which involves a myriad of spatially and temporally variable direct and indirect interactions, far more extensive than the simple interaction between parasitoids and their hosts (Vet & Dicke 1992, Godfray 1994). We realize now that plants play a prominent role in shaping the structure of higher trophic level communities, including parasitoids (van Veen et al. 2006).

Research on the interactions between plants, herbivores, and their natural enemies, what has become known as the study of MTI, aims at identifying the molecular, physiological, and behavioral processes involved and how they are shaped by natural selection, and then ultimately exploring how these affect higher-level processes such as population dynamics and the structure of ecological communities. The study of MTI is a fast developing and expanding research area that is becoming increasingly interdisciplinary, a development we see in the ecological sciences more generally. To understand fully the functioning of plant-herbivore-natural enemy communities over variable spatio-temporal scales, it is becoming imperative to connect different levels of biological organization, in other words to address and link processes operating at levels from the gene to the community. This requires broad collaborations between field-oriented ecologists and entomologists, molecular biologists, chemical and evolutionary ecologists, population geneticists, and mathematicians. New techniques such as genomics and metabolomics bring fresh opportunities to study the mechanisms underlying ecological processes. By broadening their research agenda, students of MTI are facing major new opportunities and challenges. We will discuss a range of these and look at their implications for the existing field of parasitoid behavioral ecology. We believe strongly that, as parasitoids are major players in many terrestrial food webs, the best research on MTI must address the action and reaction of this important guild of organisms.

11.2 Plant defense is central to MTI

11.2.1 Plant defense

Natural communities typically contain numerous species, some of which interact directly in trophic relationships – one species feeding on another – while others interact indirectly, for example through resource competition or shared natural enemies. In the case of direct trophic relationships, the processes of attack and defense are fundamental. In parasitoid-host interactions, numerous resistance and counter-resistance mechanisms have been identified and there is likely to be a continuing co-evolutionary interaction between the two parties (see also Chapter 14 by Kraaijeveld and Godfray). At the trophic level below, plants have evolved to become masters of the art of defense: highly necessary considering their immobility and the number of species that rely on primary producers for their resources. The consequences of successful or unsuccessful defense will cascade up through the food chain, wherever 'bottom-up' effects are significant.

Plants can defend themselves directly against attackers with physical structures, such as spines or trichomes, or with chemicals such as toxins or substances that reduce the quality and digestibility of plant tissue. Chemical defenses can be constitutively expressed, that is independent of herbivore attack, but perhaps a more efficient process is induced defense, where levels of defense are raised following attack by herbivores or pathogens (Karban & Baldwin 1997, Dicke & Vet 1999, Sabelis et al. 1999). Plants can also defend themselves indirectly through their life history strategy, for example by dispersal strategies that lead to unpredictability spatio-temporal distributions that cannot be tracked by specialist herbivores, or through seed masting that saturates local seed predator populations. Another type of indirect defense is through the recruitment of herbivore natural enemies, such as predators or parasitoids, a strategy based on the old Arab proverb 'the enemy of my enemy is my friend'. Many plant species produce herbivore-induced plant volatiles (HIPVs), which provide important information alerting natural enemies to the presence of their prey or host (see also Chapter 5 by Hilker and McNeil). Alternatively, plants may provide predators or parasitoids with food such as extra-floral nectar to enhance their attack rate on herbivores.

Both direct and indirect plant defenses are currently the subject of intensive interdisciplinary studies. They are intrinsically fascinating, but also of great potential applied (and commercial) significance. Herbivory leads to internal changes in plants, and both chemical ecologists and molecular biologists are investigating the molecular mechanisms underlying induced defense, such as the identity of herbivore elicitors, the signal transduction pathways in the plant, the role of different plant hormones, and the genes involved. At the same time, chemical and evolutionary ecologists study the evolution of plant defense, assess its costs and benefits, and its influence on species interactions in food webs. The Holy Grail is to link both approaches so as to understand both the proximate and ultimate causes and consequences of plant defense (Dicke et al. 2004, Kessler et al. 2004).

11.2.2 Plant-parasitoid mutualism: from chemical to behavioral ecology

Multitrophic thinking has fundamentally altered our views on both the foraging behavior of parasitoids and the evolution of plant defense. The bitrophic approach is far too limited to understand the full repertoire of parasitoid behaviors and we now acknowledge that plants play a significant role in the foraging behavior of parasitoids that attack herbivores (Turlings et al. 1990, Vet & Dicke 1992, Dicke & Vet 1999). There is strong evidence from studies of agricultural crops that HIPVs are important infochemicals for parasitoids, guiding them to their herbivore hosts. However, in spite of all the work that has been done with HIPVs, there is still little evidence that they play a significant role in indirect defense in natural systems (van der Meijden & Klinkhamer 2000). The vast majority of studies with HIPVs have been conducted with crops, such as cabbage, soybean, and maize, that have been artificially selected to enhance specific traits, though there is evidence that teosinte, the wild progenitor of maize, also attracts natural enemies through HIPVs (Gouinguene et al. 2001). We have no idea how the artificial selection that produced our modern crops may have modified other aspects of the plant's ecophysiological phenotype such as HIPVs production. To understand and appreciate how important HIPVs are as an active defense strategy, we need to work more with wild plants that have evolved a suite of defenses under natural selective regimes.

The often used metaphor 'plants cry for help' (Dicke et al. 1990) illustrates the hypothesis that plants are actively recruiting the enemies of their enemies (van Loon et al. 2000). It suggests an evolutionary interaction between plants and carnivores, a co-evolution of signal and response. The origin of the HIPVs may have been a mere by-product of direct defense or an unselected consequence of attack, but once a reliable signal of herbivore presence arose, selection would have operated on the natural enemy to respond to it, and then possibly on the plant to make the signal more efficient (Vet 1999a).

An active area of behavioral and evolutionary ecology that could profitably be applied to the study of HIPVs is the theory of biological signals (Maynard Smith & Harper 2003). For a signaling system to be evolutionary stable, it must neither be in the evolutionary interests of the signaler nor the receiver to change their strategy. Consider a plant attacked by an herbivore that responds by producing an HIPV. It must not be in the interest of the plant to 'cheat' and produce the signal constitutively. We could imagine circumstances when this cheating might be profitable, for example by recruiting a permanent army of natural enemies to repel future herbivores before they start feeding. Similarly, it must not be in the interests of the natural enemy to ignore the signal, as it might be if plants frequently cried wolf. The only application so far of this type of theory to plant-parasitoid signaling predicted that, for the signaling system to be stable, the production of HIPVs must be costly for the plant – if they were not, the plants would be selected to cheat by signaling constitutively – and because the signal would then lose its information content, the natural enemies would be selected to ignore it (Godfray 1995). Costly HIPVs production allows the system to be stable by providing a disincentive to the plant to cheat. This result parallels analyses of signaling in sexual selection where the tail of a peacock is a stable signal of male quality because it is too costly for poor males to produce (Grafen 1990), and parent-offspring signaling where nestling chicks provide accurate information to the parents about their levels of hunger as begging is costly in terms of energy and possibly attracting predators to the nest (Godfray 1991). Interestingly, exactly analogous theory has been developed by economists to explain why advertising works (Spence 1973).

Godfray's (1995) analysis was very simplistic and included little real biology. It would thus be interesting to see if a more realistic model accounted for unexplained aspects of tritrophic signal. This would be a move toward responding to Vet's (1999b) call for greater links between the fields of chemical ecology and behavioral ecology. Aspects of the biology that would be important to consider include:

1 the nature of the signal, whether it is qualitative (herbivore present versus herbivore absent) or quantitative (signal provides information on number of herbivore);
2 the natural variation in signal production by different plants, and whether this co-varies with factors that affect parasitoid fitness;
3 the behavioral options open to the parasitoid; and
4 how sensory and behavioral constraints limit information transfer in the system.

However, data from natural systems would be essential to test such a theory as it is dangerous to assume sufficient time has elapsed for signaling systems in crop ecosystems to have reached evolutionary equilibrium. What the theory already tells us is that selection will not just simply act on the plant to produce clear, detectable, and reliable signals, and on the natural enemy to respond automatically to the signal. Whenever there is a potential for a conflict of interest we should expect to observe only those types of signals and response, and particularly costly signals, which are parts of evolutionary stable signaling systems.

11.2.3 Plant cues mediate parasitoid foraging behavior

Parasitoids show innate responses to those plant cues that, over evolutionary time, have proved to be reliable in indicating the presence of suitable hosts (Vet et al. 1990). However, due to the considerable spatial and temporal variation in host plant use by herbivores, parasitoids require a high degree of phenotypic plasticity in their response to plant cues. Hence, responses to plant cues are generally acquired through learning, allowing parasitoids a remarkable flexibility in plant preference behavior (Turlings et al. 1993, Vet et al. 1995). Far less studied is the occurrence and importance of genetic variation in the olfactory responses of parasitoids to plant odors (Prévost & Lewis 1990, Gu & Dorn 2000, Wang et al. 2003). One of the rare studies in this area is by Wang et al. (2003), who showed genotypic variation in olfactory responses in *Cotesia glomerata* wasps to plant-host odors, which affected the efficiency of parasitism under laboratory conditions (Wang et al. 2004). Is this genetic variation a transient response to environmental changes experienced by *C. glomerata* wrought by agriculture (this parasitoid attacks *Pieris* butterflies feeding on brassicas), or does it reflect ecological and evolutionary processes promoting genetic polymorphism? There is a clear need for additional genetic studies to unravel how learning and genetic variation act together to shape parasitoid behavior in a multitrophic context, especially in non-crop systems.

In spite of the many chemical ecological studies on parasitoid attraction to HIPVs, major questions remain unsolved. First, it has proved extremely difficult to identify the key volatiles that trigger attraction in so-called naïve parasitoids that have not yet gained experience with plant volatiles (D'Alessandro & Turlings 2006). Second, plant odors are complex mixtures of compounds, and the sensory and neurological information processing of these odors is under-explored in comparison with studies of pheromonal information processing. The efficiency of a natural enemy that uses plant volatiles to find its victim is likely to depend on its ability to discriminate between 'signal' and 'noise' (Vet 1999a, Dicke et al. 2003, Perfecto & Vet 2003, Bukovinszky et al. 2005a,b, Gols et al. 2005). Parasitoids can find infested host plants within complex vegetation and studies have shown that background odors can influence the detection and response to signals (Mumm & Hilker 2005, see also Hilker and McNeil, Chapter 5, this volume), and that surrounding plants in diverse vegetation can affect the efficiency of odor-mediated foraging wasps (Perfecto & Vet 2003, Bukovinszky et al. 2005b, Gols et al. 2005). The exact mechanisms behind plant odor discrimination still need to be elucidated, and their understanding may help to develop evolutionary theories of plant-parasitoid signaling. Third, we lack insight into the information processing underlying plant odor learning. When parasitoids learn plant-volatiles, it is not known what exactly they learn and on what basis they subsequently distinguish between volatile blends. Do they learn complete odor bouquets or do they learn only differences in one or a few key compounds? To answer these questions, we need exact experimental control of the odor blends involved. D'Alessandro and Turlings (2006) advocate using selective adsorbing filters to carefully manipulate odor blends, making full use of the most advanced and latest techniques in analytical chemistry. They argue that a much stronger interdisciplinary effort between chemists and biologists is required to resolve this issue. If successful, it would create interesting possibilities for applications in which behavioral ecological interactions might be chemically manipulated to protect crops and the environment (D'Alessandro & Turlings 2006).

Another promising experimental approach has become available through a link between molecular plant biology and chemical ecology (Baldwin et al. 2001, van Poecke & Dicke 2002). When we know the mechanistic basis of the infochemical emission of plants, we can use this information to manipulate plant odor blends by silencing or over-expressing genes that are involved in the production of indirect defense chemicals (Kappers et al. 2005). Manipulation of plant traits by means of the plant's signal-transduction pathway provides us with an advanced experimental tool to study the innate and learned responses of herbivores and natural enemies to plant volatiles. It also provides experimental possibilities to assess the costs of plant volatile production, measured in terms of plant fitness. This would not only help us understand the evolution of plant defenses, but also assist in the intelligent design of the applications discussed above. Moreover, it is normally extremely hard to manipulate signaling systems, which has frustrated tests of evolutionary signaling theory. If the mechanistic basis of plant signaling turns out to be relatively simple and tractable, it might become a model system for addressing some general questions in evolutionary and behavioral ecology.

Finally, and following on from this last comment, we believe parasitoids to be excellent but under-explored model systems to study the evolution of information processing and cognition *per se*. Their diversity of ecological function, combined with relatively narrow-tuned behavioral goals, is likely to result in distinct and clearly-defined adaptations in learning and memory (Smid & Vet 2006, Smid et al. 2007). Differences between closely related species create excellent opportunities to study species-specific patterns of learning and their environmental drivers as has been shown for related *Cotesia* parasitoid species (Geervliet et al. 1998, Bleeker et al. 2006, Smid et al. 2007).

11.2.4 Cascading effects of plant defensive chemicals

In contrast to the many studies that show the effect of plant chemicals on foraging behavior in parasitoids, there is far less information about how other plant defensive chemicals affect higher trophic levels, species interactions, and community structure. The feeding ecology of the host can exert a significant effect over the biology and ecology of its parasitoids (Harvey & Strand 2002). Parasitoids of herbivores indirectly obtain their nutrition from the food plant of their host. Plant toxins that are ingested by herbivore hosts may thus cascade up through the food chain and negatively affect the performance (e.g. growth, development, survival) of parasitoids (Barbosa et al. 1986, Duffey et al. 1986, Bottrell et al. 1998) and even hyperparasitoids in the fourth trophic level (Harvey et al. 2003, 2007). There is also evidence that specialist parasitoids seem to be less affected by plant toxins than generalist parasitoids (Harvey 2005, Ode 2006). Amongst the outstanding questions in this area are:

1 resolution of the physiological mechanisms that natural enemies use to detoxify or tolerate plant defensive chemicals;
2 the effect of sequestration of plant-defense chemicals on the host acceptance behavior and fitness of parasitoids;
3 the issue of whether parasitoids themselves sequester plant compounds to defend themselves against hyperparasitoids or other natural enemies; and
4 the influence of plant chemicals that move up the food chain on host inter-specific and apparent competition, and hence on community dynamics and structure.

Understanding the factors that determine the population densities of herbivores, and how they are regulated, is interesting from a fundamental point of view but crucial in the applied context of pest control. Both bottom-up and top-down forces act together on herbivore populations and the challenge lies in understanding their interplay and relative importance. Assessing the relative roles of top-down and bottom-up forces has become a contentious and muddled issue. Part of the problem has been a failure to distinguish between key factors and regulatory processes. The population densities of probably all herbivores are influenced by bottom-up and top-down processes, though we agree it is an interesting exercise to explore the relative importance of the two. A related but different question is to understand which mortality factors vary with density and so can be regulatory. Again, these may be bottom-up or top-down, or a mixture of the two, but in general they can only be studied by detailed population analysis and manipulation. Direct defense is a classic bottom-up process while indirect defense, operating via natural enemy recruitment, can be viewed as an interesting integration of bottom-up and top-down (enemy) control. Understanding how host plant chemistry affects the functioning of MTI is a key to our ability to understand the population biology of herbivores as well as whether natural enemies can truly act as selective agents on plant investment in defense (Ode 2006).

11.3 Linking above- and below-ground interactions

Multitrophic studies generally focus on aboveground interactions and, for most students of parasitoid behavior, the soil is *terra incognita*, completely off their radar screens. However, plants do not magically stop at the soil surface and above- and belowground organisms can interact, both directly and indirectly. Comparing and linking the ecology and evolution of above- and belowground MTI, and determining how links between above- and belowground interactions affect biodiversity and ecosystem functions, is one of the greatest current challenges in multitrophic ecology (van der Putten et al. 2001, Wardle et al. 2004). As in agricultural systems, where soil-borne enemies drive the need for crop rotation, belowground interactions between plants and root herbivores can be important factors in structuring plant communities (De Deyn et al. 2003, van der Putten 2003).

Plants are not only attacked aboveground but also belowground, where a diversity of organisms including root-feeding insects, nematodes, and microbial root pathogens have a variety of local and systemic effects on plant physiology. Mycorrhizal fungi and other microbial root endophytes and epiphytes can also influence the physiological state of the plant. Above- and belowground herbivores may become dynamically but indirectly linked though mutual effects on processes such as the nutrient status of their common host plant (Bardgett & Wardle 2003), the concentration of different primary plant compounds (Masters & Brown 1997), or by triggering induced plant defense (van Dam et al. 2003, Bezemer & van Dam 2005, van Dam & Bezemer 2006). For example, in cotton belowground, feeding by root-feeding wireworms resulted in an increase in terpenoid levels in the foliage, which reduced the growth rate and food consumption of *Spodoptera exigua* caterpillars feeding aboveground (Bezemer et al. 2003). Interestingly, when induced belowground, the foliar terpenoids were distributed uniformly over all leaves while, if inducted by an aboveground herbivore, terpenoid concentrations increased only locally near the site of damage. This difference in the spatial distribution induction significantly affected the feeding behavior of the shoot-feeding caterpillars (Bezemer et al. 2004). In addition,

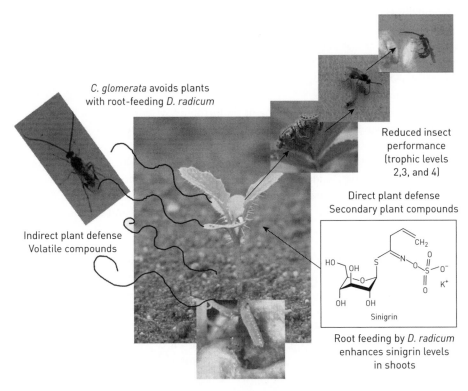

C. glomerata avoids plants
with root-feeding D. radicum

Reduced insect
performance
(trophic levels
2,3, and 4)

Direct plant defense
Secondary plant compounds

Indirect plant defense
Volatile compounds

Sinigrin

Root feeding by D. radicum
enhances sinigrin levels
in shoots

Fig. 11.1 Below-ground herbivory affects aboveground higher trophic levels in the system *Brassica nigra* (plant) – *P. brassicae* (herbivore) – *C. glomerata* (primary parasitoid) – *L. nana* (hyperparasitoid). Root feeding by *D. radicum* larvae enhances sinigrin levels in the shoots, which negatively affects the performance of the aboveground insects (after Soler et al. 2005). Foraging *C. glomerata* avoid host-infested plants with root-feeding *D. radicum* (Soler et al. 2007).

following root herbivory or mechanical root damage, the cotton plant increased the production of extrafloral nectar, a type of indirect defense, which leads to predator recruitment (Wäckers & Bezemer 2003).

The fact that changes in plant defense can mediate interactions between above- and below-ground insect assemblages has interesting implications for parasitoid behavioral ecology, as belowground processes can now influence the behavior and fitness of parasitoids. Soler et al. (2005) studied whether root herbivory by the fly *Delia radicum* influenced the development of the leaf feeder *Pieris brassicae*, its parasitoid *C. glomerata*, and its hyper-parasitoid *Lysibia nana*, through changes in primary and secondary plant compounds (Fig. 11.1). In the presence of root herbivory, the development time of the leaf herbivore and the parasitoid significantly increased, and the adult size of the parasitoid and the hyper-parasitoid were significantly reduced. Higher glucosinolate (sinigrin) levels were recorded in plants exposed to belowground herbivory, suggesting that the reduced performance of the aboveground insects was via reduced plant quality. In a further study, the authors show that root herbivory also influenced the foraging behavior of *C. glomerata* through changes in the spectrum of plant volatiles that were produced. The parasitoid responded by preferentially attacking hosts on plants without root damage (Soler et al. 2007). A possible explanation for this is that the quality of hosts or their chances of survival are lower on plants that are also attacked by root feeders. Thus, root-feeding can affect aboveground plant defense, and the consequences of induced changes in plant secondary compounds can propagate through the trophic web, affecting the behavior and development of species at higher trophic levels.

To comprehend fully the functioning of multitrophic systems, we need to understand the complex temporal and spatial variation in plant defenses that are induced by both above- and belowground herbivory, and to decipher their effects at higher trophic levels. It is a major challenge for the field of parasitoid behavioral ecology to elucidate how parasitoids deal with this complexity, to identify the proximate mechanisms and rules they use to determine their responses to an ever-changing environment, and to understand how natural selection maximizes fitness in the face of the different constraints and trade-offs involved.

11.4 Connecting different levels of biological organization

Throughout biology, there is an increasing aspiration to integrate knowledge of biological phenomena across different levels of organization, from genes to organisms and to ecosystems. The huge current interest in systems biology is a symptom of this. For evolutionary ecologists, jumping between organizational levels is not a new phenomenon: a longstanding goal of the field has been to understand population processes from the traits of individual organisms. Similarly, evolutionary ecologists are interested in how traits of organisms are influenced by population and community/ecosystem level processes. What is exciting and new is the far greater willingness of evolutionary ecologists to seek integration with the lower levels of biological organization from genes, through cells, to tissues. The revolutionary developments in molecular biology have created ample opportunities for ecologists and organismal biologists to search for the mechanistic basis of important traits, i.e. to 'search for the genes that matter'. And the trade in intellectual ideas is not just one way. Many of the genes identified in whole genome studies are of unknown role and will remain functionally unannotated unless the interaction of the organism with its biotic and abiotic environment is considered. This is 'ecology as a screen for gene function' in Ian Baldwin's memorable phrase (in a research talk).

Hence, we see an exciting increase in interdisciplinary research between biologists interested in mechanisms and researchers interested in ecology and evolution (Stearns & Magwene 2003, Thomas & Klaper 2004). The lowest gene level has become a major focus of this converging interest because, at least, potentially genes can simultaneously provide mechanistic and evolutionary insights (Boake et al. 2002). We can now link phenotype to genotype, explore how phenotypic variation is generated and explore consequences in different environments. We can also experimentally manipulate the expression of traits to study their function while a variety of new genomic tools are being developed that will allow us to investigate the transcriptional responses of organisms to their environment. Increasingly, we can even study the evolution of whole genomes in plants and animals with different evolutionary histories.

The field of MTI is rapidly embracing a molecular genetic approach and, in particular, the chemical ecology of plant defense has made major advances using *Arabidopsis thaliana* and *Nicotiana attenuata* as model species (Baldwin et al. 2001, Dicke et al. 2004, Kessler et al. 2004, Dicke 2006, Schnee et al. 2006). For example, genes encoding key enzymes in the biosynthesis of herbivore-induced terpenoids have been transferred to *Arabidopsis* by genetic engineering (Kappers et al. 2005, Schnee et al. 2006). The mutants and transgenics produced have been used in controlled laboratory and field studies to understand how plant defenses impact on the fitness of herbivores and higher trophic levels.

In addition, since the complete *Arabidopsis* genome has been sequenced and full-genome cDNA microarrays have become available, we can characterize the transcriptional response to herbivore attack involving both direct and indirect defense. Of course, not all responses to attack will involve changes in the transcriptome, and much will involve post-transcriptional modification. But, at the same time, due to considerable progress in chemical-analytical and statistical techniques, we can now characterize the plant chemical phenotype through a metabolomic approach (Fiehn et al. 2000, Sumner et al. 2003). In crucifers, for example, this reveals changes in the concentration of major secondary metabolites such as glucosinolates, alkaloids, and proteinase inhibitors, as well as changes in levels of plant volatiles such as fatty acid derivatives, phenolics, and terpenoids. Hence, we can study the effect of defense induction on both gene expression and phenotypic metabolome profiles. This research program, only in its infancy, will change completely the type of questions we can ask about MTI.

The greatest progress to date, including molecular mechanisms in studies of MTI, has occurred at the plant level, but increasingly we see similar approaches being taken with the herbivore and parasitoid. *Drosophila* was sequenced before *Arabidopsis* and, although not phytophagous, the huge progress that has been made in the functional annotation of the fly genome and the identification of cell, molecular, and behavioral genetic processes has great potential for multitrophic studies using a candidate gene approach (Fitzpatrick & Sokolowski 2004, Fitzpatrick et al. 2005). We note, in passing, that some close relatives of *Drosophila* are phytophagous, and even eat *Arabidopsis*: the drosophilid leafminer *Scaptomyza flava* is a minor pest of brassicas, and is host to both generalist and specific parasitoids. There may be rich rewards for someone willing to invest in developing this species as a model system. A number of phytophagous insects have been, or are currently being, sequenced (silk moth, *Bombyx mori*; medfly, *Ceratitis capitata*; pea aphid, *Acrythosiphon pisum*). Moreover, it is becoming increasingly easy to develop genetic resources for non-model organisms, where the full genome is not available. Classic Quantitative Trait Loci (QTL) mapping of traits involved in behavioral ecology has been possible for some time, but developing the molecular markers has been time consuming and expensive. Today, this can be contracted out to a company, and while the expenses are still non-trivial they are well within the scope of standard research grants. Similarly, Expressed Sequence Tag (EST) libraries (sequences of transcribed genes) can be constructed relatively quickly, and provide numerous markers and information on candidate genes, while techniques such as subtractive hybridization can give clues about genes transcribed in some circumstances but not others.

So far, only one parasitoid has been sequenced, the pteromalid *Nasonia vitripennis* that attacks the pupae of houseflies and other cyclorraphan Diptera (Werren et al. 2005). The *Nasonia* genome can be accessed at http://www.hgsc.bcm.tmc.edu/projects/nasonia/.

In addition, the sequence of the honey bee, *Apis mellifera*, a species that evolved from a parasitoid, is also highly relevant (http://www.hgsc.bcm.tmc.edu/projects/honeybee/). In the immediate future, behavioral and other studies with parasitoids involved in MTI will need to take the non-model organism route discussed above. We expect to see an ever-increasing development of high-density marker and EST collections, and their application in identifying functional mechanisms and genetic diversity underlying MTI.

Another connection between levels of organization that we see as both promising and necessary is that linking parasitoid behavior and community processes. An example of this is the work of Vos et al. (2001) on the parasitoid *C. glomerata* that attacks caterpillars of *Pieris* species on crucifer plants. Behavioral studies of host location and patch time allocation showed that *C. glomerata* has difficulty locating hosts on *Brassica oleracea* plants when non-host herbivores were also present due to non-specific, and therefore less informative, plant information. The parasitoids are attracted by plant volatiles from *B. oleracea* leaves induced by non-host herbivores and they spend, or rather waste, considerable amounts of time on such leaves. This lower efficiency weakens the interaction strength between parasitoid and host, which a model suggested had profound effects on the persistence of the complete parasitoid-host community. In this case, increasing herbivore diversity initially promoted the persistence of parasitoid communities. However, as herbivore diversity increased further, the model predicted a threshold would be reached when the parasitoid would become extinct because it would be unable to parasitize sufficient hosts to maintain the population.

The authors suggest that the 'wasted time' mechanism mediated by non-specific plant information may contribute to the complexity we see in natural communities. Thus, by integrating a behavioral mechanism of the parasitoid into a community model, an intriguing and testable prediction was made on a mechanism that might affect community structure and biodiversity in multitrophic systems. A similar approach may be used for other mechanisms that are now known to influence parasitoid foraging efficiency and thus parasitoid-host interactions strength, such as the effect of belowground feeding on parasitoid plant selection, as shown by Soler et al. (2007) and the effect of vegetational diversity on patch time allocation and searching efficiency, as shown by Gols et al. (2005) and Bukovinszky et al. (in press).

11.5 MTI in a changing world

Perhaps the most important aim of research on MTI in the 21st century is to predict the effects of human-induced global changes on the structure and functioning of biological communities. This is a major and increasingly significant challenge: our planet is changing at an unprecedented speed, and we are faced with the consequences of human population increase and economic growth such as the accelerating destruction of the biosphere, loss of species and natural resources, and global climate change. In addition, 'humans as the world's greatest evolutionary force' accelerate unwanted and costly evolutionary change in many other species (Palumbi 2001). Achieving true sustainable development requires a basic understanding of the structure and functioning of communities and ecosystems. It is surely the ultimate task and challenge of the ecological sciences to provide this knowledge and to make it available for society. Since insect parasitoids are an important functional guild in nearly all terrestrial food webs, the study of parasitoid behavioral ecology is embedded in this grand challenge. So what can we do to make our work more relevant to predict and mitigate the effects of global change? What questions should we ask? It would take several more chapters to fully address this issue and we restrict ourselves here to one topic: the effects of habitat fragmentation on MTI.

Environmental changes at a landscape level, such as habitat fragmentation, can have significant ecological and evolutionary consequences for community structure and trophic

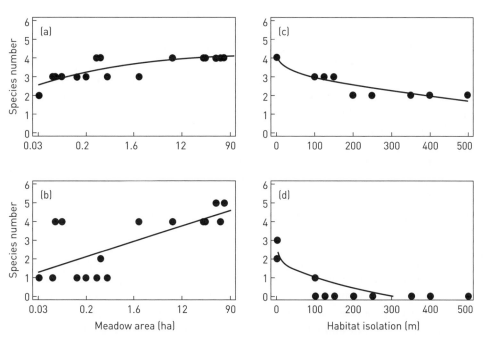

Fig. 11.2 Species richness of parasitoids and phytophagous insects are differently affected by habitat size and isolation. (a,b) Dependence of species richness on habitat size of old meadows. (a) Number of phytophagous insect species; (b) Number of parasitoid species. (c,d) Dependence of species richness on isolation of small *V. sepium* plots. (c) Number of phytophagous insect species; (d) Number of parasitoid species (from Kruess and Tscharntke 2000). See text for additional explanation.

interactions within food webs (Hoffmeister et al. 2005b). Organisms are affected by spatial processes in their own, often idiosyncratic, ways (Ritchie & Olff 1999, Bukovinszky et al. 2005c), though the scale at which these processes operate may vary systematically across trophic levels, influencing both population processes and species-area relationships (Holt et al. 1999, Thies et al. 2003). Consequently, habitat fragmentation, one aspect of habitat destruction, can have distinctly different effects on species/guilds in a community. Decreased size and connectivity between habitat fragments can disrupt trophic interactions between plants and their pollinators, or between competitors and their natural enemies (Tscharntke & Brandl 2004).

Kruess and Tscharntke (1994) argue that habitat fragmentation in the agricultural landscape poses a major threat to biological diversity. Isolation of habitat fragments results in decreased numbers of species as well as a reduction in the efficacy of natural enemies attacking pests. Insect parasitoids were far less successful in colonizing new, artificially-created patches compared with their phytophagous hosts. Hence, herbivores tended to increase in density because of reduced parasitism (which was typically only 19–60% of that experienced in non-isolated populations). Additional studies by Kruess and Tscharntke (2000) supported the differential effects of isolation on different trophic levels (Fig. 11.2). Field studies were carried out in 18 old meadows, differing in area and isolation, on the insect community (seed feeders and their parasitoids) in the pods of the bush vetch (*Vicia sepium* L.). The area of the meadows was found to be the major determinant of species diversity. The probability of being present, and the population density of endophagous insects, greatly decreased with increasing isolation. Again, parasitoids suffered more from habitat loss and isolation than their phytophagous hosts. Percent parasitism of the herbivores significantly decreased with area loss and increasing isolation of the *Vicia* plots. On the basis of these studies, the authors predict 'that conservation of large and less isolated habitat remnants enhances species diversity and parasitism of potential pest insects, i.e. the stability of ecosystem functions.' That there can indeed be a strong correlation between the stability of an ecosystem function (such as temporal variability in parasitism rates) and parasitoid species diversity was shown by subsequent work of the Tscharntke's group for naturally occurring communities of cavity nesting bees and wasps and their parasitoids in coastal Ecuador (Tylianakis et al. 2006). Parasitism rates increased with increasing parasitoid species diversity, and not simply with their abundance. Stability correlated with diversity. Plots with higher parasitoid species diversity were more stable (showed less temporal variability in parasitism rates) than lower diversity plots, independent of different habitat types (Tylianakis et al. 2006).

Results from Kruess and Tscharntke (2000), and other studies, convey an important message: local processes and larger-scale processes interact. Local species richness, interaction, and population dynamics is not only determined by local processes but also by the regional species pool (Lawton 2000). We need to connect processes happening at different scales if we want to understand communities. In addition, these results immediately point to great gaps in our knowledge, since we do not fully understand the processes and mechanisms underlying results such as those of Kruess and Tscharntke (2000). Why do processes affecting trophic levels operate at different spatial scales? What are the species characteristics responsible for these differences? Are the differences correlated with behavioral and/or life history traits?

There are now a handful of studies from the ecological literature that provides important insights into how these spatial processes affect different trophic levels. Hanski and colleagues have studied a metapopulation of Glanville Fritillary butterflies (*Mellitaea cinxia*) in the Åland archipelago in Finland for over a decade. The insect exists in scattered populations in small areas where its food plants grow, but with each population suffering a relatively high probability of extinction. Extinction rates correlate with herbivore population size, but are also influenced by parasitoids (Lei & Hanski 1997, Lei & Hanski 1998). *M. cinxia* is attacked by several ichneumonid wasps, which seem to be monophagous in Åland. Host populations of any particular size have a greater probability of extinction if they have been colonized by parasitoids. Instances of parasitoid but not host extinction, and parasitoid colonization of host patches have also been documented. The two main parasitoids have different dispersal abilities, and whether they are found on a patch, their resulting community interaction, and the probability of local extinction, are influenced by habitat size and isolation (van Nouhuys & Hanski 2002). The picture is further complicated by hyperparasitoids that are less specific. The mortality experienced by the primary parasitoids can be influenced by the presence of alternative hosts for the secondary parasitoids, an interesting example of apparent competition operating at a high tropic level (van Nouhuys & Hanski 2000).

Another important study is that of Roland and Taylor (1997) on the forest tent caterpillar (*Malacosoma disstria*) in aspen woodland in central Canada. Larvae live gregariously in silken webs, and can be abundant enough to defoliate large areas of forest. The population dynamics are not completely understood, but it appears that outbreaks are brought under control by a guild of parasitoid flies (Tachinidae and Sarcophagidae). Before European colonization of Canada, aspen forests occurred over huge areas, but today they are fragmented into stands of varying size. Roland and Taylor (1997) investigated how fragmentation and host abundance influenced the pattern of attack on *M. disstria* by the different species of parasitoid. They sampled populations across a 25×25-km grid and used statistical modeling techniques to explain rates of parasitism as a function of host density and habitat structure (the latter determined using satellite data). They found that parasitism by the three largest species of parasitoid showed significant positive spatial density dependence. The two largest species also showed significantly negative delayed density dependence. The extent of fragmentation affected parasitism by all parasitoid species but, interestingly, the spatial scale at which parasitism and fragmentation were most highly correlated differed between species. Larger species seemed to be influenced by fragmentation measured at the broadest spatial scale. Intriguingly, while the three largest flies showed lower rates of parasitism in fragmented forests, that of the smallest species was higher in small woods.

In summary, we need behavioral ecological studies to gain insight into how insects perceive their environment and respond to it (Bukovinszky et al. 2005c), how spatial processes are linked to life history evolution (Hoffmeister et al. 2005a), how trade-offs in species are related to spatial scale (Kneitel & Chase 2004), and how local and regional processes interact. These behavioral studies need to mesh with, and be interpreted within the context of, the population dynamics of the species involved. Only then can we hope to predict how different species will be affected by habitat fragmentation and other human-induced environmental changes.

11.6 General conclusions

The application of the behavioral ecological research program to parasitoids, that began in the late 1970s, has been highly successful and has taught us a huge amount about how natural selection operates on parasitoid behavior. It has helped banish much imprecise thinking, such as the automatic assumption that superparasitism is bad for the parasitoid, arguments about how natural selection acts for the benefit of the species, and the notion that coevolution may operate on host and parasitoid population dynamics so as to ensure a persistent interaction. In providing quantitative tests of sex allocation theory, parasitoid behavioral ecology has provided the classic examples for today's textbooks.

Yet, there have been areas where progress has not been so spectacular. It is natural and sensible for the first tests of any theory to be performed in the highly unnatural and artificial environment of the laboratory, yet it is still relatively rare for this work to be followed up in the field, especially in the more natural environments where the behavior probably evolved. There has also been less of a connection between behavioral ecology, in the 'Krebs-and-Davies' optimality sense, and behavioral ecology in the more mechanistic chemical-ecology sense. This second field is an area in which work with parasitoids has been at the cutting edge of the subject, for example, the recruitment of herbivore natural enemies by plants, making the disconnection even more regrettable.

A further area, where progress has not been as rapid as desired, is in linking parasitoid behavioral ecology with population dynamics (see also Chapter 13 by Bonsall and Bernstein). Part of the problem lies with population dynamicists, only a few of whom have invested in the long-term field studies that are essential to getting real insight into how populations are controlled and regulated, and how natural selection may operate on parameters that influence dynamics. But behavioral ecologists sometimes fall into the trap of automatically assuming that all behavioral ecological traits have population consequences whereas, in truth, some will just be lost in the environmental noise. Some of the same criticisms could be applied to chemical behavioral ecologists, and in particular to some of the more hyped claims about what infochemical manipulation may do for pest management. Controlling pests is an exercise in applied population dynamics, and any claims that a new manipulation is likely to have positive outcomes on the farm or in the plantation needs to be justified by a sound and specific population dynamic argument constructed at the appropriate spatial scale.

These concerns should not be taken as an argument against ignoring how behavioral ecological issues affect populations and communities, but that the links should be made carefully. Indeed, above we have given exciting examples of cases where these links have been made, and look forward to more in the future. For example, we lack a serious null-model of host–parasitoid community dynamics that we can employ to judge the degree to which real host–parasitoid food webs are structured. We can see how valuable this might be from the insights that have been gained from studying null-models in plant community ecology (Bell 2001, Hubbell 2001). No one, not least Hubbell (2001), actually believes that every individual plant is functionally identical irrespective of what species it belongs to, as his neutral models assume. Yet such models accurately predict many aspects of plant communities and more importantly do not predict others, which thus need a more biological explanation. Creating equivalent models for interactions between higher trophic levels is urgently needed, and we hope that behavioral ecological studies, both chemical and evolutionary, may be helpful in guiding the way.

Community ecologists build food webs to summarize the trophic interactions in a community, which also give insights into indirect but trophic mediated interactions such as resource exploitation and apparent competition. Community ecologists are becoming increasingly aware that other indirect interactions may be as equally important as trophic-mediated effects, in particular when the behavior of one species is modified by a third. Such trait-mediated indirect interactions are a growing area in community ecology at the moment, and one in which an alliance of chemical and population ecologists interested in parasitoids would be well placed to make significant contributions. Indeed, humans, as a visual species, are very much in a minority of organisms. Most of the animals and plants with which we share this planet receive much of their information about the environment from chemical channels, with clear consequences for population and community processes. Building infochemical interaction webs would be a natural counterpart to trophic interaction webs, and an area where parasitoid biologists could lead. Continuing technical advances in analytical chemistry can only help this.

Parasitoid behavioral ecology is an exciting field both because parasitoid behavior is intrinsically interesting, but also because it lends itself to experimental manipulation. Many interesting questions in evolutionary ecology are difficult to study experimentally (though often progress can be made using the modern comparative method) and many workers, perhaps not natural students of behavior, were attracted to the subject because of its experimental tractability. The last decade or so has seen new experimental methods entering the field, and we envisage many more entering in the next few decades. We think this will have the effect of blurring the borders of parasitoid behavioral ecology – for the better.

Examples of approaches that have already been used include population genetics such as QTR mapping (Holloway et al. 2000), the study of candidate genes (Fitzpatrick et al. 2005), and the use of artificial selection to manipulate traits that cannot be assessed directly (see also Chapter 14 by Kraaijeveld and Godfray). None of these techniques have been applied extensively to parasitoid MTI, though surely they will be in the near future.

But the major challenge and opportunity is to capitalize on the huge progress made in molecular genetics, genomics, and cell biology, to provide new tools to explore parasitoid MTI. We now have, or will shortly have, the means of identifying the mechanistic basis of many behaviors that are currently black boxes. With this knowledge, we will be able to manipulate them in precise ways that will allow both proximate and ultimate hypotheses about their operation and function to be tested in novel ways. We shall be able to identify the genes affecting phenotype, and then to look at natural variation in wild populations, and how it changes over time and in response to changed environments. Behavioral ecology has traditionally been a study of evolutionary endpoints (the assumed optimum) in contradistinction to population genetics that studied evolutionary trajectories. These boundaries will increasingly disappear.

Fundamental research on parasitoids will and should continue, though against a background of darkening environmental skies. Maintaining the fundamental research base is critical, but so is the duty of every ecologist to contribute to understanding the global change and to help mitigate its negative effects (Pimm 2001). We should simultaneously resist the impulse to claim all works are relevant to global change studies, however tangentially it impacts on policy and mitigation, but be ready to leave the comfort of the ivory tower and enter the fray when we believe our work can help with the greatest challenge facing humanity in the 21st century.

References

Baldwin, I.T., Halitschke, R., Kessler, A. and Schittko, U. (2001) Merging molecular and ecological approaches in plant-insect interactions. *Current Opinion in Plant Biology* **4**: 351–8.

Barbosa, P., Saunders, J.A., Kemper, J., Trumbule, R., Olechno, J. and Martinat, P. (1986) Plant allelochemicals and insect parasitoids. Effects of nicotine on *Cotesia congregata* (Say) (Hymenoptera: Braconidae) and *Hyposoter annulipes* (Cresson) (Hymenoptera: Ichneumonidae). *Journal of Chemical Ecology* **12**: 1319–28.

Bardgett, R.D. and Wardle, D.A. (2003) Herbivore-mediated linkages between aboveground and belowground communities. *Ecology* **84**: 2258–68.

Bell, G. (2001) Neutral macroecology. *Science* **293**: 2413–18.

Bezemer, T.M. and van Dam, N.M. (2005) Linking aboveground and belowground interactions via induced plant defenses. *Trends in Ecology & Evolution* **20**: 617–24.

Bezemer, T.M., Wagenaar, R., van Dam, N.M. and Wäckers, F.L. (2003) Interactions between above- and belowground insect herbivores as mediated by the plant defence system. *Oikos* **101**: 555–62.

Bezemer, T.M., Wagenaar, R., van Dam, N.M., van der Putten, W.H. and Wäckers, F.L. (2004) Above- and below-ground terpenoid aldehyde induction in cotton, *Gossypium herbaceum*, following root and leaf injury. *Journal of Chemical Ecology* **30**: 53–67.

Bleeker, M.A.K., Smid, H.M., Steidle, J.L.M., Kruidhof, H.M., van Loon, J.J.A. and Vet, L.E.M. (2006) Differences in memory dynamics between two closely related parasitoid wasp species. *Animal Behaviour* **71**: 1343–50.

Boake, C.R.B., Arnold, S.J., Breden, F. et al. (2002) Genetic tools for studying adaptation and the evolution of behavior. *American Naturalist* **160**: 143–59.

Bottrell, D.G., Barbosa, P. and Gould, F. (1998) Manipulating natural enemies by plant variety selection and modification: A realistic strategy? *Annual Review of Entomology* **43**: 347–67.

Bukovinszky, T., Gols, R., Posthumus, M.A., Vet, L.E.M. and van Lenteren, J.C. (2005a) Variation in plant volatiles and attraction of the parasitoid *Diadegma semiclausum* (Hellen). *Journal of Chemical Ecology* **31**: 461–80.

Bukovinszky, T., van Lenteren, J.C. and Vet, L.E.M. (2005b) Functioning of natural enemies in mixed cropping systems. In: Pimentel, D. (ed.) *Encyclopedia of Pest Management*. Marcel Dekker, New York, pp. 1–4.

Bukovinszky, T., Potting, R.P.J., Clough, Y., van Lenteren, J.C. and Vet, L.E.M. (2005c) The role of pre- and post- alighting detection mechanisms in the responses to patch size by specialist herbivores. *Oikos* **109**: 435–46.

Bukovinszky, T., Gols, R., Hemerik, L., van Lenteren, J.C. and Vet, L.E.M. (2007) Time-allocation of the parasitoid foraging in heterogeneous vegetation: implications for host-parasitoid interactions. *Journal of Animal Ecology* (in press).

D'Alessandro, M. and Turlings, T.C.J. (2006) Advances and challenges in the identification of volatiles that mediate interactions among plants and arthropods. *Analyst* **131**: 24–32.

De Deyn, G.B., Raaijmakers, C.E., Zoomer, H.R. et al. (2003) Soil invertebrate fauna enhances grassland succession and diversity. *Nature* **422**: 711–13.

Dicke, M. (2006) Chemical ecology from genes to community. In: Dicke, M. and Takken, W. (eds.) *Chemical ecology: From Gene to Ecosystem*. Springer, Dordrecht, pp. 175–89.

Dicke, M. and Vet, L.E.M. (1999) Plant-carnivore interactions: evolutionary and ecological consequences for plant, herbivore and carnivore. In: Olff, H., Brown, V.K. and Drent, R.H. (eds.) *Herbivores: Between Plants and Predators*. Blackwell Science, Oxford, pp. 483–520.

Dicke, M., Sabelis, M.W. and Takabayashi, J. (1990) Do plants cry for help? Evidence related to a tritrophic system of predatory mites, spider mites and their host plants. *Symposia Biologica Hungarica* **39**: 127–34.

Dicke, M., de Boer, J.G., Hofte, M. and Rocha-Granados, M.C. (2003) Mixed blends of herbivore-induced plant volatiles and foraging success of carnivorous arthropods. *Oikos* **101**: 38–48.

Dicke, M., van Loon, J.J.A. and de Jong, P.W. (2004) Ecogenomics benefits community ecology. *Science* **305**: 618–19.

Duffey, S.S., Bloem, K.A. and Campbell, B.C. (1986) Consequences of sequestration of plant natural products in plant-insect-parasitoid interactions. In: Boethel, D.J. and Eikenbary, R.D. (eds.) *Interactions of Plant Resistance and Parasitoids and Predators of Insects*. Horwood Press, Chichester, pp. 31–60.

Fiehn, O., Kopka, J., Dormann, P., Altmann, T., Trethewey, R.N. and Willmitzer, L. (2000) Metabolite profiling for plant functional genomics. *Nature Biotechnology* **18**: 1157–61.

Fitzpatrick, M.J. and Sokolowski, M.B. (2004) In search of food: Exploring the evolutionary link between cGMP-dependent protein kinase (PKG) and behaviour. *Integrative and Comparative Biology* **44**: 28–36.

Fitzpatrick, M.J., Ben-Shahar, Y., Smid, H.M., Vet, L.E.M., Robinson, G.E. and Sokolowski, M.B. (2005) Candidate genes for behavioural ecology. *Trends in Ecology & Evolution* **20**: 96–104.

Geervliet, J.B.F., Vreugdenhil, A.I., Vet, L.E.M. and Dicke, M. (1998) Learning to discriminate between infochemicals from different plant-host complexes by the parasitoids *Cotesia glomerata* and *C. rubecula* (Hymenoptera: Braconidae). *Entomologia Experimentalis et Applicata* **86**: 241–52.

Godfray, H.C.J. (1991) The signalling of need by offspring to their parents. *Nature* **352**: 328–30.

Godfray, H.C.J. (1994) *Parasitoids – Behavioral and Evolutionary Ecology*. Princeton University Press, Princeton.

Godfray, H.C.J. (1995) Communication between the first and third trophic levels: An analysis using biological signalling theory. *Oikos* **72**: 367–4.

Gols, R., Bukovinszky, T., Hemerik, L., Harvey, J.A., van Lenteren, J.C. and Vet, L.E.M. (2005) Reduced foraging efficiency of a parasitoid under habitat complexity: Implications for population stability and species coexistence. *Journal of Animal Ecology* **74**: 1059–68.

Gouinguene, S., Degen, T. and Turlings, T.C.J. (2001) Variability in herbivore-induced odour emissions among maize cultivars and their wild ancestors (teosinte). *Chemoecology* **11**: 9–16.

Grafen, A. (1990) Sexual selection unhandicapped by the Fisher process. *Journal of Theoretical Biology* **144**: 473–516.

Gu, H. and Dorn, S. (2000) Genetic variation in behavioral response to herbivore-infested plants in the parasitic wasp, *Cotesia glomerata* (L.) (Hymenoptera: Braconidae). *Journal of Insect Behaviour* **13**: 141–56.

Harvey, J.A. (2005) Factors affecting the evolution of development strategies in parasitoid wasps: The importance of functional constraints and incorporating complexity. *Entomologia Experimentalis et Applicata* **117**: 1–13.

Harvey, J.A. and Strand, M.R. (2002) The developmental strategies of endoparasitoid wasps vary with host feeding ecology. *Ecology* **83**: 2439–51.

Harvey, J.A., van Dam, N.M. and Gols, R. (2003) Interactions over four trophic levels: Foodplant quality affects development of a hyperparasitoid as mediated through a herbivore and its primary parasitoid. *Journal of Animal Ecology* **72**: 520–31.

Harvey, J.A., Dam, N.M.V., Witjes, L.M.A., Soler, R. and Gols, R. (2007) Effects of dietary nicotine on the development of a herbivore, its parasitoid and secondary hyperparasitoid over four trophic levels. *Ecological Entomology* **32**: 15–23.

Hoffmeister, T.S., Roitberg, B.D. and Vet, L.E.M. (2005a) Linking spatial processes to life-history evolution of insect parasitoids. *American Naturalist* **166**: 62–74.

Hoffmeister, T.S., Vet, L.E.M., Biere, A., Holsinger, K. and Filser, J. (2005b) Ecological and evolutionary consequences of biological invasion and habitat fragmentation. *Ecosystems* **8**: 657–67.

Holloway, A.K., Strand, M.R., Black, W.C. and Antolin, M.F. (2000) Linkage analysis of sex determination in *Bracon* sp near *hebetor* (Hymenoptera: Braconidae). *Genetics* **154**: 205–12.

Holt, R.D., Lawton, J.H., Polis, G.A. and Martinez, N.D. (1999) Trophic rank and the species-area relationship. *Ecology* **80**: 1495–504.

Hubbell, S.P. (2001) *The Unified Neutral Theory of Biodiversity and Biogeography*. Princeton University Press, Princeton.

Kappers, I.F., Aharoni, A., van Herpen, T., Luckerhoff, L.L.P., Dicke, M. and Bouwmeester, H.J. (2005) Genetic engineering of terpenoid metabolism attracts, bodyguards to *Arabidopsis*. *Science* **309**: 2070–2.

Karban, R. and Baldwin, I.T. (1997) *Induced Responses to Herbivory*. Chicago University Press, Chicago.

Kessler, A., Halitschke, R. and Baldwin, I.T. (2004) Silencing the jasmonate cascade: Induced plant defenses and insect populations. *Science* **305**: 665–68.

Kneitel, J.M. and Chase, J.M. (2004) Trade-offs in community ecology: Linking spatial scales and species coexistence. *Ecology Letters* **7**: 69–80.

Kruess, A. and Tscharntke, T. (1994) Habitat fragmentation, species loss, and biological control. *Science* **264**: 1581–4.

Kruess, A. and Tscharntke, T. (2000) Species richness and parasitism in a fragmented landscape: Experiments and field studies with insects on *Vicia sepium*. *Oecologia* **122**: 129–37.

Lawton, J.H. (2000) *Community Ecology in a Changing World*. Ecology Institute, Oldendorf.

Lei, G.C. and Hanski, I. (1997) Metapopulation structure of *Cotesia melitaearum*, a specialist parasitoid of the butterfly *Melitaea cinxia*. *Oikos* **78**: 91–100.

Lei, G.C. and Hanski, I. (1998) Spatial dynamics of two competing specialist parasitoids in a host metapopulation. *Journal of Animal Ecology* **67**: 422–33.

Masters, G.J. and Brown, V.K. (1997) Host-plant mediated interactions between spatially separated herbivores: Effects on community structure. In: Gange, A.C. and Brown, V.K. (eds.) *Multitrophic*

Interactions in a Changing World. 36th Symposium of the British Ecological Society. Blackwell Science, Oxford, pp. 217–32.

Maynard Smith, J. and Harper, D.G.C. (2003) *Animal Signals*. Oxford University Press, Oxford.

Mumm, R. and Hilker, M. (2005) The significance of background odour for an egg parasitoid to detect plants with host eggs. *Chemical Senses* **30**: 337–43.

Ode, P.J. (2006) Plant chemistry and natural enemy fitness: Effects on herbivore and natural enemy interactions. *Annual Review of Entomology* **51**: 163–85.

Palumbi, S.R. (2001) Evolution – Humans as the world's greatest evolutionary force. *Science* **293**: 1786–90.

Perfecto, I. and Vet, L.E.M. (2003) Effect of a nonhost plant on the location behavior of two parasitoids: The tritrophic system of *Cotesia* spp. (Hymenoptera: Braconidae), *Pieris rapae* (Lepidoptera: Pieridae), and *Brassica oleraceae*. *Environmental Entomology* **32**: 163–74.

Pimm, S.L. (2001) *The World According to Pimm*. McGraw Hill, New York.

Prévost, G. and Lewis, W.J. (1990) Genetic differences in the response of *Microplitis croceipes* to volatile semiochemicals. *Journal of Insect Behavior* **3**: 277–87.

Ritchie, M.E. and Olff, H. (1999) Spatial scaling laws yield a synthetic theory of biodiversity. *Nature* **400**: 557–60.

Roland, J. and Taylor, P.D. (1997) Insect parasitoid species respond to forest structure at different spatial scales. *Nature* **386**: 710–13.

Sabelis, M.W., van Baalen, M., Bakker, F.M. et al. (1999) The evolution of direct and indirect plant defence against herbivorous arthropods. In: Olff, H., Brown, V.K. and Drent, R.H. (eds.) *Herbivores: Between Plants and Predators*. Blackwell Science, Oxford, pp. 109–66.

Schnee, C., Kollner, T.G., Held, M., Turlings, T.C.J., Gershenzon, J. and Degenhardt, J. (2006) The products of a single maize sesquiterpene synthase form a volatile defense signal that attracts natural enemies of maize herbivores. *Proceedings of the National Academy of Sciences* **103**: 1129–34.

Smid, H.M. and Vet, L.E.M. (2006) Learning in insects: From behaviour to brain. *Animal Biology* **56**: 121–4.

Smid, H.M., Wang, G., Bukovinszky, T. et al. (2007) Species-specific acquisition and consolidation of long-term memory in parasitic wasps. *Proceedings of the Royal Society of London Series B.* **274**: 1539–46.

Soler, R., Bezemer, T.M., van der Putten, W.H., Vet, L.E.M. and Harvey, J.A. (2005) Root herbivore effects on aboveground herbivore, parasitoid and hyperparasitoid performance via changes in plant quality. *Journal of Animal Ecology* **74**: 1121–30.

Soler, R., Harvey, J.A., Kamp, A.F.D. et al. (2007) Root herbivores influence the behaviour of an aboveground parasitoid through changes in plant-volatile signals. *Oikos* **116**: 367–76.

Spence, A.M. (1973) Job market signalling. *Quarterly Journal of Economics* **87**, 355–74.

Stearns, S.C. and Magwene, P. (2003) The naturalist in a world of genomics. *American Naturalist* **161**: 171–80.

Stephens, D.W. and Krebs, J.R. (1986) *Foraging Theory*. Princeton University Press, Princeton.

Sumner, L.W., Mendes, P. and Dixon, R.A. (2003) Plant metabolomics: Large-scale phytochemistry in the functional genomics era. *Phytochemistry* **62**: 817–36.

Thies, C., Steffan-Dewenter, I. and Tscharntke, T. (2003) Effects of landscape context on herbivory and parasitism at different spatial scales. *Oikos* **101**: 18–25.

Thomas, M.A. and Klaper, R. (2004) Genomics for the ecological toolbox. *Trends in Ecology & Evolution* **19**: 439–45.

Tscharntke, T. and Brandl, R. (2004) Plant-insect interactions in fragmented landscapes. *Annual Review of Entomology* **49**: 405–30.

Turlings, T.C.J., Tumlinson, J.H. and Lewis, W.J. (1990) Exploitation of herbivore-induced plant odors by host-seeking parasitic wasps. *Science* **250**: 1251–3.

Turlings, T.C.J., Wäckers, F.L., Vet, L.E.M., Lewis, W.J. and Tumlinson, J.H. (1993) Learning of host-finding cues by hymenopterous parasitoids. In: Papaj, D.R. and Lewis, A.C. (eds.) *Insect Learning: Ecological and Evolutionary Perspectives.* Chapman & Hall, New York, pp. 51–78.

Tylianakis, J.M., Tscharntke, T. and Klein, A.-M. (2006) Diversity, ecosystem function, and stability of parasitoid-host interactions across a tropical gradient. *Ecology* **87**: 3047–57.

van Dam, N.M. and Bezemer, T.M. (2006) Chemical communication between roots and shoots: Towards an integration of aboveground and belowground induced responses in plants. In: Dicke, M. and Takken, W. (eds.) *Chemical Ecology: From Gene to Ecosystem.* Springer, Dordrecht, pp. 127–43.

van Dam, N.M., Harvey, J.A., Wäckers, F.L., Bezemer, T.M., van der Putten, W.H. and Vet, L.E.M. (2003) Interactions between aboveground and belowground induced responses against phytophages. *Basic & Applied Ecology* **4**: 63–77.

van der Meijden, E. and Klinkhamer, P.G.L. (2000) Conflicting interests of plants and the natural enemies of herbivores. *Oikos* **89**: 202–8.

van der Putten, W.H. (2003) Plant defense below ground and spatiotemporal processes in natural vegetation. *Ecology* **84**: 2269–80.

van der Putten, W.H., Vet, L.E.M., Harvey, J.A. and Wäckers, F.L. (2001) Linking above- and belowground multitrophic interactions of plants, herbivores, pathogens, and their antagonists. *Trends in Ecology & Evolution* **16**: 547–54.

van Loon, J.J.A., De Boer, J.G. and Dicke, M. (2000) Parasitoid-plant mutualism: Parasitoid attack of herbivore increases plant reproduction. *Entomologia Experimentalis et Applicata* **97**: 219–27.

van Nouhuys, S. and Hanski, I. (2000) Apparent competition between parasitoids mediated by a shared hyperparasitoid. *Ecology Letters* **3**: 82–4.

van Nouhuys, S. and Hanski, I. (2002) Colonization rates and distances of a host butterfly and two specific parasitoids in a fragmented landscape. *Journal of Animal Ecology* **71**: 639–50.

van Poecke, R.M.P. and Dicke, M. (2002) Induced parasitoid attraction by *Arabidopsis thaliana*: Involvement of the octadecanoid and the salicylic acid pathway. *Journal of Experimental Botany* **53**: 1793–9.

van Veen, F.J.F., Morris, R.J. and Godfray, H.C.J. (2006) Apparent competition, quantitative food webs, and the structure of phytophagous insect communities. *Annual Review of Entomology* **51**: 187–208.

Vet, L.E.M. (1999a) Evolutionary aspects of plant-carnivore interactions. In: Chadwick, D.J. and Goode, J.A. (eds.) *Insect-plant Interactions and Induced Plant Defence.* John Wiley & Sons, Chichester, pp. 3–13.

Vet, L.E.M. (1999b) From chemical to population ecology: Infochemical use in an evolutionary context. *Journal of Chemical Ecology* **25**: 31–49.

Vet, L.E.M. and Dicke, M. (1992) Ecology of infochemical use by natural enemies in a tritrophic context. *Annual Review of Entomology* **37**: 141–72.

Vet, L.E.M., Lewis, W.J., Papaj, D.R. and van Lenteren, J.C. (1990) A variable-response model for parasitoid foraging behavior. *Journal of Insect Behavior* **3**: 471–90.

Vet, L.E.M., Lewis, W.J. and Cardé, R.T. (1995) Parasitoid foraging and learning. In: Cardé, R.T. and Bell, W.J. (eds.) *Chemical Ecology of Insects.* Chapman & Hall, New York, pp. 65–101.

Vet, L.E.M., Hemerik, L., Visser, M.E. and Wäckers, F.L. (2002) Flexibility in host-search and patch-use strategies of insect parasitoids. In: Lewis, E.E., Cambell, J.F. and Sukhdeo, M.V.K. (eds.) *The Behavioural Ecology of Parasites.* CABI Publishing, Wallingford, pp. 39–64.

Vos, M., Berrocal, S.M., Karamaouna, F., Hemerik, L. and Vet, L.E.M. (2001) Plant-mediated indirect effects and the persistence of parasitoid-herbivore communities. *Ecology Letters* **4**: 38–45.

Wäckers, F.L. and Bezemer, T.M. (2003) Root herbivory induces an above-ground indirect defence. *Ecology Letters* **6**: 9–12.

Wajnberg, É. (2006) Time allocation strategies in insect parasitoids: From ultimate predictions to proximate behavioral mechanisms. *Behavioral Ecology & Sociobiology* **60**: 589–611.

Wajnberg, É., Gonsard, P.A., Tabone, E., Curty, C., Lezcano, N. and Colazza, S. (2003) A comparative analysis of patch-leaving decision rules in a parasitoid family. *Journal of Animal Ecology* **72**: 618–26.

Wang, Q., Gu, H. and Dorn, S. (2003) Selection on olfactory response to semiochemicals from a plant-host complex in a parasitic wasp. *Heredity* **91**: 430–5.

Wang, Q., Gu, H. and Dorn, S. (2004) Genetic relationship between olfactory response and fitness in *Cotesia glomerata* (L.). *Heredity* **92**: 579–84.

Wardle, D.A., Bardgett, R.D., Klironomos, J.N., Setälä, H., van der Putten, W.H. and Wall, D.H. (2004) Ecological linkages between aboveground and belowground biota. *Science* **304**: 1629–33.

Werren, J.H., Gadau, J., Beukeboom, L. et al. (2005) Proposal to sequence the *Nasonia* genome. http://www.genome.gov/Pages/Research/Sequencing/SeqProposals/NasoniaSeq.pdf.

12

Parasitoid sex ratios and biological control

Paul J. Ode and Ian C.W. Hardy

Abstract

In this chapter, we briefly review the major factors that influence parasitoid sex ratios and then examine theory and evidence (or lack of evidence) for how each of these affect the success of parasitoids as agents of biological pest control. We begin by assuming that ovipositing parasitoids have complete control over their sex allocation decisions and outline the major tenets of optimal sex ratio theory, particularly with respect to Fisher's theory for equal investment, host-quality based models (i.e. 'Charnov' type effects) and local mate competition (LMC) theory, and provide brief summaries of recent 'pure' empirical research in each area.

We examine how such sex allocation decisions might be expected to have important effects on the biocontrol potential of parasitoids. An example is that an understanding of host-quality effects can be used to boost female production in mass rearing facilities while keeping rearing costs under control, although it seems that such results have not been put into practice. Another example is that parasitoid density has been expected to influence host–parasitoid population dynamics via LMC because the number of female parasitoids produced per host is influenced by the density of ovipositing female wasps. We then consider factors that bring sex allocation out of the complete control of the ovipositing female. For example, we consider the effects of different sex determining mechanisms on the sex allocation strategies employed by different species of parasitoids as well as the implications for biological control efforts. Finally, we consider 'extrinsic factors' that reduce sex ratio control like infectious elements that cause mating incompatibilities and feminization of bio-control agents and discuss the consequences for biocontrol applications (e.g. choice of infected and non-infected agents). We conclude that the tremendously successful body of research into parasitoid sex ratios and sex determination has had little practical impact on biocontrol deploying parasitoid wasps, and discuss whether there is an opportunity to be better exploited.

12.1 Introduction

Sex allocation is one of the most stunningly successful areas of behavioral ecology, both in terms of the complexity of theory that has been developed and the wealth of empirical studies conducted to test the theory and stimulate further developments (Charnov 1982, Hardy 2002). Many of the empirical tests of sex allocation theory have been carried out using species in the Hymenoptera (Bourke & Franks 1995), and especially the parasitoid Hymenoptera (Godfray 1994, Ode & Hunter 2002). This is largely because sex allocation in most hymenopteran parasitoids is extremely labile, with often strong and direct links between sex allocation behavior and evolutionary fitness, facilitating examination of genetic, developmental, behavioral, and ecological factors that influence the evolution of sex ratio optima.

Hymenopteran parasitoids, of course, are also important natural enemies of insect pests and have thus been evaluated and deployed in many programs of biological pest control (Wajnberg & Hassan 1994, Wajnberg et al. 2001, Mason & Huber 2002, Jervis 2005). Parasitoid sex ratios have long been thought to be an important consideration for biological control programs. Conventional wisdom contends that production of female-biased sex ratios is largely beneficial for biological control programs because adult females, not males, are responsible for attacking (via host feeding or oviposition) individual pest insects (Clausen 1939, Waage 1982a, Hassell et al. 1983, Comins & Wellings 1985, Hall 1993, Heimpel & Lundgren 2000, Berndt & Wratten 2005, Chow & Heinz 2006). This is particularly obvious in augmentative biological control programs where large numbers of individuals are mass-reared in insectaries and released with the aim of rapidly controlling a target insect pest population. It is (mated) females that are required to build up populations in culture (Heinz 1998, Sagarra & Vincent 1999, Pérez-Lachaud & Hardy 2001, Joyce et al. 2002, Ode & Heinz 2002, Pandey & Johnson 2005, Irvin & Hoddle 2006). However, parasitoid cultures in insectaries often produce more males than typically found in natural populations and in extreme cases can come to consist entirely of males, resulting in the extinction of the culture (Wilkes 1947, Platner & Oatman 1972, Waage 1982a, van Dijken et al. 1993, Luck et al. 1999, Butcher et al. 2000a, Heimpel & Lundgren 2000, Fuester et al. 2003, Johns & Whitehouse 2004, Fig. 12.1). Sex ratio can also be a consideration for classical biological control programs where exotic natural enemies are introduced with the intention of long-term establishment from regions where the target pest is native. In this case, it is the number of female progeny produced per parasitized host, almost irrespective of the number of males, which influences host–parasitoid population dynamics (Hassell 2000, Murdoch et al. 2003). Despite the common assertion that a solid understanding of parasitoid sex ratio biology is a key element to successful biological control programs, examples of rigorous application of sex allocation theory to biological control are extremely rare.

Here, we review the many factors that influence parasitoid sex ratios and evaluate their impact (or potential for impact) on biological control programs. First, we briefly outline the major tenets of optimal sex ratio theory, initially assuming that female wasps have complete control over their sex allocation decisions. These tenets include equal investment theory, theory for host-quality related effects, and local mate competition (LMC) theory. For each, we briefly comment on empirical support for these theories and then

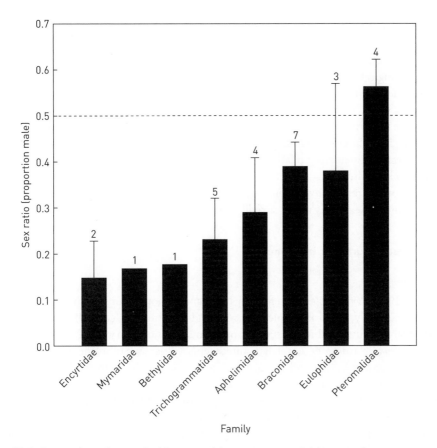

Fig. 12.1 Sex ratios of parasitoids reared from commercial insectaries. Parasitoids are grouped by hymenopteran family, numbers above columns indicate the number of species sampled per family, and bars denote standard errors. Note that there is a large inter-specific variation in sex ratios of parasitoids deployed as biological control agents, but most have sex ratios that are more female biased than would be expected under Fisherian sex allocation (dashed line). Redrawn, with modification, from Fig. 1 in Heimpel and Lundgren (2000), with permission from Elsevier.

examine how sex allocation decisions are expected to affect the biological control potential of parasitoids. We then relax the above assumption and consider various factors that remove sex allocation from complete control of the ovipositing female. We introduce genetic considerations (particularly the role of sex determining mechanisms on the sex allocation patterns of different parasitoid species) and their implications for biological control, together with their effects on sex allocation and population biology. We further examine various infectious elements that result in mating incompatibilities, death of male offspring, and feminization of parasitoids and also discuss their consequences for biological control. We conclude with an overall discussion of whether the tremendously rich body of research into parasitoid sex ratios and sex determination has had any practical impact on biological control efforts and, if not, whether there is an opportunity to be exploited.

12.2 Sex ratios under maternal control

Sex allocation theory assuming maternal control over sex allocation is now well developed. The following relatively recent reviews present the intricacies of current sex ratio research in more detail than is possible here (Hardy 1992, 1994, Godfray 1994, Hardy 1997, Godfray & Cook 1997, Hardy 2002, Ode & Hunter 2002, West et al. 2002, Hardy et al. 2005). The majority of theoretical and empirical studies of parasitoid sex allocation have dealt with either the effect of host quality on the sex allocation decisions of individual females or the effects of interacting relatives in spatially-structured populations on sex allocation optima. In both situations, sex ratio effects are predicted due to sexually differential fitness returns from parental investment in offspring. Before examining these scenarios, we briefly consider sex ratios under sexually identical investment returns.

12.2.1 Equal returns for investment

Fisher's theory
Fisher's ([1930] 1999) explanation for why many animal populations exhibit equal sex ratios or, more correctly, equal investment in sons and daughters, is generally accepted as the conceptual and historical foundation of modern sex ratio theory (Bull & Charnov 1988, Godfray 1994, Hardy 1997), yet has antecedents (Seger & Stubblefield 2002). Fisher's explanation assumes that individuals find mates 'at random' from throughout large populations (i.e. there is no population structure in regard to mate choice, a situation called panmixis). When sons and daughters are equally costly to produce and the population sex ratio is unbiased, each son will on average mate with one female and an ovipositing mother will realize equal fitness gains from investing in a son or a daughter. If the population becomes biased toward either males or females, mothers that preferentially produce the rarer sex will realize greater fitness gains, until the population sex ratio returns to equality via frequency dependent selection. If one sex is more costly to produce than the other, selection for equal investment will lead to a sex ratio biased in favor of the less costly sex.

Under Fisherian assumptions, investment optima are unaffected by sexually differential post-investment mortality or whether mating is monogamous or polygamous. These factors affect the variance in male or female mating success, but not their means (Leigh 1970, Bull & Charnov 1988, Hardy 1997). Further, Fisherian sex ratios apply to population investment ratios rather than the sex allocation decisions made by individuals. In large populations, equal investment ratios can be arrived at in a large number of ways, ranging from all mothers investing equally in sons and daughters to half of the mothers producing only daughters and the other half producing only sons. While the latter may seem unlikely in many animals, parasitoids have haplo-diploid sex determination (Section 12.3.2) with the consequence that unmated females can reproduce but are constrained to produce male offspring only. If such constrained reproduction is common in panmictic populations, frequency dependent selection will lead to mated females producing female based sex ratios leading to an unbiased population sex ratio (Godfray 1990, Godfray & Hardy 1993, Godfray 1994, Ode et al. 1997, Henter 2004).

If Fisher's assumption of a large population is violated (Section 12.2.3), mothers are selected to match the predicted population average by reducing variance in offspring sex ratios (Hardy 1992, 1997, Krackow et al. 2002) and mated females should make little adjustment to their sex ratio in the presence of reproducing but unmated females (Godfray 1990, Godfray & Hardy 1993).

Evidence
Although populations of many parasitoid and other animal species have unbiased sex ratios (Godfray 1994, Godfray & Cook 1997), this is only weak evidence for the operation of Fisherian frequency dependent selection (Bull & Charnov 1988, Hardy 1997), because many other factors provide candidate alternative explanations. Further, a numerical bias may also be expected under sex allocation equilibrium. The most convincing test would thus be a 'dynamic test' evaluating whether sex allocation returns to equality after a perturbation (Bull & Charnov 1988), coupled with empirical evaluation of the assumptions on which the theory is based, such as mating in large swarms suggesting panmixis (Hardy 1994, Godfray & Cook 1997, Hardy et al. 2005). We know of no such tests employing parasitoids, although Conover and van Voorhees (1990) provide a perturbation test using fish.

Biological control relevance
We know of no direct usage of Fisher's theory in biological control programs, beyond the fact that it provides a baseline set of expectations. Biocontrol practitioners can at least expect that the sex ratios of parasitoids they mass-rear or deploy should approximate equality (Fig. 12.1). If a species has biased sex ratios there are numerous possible explanations. A bias toward females is possibly an advantage that can be exploited, but a bias toward males likely constitutes a production problem. The exploitation of such advantages and the solution of such problems are discussed in the sections below.

12.2.2 Differential returns due to host quality

Charnov's theory
The Fisherian assumption of identical fitness returns from investment in male and female offspring breaks down in a number of circumstances, leading to different predicted sex ratio strategies. If there are sexually differential returns from resource investment, mothers are able to influence more strongly the success of one sex of offspring than the other. A mother with a large amount of resource to invest should produce the sex that benefits more from investment and a mother with little resource to invest should produce the sex whose fitness is less sensitive to investment. Such arguments were first formally proposed by Trivers and Willard (1973) in the context of maternal investment patterns in mammals, but are readily applicable to a much wider range of situations (Charnov 1979, Bull 1981), including parasitoid sex allocation decisions when host quality varies (Charnov et al. 1981, reviewed by King 1993, Godfray 1994). Because Charnov and his colleagues were the first to explicitly consider sex allocation responses to host quality in parasitoids, Trivers–Willard type ideas are often known as 'Charnov's theory' by parasitoid researchers.

The host quality model (Charnov et al. 1981) was developed specifically for solitary parasitoids (where at most one offspring develops per host), where body size is strongly correlated with the amount of resources available in the host. In many solitary parasitoids, host size is thought to more strongly affect female fitness than male fitness (King 1993, see Fig. 12.2). In this case, the ovipositing female would realize greater fitness returns by allocating daughters to larger, higher-quality hosts, and sons to smaller, poorer-quality hosts.

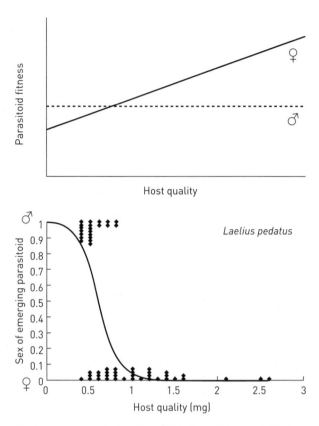

Fig. 12.2 Sex ratio in response to host quality. The top panel illustrates the assumption that female fitness is more strongly affected than male fitness by variation in host quality (after King 1993). The lower panel shows a sex ratio response to host quality (mass) in *Laelius pedatus* (Hymenoptera: Bethylidae) when laying just one egg onto a host. This matches expectations from Charnov's theory. (On even higher quality hosts *L. pedatus* tend to lay larger numbers of eggs with sex allocation patterns apparently influenced by both host quality effects and LMC, Section 12.2.3). Data are derived from Mayhew and Godfray's (1997) study and were re-analysed using logistic regression. The probability of the developed offspring being male decreased significantly with increase in host quality ($G_1 = 31.60$, $p < 0.001$). Data are shown displaced from their binary positions to illustrate sample sizes.

Charnov's host quality model makes two additional predictions. First, a size threshold exists, above which only daughters are laid and below which only sons are laid. Second, this size threshold is relative rather than absolute. If an ovipositing female encounters a medium-sized host that is in a population of mostly large-sized hosts, she should lay a son in this host but, if this same medium-sized host were encountered in a population of mostly small-sized hosts, she should lay a daughter in this host. Many of the predictions of Charnov's host quality model are based on the implicit assumption that ovipositing females have complete knowledge of the size distribution of the hosts available. More realistically, females are likely to update their perceptions of host-size distributions by comparing the sizes of currently encountered hosts with those of previously encountered hosts.

Population level consequences of host-quality effects

In opposite vein to Fisher's theory, Charnov's theory was developed to predict sex allocation decisions made by individuals rather than population-level investment ratios. However, further modeling has predicted that under host-quality dependent sex allocation, a small numerical bias toward males may be expected in the population (Frank & Swingland 1988, Charnov & Bull 1989, Pen & Weissing 2002).

Host-quality effects on sex allocation have been incorporated into host–parasitoid population dynamic theory by Murdoch and co-workers (Murdoch et al. 1992, 2003, 2005, Gutierrez et al. 1993). These studies have primarily been aimed at exploring the dynamic influences of host-stage dependent behavior (and consequent female offspring production) in general rather than sex allocation *per se*, and therefore often include consideration of aspects such as host-size and parasitoid-state-dependent host-feeding (Briggs et al. 1999, Murdoch et al. 2003). Such models generally treat hosts as discrete (size and/or developmental stage) classes and often assume that a given class of host always receives a given sex of egg on parasitism (Murdoch et al. 1992, 2003), or that female eggs are laid with host-class dependent probability (Murdoch et al. 2005). Such assumptions capture the basic pattern of size-dependent sex allocation expected under Charnov's model, but exclude the possibility of foraging females adjusting their sex allocation decisions according to a relative size threshold (Murdoch et al. 1992). In the model outlined by Murdoch et al. (2003), an increase in the density of foraging females leads to increased attack on young hosts, and male parasitoid production, leaving subsequently fewer older hosts and thus a reduction in the production of female parasitoid offspring. These delayed density dependent effects can potentially destabilize initially stable host–parasitoid equilibria, although stabilizing effects are also possible (Murdoch et al. 1992). However, the conclusions from a particular model may be different if the assumptions were refined to include a more flexible sex-allocation strategy, such that under increased density foraging parasitoids laid female eggs in some young hosts. Murdoch et al. (1992) further point out that optimal sex allocation theory usually (and often implicitly) assumes stable population dynamics (exceptions include Werren & Charnov 1978, Werren & Taylor 1984, West & Godfray 1997) and that further analyses are needed to consider simultaneously the evolutionary and population dynamics associated with host-quality effects.

Evidence

Researchers have long known that host size, the most frequent measure of host quality, influences the sex allocation patterns observed in many parasitoids. Ovipositing females tend to lay sons in smaller hosts and daughters in larger hosts (Clausen 1939, Charnov 1982, King 1987, 1993, Godfray 1994, Luck & Nunney 1999, see Fig. 12.2). The tendency is also stronger among parasitoids whose hosts cease growth upon parasitism (idiobionts) than among species whose hosts continue to develop (koinobionts), where maternal estimation of the amount of resource available to offspring is more difficult (West & Sheldon 2002). While the broadest prediction of Charnov's theory is well supported empirically, only a few studies have tested the prediction that ovipositing females adjust their allocation decisions according to the host size distributions in the population that they encounter. Several of these studies have found support for assessment of population host-size

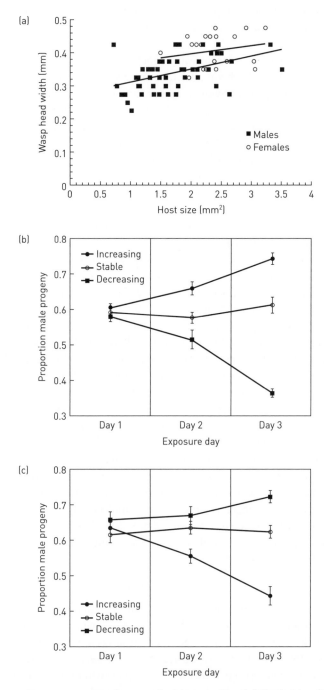

Fig. 12.3 Sex ratio response to changes in host quality. (a) Body size (head width) of male and female *D. isaea* as a function of host size (test of different slopes: t = 0.19, d.f. = 1, P = 0.8532). (b, c) Sex allocation patterns (means ± SE) of individual females (b) and groups of 30 females (c) exposed to increasing (closed circles), decreasing (closed squares), or stable (open circles) host size distributions over 3 days. Reprinted from Figs 1–3 in Ode and Heinz (2002), with permission from Elsevier.

distributions, with according adjustment of sex allocation behavior (Charnov et al. 1981, King 1987, Heinz & Parrella 1989, 1990, Ode & Strand 1995), but other parasitoid species do not appear to be able to shift sex allocation patterns within a generation (Jones 1982, van Dijken et al. 1991, Hare & Luck 1991, Bernal et al. 1998).

The crucial assumption that female fitness is more positively affected by increasing host size (via a positive correlation between host size and parasitoid adult body size, Fig. 12.3a) than is male fitness has also been evaluated by relatively few studies. There is evidence to support the assumption in some but not all species studied (Charnov et al. 1981, Jones 1982, King 1988, van den Assem et al. 1989, Heinz 1991, King & King 1994, King & Lee 1994, Ode et al. 1996, Napoleon & King 1999, Ueno 1999). There are several interrelated practical constraints that make the assessment of the crucial relationship(s) between fitness and host size and/or adult body size challenging (discussed by Karsai et al. 2006, and references therein):

1 parasitoid fitness is likely to have many components, each of which may be affected differently by host or body size;
2 male and female fitness will inevitably have different components and will require different assessment procedures;
3 evaluations of how fitness is affected by host or body size will in general be easier to carry out in the laboratory than in the field, but laboratory-estimated relationships may not accurately reflect their natural form.

Biological control relevance

Aside from the general awareness that maintaining high host quality for biological control agents is important for the production of female parasitoids (Lampson et al. 1996), Charnov's model has rarely been applied to the improvement of biological control programs. However, there are some notable exceptions. Multi-trophic field studies have shown that biological control of the cassava mealybug by the encyrtid parasitoid *Apoanagyrus lopezi*, which exhibits host-size dependent sex allocation (Kraaijeveld & van Alphen 1986, van Dijken et al. 1991, Gutierrez et al. 1993), can be improved by enhancing the nutrient content of soil. Better soil leads to stronger plants bearing larger mealybugs, which in turn leads to an increased proportion of females among the emergent parasitoids and improved pest control (Schulthess et al. 1997, Neuenschwander 2001). Similarly, citrus cultivar can affect the size of California red scale (*Aonidiella aurantii*, Homoptera) and the subsequent sex ratios of its biological control agent *Aphytis melinus* (Aphelinidae, Hare & Luck 1991, Luck & Nunney 1999). Further, field studies of the interactions between two congeneric parasitoids of California red scale have shown that host-size dependent sex allocation can explain patterns of competitive exclusion (Luck & Podoler 1985, Reeve 1987). Indeed, in interior regions of southern California, *A. melinus* competitively excludes *A. lingnanensis* in part because it accepts smaller, younger hosts for the oviposition of daughters thereby pre-empting suitable hosts for *A. lingnanensis* (Luck & Podoler 1985, Opp & Luck 1986, Luck & Nunney 1999). The most explicit cases where an understanding of the theory of host-size dependent sex allocation has been used to improve biological control potential, however, concern mass-rearing procedures in insectaries. We examine two such cases in detail here before briefly discussing some population dynamic considerations.

Tolerance for arthropod damage is very low in many horticultural and agricultural crops. While biological control can provide effective levels of control in many situations, it is often prohibitively expensive, in terms of the production and application costs, relative to chemical control (Parrella et al. 1992, van Lenteren et al. 1997). One cost associated with mass-reared natural enemies is the production of more males than are necessary to mate females that are produced. The sex ratios of many parasitic Hymenoptera used in biological control are male biased, suggesting that this cost is widespread in biological control programs (Heimpel & Lundgren 2000). The male-biased sex ratios reported in many laboratory and mass-rearing studies are thought to be due to host quality or other rearing conditions (see references cited in Heimpel & Lundgren 2000). If mass-rearing programs used by insectaries incorporate techniques to increase the proportion of females produced, this may allow insectaries to produce a higher-quality product (i.e. more females) at equal or nearly equal cost to the original production techniques. While the price per individual parasitoid (irrespective of gender) may not decrease, biocontrol practitioners may be able to realize the same degree of pest control with fewer natural enemy releases. This would reduce the overall cost of pest control via biological control.

Heinz (1998) was the first explicitly to use insights generated by the host quality model to modify the sex ratio produced by a natural enemy important in biological control. *Catolaccus grandis* (Hymenoptera: Pteromalidae) is a solitary, idiobiont ectoparasitoid of the boll weevil, *Anthonomus grandis*, an important pest of cotton. Several aspects of the reproductive biology of this species match the assumptions and predictions of Charnov's model (Heinz 1998). Female body size is more strongly affected than male body size as host size increases, suggesting that host quality affects female fitness more strongly than male fitness. In laboratory cultures, females tend to be produced on larger hosts and sons on smaller hosts, and egg-transfer experiments have eliminated sexually differential mortality as a candidate explanation for this pattern. This trend was observed despite fluctuation in average host size in the laboratory population over time, suggesting that females make sex allocation decisions based on relative, not absolute, host size (Heinz 1998). Presenting individual females with small- and medium-sized hosts on day one and then medium- and large-sized hosts on day two further supported the observation that relative rather than absolute host size influences sex allocation decisions. Finally, because it would be uneconomic for mass-rearing programs to handle parasitoids on an individual basis, the method was extended to groups of one hundred *C. grandis* females exposed to one of three host size treatments over four consecutive days: (i) successively increasing; (ii) decreasing; or (iii) stable host size distributions. Groups exposed to successively larger hosts exhibited a shift in sex ratio from 0.33 to 0.23 (proportion male), whereas groups exposed to successively smaller hosts produced more male biased sex ratios; at the end of the four-day experiment 0.35 shifted to 0.44 (Heinz 1998). The ability to increase female bias in *C. grandis* sex ratios was calculated to save US$50 per hectare in 1998.

A second example of the use of Charnov's model to improve biological control involves *Diglyphus isaea* (Hymenoptera: Eulophidae), a solitary idiobiont ectoparasitoid of agromyzid leafminers, including *Liriomyza langei*, which are important, highly polyphagous pests in agriculture and horticulture. Sex ratios of *D. isaea* are extremely male biased (up to 0.77, Heimpel & Lundgren 2000). At current prices, biocontrol practitioners pay up to US$95 for 200–250 adult *D. isaea* (which equates to *ca.* US$1.65–2.07 per adult female) (Chow & Heinz 2006), making this an expensive biological control agent relative to many other mass-reared species (van Lenteren et al. 1997) and reducing its adoption by growers as a method of control of leafminer pests in greenhouses. As for *C. grandis*, daughters are laid on larger than average hosts and sons on smaller than average hosts, and differential mortality cannot account for the pattern (Ode & Heinz 2002).

Using a modified temporal weighting rule model (based on Devenport & Devenport 1994), that weights the influence of a given host encounter based on its recency, Ode and Heinz (2002) empirically determined the degree to which female *D. isaea* weight current and past determinations of host size. The empirically determined weighting parameter indicates the threshold value a female is likely to use to determine whether a given host is small or large. If the average host size encountered by females changes more rapidly than females are able to adjust their perception of the population host-size distribution, and if the hosts presented are always larger than the calculated threshold size, females will consider each host to be larger than average and produce a daughter.

Following methods developed for *C. grandis*, individual female *D. isaea* were presented with hosts in one of three treatments over a three-day period: (i) progressively larger hosts; (ii) same size hosts; and (iii) progressively smaller hosts (Ode & Heinz 2002). While daughters were allocated to larger hosts, the relationship between host size and body size did not differ significantly between males and females (Ode & Heinz 2002, Fig. 12.3a). On day one, females in all three treatments encountered intermediate-sized hosts and produced sex ratios of approximately 0.60 (proportion male). By day three, females given progressively larger hosts produced sex ratios of 0.36, whereas females given progressively smaller hosts produced sex ratios of 0.74 and females that received only intermediate-sized hosts consistently produced sex ratios around 0.60 (Ode & Heinz 2002, see Fig. 12.3b). Groups of thirty females exposed to the same three treatments produced similar, albeit less extreme, shifts in sex ratio. After three days, sex ratios were respectively 0.45, 0.73, and 0.63 (Fig. 12.3c).

Chow and Heinz (2005) subsequently observed similar shifts in sex ratio when combinations of large and small hosts were presented to both individual and groups of females. Under simulated mass-rearing conditions over eight consecutive weeks, presentations of a combination of host sizes were found to produce a significantly higher proportion of females compared to presentations consisting of only large hosts. When given only large hosts over two days, ovipositing *D. isaea* females produced significantly more male-biased sex ratios on the second day (Chow & Heinz 2005). This production method was estimated to reduce the production cost per female by approximately 50%. Wasps produced by this modified method provided similar levels of leafminer control (in terms of numbers of dead leafminer larvae (Fig. 12.4a), numbers of unparasitized leafminer pupae (Fig. 12.4b), and number of chrysanthemum flower buds (Fig. 12.4c), in chrysanthemum under simulated commercial production practices compared to control provided by wasps produced in the conventional mass-rearing method when released in similar numbers and sex ratio (Chow & Heinz 2006). This suggests that there is no reduction in quality of wasps produced by this modified method.

This 'host size distribution manipulation technique' clearly has great potential to be applied in many biological control programs involving the mass-rearing of solitary idiobiont parasitoids. Many such parasitoids have been shown to exhibit host-size-dependent sex allocation and, of these, many (but not all) are known to assess host size on a relative basis. Those that do are expected to be prime candidates for manipulating mass-reared sex ratios following methods developed by Heinz and his colleagues. However, we are not aware of any commercial insectaries that have adopted these modified rearing techniques. Even if parasitoid sex allocation behavior does not match exactly that predicted by Charnov's theory, female production may still be significantly enhanced by consideration of host size effects in parasitoid cultures (Lampson et al. 1996, Joyce et al. 2002, Paine et al. 2004, Pandey & Johnson 2005).

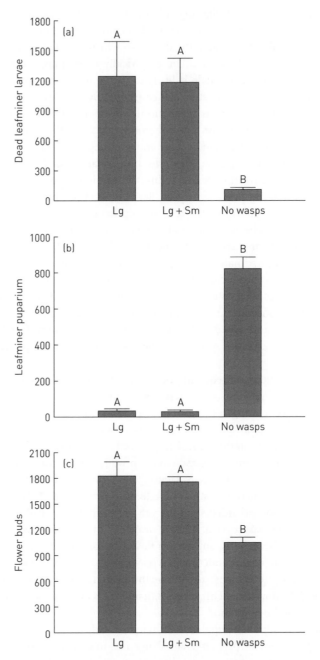

Fig. 12.4 Effect of rearing technique on realized pest control. Numbers (mean + SE) of dead *L. langei* larvae (a), unparasitized *L. langei* pupae (b), and number of flower buds per chrysanthemum plant (c) when exposed to: wasps reared from 8-day-old host (Lg: conventional rearing method), wasps reared from mixtures of 8-day and 6-day-old hosts (Lg + Sm: modified rearing method), or no wasps. Bars with the same letter are not significantly different (p > 0.05: Bonferroni multicomparison test). Reprinted from Figs 4–6 in Chow and Heinz (2006), with permission from Blackwell Publishing.

Host-quality effects on sex ratios are likely to operate after parasitoids have been released and influence biological control success in the field, as in the cassava mealybug (Schulthess et al. 1997, Neuenschwander 2001) and California red scale (Luck & Podoler 1985) examples already mentioned above. In both these biological control systems, host-stage dependent sex allocation has been incorporated into theoretical models aimed at understanding host population regulation. The suite of models explored by Murdoch and co-workers (Section 12.2.2) has been developed with particular reference to California red scale control (Murdoch et al. 2003, 2005, Luck & Nunney 1999). These models treat sex allocation patterns as phenomenological rather than explicitly adaptive and also as just one part of host-stage dependent parasitoid behavior, but their results imply that sex allocation and related host-stage dependent behaviors play an important role in pest population suppression and in the stability of biological control. Similarly, Gutierrez et al. (1993) assume (with empirical support) that the size and age of cassava mealybug hosts affect sex allocation decisions by two exotic biological control agents, A. lopezi and A. diversicornis. Their model predicts that parasitoid sex ratios are indirectly affected by influences of plant growth via their effects on host size and that inter-specific differences in host-size dependent sex allocation contribute greatly to the competitive displacement of E. diversicornis (which has a higher threshold host-size for female production).

As a final example of the potential importance for host-quality effects for biological control, Godfray and Waage (1991) developed a model predicting the biological control success of two encyrtid parasitoids that attack different life-history stages of the mango mealybug, Rastrococcus invadens (Hempitera: Pseudococcidae). They concluded that only one species should be introduced. Subsequent empirical work, however, suggests that deployment of both species may result in the best biological control. The model assumed that parasitoid sex ratios are independent of host-size within the host-stage they attack, but sex ratios of both species are in fact host-size dependent (larger hosts tend to receive female eggs). The incorporation of such patterns into the assumptions of future models may improve the accuracy of predictions (Godfray & Waage 1991, Boavida et al. 1995, Bokonon-Ganta et al. 1995), particularly because the sensitivity analysis carried out by Godfray and Waage (1991) suggests that overall sex ratios of the species considered have a large effect on population equilibria.

We conclude by noting that it is now nearly 26 years since Waage (1982a) highlighted the potential value of formally linking host-quality dependent sex allocation to the dynamics of host and parasitoid populations in the context of biological control. To our knowledge, the above-mentioned models constitute nearly all of the surprisingly few attempts to implement his suggestions (Section 12.2.3).

12.2.3 Differential returns due to structured populations

Hamilton's theory and its extensions
When populations are structured, either spatially or temporally, relatives are more likely to interact with one another than they would if interactions were distributed randomly throughout the population. Maternal sex ratio (MSR) optima are biased when the effects of such interactions are sexually differential. The best known and most studied case is Hamilton's (1967, 1979) theory of LMC, which provides an adaptive explanation for why so many parasitoid populations exhibit female-biased sex ratios (Luck et al. 1993, Godfray 1994). In many parasitoid species, one or more 'foundress' females contribute offspring

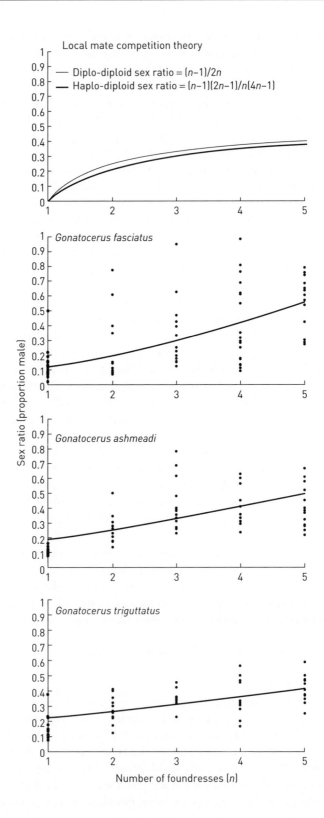

Local mate competition theory

— Diplo-diploid sex ratio = $(n-1)/2n$
— Haplo-diploid sex ratio = $(n-1)(2n-1)/n(4n-1)$

Gonatocerus fasciatus

Gonatocerus ashmeadi

Sex ratio (proportion male)

Gonatocerus triguttatus

Number of foundresses (n)

to a patch, which typically consists of a single host or a clustered group of hosts (see also Chapter 8 by van Alphen and Bernstein). Hamilton assumed that once offspring mature, all mating occurs within the patch. Mated females then disperse to forage for oviposition opportunities in unexploited patches. When only one female contributes offspring to a patch, the evolutionarily stable strategy (Maynard Smith 1982) is to produce only as many sons as are needed to mate with all the daughters. Because males are usually capable of polygyny, female biased sex ratios are predicted (Fig. 12.5). Such female-biased sex ratios reduce the amount of competition between sons for mates. Sib-mating also has the effect of increasing mother-offspring relatedness, which in itself selects for a small degree of female bias. As the number of foundresses increases, the population becomes progressively less structured with regard to mating opportunities (i.e. although all mating occurs locally within the natal patch, the population on the natal patch is large) and the optimal progeny sex ratio for each foundress asymptotically approaches equality (Fig. 12.5). This is because sons increasingly compete with non-siblings rather than siblings, resulting in more equal fitness returns through sons and daughters for each foundress (as assumed under Fisherian sex allocation).

Hamilton's initial LMC model has been modified to incorporate numerous more realistic or more specific assumptions about parasitoid biology, including the consideration of host-quality effects (Section 12.2.2) operating simultaneously with population structure (Werren & Simbolotti 1989, Luck et al. 1993, Godfray 1994, Hardy 1994, Hardy et al. 1998, 2005, for recent developments see Courteau & Lessard 1999, 2000, Abe et al. 2003a, Shuker et al. 2005, Wakano 2005, Gardner et al. 2007, Shuker et al. 2007). One modification is to treat broods as being composed of discrete (and relatively small) numbers of offspring, rather than assuming infinite brood size. This has lead to a number of theoretical advances concerning selection for low sex ratio variance and the effect of developmental mortality on sex allocation strategies (Green et al. 1982, Hardy 1992, Heimpel 1994, Nagelkerke & Hardy 1994, Nagelkerke 1996, Hardy et al. 1998, Krackow et al. 2002). Developmental mortality brings the offspring sex ratio at maturity (the time at which offspring mate) out of complete maternal control, even though mothers may have complete control over the sex of the eggs they lay (Nagelkerke & Hardy 1994), while these factors have no influence on sex allocation optima under Fisherian assumptions (Leigh

Fig. 12.5 (*Opposite*) Sex ratio in response to foundress number. The top panel shows the predictions of Hamilton's (1967) theory of LMC in its original formulation for diplo-diploid genetics and the modified prediction (Hamilton 1979) assuming haplo-dipolid inheritance, which is appropriate for the parasitoid Hymenoptera. Note that haplo-diploidy leads to a slight increase in predicted bias when foundress number is greater than 1. The lower three panels show offspring-group sex ratios in response to foundress number in three egg-parasitic species of *Gonatocerus* (Hymenoptera: Mymaridae): *G. fasciatus* is gregarious, while *G. ashmeadi* and *G. triguttatus* are solitary (quasi-gregarious). Data are derived from Irvin and Hoddle's (2006) study and were re-analyzed using logistic regression, with the scale parameter estimated empirically to account for overdispersion (Wilson & Hardy 2002). Sex ratios increased significantly with increase in foundress number in all 3 species: (*G. fasciatus*: $F_{(1,83)} = 38.93$, $p < 0.001$; *G. ashmeadi*: $F_{(1,57)} = 31.76$, $p < 0.001$; *G. triguttatus*: $F_{(1,54)} = 24.39$, $p < 0.001$). Further, the sex ratio response to foundress number did not differ between the 3 species ($F_{(2,196)} = 0.13$, $p = 0.88$).

1970). Mothers are predicted to adjust the sex ratio at oviposition according to the probability, and distribution across broods and sexes, of pre-adult mortality among their offspring.

A particularly useful modification of LMC theory has been to formally consider mating structures that are between the extremes of strictly local and fully panmictic, known as 'partial local mating' (i.e. considering that some matings can occur after dispersal from the natal patch). As might be intuitively expected, sex ratio optima are generally less female biased when local mating is less common because populations become increasingly Fisherian (Luck et al. 1993, Hardy 1994, Hardy et al. 2005, Courteau & Lessard 2000, Gardner et al. 2007).

Population level consequences of LMC

A different way of extending LMC theory has been to incorporate it as an assumption into models of host–parasitoid population dynamics. Identifying factors that promote stability or persistence, and hence pest population regulation, in host–parasitoid systems has been a major preoccupation of population ecologists since the work of Nicholson and Bailey (1935) (Bernstein 2000, Hassell 2000). Several formal models have been developed following suggestions by Waage (1982a) that density-dependent sex ratios, with LMC as one of several possible density-dependent mechanisms, might act to stabilize population dynamics. Models by Hassell et al. (1983) and Comins and Wellings (1985) both predicted that sex ratio shifts in response to changes in host density can have a strong stabilizing effect on host–parasitoid dynamics. However, these studies did not distinguish the effects of LMC from other density-dependent influences on the sex ratio, such as differential mortality (Hassell 2000, Meunier & Bernstein 2002). As a result, sex ratio is free to vary beyond the bounds predicted under LMC (e.g. sex ratios become male-biased when host densities become high). Meunier and Bernstein (2002) modeled explicitly the influence of LMC on the stability of host–parasitoid population dynamics. Their models predict that, while LMC can contribute to the stability of host–parasitoid systems (along with host density-dependent effects on larval mortality, host size, and parasitoid aggregation), LMC will not contribute to stability as much as previously thought.

Evidence

Many parasitoid species have female biased sex ratios (Fig. 12.1) and a large number of laboratory and field studies have provided support that LMC conditions can result in female bias (Hamilton 1967, Godfray 1994). However, because LMC is the most obvious explanation for the production of female-biased parasitoid sex ratios, it has often been assumed to occur without accompanying empirical observation (Orzack 1993, 2002). It is an exemplary caution that female bias has been observed in gregarious parasitoids in which sibling mating has been shown to be avoided (Ode et al. 1995, Heimpel 1997, Gu & Dorn 2003). More generally, it is increasingly recognized that the mating systems of many parasitoid species are intermediate between panmixis and strict LMC (Hardy 1994, Godfray & Cook 1997, Hardy et al. 2005), together with an accumulation of empirical support for predicted sex allocation patterns under partial-LMC in parasitoids and other species with analogous life-histories (Hardy & Mayhew 1998, Drapeau & Werren 1999, Fellowes et al. 1999, Peer & Taborsky 2004). Many of the predictions of other modifications of basic LMC theory are also supported empirically. For instance, low sex ratio variances are commonly observed and developmental mortality appears to influence sex allocation decisions in at least some species (Hardy 1992, Hardy et al. 1998). Current theoretically-oriented empirical work on parasitoid sex ratios under LMC includes the use of sex allocation behavior to address general questions in behavior and evolution, such as how individuals process information about their environment and whether they recognize kin or the presence of unmated females (Ode et al. 1997, West & Herre 2002, King & D'Souza 2004, Reece et al. 2004, Shuker & West 2004, Shuker et al. 2006a) and the influence of male-male competition on sex allocation strategies (Abe et al. 2003b, 2005).

Biological control relevance
The influence of spatially-structured populations, and LMC in particular, on population sex ratios is of great potential importance for biological control programs because beneficial strong female biases can occur. Despite this, we know of only one explicit attempt to marry the understanding provided by LMC theory with sex ratio manipulation of parasitoids for use in biological control (Irvin & Hoddle 2006).

Since the major prediction of LMC models is that sex ratios will be more female biased when fewer females contribute offspring to a patch, the most obvious possibility is to alter mass-rearing programs such that females encounter host patches in isolation rather than in the presence of other females. Irvin and Hoddle (2006) studied sex ratios of three mymarid species, *Gonatocerus ashmeadi*, *G. triguttatus*, and *G. fasciatus* (Mymaridae), egg parasitoids of the glassy-winged sharpshooter (Hemiptera), which is a pest of numerous crop plants in the USA. *G. ashmeadi* and *G. triguttatus* are solitary (i.e. quasi-gregarious) parasitoids, while *G. fasciatus* is gregarious. Overall, all three species have female-biased sex ratios (proportion male of about 0.32), but Irvin and Hoddle found that the sex ratio produced depends strongly on the number of females ovipositing on a patch of host in a manner generally consistent with the predictions of LMC theory (Fig. 12.5). Further, consistent with scenarios assumed by LMC theory, male protandry (males emerge before females) followed by local (on patch) mating has also been observed in all three species (Triapitsyn et al. 2003, Irvin & Hoddle 2006). These results indicate that the efficiency of mass rearing facilities can be significantly improved by minimizing contact between females presented with hosts. Estimated economic costs of inefficiently rearing these parasitoids under conditions that promoted male-biased sex ratios increased the cost per female from about US$1.42 to US$4.51. Irvin and Hoddle (2006) further showed that sex allocation behavior is unaffected by a female's experience of conspecifics prior to presentation with hosts. This has the practical advantage that there is no need to isolate females from each other in holding cages or during transport to the site of field release, provided they are separated or held at very low density once in the presence of hosts.

A further use of LMC ideas to increase female production of gregarious parasitoids in culture is to increase the size of the host presented to each female. Increasing host size will generally increase the absolute number of female offspring produced (since more eggs are generally laid on larger hosts, Godfray 1994) but also the proportion of females increases among the offspring. For instance, many species of bethylid wasps (parasitoids of a variety of important lepidopteran and coleopteran pest of crops and stored products) are gregarious and, generally, only one female oviposits on each host (Griffiths & Godfray 1988). The sex ratios of emergent broods tend to be distinctly female biased, as would be predicted by LMC theory for the single foundress case. Bethylid broods are typically relatively small (less than 20 individuals) with larger broods produced on larger hosts. Theory assuming discrete brood (or clutch) sizes (Section 12.2.3) predicts that sex ratio

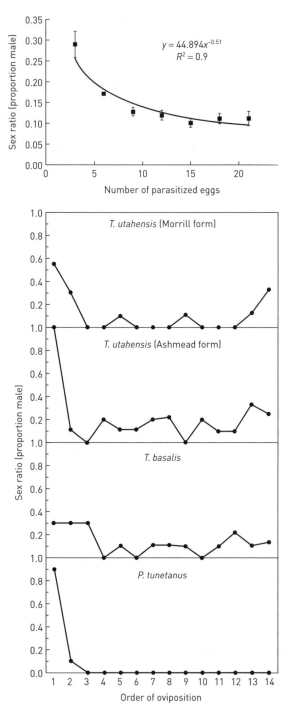

Fig. 12.6 Sex ratio in response to host number in quasi-gregarious parasitoids. The top panel shows the sex ratios (means and standard errors) produced by individual females of the mymarid wasp *G. ashmeadi* presented with egg masses of different sizes and the fitted line from regression analysis. Reprinted from Fig. 4 of Irvin and Hoddle (2006), with permission from Elsevier. The lower four panels show mean sex ratios of progeny emerging from each of 14 host eggs presented to individual females of the scelionids *Psix tunetanus* and *Trissolcus* spp. Reprinted from Fig. 1 of Weber et al. (1996), with permission from Elsevier.

should become more female biased as brood sizes increases, because one or only a few males are needed to mate with all females in a brood (Griffiths & Godfray 1988). Such patterns are clearly observed across species (Griffiths & Godfray 1988, Hardy & Mayhew 1998) and within some of the bethylid species examined (for a discussion of theoretically-based explanations for the absence of this trend, see Hardy et al. 1998, 2000). Laboratory cultures of the gregarious eulophid *Colpoclypeus florus*, a parasitoid of tortricid leaf rollers, also show increased female bias when larger hosts are provided (Hardy et al. 1998). By the same logic, increasing the number of hosts in clusters presented to individual quasi-gregarious parasitoids may also increase female bias, as observed in *G. ashmeadi* (Irvin & Hoddle 2006, Weber et al. 1996, see Fig. 12.6) but the opposite effect may also be observed if there is, for the species in culture, a natural positive correlation between the number of hosts in a patch and foundress number (a cross-species example is provided by Waage 1982b).

The above cases illustrate that relatively straightforward explorations of the major influences on sex ratio predicted by LMC theory, foundress number, and host size, can lead directly to practical recommendations to increase female production in parasitoid cultures, which should be adopted, provided that any changes to current practice are cost effective (Luck et al. 1999, Irvin & Hoddle 2006).

Finally, we know of no biological control-oriented studies that have explored potential relationships between LMC and host–parasitoid population dynamics. The nearest example is provided by Reeve and Murdoch (1986), who assessed the potential that density-dependence on sex ratios (Section 12.2.3) could account for the successful and stable biological control of California red scale by *A. melinus*. However, these authors considered that local mating in *A. melinus* is unlikely as the males have high dispersal capabilities. They also found no effect of density on parasitoid sex ratio and concluded that density-dependent sex ratios thus cannot explain the success of this biological control agent (Murdoch et al. 2005).

12.3 Sex ratios under compromised maternal control

In the preceding sections of this chapter we have discussed sex ratios in the context of theoretical models that take the 'behavioral ecology approach' of assuming that mothers are in control of sex allocation, and the sex ratios of their progeny thus reflect maternal optima in terms of maximizing fitness. These issues form the major focus of our chapter. However, there are factors that bring parasitoid sex ratios out of complete maternal control. Such factors are largely genetic, and there is a similarly well developed, and complementary, 'genetic approach' to sex ratio research, much of which has relevance for biological control programs (Luck et al. 1993, Godfray & Cook 1997, Hunter 1999, Luck et al. 1999, Cook 2002, Stouthamer et al. 2002, Majerus 2003, Beukeboom & Zwaan 2005, Burt & Trivers 2006). Here we first provide a brief background about what is known regarding genetic variation for the sex ratio in parasitoids, as genetic variation provides the opportunity for sex ratio to respond to selection (Section 12.3.1). We then discuss sex determination mechanisms (Section 12.3.2), with particular emphasis on complementary sex determination (CSD), and their impact on biological control programs. Finally, we discuss a variety of selfish genetic elements, such as extrachromosomal elements and endo-symbionts and their effects on sex allocation patterns as well as their implication for biocontrol programs (Section 12.3.3).

12.3.1 Genetic variation for the sex ratio

Sex allocation theory presumes that natural selection results in changes in the sex ratio, defined as the proportion of offspring that are male. If sex ratio responds to selection, genetic variation for the sex ratio must exist in the population. Surprisingly, however, few studies have sought to demonstrate genetic variance for the sex ratio in parasitoids. However, additive genetic variation in sex ratio has been shown for the solitary trichogrammatid wasp *Uscana semifumipennis* (Henter 2004), and a series of studies on the gregarious ptero-malid wasp *Nasonia vitripennis* has shown sex ratio differences among isofemale lines and changes in sex ratio among inbred lines during artificial selection experiments (Parker & Orzack 1985, Orzack 1990, Orzack & Parker 1990). Sex allocation effects of (genetic) iden-tity of the male that the female mated with have also been found in both of these species (Henter 2004, Shuker et al. 2006b).

While sex allocation theory often assumes that selection acts on the sex ratio itself, this is not necessarily the case as sex ratio is a complex trait. Although sex ratio is straightforward to measure (simply count the numbers of male and female offspring), it is frequently correlated with factors such as fecundity or longevity and a variety of envi-ronmental conditions including host quality and the presence of other ovipositing females (Godfray 1994, Henter 2003, 2004). Selection on a correlated trait may result in indirect selection on the sex ratio. For instance, genetic variation for the sex ratio may be limited by the amount of variation present for fecundity. Realized fecundity (the number of eggs laid) is often correlated with sex ratio (Henter 2004), particularly in those species that lay sons and daughters in a stereotypic sequence and/or that tend to produce an almost invariant number of males in each offspring batch (Hardy 1992, Hardy et al. 1998). In a pair of studies, Antolin (1992a,b) used reaction norms and diallel crosses to examine genetic variation in several strains of the pteromalid *Muscidifurax raptor*. While limited among-strain variation exists for sex ratio, sex ratio is correlated strongly with fecundity (Antolin 1992b, Henter 2004). Selection on fecundity may constrain selection on the sex ratio, given that selection on sex ratio is frequency dependent, whereas selection on fecundity is direc-tional (Antolin 1992b).

Because sex ratio is a straightforwardly measured trait, many biological control practi-tioners have been tempted to manipulate parasitoid sex ratios, via artificial selection, in an effort to enhance the efficacy of biological control agents. Indeed, artificial selection for improving a variety of natural enemy traits has been attempted (Whitten & Hoy 1999, Wajnberg 2004). While sex ratio is straightforward to measure, the responses of sex ratio to selection experiments may not be easily predictable given the potential correlations described above. Attempts to alter parasitoid sex ratios in an effort to enhance the qual-ity of biological control agents have met with mixed success (Simmonds 1947, Wilkes 1947, Ram & Sharma 1977). Correlated factors such as fecundity, response to the presence of other females, and even the sex determination mechanism (Section 12.3.2) may all act as constraints on the response to artificial selection on the sex ratio. Instead, a more fruitful approach may involve developing an understanding of the mechanisms responsible for sex allocation patterns and manipulating the responsible mechanisms to alter the sex ratio.

12.3.2 Sex determination mechanisms

From the perspective of behavioral ecology, the sex determination mechanism of a stud-ied parasitoid species may be of interest if it influences or constrains aspects such as sex allocation optima and mating decisions. From the perspective of biological control, sex determination mechanisms are of interest if they affect factors such as mass rearing efficiency or post-release population establishment and dynamics. Detailed discussions of parasitoid sex determination mechanisms have been provided elsewhere (Cook 1993a, 2002, Beukeboom 1995, Cook & Crozier 1995, Godfray & Cook 1997, Dobson & Tanouye 1998, Luck et al. 1999, Majerus 2003, Beukeboom & Zwaan 2005. Here we restrict ourselves to a brief overview of mechanisms, and the degree of control of sex that they provide, followed by a more detailed discussion of one mechanism, CSD, that has been the most explored with regard to its consequences for sex allocation optima and for biological control.

Parasitoid wasps are haplo-diploid. With some exceptions, females develop from fertilized (diploid) eggs and males from unfertilized (haploid) eggs. An understanding of both the broad relationship between ploidy and gender and the exceptions requires an understanding of the genetic mechanism of sex (gender) determination. However, there are several sex determination mechanisms operating within the parasitoid Hymenoptera. Some of these are relatively well understood and others are only beginning to be explored. Almost irrespective of the particular mechanism by which haploids become males and diploids become females, it has commonly been assumed that parasitoid wasps have a higher degree of control over the sex of their progeny than organisms with, for example, diplo-diploid (chromosomal) sex determination mechanisms (because a mated female can control progeny gender by controlling the act of egg fertilization). This view is challenged by particular case studies showing remarkable sex ratio adjustment in several higher vertebrate species, such as the Seychelles warbler (*Acrocephalus sechellensis*, Komdeur 1996), and by meta-analyses examining sex ratio patterns across wide ranges of taxa with different sex determination mechanisms (West & Sheldon 2002, West et al. 2005). While vertebrates appear to have generally less sex ratio control than invertebrates, there is no clear difference in sex ratio adjustment between haplo-diploids and invertebrate species with supposedly constraining sex determination mechanisms (West et al. 2005). Further, in a meta-analysis focused on differential returns due to host quality (Section 12.2.2) and its avian analogs, West and Sheldon (2002) found that environmental predictability is more important than taxon (i.e. sex determination mechanism) in explaining the degree of sex ratio adjustment achieved by parasitoids and birds.

Mechanism gallery

In haplo-diploids, there are no heteromorphic sex chromosomes, such as the mammalian X and Y. Males and females have chromosomes in the same proportions and the only difference is the number of chromosome sets (Cook 1993a, Beukeboom & Kamping 2006). Five different mechanisms for parasitoid sex determination have been suggested, none of these operate in all species studied, and most are not supported by current evidence, with further modifications required to fully explain observed gender patterns (Dobson & Tanouye 1998, Beukeboom & Kamping 2006).

Under fertilization sex determination (FSD), sex is determined by the fertilization event rather than ploidy. Although FSD was proposed following studies of a polyploidy strain of *Nasonia vitripennis*, more recent studies using this species have excluded FSD as a candidate mechanism (Dobson & Tanouye 1998, Beukeboom & Kamping 2006). Another hypothesized mechanism is genic balance sex determination (GBSD) under which sex is determined by a set of dosage-compensated maleness genes and a set of dosage-dependent femaleness genes. The effects of maleness genes are the same in haploid and diploid offspring, while the femaleness genes have twice the effect in diploids than in haploids.

If maleness gene effects outweighed femaleness effects in haploids but not diploids, haploids would develop as males and diploid as females. However, there is no good evidence that GBSD operates and it can be excluded in several dozen hymenopteran species in which diploid males have been observed (Cook 1993a, Dobson & Tanouye 1998). Another balance-based hypothesis is maternal effect sex determination (MESD). Under MESD, a maternally derived cytoplasmic component leads to male development in haploids but is outweighed by feminizing nuclear components (loci) in diploids. Like GBSD, MESD is shown not to operate in some species by the presence of diploid males (Dobson & Tanouye 1998).

The most recent proposal is genomic imprinting sex determination (GISD) (Beukeboom 1995). Under GISD, one or more loci are differentially imprinted in maternal and paternal development. Unfertilized eggs contain only maternally derived and imprinted loci, which are unable to bind with, or activate, an unknown 'product', and develop into males. Fertilized eggs contain paternally imprinted loci, which bind with and activate the product, resulting in female development. While GISD has been supported as the mechanism operating in *Nasonia vitripennis* (Dobson & Tanouye 1998, Trent et al. 2006), the most recent evidence shows that paternally-derived loci are not required for female development and thus that, in *N. vitripennis* at least, sex determination operates by a yet more complex mechanism (Beukeboom & Kamping 2006). The remaining suggested mechanism is CSD, which is considered below.

Complementary sex determination

Under the simplest form of CSD, sex is determined by a single genetic locus with multiple alleles (single-locus-CSD or sl-CSD). Unfertilized eggs develop into haploid males (hemizygous at the sex determination locus), whereas fertilized eggs develop into either females or males, depending on the zygosity at the sex locus. Individuals heterozygous at the sex locus develop into females, whereas homozygous individuals develop into diploid males. CSD differs from the sex determination mechanisms discussed above in that there is now convincing evidence that it operates in many species (Beye et al. 2003). Indeed, since its first description in *Habrobracon hebetor* (Whiting 1943), sl-CSD and/or the production of diploid males has been documented in at least 40 species of Hymenoptera, and appears to be the ancestral sex determination mechanism of the Hymenoptera (Cook 1993a, Periquet et al. 1993, Cook & Crozier 1995, Luck et al. 1999, Butcher et al. 2000a,b, Noda 2000, Beukeboom 2001, Wu et al. 2003, Beukeboom & Zwaan 2005, Zhou et al. 2006, de Boer et al. 2007). However, the sex determination mechanism must be evolutionarily labile, as within families, and even in genera, CSD may be present in one species and absent in another (Beukeboom et al. 2000, Niyibigira et al. 2004a,b, Wu et al. 2005, Zhou et al. 2006, de Boer et al. 2007), and appears to be absent from the Chalcidoidea (Butcher et al. 2000a, Niyibigira et al. 2004a,b, Beukeboom & Kamping 2006).

Single-locus CSD may represent a special case of multi-locus-CSD (ml-CSD) where heterozygosity at any one of several loci involved in sex determination is sufficient to result in female development (Cook 1993a, Cook & Crozier 1995). While the existence of ml-CSD has been conjectured for some time (Crozier 1977), its detection may be problematic (Cook 1993a) and it has been tested for in only a few species. For instance, ml-CSD and sl-CSD have been ruled out in the bethylid wasp *Goniozus nephantidis* (Cook 1993b), and six species of *Diadegma* (Ichneumonidae) have been shown to have sl-CSD and not two-locus-CSD (tl-CSD) or ml-CSD (Butcher et al. 2000a,b, Niyibigira et al. 2004a,b). However, there have been unconfirmed suggestions of tl-CSD in several species (Luck et al. 1999) and recent work with the braconid *Cotesia vestalis* (= *C. plutellae*) provides sex ratio and diploid male production data consistent with tl-CSD (de Boer et al. 2007).

In some species possessing CSD, the majority of diploid males fail to mature into viable adults. In other species, diploid males develop with normal survivorship and produce unreduced, diploid sperm (Whiting 1943, Elagoze & Periquet 1993, Holloway et al. 1999, Luck et al. 1999, Cowan & Stahlhut 2004, Niyibigira 2004a, Zhou et al. 2006). If diploid males survive and mate, there are potential fitness consequences in terms of, for instance, the ploidy of their offspring. However, these fitness costs may be absent in some species (Cowan & Stahlhut 2004, Wu et al. 2005). Under CSD, the production of diploid males increases with inbreeding, since inbreeding increases the levels of homozygosity at all loci including the sex locus. Diploid male production is also considerably less likely under ml-CSD than under sl-CSD. Importantly, in species possessing CSD, females lose a degree of control over their sex allocation decisions (Cook & Crozier 1995, Ode et al. 1995, Godfray & Cook 1997). The degree of loss depends on a variety of factors, including the allelic diversity at the sex locus (or loci) in the population, the level of inbreeding, the survivorship of diploid males, and the likelihood of remating or multiple mating by females.

CSD and biological control

One outcome of diploid male production under CSD is the concomitant reduction in the number of females that are produced (each diploid male would have been a female had the sperm and egg not borne the same allele at the sex locus). In a matched mating, where the father shares one of the mother's two alleles at the sex determination locus, half of the fertilized eggs on average will become diploid males. From a biological control standpoint, mass-rearing techniques that inadvertently promote inbreeding may result in the production of excessive levels of diploid males among parasitoids with CSD. Indeed, laboratory cultures of many braconid and ichneumonids are often male-biased (Platner & Oatman 1972, Rappaport & Page 1985, Smith et al. 1990, Grinsberg & Wallner 1991, Johns & Whitehouse 2004) suggesting the production of diploid males under conditions of inbreeding (as, for instance, found by Noda 2000). Further, biocontrol programs using parasitoids in the families Ichneumonidae and Braconidae are approximately twice as likely to fail to establish compared to programs using wasps from the Chalcidoidea (Stouthamer et al. 1992) and it is therefore encouraging when it is established that a biological control agent in one of these families does not have CSD or male-biased sex ratios in mass culture (Niyibigira 2004a).

For those programs deploying parasitoids with, or suspected of having, CSD, an important consideration is the field-collection of sufficient numbers of individuals during exploration so as to collect as many alleles as possible from the original field population. Estimates of allelic diversity at the sex locus in natural populations range from 3 to 86, and may vary within a species' geographic range (Cook & Crozier 1995, Butcher et al. 2000a,b, Wu et al. 2003). Mass-rearing techniques in insectaries are another potential cause of allele loss that would increase the rate at which diploid males are produced (Fig. 12.7). Two recommendations have been put forward to reduce the loss of allelic diversity at the sex locus (loci) under mass rearing conditions (Stouthamer et al. 1992, Cook 1993c). First, the mass-reared parasitoids can be maintained as a single large population to minimize the rate at which alleles are lost (Stouthamer et al. 1992). However this approach is

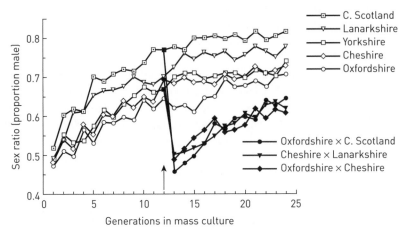

Fig. 12.7 Sex ratio changes under CSD during mass rearing. The sex ratio of lines of the solitary parasitoid *Diadegma fabricianae* (Hymenoptera: Ichneumonidae), which has single-locus CSD, initiated with around 60 adults from each of 5 sites around the UK, became progressively male biased due to inbreeding, and female fecundity also decreased. Sex ratio and fecundity were both restored by outbreeding (crosses between lines) at generation 12 (shown by arrow) but continued inbreeding again lead to sex ratio changes. Sex ratios are shown as mean proportion males across 20 cultures per line. Reprinted, with modification, from Fig. 1A from Butcher et al. (2000), with permission from Blackwell Publishing.

susceptible to periodic fluctuations in population sizes (and consequent allele loss) as a result of difficulties in colony maintenance (Cook 1993c).

The other approach is to maintain multiple, inbred lines (Cook 1993c). Each inbred line will contain a minimum of two alleles. As long as each line produces 10 fertilized eggs each generation, the probability of an individual line (and the 2 alleles it carries) going extinct is less than 1 in 1000. In species possessing sl-CSD where females mate only once (e.g. *Habrobracon hebetor*, Ode et al. 1995), females that mate with diploid males are constrained to produce only sons. Even in cases where diploid male survival is high (e.g. *H.* sp. near *hebetor*, *H. serinopae*, *H. brevicornis*, Holloway et al. 1999), a modest increase in the number of females produced per line per generation will ensure the survival of these inbred lines. For instance, *H. hebetor* females fertilize approximately two-thirds of the eggs they lay. Therefore, a female has a 50% chance of mating with a diploid male (provided that diploid males have survival rates similar to haploid males, although this happens not to be the case with *H. hebetor* (Petters & Mettus 1980)). Prior to release in the field for biological control, offspring from all the separately maintained lines can be combined to reconstitute a large culture. While each technique has its disadvantages, perhaps the best strategy would be to employ both approaches (Cook 1993c) and a further possibility is to periodically augment laboratory cultures with wild caught individuals (Johns & Whitehouse 2004).

Inbreeding under non-CSD mechanisms

We conclude our discussion of sex determination mechanisms by noting that, while the potential effects of inbreeding in species with CSD are well recognized (i.e. diploid male production, Sections 12.3.2 and 12.3.2, Butcher et al. 2000a, see Fig. 12.7), it has generally been assumed that, due to the purging of genetic load in haploid males, there will be little inbreeding depression in parasitoids with other sex determination mechanisms (Section 12.3.2, Werren 1993, Butcher et al. 2000a, Zayed & Packer 2005). However, substantial inbreeding depression has been observed in the normally outbreeding trichogrammatid wasp *Uscana semifumipennis* (Henter 2003) and meta-analysis has found no difference in (experimentally imposed) inbreeding depression between gregarious (likely to have a history of inbreeding) and solitary (generally outbreeding) haplodiploid species, with known CSD-species excluded (Henter 2003). While haplodiploid species can clearly suffer inbreeding depression, the severity of inbreeding depression is greater in diploid species (Henter 2003).

12.3.3 Sex ratio distorters

Another important way that ovipositing female hymenopterans do not have complete control over sex allocation decisions is through the influence of sex ratio distorters. For more detailed accounts of sex ratio distorters than we can provide here, we direct the reader to one of the many available reviews (Godfray 1994, O'Neill et al. 1997, Werren 1997, Hunter 1999, Luck et al. 1999, Stouthamer et al. 1999, 2002, Majerus 2003, Butcher & Jervis [Section 6.5 in Sunderland et al. 2005], Burt & Trivers 2006). As a rule, sex ratio distorters bias the sex ratio in favor of the sex through which they are transmitted. Most increase the proportion of females in offspring of infected individuals because these factors are transmitted through the cytoplasm of the egg.

Gallery of sex ratio distorters
One sex ratio distorter is 'paternal sex ratio' (PSR), a supernumerary (or B) chromosome (Werren et al. 1987, Nur et al. 1988). Like other supernumerary chromosomes, PSR generally does not survive meiotic reduction divisions. Instead, PSR is transmitted through haploid sperm, which do not undergo meiosis. After fertilization, PSR destroys the paternal genome leaving only the maternal genome and resulting in the production of a son. PSR has been found in *Nasonia vitripennis* (Werren et al. 1987), *Encarsia formosa* (Hunter et al. 1993), and *Trichogramma kaykai* (Werren & Stouthamer 2003). In contrast, MSR is a cytoplasmically inherited sex ratio distorter (Skinner 1982). Females infected with MSR produce extremely female-biased sex ratios by fertilizing all of their eggs. MSR is predicted to spread through the population, ultimately driving it to extinction. In populations of *N. vitripennis*, where both PSR and MSR are present, PSR can prevent the fixation of MSR, at least in highly subdivided populations (Werren & Beukeboom 1993).

Many arthropods, including many parasitic wasps and coccinellid beetles, harbor symbiotic bacteria from a wide variety of bacterial families, of which *Wolbachia* (α-proteobacteria) is the best studied genus (Hurst et al. 1997, Werren 1997). *Wolbachia* are known from at least 70 species of parasitic wasps (Cook & Butcher 1999). It is estimated that 18–30% of all insect species may be infected with *Wolbachia* (Cook & Butcher 1999). These symbiotic bacteria influence sex allocation patterns in a variety of ways including (Beukeboom & Zwaan 2005):

1 killing male offspring;
2 inducing cytoplasmic incompatibility (CI), resulting in reproductive isolation under some conditions;
3 inducing thelytoky (parthenogenesis induction, PI); and
4 feminizing males.

Male-killing bacteria are known to occur in at least 19 species across 5 insect orders (Hurst et al. 1997) and are particularly common among aphidophagous coccinellids (Majerus & Hurst 1997). The mechanism by which these bacteria differentially kill males is not known. In *Nasonia vitripennis*, the bacterium *Arsenophonus nasoniae* (Enterobacteriaceae) results in the death of unfertilized eggs of infected females. Further, if superparasitism occurs, the male offspring of uninfected females can also be killed, meaning that both vertical and horizontal transmission can occur (Skinner 1985). Male-killing results in a decrease in the number of offspring successfully produced by an infected female. An adaptive advantage to male-killing is open to debate, although male-killing may be advantageous to the bacterium if the fitness of remaining females is enhanced through increased access to resources in the form of male eggs that are consumed, decreased rates of inbreeding, or decreased competition for food (Hurst et al. 1997).

CI is well known from a wide variety of insects including mosquitoes (*Culex pipiens*), *Drosophila* spp. *Tribolium confusum*, *Nasonia vitripennis*, and others (Hoffmann & Turelli 1997). In failed crosses, the paternal set of chromosomes fails to condense and is lost. In many species, incompatibility is unidirectional, that is, an infected male is unable to successfully fertilize an uninfected female, although the reverse cross is successful. Although much rarer (Breeuwer & Werren 1990), bi-directional incompatibility has been observed between different populations of *N. vitripennis*, each containing different strains of *Wolbachia*. Other *N. vitripennis* populations containing both *Wolbachia* strains were incompatible with populations containing single strains (Perrot-Minnot et al. 1996). In diploid species, incompatibility results in the production of no offspring whereas in haplodiploid species only sons are produced.

Wolbachia are associated with parthenogenesis induction in at least 33 species of parasitic Hymenoptera from 7 families, especially wasps in the genus *Trichogramma* (Stouthamer 1997). Thelytoky, the parthenogenetic production of daughters, is considerably more common than the number of confirmed *Wolbachia* cases, raising the possibility that *Wolbachia* remains to be detected in many more thelytokous species (Stouthamer et al. 1990). *Wolbachia* prevents the segregation of the two sets of chromosomes during the first meiotic division, restoring diploidy by gamete duplication (Stouthamer & Kazmer 1994). Gamete duplication is expected to be incompatible with CSD. Indeed, none of the nine thelytokous species in the Ichneumonoidea tested for *Wolbachia* tested positive (Butcher cited in Cook & Butcher 1999).

Finally, feminization by *Wolbachia* is known from crustaceans (*Armadillidium* spp., Rigaud 1997) and one lepidopteran species (*Ostrinia furnacalis*, Kageyama et al. 1998). Interestingly, females in these species are heterogametic.

Sex ratio distorters and biological control

Wolbachia-induced effects, such as CI and thelytoky, are common in many parasitoids important in biological control (Pijls et al. 1996, Mochiah et al. 2002, Rincon et al. 2006), most notably several species of *Trichogramma* (Silva et al. 2000, Lundgren & Heimpel 2002, Gonçlaves et al. 2006). In many ways, the advantages and disadvantages of using thelytokous organisms as biological control agents are the same arguments regarding the costs and benefits of sexual versus asexual reproduction (Williams 1975, Maynard Smith 1978, Michod & Levin 1988). Sexual reproduction has two distinct disadvantages compared to asexual reproduction with only females. First, if there are any advantageous combinations of genes at different loci, sexual recombination will often destroy these combinations. Second, and perhaps more important, is that, all else being equal, asexual populations will have a rate of increase twice that of sexual populations. Countering this is the fact that sexual populations produce genetically variable offspring, which are better able to adapt to novel environments. Sexual populations are also able to generate mutation-free genotypes through recombination.

Thelytokous biological control agents are, in principle, able to produce more females than sexual lines. When population densities are low, males and females may have difficulty in locating each other. Asexual lines obviously do not have this constraint (Stouthamer 1993). Of course, these putative attributes of thelytoky in biological control depend on the assumption that the fecundity of thelytokous females is not reduced by *Wolbachia*. This is clearly not the case in some, and possibly most, thelytokous species infected with *Wolbachia*. Laboratory comparisons of females from thelytokous (infected) and arrhenotokous (uninfected) lines of *Trichogramma cordubensis* and *T. deion* have shown that thelytokous females had lower fecundity and lower dispersal ability, suggesting that *Wolbachia* infection bears a fitness cost (Silva et al. 2000). However, the advantage of producing only females appears to outweigh the per capita fitness cost of being infected with *Wolbachia*. In greenhouse experiments where sex ratio was taken into account, despite arrhenotokous females still parasitizing more eggs than their thelytokous counterparts, a release of 100 arrhenotokous wasps resulted in a lower overall parasitism rate than the release of 100 thelytokous wasps. Both Aeschlimann (1990) and Stouthamer (1993) suggest using both sexual and asexual strains simultaneously, where possible, to take advantage of the different attributes of sexual and asexual lines. Some researchers have suggested taking *Wolbachia* or the eubacteria, *Rickettsia*, from thelytokous species and transfecting other confamilial parasitoids in an effort to improve biocontrol effectiveness (Tagami et al. 2006).

12.4 Conclusions

As noted above, and in numerous reviews, the study of sex ratios has had a central importance in behavioral evolutionary ecology, in that it has provided some of the clearest examples of adaptation (Godfray 1994, West et al. 2000, Hardy 2002). One of the most remarkable features of the field of sex ratio biology is the depth to which theory and empirical support (both phenotypically and genetically based) have been developed, assisted by their strong mutual interplay. Further, it has long been acknowledged that sex ratio biology has important applications in terms of parasite and disease epidemiology (Read et al. 2002), conservation biology (e.g. for an excellent example of how Trivers–Willard theory has been used to effectively manage the critically endangered parrot, the kakapo (Clout et al. 2001, Tella 2001, Robertson et al. 2006)), and especially biological pest control. However, we consider that meaningful application of sex ratio theory and empirical findings to biological control has largely fallen short of its potential.

Encouragingly, consideration of 'genetic' mechanisms, particularly the effects of CSD and infection by *Wolbachia*, for biological control seems to be increasing. Less positive is that the incorporation of sex ratio responses into host–parasitoid population dynamic theory is in its infancy and has not yielded many results readily or reliably applicable by biocontrol practitioners. This is despite the passage of a quarter century since Waage's (1982a) call for developments in this area. Our primary focus has, however, been on the potential application of the 'behavioral ecology' based understanding of sex allocation. Here again, we are forced to bemoan the state of progress. For instance, biological control practitioners and researchers have long been aware of the relationship between host size and sex ratio. Indeed, many of the early examples used by theory were derived from biological control studies. Yet, the vast majority of the multitude of biological control studies that have assessed host-size effects on sex ratio with a view toward improvements in mass rearing rarely mention the underlying Charnov host quality models (which have been known to parasitoid biologists for several decades), let alone attempt to use an understanding of theory to modify sex ratios. While there are few important exceptions to this (e.g. the work by Heinz and colleagues during the last decade), these approaches have yet to be adopted by biocontrol practitioners. The approaches described here can be potentially used for any solitary parasitoid biocontrol agent that exhibits host-size dependent sex allocation. Likewise, LMC can result in extremely female-biased, population-wide sex ratios, which can be potentially manipulated for the purposes of biological control. A theoretical understanding of this phenomenon has been available to parasitoid biologists for nearly 40 years (Hamilton 1967) and is one of the best known theories in the whole field of evolutionary ecology. Irvin and Hoddle's (2006) application of LMC theory to manipulate sex ratios is one study that shows the plausibility of this approach for biological control. Yet, again, this technique has yet to be adopted in biocontrol programs. This may be that Irvin and Hoddle's (2006) work is very new, but the history of the adoption of Heinz and colleagues work does not give us any great hope. The situation is frustrating because simple, relatively inexpensive techniques, well-grounded in sex ratio theory, have been shown to result in significantly improved sex ratios, which in turn should reduce production costs making biological control an attractive alternative to chemical control.

Acknowledgments

We contributed approximately equally to this chapter: order of authorship was determined by closest proximity to an unbiased progeny sex ratio. We thank Éric Wajnberg, Carlos Bernstein, and Jacques van Alphen for inviting this contribution, and our many colleagues in sex ratio and/or biocontrol research for the part they have wittingly or unwittingly played. We thank Nic Irvin, Mark Hoddle, and Peter Mayhew for providing raw data for some of the graphs. This work was partially supported by National Science Foundation (US) IBN 0344665.

References

Abe, J., Kamimura, Y., Ito, H., Matsuda, H. and Shimada, M. (2003a) Local mate competition with lethal male combat: effects of competitive asymmetry and information availability on a sex ratio game. *Journal of Evolutionary Biology* **16**: 607–13.

Abe, J., Kamimura, Y., Kondo, N. and Shimada, M. (2003b) Extremely female-biased sex ratio and lethal male-male combat in a parasitoid wasp, *Melittobia australica* (Eulophidae). *Behavioral Ecology* **14**: 34–9.

Abe, J., Kamimura, Y. and Shimada, M. (2005) Individual sex ratios and offspring emergence patterns in a parasitoid wasp, *Melittobia australica* (Eulophidae), with superparasitism and lethal combat among sons. *Behavioral Ecology & Sociobiology* **57**: 366–73.

Aeschlimann, J.P. (1990) Simultaneous occurrence of thelytoky and bisexuality in hymenopteran species, and its implications for the biological control of pests. *Entomophaga* **35**: 3–5.

Antolin, M.F. (1992a) Sex ratio variation in a parasitic wasp. I. Reaction norms. *Evolution* **46**: 1496–510.

Antolin, M.F. (1992b) Sex ratio variation in a parasitic wasp. II. Diallel cross. *Evolution* **46**: 1511–24.

Bernal, J.S., Luck, R.F. and Morse, J.G. (1998) Sex ratios in field populations of two parasitoids (Hymenoptera: Chalcidoidea) of *Coccus hesperidum* L. (Homoptera: Coccidae). *Oecologia* **116**: 510–18.

Berndt, L.A. and Wratten, S.D. (2005) Effects of alyssum flowers on the longevity, fecundity, and sex ratio of the leafroller parasitoid *Dolichogenidea tasmanica*. *Biological Control* **32**: 65–9.

Bernstein, C. (2000) Host-parasitoid models: the story of a successful failure. In: Hochberg, M.E. and Ives, A.R. (eds.) *Parasitoid Population Biology*. Princeton University Press, Princeton, pp. 41–57.

Beukeboom, L.W. (1995) Sex determination in Hymenoptera: A need for genetic and molecular studies. *Bioessays* **17**: 813–17.

Beukeboom, L.W. (2001) Single-locus complementary sex determination in the ichneumonid *Venturia canescens* (Gravenhorst) (Hymenoptera). *Netherlands Journal of Zoology* **51**: 1–15.

Beukeboom, L.W. and Kamping, A. (2006) No patrigenes required for femaleness in the haplodiploid wasp *Nasonia vitripennis*. *Genetics* **172**: 981–9.

Beukeboom, L.W. and Zwaan, B.J. (2005) Genetics. In: Jervis, M.A. (ed.) *Insects as Natural Enemies: a Practical Perspective*. Springer, Dordrecht, pp. 167–218.

Beukeboom, L.W., Ellers, J. and van Alphen, J.J.M. (2000) Absence of single-locus complementary sex determination in the braconid wasps *Asobara tabida* and *Alysia manducator*. *Heredity* **84**: 29–36.

Beye, M., Hasselmann, M., Fondrk, M.K., Page, R.E. and Omholt, S.W. (2003) The gene *csd* is the primary signal for sexual development in the honeybee and encodes an SR-type protein. *Cell* **114**: 419–29.

Boavida, C., Ahounou, M., Vos, M., Neuenschwander, P. and van Alphen, J.J.M. (1995) Host stage selection and sex allocation by *Gyranusoidea tebygi* (Hymenoptera: Encyrtidae), a parasitoid of the mango mealybug, *Rastrococcus invadens* (Homoptera: Pseudococcidae). *Biological Control* **5**: 487–96.

Bokonon-Ganta, A.H., Neuenschwander, P., van Alphen, J.J.M. and Vos, M. (1995) Host stage selection and sex allocation by *Anagyrus mangicola* (Hymenoptera: Encyrtidae), a parasitoid of the mango mealybug, *Rastrococcus invadens* (Homoptera: Pseudococcidae). *Biological Control* **5**: 479–86.

Bourke, A.F.G. and Franks, N.R. (1995) *Social Evolution in Ants*. Princeton University Press, Princeton.

Breeuwer, J.A.J. and Werren, J.H. (1990) Microorganisms associated with chromosome destruction and reproductive isolation between two insect species. *Nature* **346**: 558–60.

Briggs, C.J., Murdoch, W.W. and Nisbet, R.M. (1999) Recent developments in theory for biological control of insect pests by parasitoids. In: Hawkins, B.A. and Cornell, H.V. (eds.) *Theoretical Approaches to Biological Control*. Cambridge University Press, Cambridge, pp. 22–42.

Bull, J.J. (1981) Sex ratio evolution when fitness varies. *Heredity* **46**: 9–26.

Bull, J.J. and Charnov, E.L. (1988) How fundamental are Fisherian sex ratios? *Oxford Surveys in Evolutionary Biology* **5**: 96–135.

Burt, A. and Trivers, R.L. (2006) *Genes in Conflict: the Biology of Selfish Genetic Elements*. Belknap Harvard, Cambridge.

Butcher, R.D.J., Whitfield, W. and G.F. and Hubbard, S.F. (2000a) Complementary sex determination in the genus *Diadegma* (Hymenoptera: Ichneumonidae). *Journal of Evolutionary Biology* **13**: 593–606.

Butcher, R.D.J., Whitfield, W.G.F. and Hubbard, S.F. (2000b) Single-locus complementary sex determination in *Diadegma chrysostictos* (Gmelin) (Hymenoptera: Ichneumonidae). *Journal of Heredity* **91**: 104–11.

Charnov, E.L. (1979) The genetical evolution of patterns of sexuality: Darwinian fitness. *American Naturalist* **113**: 465–80.

Charnov, E.L. (1982) *The Theory of Sex Allocation*. Princeton University Press, Princeton.

Charnov, E.L. and Bull, J.J. (1989) The primary sex ratio under environmental sex determination. *Journal of Theoretical Biology* **139**: 431–6.

Charnov, E.L., los-den Hartogh, R.L., Jones, W.T. and van den Assem, J. (1981) Sex ratio evolution in a variable environment. *Nature* **289**: 27–33.

Chow, A. and Heinz, K.M. (2005) Using hosts of mixed sizes to reduce male-biased sex ratio in the parasitoid wasp, *Diglyphus isaea*. *Entomologia Experimentalis et Applicata* **117**: 193–9.

Chow, A. and Heinz, K.M. (2006) Control of *Liriomyza langei* on chrysanthemum by *Diglyphus isaea* produced with a standard or modified parasitoid rearing technique. *Journal of Applied Entomology* **130**: 113–21.

Clausen, C.P. (1939) The effect of host size upon the sex ratio of hymenopterous parasites and its relation to methods of rearing and colonization. *Journal of the New York Entomological Society* **47**: 1–9.

Clout, M.N., Elliot, G.P. and Robertson, B.C. (2001) Effects of supplementary feeding on the offspring sex ratio of the kakapo: A dilemma for the conservation of a polygynous parrot. *Biological Conservation* **107**: 13–18.

Comins, H.N. and Wellings, P.W. (1985) Density-related parasitoid sex ratio: Influence on host-parasitoid dynamics. *Journal of Animal Ecology* **54**: 583–94.

Conover, D.O. and van Voorhees, D.A. (1990) Evolution of a balanced sex-ratio by frequency dependent selection in a fish. *Science* **250**: 1556–8.

Cook, J.M. (1993a) Sex determination in the Hymenoptera: a review of models and evidence. *Heredity* **71**: 421–35.

Cook, J.M. (1993b) Experimental test of sex determination in *Goniozus nephantidis*. *Heredity* **71**: 130–7.

Cook, J.M. (1993c) Inbred lines as reservoirs of sex alleles in parasitoid rearing programs. *Environmental Entomology* **22**: 1213–16.

Cook, J.M. (2002) Sex determination in invertebrates. In: Hardy, I.C.W. (ed.) *Sex Ratios: Concepts and Research Methods*. Cambridge University Press, Cambridge, pp. 178–94.

Cook, J.M. and Crozier, R.H. (1995) Sex determination and population biology in the Hymenoptera. *Trends in Ecology & Evolution* **10**: 281–6.

Cook, J.M. and Butcher, R.D.J. (1999) The transmission and effects of *Wolbachia* bacteria in parasitoids. *Researches on Population Ecology* **41**: 15–28.

Courteau, J. and Lessard, S. (1999) Stochastic effects in LMC models. *Theoretical Population Biology* **55**: 127–36.

Courteau, J. and Lessard, S. (2000) Optimal sex ratios in structured populations. *Journal of Theoretical Biology* **207**: 159–75.

Cowan, D.P. and Stahlhut, J.K. (2004) Functionally reproductive diploid and haploid males in an inbreeding hymenopteran with complementary sex determination. *Proceedings of the National Academy of Sciences USA* **101**: 10374–9.

Crozier, R.H. (1977) Evolutionary genetics of the Hymenoptera. *Annual Review of Entomology* **22**: 263–88.

de Boer, J.G., Ode, P.J., Vet, L.E.M., Whitfield, J. and Heimpel, G.E. (2007) Complementary sex determination in the parasitoid wasp *Cotesia vestalis* (*C. plutellae*). *Journal of Evolutionary Biology* **20**: 340–8.

Devenport, L.D. and Devenport, J.A. (1994) Time-dependent averaging of foraging information in least chipmunks and golden-mantled ground squirrels. *Animal Behaviour* **47**: 787–802.

Dobson, S.L. and Tanouye, M.A. (1998) Evidence for a genomic imprinting sex determination mechanisms in *Nasonia vitripennis* (Hymenoptera; Chalcidoidea). *Genetics* **149**: 233–42.

Drapeau, M.D. and Werren, J.H. (1999) Differences in mating behaviour and sex ratio between three sibling species of *Nasonia*. *Evolutionary Ecology Research* **1**: 223–34.

Elagoze, M. and Periquet, G. (1993) Viability of diploid males in the parasitic wasp *Diadromus pulchellus* (Hym, Ichneumonidae). *Entomophaga* **38**: 199–206.

Fellowes, M.D.E., Compton, S.G. and Cook, J.M. (1999) Sex allocation and local mate competition in Old World non-pollinating fig wasps. *Behavioral Ecology & Sociobiology* **46**: 95–102.

Fisher, R.A. ([1930] 1999) *The Genetical Theory of Natural Selection*. A Complete Variorum Edition, Bennett, J.H. (ed.), Oxford University Press, Oxford.

Frank, S.A. and Swingland, I.R. (1988) Sex ratio under conditional sex expression. *Journal of Theoretical Biology* **135**: 415–18.

Fuester, R.W., Swan, K.S., Dunning, K., Taylor, P.B. and Ramaseshiah, G. (2003) Male-biased sex ratios in *Glyptapanteles flavicoxis* (Hymenoptera : Braconidae), a parasitoid of the gypsy moth (Lepidoptera: Lymantriidae). *Annals of the Entomological Society of America* **96**: 553–9.

Gardner, A., Hardy, I.C.W., Taylor, P.D. and West, S.A. (2007) Spiteful soldiers and sex ratio conflict in polyembryonic parasitoid wasps. *American Naturalist* **169**: 519–33.

Godfray, H.C.J. (1990) The causes and consequences of constrained sex allocation in haplodiploid animals. *Journal of Evolutionary Biology* **3**: 3–17.

Godfray, H.C.J. (1994) *Parasitoids: Behavioral and Evolutionary Ecology*. Princeton University Press, Princeton.

Godfray, H.C.J. and Waage, J.K. (1991) Predictive modeling in biological control – the mango mealy bug (*Rastrococcus invadens*) and its parasitoids. *Journal of Applied Ecology* **28**: 434–53.

Godfray, H.C.J. and Hardy, I.C.W. (1993) Sex ratio and virginity in haplodiploid insects. In: Wrensch, D.L. and Ebbert, M.A.(eds.) *Evolution and Diversity of Sex Ratios in Insects and Mites*. Chapman & Hall, Inc., New York, pp. 402–17.

Godfray, H.C.J. and Cook, J.M. (1997) Mating systems of parasitoid wasps. In: Choe, J.C. and Crespi, B.J. (eds.) *The Evolution of Mating Systems in Insects and Arachnids*. Cambridge University Press, Cambridge, pp. 211–25.

Gonçalves, C.I., Huigens, M.E., Verbaarschotb, P., Duartea, S., Mexiaa, A. and Tavaresc, J. (2006) Natural occurrence of *Wolbachia*-infected and uninfected *Trichogramma* species in tomato fields in Portugal. *Biological Control* **37**: 375–81.

Green, R.F., Gordh, G. and Hawkins, B.A. (1982) Precise sex ratios in highly inbred parasitic wasps. *American Naturalist* **120**: 653–65.

Griffiths, N.T. and Godfray, H.C.J. (1988) Local mate competition, sex ratio and clutch size in bethylid wasps. *Behavioral Ecology & Sociobiology* **22**: 211–17.

Grinsberg, P.S. and Wallner, W.E. (1991) Long-term laboratory evaluation of *Rojas lymantriae*: a braconid endoparasite of the gypsy moth. *Entomophaga* **36**: 205–12.

Gu, H.N. and Dorn, S. (2003) Mating system and sex allocation in the gregarious parasitoid *Cotesia glomerata*. *Animal Behaviour* **66**: 259–64.

Gutierrez, A.P., Neuenschwander, P. and van Alphen, J.J.M. (1993) Factors affecting biological control of cassava mealybug by exotic parasitoids: A ratio-dependent supply-demand driven model. *Journal of Applied Ecology* **30**: 706–21.

Hall, R.W. (1993) Alteration of sex ratios of parasitoids for use in biological control. In: Wrensch, D.L. and Ebbert, M.A. (eds.) *Evolution and Diversity of Sex Ratios in Insects and Mites*. Chapman & Hall, Inc., New York, pp. 542–7.

Hamilton, W.D. (1967) Extraordinary sex ratios. *Science* **156**: 477–88.

Hamilton, W.D. (1979) Wingless and fighting males in fig wasps and other insects. In: Blum, M.S. and Blum, N.A. (eds.) *Sexual Selection and Reproductive Competition in Insects*. Academic Press, London, pp. 167–220.

Hardy, I.C.W. (1992) Non-binomial sex allocation and brood sex ratio variances in the parasitoid Hymenoptera. *Oikos* **65**: 143–50.

Hardy, I.C.W. (1994) Sex ratio and mating structure in the parasitoid Hymenoptera. *Oikos* **69**: 3–20.

Hardy, I.C.W. (1997) Possible factors influencing vertebrate sex ratios: An introductory overview. *Applied Animal Behaviour Science* **51**: 217–41.

Hardy, I.C.W. (ed.) (2002) *Sex Ratios: Concepts and Research Methods*. Cambridge University Press, Cambridge.

Hardy, I.C.W. and Mayhew, P.J. (1998) Sex ratio, sexual dimorphism and mating structure in bethylid wasps. *Behavioral Ecology & Sociobiology* **42**: 383–95.

Hardy, I.C.W., Dijkstra, L.J., Gillis, J.E.M. and Luft, P.A. (1998) Patterns of sex ratio, virginity and developmental mortality in gregarious parasitoids. *Biological Journal of the Linnean Society* **64**: 239–70.

Hardy, I.C.W., Stokkebo, S., Bønløkke-Pedersen, J. and Sejr, M.K. (2000) Insemination capacity and dispersal in relation to sex allocation decisions in *Goniozus legneri* (Hymenoptera: Bethylidae): Why are there more males in larger broods? *Ethology* **106**: 1021–32.

Hardy, I.C.W., Ode, P.J. and Siva-Jothy, M.T. (2005) Mating systems. In: Jervis, M.A. (ed.) *Insects as Natural Enemies: a Practical Perspective*. Springer, Dordrecht, pp. 261–98.

Hare, J.D. and Luck, R.F. (1991) Indirect effects of citrus cultivars on life history parameters of a parasitic wasp. *Ecology* **72**: 1576–85.

Hassell, M.P. (2000) *The Spatial and Temporal Dynamics of Host-Parasitoid Interactions*. Oxford University Press, Oxford.

Hassell, M.P, Waage, J.K. and May, R.M. (1983) Variable parasitoid sex ratios and their effect on host-parasitoid dynamics. *Journal of Animal Ecology* **52**: 889–904.

Heimpel, G.E. (1994) Virginity and the cost of insurance in highly inbred Hymenoptera. *Ecological Entomology* **19**: 299–302.

Heimpel, G.E. (1997) Extraordinary sex ratios for extraordinary reasons. *Trends in Ecology & Evolution* **12**: 298–9.

Heimpel, G.E. and Lundgren, J.G. (2000) Sex ratios of commercially reared biological control agents. *Biological Control* **19**: 77–93.

Heinz, K.M. (1991) Sex-specific reproductive consequences of body size in the solitary ectoparasitoid, *Diglyphus begini*. *Evolution* **45**: 1511–15.

Heinz, K.M. (1998) Host size-dependent sex allocation behaviour in a parasitoid: Implications for *Catolaccus grandis* (Hymenoptera: Pteromalidae) mass rearing programmes. *Bulletin of Entomological Research* **88**: 37–45.

Heinz, K.M. and Parrella, M.P. (1989) Attack behavior and host size selection by *Diglyphus begini* on *Liriomyza trifolii* in chrysanthemum. *Entomologia Experimentalis et Applicata* **53**: 147–56.

Heinz, K.M. and Parrella, M.P. (1990) The influence of host size on sex ratios in the parasitoid *D. begini* (Hymenoptera: Eulophidae). *Ecological Entomology* **15**: 391–9.

Henter, H.J. (2003) Inbreeding depression and haplodiploidy: experimental measures in a parasitoid and comparisons across diploid and haplodiploid insect taxa. *Evolution* **57**: 1793–1803.

Henter, H.J. (2004) Constrained sex allocation in a parasitoid due to variation in male quality. *Journal of Evolutionary Biology* **17**: 886–96.

Hoffman, A.A. and Turelli, M. (1997) Cytoplasmic incompatibility in insects. In: O'Neill, S.L., Hoffmann, A.A. and Werren, J.H. (eds.) *Influential Passengers: Inherited Microorganisms and Arthropod Reproduction*. Oxford University Press, Oxford, pp. 42–80.

Holloway, A.K., Heimpel, G.E., Strand, M.R. and Antolin, M.F. (1999) Survival of diploid males in *Bracon* sp. near *hebetor* (Hymenoptera: Braconidae). *Annals of the Entomological Society of America* **92**: 110–16.

Hunter, M.S. (1999) Genetic conflict in natural enemies: a review, and consequences for the biological control of arthropods. In: Hawkins, B.A. and Cornell, H.V. (eds.) *Theoretical Approaches to Biological Control*. Cambridge University Press, Cambridge, pp. 231–58.

Hunter, M.S., Nur, U. and Werren, J.H. (1993) Origin of males by genome loss in an autoparasitoid wasp. *Heredity* **70**: 162–71.

Hurst, G.D.D., Hurst, L.D. and Majerus, M.E.N. (1997) Cytoplasmic sex-ratio distorters. In: O'Neill, S.L., Hoffmann, A.A. and Werren, J.H. (eds.) *Influential Passengers: Inherited Microorganisms and Arthropod Reproduction*. Oxford University Press, Oxford, pp. 125–54.

Irvin, N.A. and Hoddle, M.S. (2006) The effect of intraspecific competition on progeny sex ratio in *Gonatocerus* spp. for *Homalodisca coagulata* egg masses: Economic implications for mass rearing and biological control. *Biological Control* **39**: 162–70.

Jervis, M. (ed.) (2005) *Insects as Natural Enemies: a Practical Perspective.* Springer, Dordrecht.

Johns, C.V. and Whitehouse, M.E.A. (2004) Mass rearing of two larval parasitoids of *Helicoverpa* spp. (Lepidoptera: Noctuidae): *Netelia producta* (Brulle) and *Heteropelma scaposum* (Morley) (Hymenoptera: Ichneumonidae) for field release. *Australian Journal of Entomology* **43**: 83–7.

Jones, W.T. (1982) Sex ratio and host size in a parasitic wasp. *Behavioral Ecology & Sociobiology* **10**: 207–10.

Joyce, A.L., Millar, J.G., Paine, T.D. and Hanks, L.M. (2002) The effect of host size on the sex ratio of *Syngaster lepidus*, a parasitoid of Eucalyptus longhorned borers (*Phoracantha* spp.). *Biological Control* **24**: 207–13.

Kageyama, D., Hoshizaki, S. and Ishikawa, Y. (1998) Female-biased sex ratio in the Asian corn borer, *Ostrinia furnacalis*: Evidence for the occurrence of feminizing bacteria in an insect. *Heredity* **81**: 311–16.

Karsai, I., Somogyi, K. and Hardy, I.C.W. (2006) Body size, host choice and sex allocation in a spider hunting pompilid wasp. *Biological Journal of the Linnean Society* **87**: 285–96.

King, B.H. (1987) Offspring sex ratios in parasitoid wasps. *Quarterly Review of Biology* **62**: 367–96.

King, B.H. (1988) Sex-ratio manipulation in response to host size by the parasitic wasp *Spalangia cameroni*: A laboratory study. *Evolution* **42**: 1190–8.

King, B.H. (1993) Sex ratio manipulation by parasitoid wasps. In: Wrensch, D.L. and Ebbert, M.A. (eds.) *Evolution and Diversity of Sex ratio in Insects and Mites.* Chapman & Hall, New York, pp. 418–41.

King, B.H. and King, R.B. (1994) Sex ratio manipulation in relation to host size in the parasitic wasp *Spalangia cameroni*: Is it adaptive? *Behavioral Ecology* **5**: 448–54.

King, B.H. and Lee, H.E. (1994) Test of the adaptiveness of sex ratio manipulation in a parasitoid wasp. *Behavioral Ecology & Sociobiology* **35**: 437–43.

King, B.H. and D'Souza, J.A. (2004) Effects of constrained females on offspring sex ratios of *Nasonia vitripennis* in relation to local mate competition theory. *Canadian Journal of Zoology* **82**: 1969–74.

Komdeur, J. (1996) Facultative sex ratio bias in the offspring of Seychelles warblers. *Proceedings of the Royal Society of London Series B Biological Science* **263**: 661–6.

Kraaijeveld, A.R. and van Aphen, J.J.M. (1986) Host-stage selection and sex allocation by *Epidinocarsis lopezi* (Hymenoptera; Encrytidae), a parasitoid of the cassava mealybug, *Phenacoccus manihoti* (Homoptera; Pseudococcidae). *Mededelingen van de Faculteit Landbouwwetenschappen Rijksuniversiteit Gent* **51**: 1067–78.

Krackow, S., Meelis, E. and Hardy, I.C.W. (2002) Analysis of sex ratio variances and sequences of sex allocation. In: Hardy, I.C.W. (ed.) *Sex Ratios: Concepts and Research Methods.* Cambridge University Press, Cambridge, pp. 112–31.

Lampson, L.J., Morse, J.G. and Luck, R.F. (1996) Host selection, sex allocation, and host feeding by *Metaphycus helvolus* (Hymenoptera: Encyrtidae) on *Saissetia oleae* (Homoptera: Coccidae) and its effect on parasitoid size, sex, and quality. *Environmental Entomology* **25**: 283–94.

Leigh, E.G. Jr. (1970) Sex ratio and differential mortality between the sexes. *American Naturalist* **104**: 205–10.

Luck, R.F. and Podoler, H. (1985) The potential role of host size in the competitive exclusion of *Aphytis lingnanensis* by *A. melinus*. *Ecology* **66**: 904–13.

Luck, R.F. and Nunney, L.P. (1999) A Darwinian view of host selection and its practical applications. In: Hawkins, B.A. and Cornell, H.V. (eds.) *Theoretical Approaches to Biological Control.* Cambridge University Press, Cambridge, pp. 283–303.

Luck, R.F., Stouthamer, R. and Nunney, L.P. (1993) Sex determination and sex ratio patterns in parasitic Hymenoptera. In: Wrensch, D.L. and Ebbert, M.A. (eds.) *Evolution and Diversity of Sex Ratio in Insects and Mites.* Chapman & Hall, New York, pp. 442–76.

Luck, R.F., Nunney, L.P. and Stouthamer, R. (1999) Sex ratio and quality in the culturing of parasitic Hymenoptera: a genetic and evolutionary perspective. In: Bellows, S.T. Jr and Fisher, T.W.

(eds.) *Handbook of Biological Control: Principles and Applications of Biological Control.* Academic Press, San Diego, pp. 653–2.

Lundgren, J.G. and Heimpel, G.E. (2002) Quality assessment of three species of commercially produced *Trichogramma* and the first report of thelytoky in commercially produced *Trichogramma. Biological Control* **26**: 68–73.

Majerus, M.E.N. (2003) *Sex Wars: Genes, Bacteria, and Biased Sex Ratios.* Princeton University Press, Princeton.

Majerus, M.E.N. and Hurst, G.D.D. (1997) Ladybirds as a model system for the study of male-killing symbionts. *Entomophaga* **42**: 13–20.

Mason, P.G. and Huber, J.T. (eds.) (2002) *Biological Control Programmes in Canada, 1981–2000.* CABI Publishing, Wallingford.

Mayhew, P.J. and Godfray, H.C.J. (1997) Mixed sex allocation strategies in a parasitoid wasp. *Oecologia* **110**: 218–21.

Maynard Smith, J. (1978) *The Evolution of Sex.* Cambridge University Press, Cambridge.

Maynard Smith, J. (1982) *Evolution and the Theory of Games.* Cambridge University Press, Cambridge.

Meunier, J. and Bernstein, C. (2002) The influence of local mate competition on host-parasitoid dynamics. *Ecological Modelling* **152**: 77–88.

Michod, R.E. and Levin, B.R. (1988) *The Evolution of Sex: An Examination of Current Ideas.* Sinauer Associates, Sunderland.

Mochiah, M.B., Ngi-Song, A.J., Overholt, W.A. and Stouthamer, R. (2002) *Wolbachia* infection in *Cotesia sesamiae* (Hymenoptera: Braconidae) causes cytoplasmic incompatibility: Implications for biological control. *Biological Control* **25**: 74–80.

Murdoch, W.W., Nisbet, R.M., Luck, R.F., Godfray, H.C.J. and Gurney, W.S.C. (1992) Size-selective sex-allocation and host feeding in a parasitoid-host model. *Journal of Animal Ecology* **61**: 533–41.

Murdoch, W.W., Briggs, C.J. and Nisbet, R.M. (2003) *Consumer-Resource Dynamics.* Princeton University Press, Princeton.

Murdoch, W., Briggs, C.J. and Swarbrick, S. (2005) Host suppression and stability in a host-parasitoid system: Experimental demonstration. *Science* **309**: 610–13.

Nagelkerke, C.J. (1996) Discrete clutch sizes, local mate competition, and the evolution of precise sex allocation. *Theoretical Population Biology* **49**: 314–43.

Nagelkerke, C.J. and Hardy, I.C.W. (1994) The influence of developmental mortality on optimal sex allocation under local mate competition. *Behavioral Ecology* **5**: 401–11.

Napoleon, M.E. and King, B.H. (1999) Offspring sex ratio response to host size in the parasitoid wasp *Spalangia endius. Behavioral Ecology & Sociology* **46**: 325–32.

Neuenschwander, P. (2001) Biological control of cassava mealybug in Africa: A review. *Biological Control* **21**: 214–29.

Nicholson, A.J. and Bailey, V.A. (1935) The balance of animal populations. Part I. *Proceedings of the Zoological Society of London* **3**: 551–98.

Niyibigira, E.I., Overholt, W.A. and Stouthamer, R. (2004a) *Cotesia flavipes* Cameron and *Cotesia sesamiae* (Cameron) (Hymenoptera: Braconidae) do not exhibit complementary sex determination: Evidence from field populations. *Applied Entomology & Zoology* **39**: 705–15.

Niyibigira, E.I., Overholt, W.A. and Stouthamer, R. (2004b) *Cotesia flavipes* Cameron (Hymenoptera: Braconidae) does not exhibit complementary sex determination (ii) Evidence from laboratory experiments. *Applied Entomology & Zoology* **39**: 717–25.

Noda, T. (2000) Detection of diploid males and estimation of sex determination system in the parasitic wasp *Diadegma semiclausum* (Hellen) (Hymenoptera: Ichneumonidae) using an allozyme as a genetic marker. *Applied Entomology & Zoology* **35**: 41–4.

Nur, U., Werren, J.H., Eickbush, D.G., Burke, W.D. and Eickbush, T.H. (1988) A 'selfish' B chromosome that enhances its transmission by eliminating the paternal genome. *Science* **240**: 512–14.

Ode, P.J. and Strand, M.R. (1995) Progeny and sex allocation decisions of the polyembryonic wasp *Copidosoma floridanum. Journal of Animal Ecology* **64**: 213–24.

Ode, P.J. and Heinz, K.M. (2002) Host-size-dependent sex ratio theory and improving mass-reared parasitoid sex ratios. *Biological Control* **24**: 31–41.

Ode, P.J. and Hunter, M.S. (2002) Sex ratios of parasitic Hymenoptera with unusual life-histories. In: Hardy, I.C.W. (ed.) *Sex Ratios: Concepts and Research Methods*. Cambridge University Press, Cambridge, pp. 218–34.

Ode, P.J., Antolin, M.F. and Strand, M.R. (1995) Brood-mate avoidance in the parasitic wasp *Bracon hebetor* Say. *Animal Behaviour* **49**: 1239–48.

Ode, P.J., Antolin, M.F. and Strand, M.R. (1996) Sex allocation and sexual asymmetries in intra-brood competition in the parasitic wasp *Bracon hebetor. Journal of Animal Ecology* **65**: 690–700.

Ode, P.J., Antolin, M.F. and Strand, M.R. (1997) Constrained oviposition and female-biased sex allocation in a parasitic wasp. *Oecologia* **109**: 547–55.

O'Neill, S.L., Hoffmann, A.A. and Werren, J.H. (1997) *Influential Passengers: Inherited Micro-organisms and Arthropod Reproduction*. Oxford University Press, Oxford.

Opp, S.B. and Luck, R.F. (1986) Effects of host size on selected fitness components of *Aphytis melinus* and *A. lingnanensis* (Hymenoptera: Aphelinidae). *Annals of the Entomological Society of America* **79**: 700–4.

Orzack, S.H. (1990) The comparative biology of second sex ratio evolution within a natural population of a parasitic wasp, *Nasonia vitripennis. Genetics* **124**: 385–96.

Orzack, S.H. (1993) Sex ratio evolution on parasitic wasps. In: Wrensch, D.L. and Ebbert, M.A. (eds.) *Evolution and Diversity of Sex Ratio in Insects and Mites*. Chapman & Hall, New York, pp. 477–511.

Orzack, S.H. (2002) Using sex ratios: The past and the future. In: Hardy, I.C.W. (ed.) *Sex Ratios: Concepts and Research Methods*. Cambridge University Press, Cambridge, pp. 383–98.

Orzack, S.H. and Parker, E.D. (1990) Genetic variation for sex ratio traits within a natural population of a parasitic wasp. *Genetics* **124**: 373–84.

Paine, T.D., Joyce, A.L., Millar, J.G. and Hanks, L.M. (2004) Effect of variation in host size on sex ratio, size, and survival of *Syngaster lepidus*, a parasitoid of Eucalyptus longhorned beetles (*Phoracantha* spp.): II. *Biological Control* **30**: 374–81.

Pandey, R.R. and Johnson, M.W. (2005) Effect of pink pineapple mealybug hosts on *Anagyrus ananatis* Gahan size and progeny production. *Biological Control* **35**: 1–8.

Parker, E.D. and Orzack, S.H. (1985) Genetic variation for the sex ratio in *Nasonia vitripennis. Genetics* **110**: 93–105.

Parrella, M.P., Heinz, K.M. and Nunney, L. (1992) Biological control through augmentative releases of natural enemies: A strategy whose time has come. *American Entomologist* **38**: 172–9.

Peer, K. and Taborsky, M. (2004) Female ambrosia beetles adjust their offspring sex ratio according to outbreeding opportunities for their sons. *Journal of Evolutionary Biology* **17**: 257–64.

Pen, I. and Weissing, F.J. (2002) Optimal sex allocation: steps towards a mechanistic theory. In: Hardy, I.C.W. (ed.) *Sex Ratios: Concepts and Research Methods*. Cambridge University Press, Cambridge, pp. 26–45.

Pérez-Lachaud, G. and Hardy, I.C.W. (2001) Alternative hosts for bethylid parasitoids of the coffee berry borer, *Hypothenemus hampei* (Coleoptera: Scolytidae). *Biological Control* **22**: 265–77.

Periquet, G., Hedderwick, M.P., Elagoze, M. and Poirié, M. (1993) Sex determination in the hymenopteran *Diadromous pulchellus* (Icheneumonidae) – Validation of the one-locus multi-allele model. *Heredity* **70**: 420–7.

Perrot-Minnot, M.-J., Guo, L.R. and Werren, J.H. (1996) Single and double infections with *Wolbachia* in the parasitic wasp *Nasonia vitripennis*: effects on compatibility. *Genetics* **143**: 961–72.

Petters, R.M. and Mettus, R.V. (1980) Decreased diploid male viability in the parasitic wasp, *Bracon hebetor. Journal of Heredity* **71**: 353–6.

Pijls, J.W.A.M., van Steenbergen, H.J. and van Alphen, J.J.M. (1996) Asexuality cured: The relations and differences between sexual and asexual *Apoanagyrus diversicornis*. *Heredity* **76**: 506–13.

Platner, G.R. and Oatman, E.R. (1972) Techniques for culturing and mass producing parasites of the potato tuberworm. *Journal of Economic Entomology* **65**: 1336–8.

Ram, A. and Sharma, A.K. (1977) Selective breeding for improving the fecundity and sex ratio of *Trichogramma fasciatum* (Perkins) (Trichogrammatidae: Hymenoptera), an egg parasite of lepidopterous hosts. *Entomon* **2**: 133–7.

Rappaport, N. and Page, M. (1985) Rearing *Glypta fumiferanae* on a multivoltine laboratory colony of the western spruce budworm. *Journal of Economic Entomology* **65**: 1336–8.

Read, A.F., Smith, T.G., Nee, S. and West, S.A. (2002) Sex ratios of malaria parasites and related protozoa. In: Hardy, I.C.W. (ed.) *Sex Ratios: Concepts and Research Methods.* Cambridge University Press, Cambridge, pp. 314–32.

Reece, S.E., Shuker, D.M., Pen, I. et al. (2004) Kin discrimination and sex ratios in a parasitoid wasp. *Journal of Evolutionary Biology* **17**: 208–16.

Reeve, J.D. (1987) Foraging behavior of *Aphytis melinus*: Effects of patch density and host size. *Ecology* **68**: 530–8.

Reeve, J.D. and Murdoch, W.W. (1986) Biological control of the parasitoid *Aphytis melinus*, and population stability of the California red scale. *Journal of Animal Ecology* **55**: 1069–82.

Rigaud, T. (1997) Inherited microorganisms and sex determination of arthropod hosts. In: O'Neill, S.L., Hoffmann, A.A. and Werren, J.H. (eds.) *Influential Passengers: Inherited Microorganisms and Arthropod Reproduction.* Oxford University Press, Oxford, pp. 81–101.

Rincon, C., Bordat, D., Löhr, B. and Dupas, S. (2006) Reproductive isolation and differentiation between five populations of *Cotesia plutellae* (Hymenoptera: Braconidae), parasitoid of *Plutella xylostella* (Lepidoptera: Plutellidae). *Biological Control* **36**: 171–82.

Robertson, B.C., Elliot, G.P., Eas on, D.K., Clout, M.N. and Gemmell, N.J. (2006) Sex allocation theory aids species conservation. *Biology Letters* **2**: 229–31.

Sagarra, L.A. and Vincent, C. (1999) Influence of host stage on oviposition, development, sex ratio, and survival of *Anagyrus kamali* Moursi (Hymenoptera: Encrytidae), a parasitoid of the hibiscus mealybug, *Maconellicoccus hirsutus* Green (Homptera: Pseudococcidae). *Biological Control* **15**: 51–6.

Schulthess, F., Neuenschwander, P. and Gounou, S. (1997) Multitrophic interactions in cassava, *Manihot esculenta*, cropping systems in the subhumid tropics of West Africa. *Agriculture, Ecosystems & Environment* **66**: 211–22.

Seger, J. and Stubblefield, J.W. (2002) Models of sex ratio evolution. In: Hardy, I.C.W. (ed.) *Sex Ratios: Concepts and Research Methods.* Cambridge University Press, Cambridge, pp. 2–25.

Shuker, D.M. and West, S.A. (2004) Information constraints and the precision of adaptation: Sex ratio manipulation in wasps. *Proceedings of the National Academy of Sciences USA* **101**: 10363–7.

Shuker, D.M., Pen, I., Duncan, A.B., Reece, S.E. and West, S.A. (2005) Sex ratios under asymmetrical local mate competition: Theory and a test with parasitoid wasps. *American Naturalist* **166**: 301–6.

Shuker, D.M., Pen, I. and West, S.A. (2006a) Sex ratios under asymmetrical local mate competition in the parasitoid wasp *Nasonia vitripennis*. *Behavioral Ecology* **17**: 345–52.

Shuker, D.M., Sykes, E.M., Browning, L.E., Beukeboom, L.W. and West, S.A. (2006b) Male influence on sex allocation in the parasitoid wasp *Nasonia vitripennis*. *Behavioral Ecology & Sociobiology* **59**: 829–35.

Shuker, D.M., Reece, S.E., Lee, A., Graham, A., Duncan, A.B. and West, S.A. (2007) Information use in space and time: Sex allocation behaviour in the parasitoid wasp *Nasonia vitripennis*. *Animal Behaviour* **73**: 971–7.

Silva, I.M.M.S., van Meer, M.M.M., Roskam, M.M., Hoogenboom, A., Gort, G. and Stouthamer, R. (2000) Biological control potential of *Wolbachia*-infected versus uninfected wasps: Laboratory and greenhouse evaluation of *Trichogramma cordubensis* and *T. deion* strains. *Biocontrol Science & Technology* **10**: 223–38.

Simmonds, F.J. (1947) Improvement of the sex-ratio of a parasite by selection. *Canadian Entomologist* **79**: 41–4.

Skinner, S.W. (1982) Maternally inherited sex ratio in the parasitoid wasp, *Nasonia vitripennis*. *Science* **215**: 1133–4.

Skinner, S.W. (1985) Son-killer: A third extrachromosomal factor affecting sex ratios in the parasitoid wasp *Nasonia vitripennis*. *Genetics* **109**: 745–54.

Smith, Jr. J.W., Rodriguez-del-Bosque, L.A., Agnew, C.W. (1990) Biology of *Mallochia pyralidis*, an ectoparasite of *Eoreuma loftini* from Mexico. *Annals of the Entomological Society of America* **83**: 961–6.

Stouthamer, R. (1993) The use of sexual versus asexual wasps in biological control. *Entomophaga* **38**: 3–6.

Stouthamer, R. (1997) *Wolbachia*-induced parthenogenesis. In: O'Neill, S.L., Hoffmann, A.A. and Werren, J.H. (eds.) *Influential Passengers: Inherited Microorganisms and Arthropod Reproduction.* Oxford University Press, Oxford, pp. 102–24.

Stouthamer, R. and Kazmer, D.J. (1994) Cytogenetics of microbe-associated parthenogenesis and its consequences for gene flow in *Trichogramma* wasps. *Heredity* **73**: 317–27.

Stouthamer, R., Pinto, J.D., Platner, G.R. and Luck, R.F. (1990) Taxonomic status of thelytokous forms of *Trichogramma* (Hymenoptera: Trichogrammatidae). *Annals of the Entomological Society of America* **83**: 475–81.

Stouthamer, R., Luck, R.F. and Werren, J.H. (1992) Genetics of sex determination and the improvement of biological control using parasitoids. *Environmental Entomology* **21**: 427–35.

Stouthamer, R., Breeuwer, J.A.J. and Hurst, G.D.D. (1999) *Wolbachia pipientis*: Microbial manipulator of arthropod reproduction. *Annual Review of Microbiology* **53**: 71–102.

Stouthamer, R., Hurst, G.D.D. and Breeuwer, J.A.J. (2002) Sex ratio distorters and their detection. In: Hardy ICW (ed.) *Sex Ratios: Concepts and Research Methods.* Cambridge University Press, Cambridge, pp. 195–215.

Sunderland, K.D., Powell, W. and Symondson, W.O.C. (2005) Populations and communities. In: Jervis, M.A. (ed.) *Insects as Natural Enemies: a Practical Perspective.* Springer, Dordrecht, pp. 299–434.

Tagami, Y., Doi, M., Sugiyami, K., Tatara, A. and Saito, T. (2006) Survey of leafminers and their parasitoids to find endosymbionts for improvement of biological control. *Biological Control* **38**: 210–16.

Tella, J.L. (2001) Sex-ratio theory in conservation biology. *Trends in Ecology & Evolution* **16**: 76–7.

Trent, C., Crosby, C. and Eavey, J. (2006) Additional evidence for the genomic imprinting model of sex determination in the haplodiploid wasp *Nasonia vitripennis*: Isolation of biparental diploid males after X-ray mutagenesis. *Heredity* **96**: 368–76.

Triapitsyn, S.V., Morgan, D.J.W., Hoddle, M.S. and Berezovskiy, V.V. (2003) Observations on the biology of *Gonatocerus fasciatus* Girault (Hymenoptera: Mymaridae), egg parasitoid of *Homalodisca coagulata* (Say) and *Oncometopia orbona* (Fabricius) (Hemiptera: Clypeorrhyncha: Cicadellidae) *Pan-Pacific Entomologist* **79**: 75–6.

Trivers, R.L. and Willard, D.E. (1973) Natural selection of parental ability to vary the sex ratio of offspring. *Science* **179**: 90–2.

Ueno, T. (1999) Host-size-dependent sex ratio in a parasitoid wasp. *Researches in Population Ecology* **41**: 47–7.

van den Assem, J., van Iersal, J.A. and los-den Hartogh, R.L. (1989) Is being large more important for female than male parasitic wasps? *Behaviour* **108**: 160–95.

van Dijken, M.J., Neuenschwander, P., van Alphen, J.J.M. and Hammond, W.N.O. (1991) Sex ratios in field populations of *Epidinocarsis lopezi*, an exotic parasitoid of the cassava mealybug, in Africa. *Ecological Entomology* **16**: 233–40.

van Dijken, M.J., van Stratum, P. and van Alphen, J.J.M. (1993) Superparasitism and sex ratio in the solitary parasitoid, *Epidinocarsis lopezi*. *Entomologia Experimentalis et Applicata* **68**: 51–8.

van Lenteren, J.C., Roskam, M.M. and Timmer, R. (1997) Commercial mass production and pricing of organisms for biological control of pests in Europe. *Biological Control* **10**: 143–9.

Waage, J.K. (1982a) Sex ratio and population dynamics of natural enemies – some possible interactions. *Annals of Applied Biology* **101**: 159–64.

Waage, J.K. (1982b) Sib-mating and sex ratio strategies in scelionid wasps. *Ecological Entomology* **7**: 103–12.

Wajnberg, É. (2004) Measuring genetic variation in natural enemies used for biological control: Why and how? In: Ehler, L., Sforza, R. and Mateille, Th. (eds.) *Genetics, Evolution and Biological Control*. CABI Publishing, Wallingford, pp. 19–37.

Wajnberg, É. and Hassan, S.A. (eds.) (1994) *Biological Control with Egg Parasitoids*. CABI Publishing, Wallingford.

Wajnberg, É., Scott, J.K. and Quimby, P.C. (eds.) (2001) *Evaluating Indirect Ecological Effects of Biological Control*. CABI Publishing, Wallingford.

Wakano, J.Y. (2005) Evolution of extraordinary female-biased sex ratios: The optimal schedule of sex ratio in local mate competition. *Journal of Theoretical Biology* **237**: 193–202.

Weber, C.A., Smilanick, J.M., Ehler, L.E. and Zalom, F.G. (1996) Ovipositional behavior and host discrimination in three scelionid egg parasitoids of stink bugs. *Biological Control* **6**: 245–52.

Werren, J.H. (1993) The evolution of inbreeding in haplodiploid organisms. In: Thornhill, N.W. (ed.) *The Natural History of Inbreeding and Outbreeding*. Chicago University Press, Chicago, pp. 42–59.

Werren, J.H. (1997) Biology of *Wolbachia*. *Annual Review of Entomology* **42**: 587–609.

Werren, J.H. and Charnov, E.L. (1978) Facultative sex ratios and population dynamics. *Nature* **272**: 349–50.

Werren, J.H. and Taylor, P.D. (1984) The effects of population recruitment on sex ratio selection. *American Naturalist* **124**: 143–8.

Werren, J.H. and Simbolotti, G. (1989) Combined effects of host quality and local mate competition on sex allocation in *Lariophagous distinguendus*. *Evolutionary Ecology* **3**: 123–43.

Werren, J.H. and Beukeboom, L. (1993) Population genetics of a parasitic chromosome: Theoretical analysis of PSR in subdivided populations. *American Naturalist* **142**: 224–41.

Werren, J.H. and Stouthamer, R. (2003) PSR (paternal sex ratio) chromosomes: The ultimate selfish genetic elements. *Genetica* **117**: 85–101.

Werren, J.H., Nur, U. and Eickbush, D. (1987) An extrachromosomal factor causing loss of paternal chromosomes. *Nature* **327**: 75–6.

West, S.A. and Godfray, H.C.J. (1997) Sex ratio strategies after perturbation of the stable age distribution. *Journal of Theoretical Biology* **186**: 213–21.

West, S.A. and Herre, E.A. (2002) Using sex ratios: Why bother? In: Hardy, I.C.W. (ed.) *Sex Ratios: Concepts and Research Methods*. Cambridge University Press, Cambridge, pp. 399–413.

West, S.A. and Sheldon, B.C. (2002) Constraints in the evolution of sex ratio adjustment. *Science* **295**: 1685–8.

West, S.A., Herre, E.A. and Sheldon, B.C. (2000) The benefits of allocating sex. *Science* **290**: 288–90.

West, S.A., Reece, S.E. and Sheldon, B.C. (2002) Sex ratios. *Heredity* **88**: 117–24.

West, S.A., Shuker, D.M. and Sheldon, B.C. (2005) Sex-ratio adjustment when relatives interact: A test of constraints on adaptation. *Evolution* **59**: 1211–28.

Whiting, P.W. (1943) Multiple alleles in complementary sex determination of *Habrobracon*. *Genetics* **28**: 365–82.

Whitten, M.J. and Hoy, M.A. (1999) Genetic improvement and other genetic considerations for improving the efficacy and success rate of biological control. In: Bellows, T.S. and Fisher, T.W. (eds.) *Handbook of Biological Control: Principles and Applications of Biological Control*. Academic Press, San Diego, pp. 271–96.

Wilkes, A. (1947) The effects of selective breeding on the laboratory propagation of insect parasites. *Proceedings of the Royal Society of London Series B Biological Science* **134**: 227–45.

Williams, G.C. (1975) Sex and Evolution. Princeton University Press, Princeton.

Wilson, K. and Hardy, I.C.W. (2002) Statistical analysis of sex ratios: an introduction. In: Hardy, I.C.W. (ed.) *Sex Ratios: Concepts and Research Methods. Cambridge* University Press, Cambridge, pp. 48–92.

Wu, Z., Hopper, K.R., Ode, P.J., Fuester, R.W., Chen, J. and Heimpel, G.E. (2003) Complementary sex determination in hymenopteran parasitoids and its implications for biological control. *Entomologia Sinica* **10**: 81–93.

Wu, Z., Hopper, K.R., Ode, P.J., Fuester, R.W., Tuda, M. and Heimpel, G.E. (2005) Single-locus complementary sex determination absent in *Heterospilus prosopidis* (Hymenoptera: Braconidae). *Heredity* **95**: 228–34.

Zayed, A. and Packer, L. (2005) Complementary sex determination substantially increases extinction proneness of haplodiploid populations. *Proceedings of the National Academy of Sciences USA* **102**: 10742–6.

Zhou, Y., Gu, H. and Dorn, S. (2006) Single-locus sex determination in the parasitoid wasp *Cotesia glomerata* (Hymenoptera: Braconidae). *Heredity* **96**: 487–92.

13

Linking foraging and dynamics

Michael B. Bonsall and Carlos Bernstein

Abstract

Behavioral ecology is rich in theories about how parasitoid behavior should affect host–parasitoid dynamics. For instance, models have explored the influence of ideal free and ideal despotic distributions or decisions such as when to accept superparasitized hosts, when to host feed, or when to search for alternative (non-host) resources. However, population dynamics predictions with different aspects of behavior are often in conflict, and different representations of the same behaviors or changes in the behavioral currency often give quite different or opposing results. To explore these discrepancies, we review the contemporary literature, and examine the role of individual behaviors to consider whether phenomenological models capture the broad ideas of behavior that are appropriate for understanding the population dynamics of host–parasitoid interactions. We will also consider new conceptual and methodological developments. For instance, we discuss the use of probabilistic based rules for behavior (e.g. reproductive queues, distributions of attack, leaving times) and how these scale up to affect the population and evolutionary dynamics. Over-arching all of this is the explicit consideration that pluralistic approaches of combining behavior, population dynamics, and evolution are necessary to understand host–parasitoid systems.

13.1 Introduction

As a discipline, population ecology fundamentally describes how the births, deaths, and dispersal of individuals affect changes in the distribution and abundance of species. Individual differences are often overlooked and populations are treated as a homogeneous collection of individuals. In contrast, behavioral ecology focuses on how individual decisions are shaped by evolution but not necessarily how this behavior translates to the population level. Even though one of the first aims of behavioral ecology was to link individual and population level phenomena, this aim was forgotten for many years but now is a contemporary challenge in evolutionary ecology. In this chapter, we will explore this theme of linking individual foraging behaviors to the population and evolutionary dynamics in host–parasitoid interactions.

Associations between insect hosts and their hymenopteran, coleopteran, or dipteran parasitoids have biological and ecological features that allow detailed investigation of their individual behavioral, population, and evolutionary ecology. In contrast to predators, where male, female, and juvenile stages must locate and consume prey, it is only adult female parasitoids that search for hosts. Describing parasitoid searching and oviposition behavior defines this type of interaction, and relating these behaviors to the population level dynamics is a venerable problem and a contemporary challenge (van Alphen et al. 2003). Further, the close relationship between oviposition behavior and fitness gains, at least for solitary parasitoids, provides a unique way to linking individual behavior and population dynamics to evolutionary processes.

One of the first approaches in exploring the link between individual parasitoid behavior and population dynamics was a model developed by Thompson (1924), who was concerned with biological control and how a parasitoid might affect the abundance of a pest population. In his model, he assumed that parasitoids would lay a single egg in each host, the number of eggs laid was directly equated with the number of host parasitized, and the encounters between hosts and parasitoids were assumed to occur at random. Although accurately describing the interaction when parasitoids were released into a region of high host abundance, this model of individual wasp foraging behavior was unable to provide an appropriate description of the population-level interaction when hosts were rare and thus parasitoid searching efficiency becomes important.

A more formal treatment of the link between individual parasitoid behavior and population dynamics was provided by Nicholson (1933) and Nicholson and Bailey (1935). Using a similar approach to Thompson (1924), these authors made two distinct and important assumptions in linking behavior to dynamics:

1 that the number of encounters with hosts by parasitoids occurs in direct proportion to host density; and

2 that these encounters occur at random.

Implicit in these assumptions is the idea of individual behaviors: individual parasitoids forage for hosts (in direct proportion to host density), and each encounter between a parasitoid and its host is a probabilistic event. What this describes then is a simple, albeit implicit, scaling from individual behavior to the population level:

$$\begin{cases} H_{t+1} = exp(r) \cdot H_t \cdot exp(-a \cdot P_t) \\ P_{t+1} = H_t \cdot [1 - exp(-a \cdot P_t)] \end{cases} \tag{13.1}$$

where r is the host (H) intrinsic rate of increase and a is the searching efficiency of the parasitoids (P). The individual foraging behavior in the Nicholson and Bailey (1935) model (Equation 13.1) is expressed through the probability that a particular host is not attacked by randomly searching wasps, and this is described by the zero term of the Poisson distribution ($exp(-a \cdot P_t)$). These early theoretical models (Thompson 1924, Nicholson & Bailey

1935) set the scene for the elaboration of how different host or parasitoid behaviors might affect the population dynamics and, over the last few decades, elaborations of these models has led to the rapid development of ecological theory describing host–parasitoid interactions (Hassell 1978, Hassell 2000, Murdoch et al. 2003).

In this chapter, we review the ideas and concepts associated with parasitoid foraging. In the next section, we explore how spatial heterogeneity, optimal foraging, and behavioral hierarchies are shaping our understanding of the details of parasitoid foraging. We go on to discuss the relevance and implications of these detailed findings for the population and evolutionary dynamics of host–parasitoid interactions. We argue that, while the detailed individual behaviors have been shaped by natural selection and is important for the dynamics of populations, the reductionist approach of the finer and finer dissection of these behavior may be difficult (but not impossible) to reconcile with the broader population-level interactions. Yet, the details of the behavior are fundamentally important to understanding the evolutionary dynamics, which do depend on understanding the population dynamics. We discuss these topics with reference to recent developments in parasitoid foraging biology.

13.2 Foraging theories

Foraging and foraging theory can be defined in terms of three criteria: (i) decision assumptions; (ii) currency assumptions; and (iii) constraints assumptions (Stephens & Krebs 1986). Decision assumptions can define the broad strategy sets but, for straightforward foraging theory, decisions center on which hosts to attack and when to leave a patch of hosts (see also Chapter 8 by van Alphen and Bernstein). Currency assumptions are used to compare the outcome of different decisions, and constraint assumptions define the factors that limit the relationship between currency and decision (Stephens & Krebs 1986). In this chapter, a central goal is to explore different descriptions of parasitoid foraging behaviors in terms of decisions, currencies, and constraints, and we illustrate how these can lead to different and often opposing population and evolutionary dynamical predictions.

Standard statistical theory predicts that individual behaviors will lead to nothing more than the sum of their parts. For instance, if individual parasitoids act independently in distributing attacks on hosts, then the overall distribution of parasitism will be predicted by the central limit theorem: the distribution of parasitism will follow a random distribution. This is our null hypothesis and we ask: How do departures from randomly distributed attacks scale up to affect the collective dynamics at the level of the population in terms of the dynamics and evolution?

The effects of randomly distributed attacks on hosts are known to lead to unstable and often divergent oscillations between hosts and parasitoids (Nicholson & Bailey 1935) and central to these dynamics is the functional response (the number of prey attacked per predator per unit of time). Functional response models (Solomon 1949, Holling 1959) are fundamental to the way in which we think about foraging and how behaviors might translate to affect population dynamics. Simple functional responses assume that parasitoids encounter hosts sequentially and each host provides a fixed fitness (offspring) gain. If a parasitoid decides to attack a host, then a fixed amount of time must be invested in pursuing, capturing, and/or ovipositing in the host. This handling time forms a component of the functional response models relating behavior in a broad way to population dynamics.

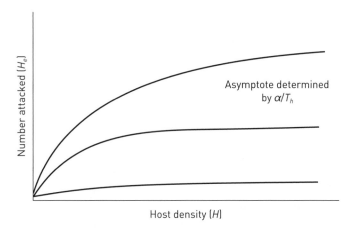

Number attacked (H_e)

Asymptote determined
by α/T_h

Host density (H)

Fig. 13.1 Type II Functional Response (Equation 13.2). Upper asymptote determined by the ratio between searching efficiency and handling time. Increases in handling time lead to low absolute values in this ratio (characterized by the different lines). This behavioral relationship between the number of hosts attacked and host density depends on the time necessary to handle each host and is known to lead to unstable population dynamics.

A simple derivation of Holling's (1959) disc equation follows from considering that the number of hosts subdued and attacked (H_e) is a function of: (i) the attack coefficient a which has units of hosts/time; (ii) the availability of hosts (H); (iii) the time available for searching (T); and (iv) the time necessary to handle each host (T_h). Thus:

$$H_e = \frac{\alpha \cdot T \cdot H}{1 + \alpha \cdot T_h \cdot H}.$$
(13.2)

The relationship between the number of hosts attacked (H_e) and host population size (H) is a convex function, with the asymptote determined by the ratio between the searching efficiency and the handling time (Fig. 13.1). Expressing hosts encountered as a function of the proportion of hosts attacked illustrates that, as the host population size increases, the proportion of hosts attacked declines. This effect of a decreasing death rate with increasing density is known to destabilize population dynamics and in particular this functional response model is expected to lead to unstable (and sometimes non-persistent) host–parasitoid dynamics (see below).

13.3 Spatial heterogeneity

Within heterogeneous environments, where hosts are distributed in patches of differing density, some hosts, by virtue of their spatial distribution, are less susceptible to attack than others. This presumes that parasitoids forage in a non-random way. Hassell (1978, 2000), for example, discusses in some detail how parasitoids tend to aggregate to patches of higher host density. A simple description of non-random foraging leading to differential distribution of parasitism is a power-law relationship of the form:

$$\beta_i = c \cdot \alpha_i^\mu \tag{13.3}$$

Here, β_i is the proportion of parasitoids relative to the proportion of hosts α_i in the ith patch, c is a normalization constant, and μ describes the strength of the aggregative response (Hassell & May 1973, 1974). If $\mu = 1$, parasitoids respond in a linear fashion to changes in host density while, if $\mu > 1$, parasitoids aggregate to high density patches and, as $\mu \rightarrow \infty$, parasitoids preferentially enter the highest density patches. Such non-random foraging leads to differential levels of parasitism amongst patches and is predicted to give rise to various patterns of parasitism. In general, correlations between host density and intensity of parasitism per patch may be positive, negative, or independent of density (Hassell 2000). Various attempts have been made to catalog the multitude of studies on patterns of parasitism. Reviews by Lessells (1985), Stiling (1987), Walde and Murdoch (1988), and Hassell and Pacala (1990) estimate the frequency with which these three patterns arise, and their consequences for the resulting population dynamics.

The original theory developed by Hassell and May (1973, 1974) proposed that aggregative behaviors and non-random distribution of parasitism would lead to positive patterns in the relationship between host density and the proportion of hosts parasitized. Development of this theory illustrated that different behaviors could lead to different patterns of parasitism and predictions about the local population dynamics. For example, Hassell (1982) illustrated that the range of patterns of parasitism can arise if parasitoids show a behavioral (aggregative) response to patchily distributed hosts (Equation 13.3) and if exploitation within a patch is limited by factors such as handling time. This patch-to-patch exploitation model (Hassell 1982) illustrates the necessity to consider multiple (and often conflicting) behaviors in determining how non-random foraging might affect the dynamical interaction between hosts and parasitoids.

13.4 Optimal foraging

Appreciating that consumers (such as parasitoids) are more likely to forage non-randomly, and that searching for hosts leads to differential patch-time allocation strategies, was instrumental in the development of the ideas of optimal foraging (MacArthur & Pianka 1966, Emlen 1973, Charnov 1976). In terms of foraging theory, optimal foraging describes the ideal allocation of searching time to maximize resource capture rate, and hence reproductive fitness. The individual-level patch exploitation theory (i.e. the marginal value theorem, Charnov 1976) is based on travel times amongst patches and the marginal gains achieved by differential patch residence times (see also Chapter 8 by van Alphen and Bernstein). Long travel times between patches and highly profitable patches tend to promote long residence times. Very often, parasitoids deplete patches in which they forage. Non-ovicidal idiobiont parasitoids, for instance, paralyze hosts and these are then no longer accessible to parasitoids which forage for actively moving hosts. Through time, patch profitability declines and the optimal parasitoid has to choose when on the curve of diminishing returns to leave the patch and forage elsewhere. In order to maximize foraging and oviposition rate, a parasitoid chooses to attack hosts in a patch until the tangent to the diminishing returns curve is as steep as possible (Fig. 13.2). If the habitat contains a variety of patches (of differing host density or quality) the optimal solution is then to stay in each patch until the intake rate is equivalent to the net rate of food intake across

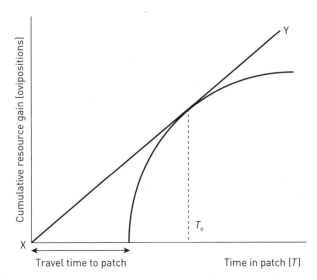

Fig. 13.2 Optimal foraging time in a patch in which food is depleted (from Krebs 1978). The curve represents the cumulative gain in resources as function of time on a patch. Optimal parasitoids should choose to stay long enough to maximize the tangent (line XY). Parasitoids should do this by leaving a patch at T_o.

the whole habitat (see also Chapter 8 by van Alphen and Bernstein). Then, the overall tendency is for individual wasps to deplete patches so that all patches have the same rate of host capture.

Determining the intake curve within a patch and the travel time between patches allows predictions on the time spent by an individual in each patch, the number of hosts attacked per patch, or the marginal oviposition rate to be made (Krebs 1978). By extending a version of the disc equation (Rogers 1972), Cook and Hubbard (1977) show that this version of optimal foraging may hold for insect parasitoids. Their depletion model predicted that all host patches should be reduced to the same threshold encounter rate, and that the time allocation between patches is dependent on the availability of patch types, the density of foraging wasps, and the time available for foraging. Using the ichneumonid parasitoid, *Venturia* (*Nemeritis*) *canescens*, Hubbard and Cook (1978) tested these predictions. They showed that all the patches that wasps used are reduced to the same rate of host encounter per unit of time and that giving-up times (patch leaving times) are modified in response to the availability of hosts. However, one of the main differences in the predictions of Cook and Hubbard's (1977) model, and other models of optimal foraging, is that the individual time budgets depend upon the level of exploitation during the search time: the amount of time spent on high density patches declines as exploitation increases. It should be pointed out that, although they apply their model to groups of foraging parasitoids, it is not shown whether or not the optimal solution is evolutionary stable.

A population level concept related to optimal foraging is the ideal free distribution (IFD) in which parasitoids distribute their foraging activities in an evolutionary stable manner. Under the IFD (Fretwell & Lucas 1970, Harper 1982) individuals are free to distribute their foraging activities across a range of patches and there is a preferential tendency for individuals to go to patches where rewards are highest. IFDs assume that foragers are omniscient, travel costs are limited, and average fitness benefits should be equal for individuals in different patches. If one patch has high resources, then individuals should move to balance out the costs of resource depletion (by conspecific consumers) and the benefits of high host availability. In making the explicit link between the effects of conspecifics (through interference) and resource maximization, Sutherland (1983) argued that the IFD involves an alteration in the functional response that could lead to parasitoids being distributed as a function of not only host density but also due to the strength of interference. Interference can be defined as a decline in foraging efficiency as conspecific density increases. As parasitoids aggregate, they are more likely to interact with conspecifics searching for hosts in the same area or on the same patch (Hassell 1978). The common effect of this behavior is to reduce the available search time in direct proportion to the frequency of encounters.

Field and laboratory experimental tests of this theory have been developed using a variety of host–parasitoid interactions. For instance, using the parasitoid *V. canescens*, Tregenza et al. (1996a,b) explored the predictions of the effects of interference and ideal free foraging. They showed that parasitoid fitness (gain rate) declined as density of parasitoids on a patch increased (due to the effects of interference). More importantly, they showed that the effects of sampling behavior and perceptual constraints also affected the gain rate across patches of different host densities, and that wasp movement was high during the initial period of the experiments. Further, field studies on host–parasitoid interactions have shown that the distribution of foraging activity of parasitoids may not necessarily follow a pure IFD (Williams et al. 2001). Foraging activity of the parasitoids (*Pteromalus elevatus*, *P. albipennis*, *Torymus* sp. A, *Torymus* sp. B) of the tephritid, *Terellia ruficauda*, are more heavily distributed on hosts in isolated, as opposed to crowded, patches (Williams et al. 2001). It was argued that this can result in an inverse density-dependent pattern of parasitism. Although the mechanisms behind this pattern of parasitism are well understood from the behavior of the parasitoids through handling time limitations (Hassell 1982), the interplay between interference and the distribution of foragers among patches is more likely to involve combinations and mixtures of different behaviors. In fact, models by Bernstein et al. (1988, 1991) have revealed that foraging by non-omniscient animals relying on their individual experience and learning can also lead to an IFD. In these models, parasitoids base patch-leaving decisions on private estimations of overall host availability environment, and this leads to a mixture of migration patch exploitation patterns.

If ideal free strategies are more likely to involve a mixture of behaviors, then this has a range of implications for the local distribution of individuals and the resulting population dynamics. Parasitoids are well-known to use hosts for feeding on as well as for ovipositing in (see also Chapter 7 by Bernstein and Jervis). This necessity of resources to develop eggs (synovigeny) can affect the expected lifespan and total reproductive success. For instance, using a stochastic dynamic approach, Sirot and Bernstein (1996) have shown that, when metabolic resources and hosts are in different parts of the environment, a state-dependent IFD occurs and is determined by the number of conspecifics and the level of interference (see also Chapter 7 by Bernstein and Jervis). This has implications for the population dynamics of host–parasitoid interactions.

A more likely, but less explored, foraging distribution is that described by the ideal despotic distribution. This distribution of foragers can occur in species with structured hierarchies. Here, the IFD may no longer apply as individuals are not free to move amongst all patches. The first arriving at a patch may gain a hold of the resources and obtain a highest fitness rewards (e.g. all hosts are susceptible and the costs of superparasitism are low) and average fitness may vary between patches. Acquiring information on the presence of conspecifics, and interacting with them, is often a more reliable indicator of foraging success (van Alphen et al. 2003, Wajnberg et al. 2004) and the ideal despotic distribution might be a more realistic framework in which to link individual behaviors to the broader collective dynamics.

13.5 Queues, individual behaviors, and collective dynamics

A collorary of this despotic distribution and structured hierarchies is the activity where individuals form some sort of queue (or there is reproductive skew – differential ovi-position rates amongst individual parasitoids) such that some individuals have different residence times on patches to achieve higher fitness. Application of queuing theory (Newell 1971) and its optimization to foraging theory remain relatively undeveloped in behavioral ecology but may provide useful ways in which we might think about how individual beha-viors translate to broader collective patterns. One way to approach this is to measure and analyse features of foraging parasitoids on a scale where a single foraging individual is a negligibly small quantity. This assumption might not be too far-fetched as foraging across an environment for hosts is likely to involve large numbers of interacting wasps. The strat-egy in understanding these ideas is to think of foraging and the distribution of foragers as a queue. Each individual parasitoid arriving on a patch is serviced by that patch (oviposits) and then leaves. This might describe a hierarchical structure (e.g. reproductive skew) or different residence times on patches. As such, the details of the arrival and departure rules might differ and we could postulate a stochastic model to represent this queue and derive some underpinning probability distribution (with unknown parameters). On the other hand, we might postulate that if there are large numbers of foragers, the relative change in the number of parasitoids through arrival and departure rules might be negligible: uncertainty in the random fluctuation of foragers arriving on a patch will be small compared to the observed number of arrivals. The size of the queue ($Q(t)$) on a particular patch can be described extremely simply as

$$Q(t) \approx Q(0) + (\lambda - \mu) \cdot t \tag{13.4}$$

where λ and μ are the arrival and departure rates per unit time t. If $\lambda < \mu$, the queue decreases and, if $\lambda > \mu$, the queue increases. For some smooth cumulative arrival function (A), the instantaneous arrival rate can be defined as

$$\frac{dA(t)}{dt} = \lambda(t) \tag{13.5}$$

Figure 13.3 shows a typical arrival and departure curve, where arrival and departure rates are constant through time. The difference at any point between $A(t)$ and $D(t)$ gives the instantaneous delay (instantaneous patch residence time), and the total delay (total patch residence time) is obtained from the area between $A(t)$ and $D(t)$, or by integration:

$$Q(t) = A(t) - D(t) = \int_{t\,min}^{t\,max} [\lambda(s) - \mu] ds \tag{13.6}$$

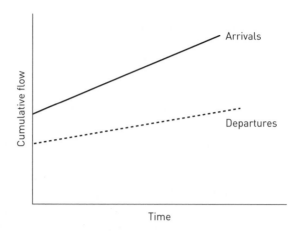

Fig. 13.3 Schematic representation of the flow through a queue due to time-dependent arrivals (A(t)) and time-independent departures (D). Differences between the two curves give the instantaneous delay (patch residency) and the area between the two lines is the total time delay.

It follows that the number of foragers waiting to oviposit is a maximum if $dQ/dt = 0$, and this occurs when $\lambda(t) = \mu$. The implications of this, from an adaptive perspective, are that individuals should adapt their foraging decisions based on the number of conspecifics in the local environment (van Alphen et al. 2003): long queues are prohibitively expensive in terms of future reproductive success. Stochastic versions of this basic theory can also be derived. For instance, if the arrivals and departures in time t are approximately normal and statistically independent, then the difference in the size of the forager queue in time t is the difference between $A(t)$ and $D(t)$. While the mean change in queue length is simply the difference between the expectation of arrivals and the expectation of departures, the variance in change in queue length is

$$Var[A(t) - D(t)] = Var[A(t)] + Var[D(t)] \tag{13.7}$$

If, for example, arrival and departure rates are constant (a Poisson process):

$$E[A(t)] = \lambda \cdot t \tag{13.8}$$

$$E[D(t)] = \mu \cdot t \tag{13.9}$$

then, the variance in queue length is

$$Var[Q(t)] = t \cdot (\lambda + \mu) \tag{13.10}$$

Queue length increases linearly with time and with the difference between arrivals and departures (Equation 13.6), while the standard deviation increases as $\sqrt{t}$ and as the square-root of the sum of the arrivals and departures. The implications of this are that understanding the variance in queue length, and consequently previously levels of patch exploitation, is critically important to foraging decisions: while prohibitively long queues might restrict foraging activities, so might previous arrivals and departures from the local environment. Understanding patch residence times or foraging activities, in terms of reproductive skew, might be profitably understood by the application and development of these sorts of stochastic models.

Understanding the patch residence time or rules about when to give up and search elsewhere are of fundamental importance to linking behaviors, dynamics, and evolution. As already mentioned, time invested in non-oviposition activities can affect individual fitness and the collective effects on population dynamics (see also Chapter 7 by Bernstein and Jervis; and Chapter 9 by Haccou and van Alphen). The notion of the time spent on a patch, the determinants of such behaviors, and the mechanisms associated with leaving have been thoroughly explored using insect parasitoids (Wajnberg et al. 2006). Waage (1979), using the parasitoid wasp *V. canescens*, suggested that residence time on a patch is determined by the density of hosts and the rate of oviposition, both of which lead to longer residence times. Patch leaving is influenced not simply by the time since the last oviposition but by all ovipositions on a patch (van Alphen et al. 2003, Wajnberg 2006, see also Chapter 8 by van Alphen and Bernstein). This behavioral assessment of patch quality provides a more accurate assessment response by the forager than a simple threshold model or stochastic variants (Oaten 1977, see also Chapter 16 by Pierre and Green).

More thoroughly, these leaving rules are best couched as probability models that describe the chance that an individual parasitoid leaves a patch per unit of time given that the individual is in the patch. Haccou et al. (1991) considered how the patch leaving tendency in a *Drosophila* parasitoid, *Leptopilina heterotoma*, is reset after each oviposition event and after each time a patch is left and re-entered. As might be expected, there is an effect that the leaving tendency decreases with each additional oviposition event (Waage 1979, Haccou et al. 1991). Long foraging bouts between ovipositions increase the likelihood that a wasp will leave a patch, although tendency to return has little effect on subsequent leaving times (Haccou et al. 1991).

As noted by Waage (1979), host density can also affect patch residence times. As patch density increases, leaving probability is expected to decline as the expected fitness return from increased encounters with healthy hosts offsets the travel costs of searching for new patches. In low host density or uniform density environments, oviposition events by parasitoids such as *V. canescens* decrease the amount of time spent in a patch (Iwasa et al. 1981, Driessen et al. 1995). In fact, the key factor here is how reliable is the information that a parasitoid has on host availability in the patch? Reliable information requires a countdown mechanism and a uniform distribution is a particular case through which such information can be obtained. As ovipositions accumulate, each oviposition event leads to a decrease in the current level of availability and this countdown mechanism allows individual wasps to adjust patch persistence to the initial profitability of the patch and realized fitness gains. It has been argued that these patch leaving rules are under significant genetic variation (Wajnberg et al. 1999, 2004). Exploring interactions amongst conspecifics, Wajnberg et al. (2004) also illustrate that detailed interaction amongst individuals are important.

Knowledge of the competitive dominance, which is under genetic control, of other individuals appears to be crucial to determining patch residence strategies and leaving rules. Proximate descriptions of all of these foraging patterns can be achieved by considering the adoption of particular behaviors or strategies as state-dependent processes. Whether to forage in a particular patch depends on host density, the number of ovipositions already delivered in a patch, the presence of conspecifics, and how many times a patch as been re-entered. This state-dependency determines successful foraging and in emphazing this, Wäckers (1994) has shown that the nutritional state of the parasitoid *Cotesia rubecula* influences its choice between finding hosts for oviposition and resources to survive (see also Chapter 7 by Bernstein and Jervis). This state-dependent behavioral information introduces an explicit, additional temporal component (Haccou & Meelis 1994) to foraging that has important collective population dynamic consequences.

13.6 Population dynamics

Theory on the dynamics of host–parasitoid systems predicts that these interactions have an inherent tendency to lead to divergent oscillations. When the parasitoid is initially rare, then the host population increases exponentially (in the absence of any other regulating factor). As the hosts become more abundant, the size of the parasitoid population increases. This has a detrimental effect on the host population leading to a decline in host numbers. As the host becomes scarce, then the parasitoid population declines and due to the generation time lag the cycle starts again with ever increasing amplitude. This intuitive description of the discrete-time dynamics is captured by the Nicholson and Bailey (1935) model (Equation 13.1) and a similar continuous-time version was proposed independently by Lotka (1925) and Volterra (1926). These models describe the simplest behavior of a foraging parasitoid – the distribution of attacks is proportional to prey density and it is this type of general functional response that leads to the unstable population dynamics.

The interest in linking behavior to dynamics has led to the development of a strong theoretical basis for investigating how host and parasitoid behavior might modulate, mitigate, or alter the inherent tendency of these predator-prey interactions to show diverging oscillations (Bailey et al. 1962, Hassell & May 1973, 1974, Chesson & Murdoch 1986, Hassell et al. 1991, Pacala & Hassell 1991). One broad consequence for the population dynamics of these behaviors is that the time available for searching declines as density of foragers increases. This density-dependent response has been described in a number of ways. For instance, Hassell and Varley (1969) proposed that one manifestation of this effect would be through mutual interference. Interactions amongst wasps leads to a reduction in searching efficiency as wasp density increases (Hassell & Huffaker 1969) and the stronger this interference effect, the more likely it is that the dynamical interaction between a host and its parasitoid will be stable (Hassell & May 1973).

However, this reduction in searching efficiency with increasing density can arise without any direct behavioral interference. From the Nicholson and Bailey (1935) model (Equation 13.1), the per capita searching efficiency is

$$a = \frac{1}{P_t} \cdot ln\left[\frac{H_t}{H_t - H_e}\right] \tag{13.11}$$

where H_e is the number of hosts attacked, P_t is the density of searching parasitoids, and H_t is the abundance of susceptible hosts. Other things being equal, a declines as P_t increases. This effect is known as pseudo-interference (Free et al. 1977) and arises when the risk of parasitism (for whatever reason – captured by Equation 13.11 through the

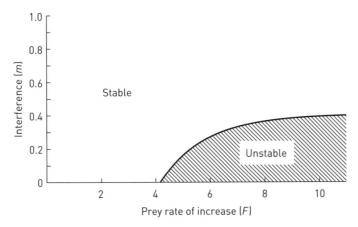

Fig. 13.4 Stability boundaries for a model in which interference and aggregation affect the dynamics of a host–parasitoid interaction (from Sutherland 1983).

difference between susceptible host density and the number of hosts attacked) varies between host individuals. The critical condition for population stability is how the fraction of hosts parasitized changes with alterations in parasitoid density. If the per capita parasitoid foraging efficiency declines as parasitoid density increases, the effects of density dependence are experienced (birth rate declines as density increases) and the parasitoid population growth rate is constrained.

Pseudo-interference predicts stable population dynamic interactions between a host and its parasitoid. The effects of interference contribute to population stability through a non-random distribution in parasitism. On a similar theme, Sutherland (1983) shows that coupling interference and aggregation (leading to an IFD) can also affect the stability of host–parasitoid interactions. In particular, avoidance of interference through parasitoid redistribution leads to a wide set of stability conditions (Fig. 13.4) as interference and aggregation interact to affect the strength of density dependence acting on the parasitoid population. Similar effects are also observed by considering more explicit distributions of foraging parasitoids. For example, from a theoretical perspective, Comins and Hassell (1979) illustrate that optimal foraging can contribute to the stability of a host–parasitoid interaction if the host is sufficiently unevenly distributed amongst patches. As noted by Comins and Hassell (1979), these effects of optimal foraging on the persistence and stability of host–parasitoid interactions are in broad agreement with the effects of fixed parasitoid aggregation strategies (Hassell & May 1973, 1974).

The effects of foraging strategy on the dynamics of host–parasitoid interactions are also likely to be influenced by other factors than by simply maximizing the rate of encounters with healthy hosts. In particular, avoiding other parasitoids or acquiring resources or mates are all equally likely attributes that might be maximized to achieve maximum fitness returns. The dependency of current state on future reproductive or foraging decisions is an important aspect in relating behavior to dynamics (Mangel & Clark 1988, Clark & Mangel 2000). As mentioned above, Sirot and Bernstein (1996) explored a situation in which the distribution of parasitoids is determined by the energy state of the wasps (see also Chapter 7 by Bernstein and Jervis). This state-dependent IFD could promote stable host–parasitoid interactions as it affects the density of parasitoids searching for hosts at any one instance in time. Křivan (1997) considers a more dynamic interaction between the (ideal free) distribution of foragers and the population dynamics of a host–parasitoid interaction. By assuming parasitoids and/or hosts are free to move, then the dynamics of interaction can be determined from the maximization of per capita growth rate functions for hosts:

$$max\left[\sum_i (r_i - f(H,P)_i) \cdot Pr[H_i]\right] \tag{13.12}$$

and for parasitoids:

$$max\left[\sum_i (f(H,P)_i - g(P_i)) \cdot Pr[P_i]\right] \tag{13.13}$$

respectively. Here, r_i is the host growth rate, $f(H,P)_i$ is the function for parasitism (i.e. the functional response), and $g(P_i)$ is the death rate of the parasitoid in the ith patch. $Pr[H_i]$ is the probability of a host being in the ith patch, and similarly, $Pr[P_i]$ is the probability that a parasitoid is in the ith patch. Optimal control, with respect to the distribution of host and parasitoids, reveals that the adaptive response, if parasitoid abundance is low, is for both the hosts and parasitoids to aggregate in the patch with the highest host growth rate. Hosts aggregate in this patch due to low parasitoid abundance, while the parasitoids preferentially prefer this patch as host density is increasing (Křivan 1997). Even though an IFD is achieved, the population dynamics of this sort of interaction in an environment with a small number of patches tend to predict limit cycles (Křivan 1997). However, it remains entirely plausible that the adaptive response (through optimal control) and the IFD of hosts and parasitoids in complex, multi-patch environments might interplay with the population dynamics to promote other types of population dynamics such as stable equilibrium or highly non-linear dynamics. Bernstein et al. (1999) developed this theme to explore how the influence of parasitoid movement can affect the IFD and population stability. Using a set of differential equations to describe a three-patch system, these authors showed that the problem of different timescales between individual level phenomena and population processes can be disentangled. Different cognitive or perceptual abilities (e.g. parasitoids are omniscient, limitations in making optimal decisions or constraints on patch intake rate assessment), and movement rules (migrate whenever current intake rate is less that the expected value for the whole environment), affect the IFD and stability of the system. When migration rates are high, the parasitoid population converges to the IFD and parasitoid movement plays no role in the stability of the trophic interaction. If migration rates are lower, or decisions are less than optimal, then the factors that hinder the convergence of the population to the IFD also affect the stability of the system by imposing levels of spatial heterogeneity.

Cressman and Křivan (2006) have shown that movement dynamics are central to understand the IFDs. By using a series of theoretical models, these authors make a number of predictions including that foraging by non-omniscient individuals and stable population dynamics can both lead to an IFD.

To explore how patch selection affects the ecological stability of host–parasitoid interactions, van Baalen and Sabelis (1993, 1999) have also developed a series of evolutionary and co-evolutionary models. If selection acts on parasitoids alone, then the evolutionary stable strategy (ESS) is that parasitoids are expected to develop an aggregative response to unevenly distributed prey. Here, the ESS specifies an IFD since no parasitoid does any better by visiting alternative patches. Under co-evolutionary host–parasitoid models, if patch quality is homogeneous, then it is expected that selection on host foraging will be to choose low density patches leading to a more evenly distributed population. However, variability in patch quality leads to differential host foraging and uneven (aggregated) distribution of hosts. Again, neither parasitoids nor hosts gain by changing patch selection strategies, and this rather restrictive ESS condition, where both hosts and parasitoids aggregate across patches, might lead to ecological stability in the trophic interaction (van Baalen & Sabelis 1993).

In contrast to this work, Schreiber et al. (2000) have explored how non-congruent patch selection strategies by hosts and parasitoids affect the ecological dynamics of these trophic interactions. If life history traits of hosts and parasitoids vary amongst patches, then a range of effects can be expected depending on whether hosts or parasitoids are maximizing their fitness. Under fixed parasitoid strategies, hosts are expected to lay eggs in patches that suffer no or low levels of parasitism. Under co-evolving host and parasitoid strategies, the joint ESS occurs when hosts and parasitoids preferentially select patches that lead to an IFD of hosts and parasitoids, and thus higher population equilibria. Patch heterogeneity can lead to contrary choices in which hosts and parasitoids prefer different patches. Such effects can lead to the full range (positive, independent, and inverse) patterns of parasitism (Schreiber et al. 2000).

It is the distribution of parasitoids that is central to linking foraging and dynamics and, as mentioned above, initial theory focused on the key assumption that parasitoid aggregation is critical for stabilizing host–parasitoid interactions (Hassell & May 1973, 1974, Murdoch & Oaten 1975, Murdoch 1977). Of over-riding importance is the distribution of parasitism by non-randomly foraging parasitoids. Theory shows that aggregation to patches of higher host density can stabilize the otherwise non-persistent Nicholson and Bailey (1935) model (Hassell & May 1973, 1974). Following these initial theoretical investigations, a series of papers (Beddington et al. 1978, May 1978, Hassell, 1980, May & Hassell 1981, Hassell 1982) confirmed that parasitoid aggregation on patch of high host density (leading to positive patterns of parasitism) can contribute to the stability of the population interaction between a host and its parasitoid. However, based on laboratory and field data on parasitism, Pacala and Hassell (1991) suggested that, in spite of its stabilizing potential, parasitoid aggregation in patchy environments is often not sufficient to stabilize host–parasitoid systems on its own. Additional aspects of host and/or parasitoid ecology need to be factored into determining the processes leading to stable host–parasitoid dynamics.

As originally highlighted by Reeve et al. (1989), and explored more fully by Gross and Ives (1999), inferring the temporal dynamics of host–parasitoid interactions from static spatial snapshots of the patterns of parasitism is complicated by a number of processes such as the foraging activity of wasps within and between patches. Scaling up from patches to dynamics may be thwarted with difficulties and an alternative, potentially more robust way to test the link between behavior and dynamics is to examine the population dynamic consequences of patchily-distributed foraging activities in well-replicated experiments. In this approach, the behavior of how the natural enemy might affect dynamics can be explored by investigating the dynamical patterns observed in time series and contrasting them with predictions from relevant behavioral ecological models.

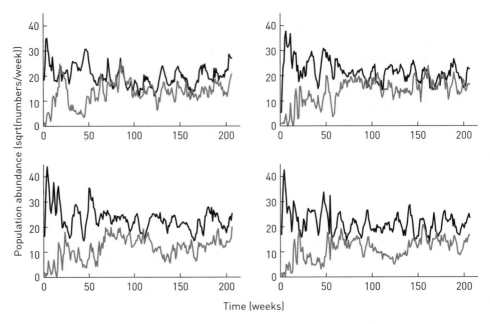

Fig. 13.5 Replicate time series of *D. ananassae* and *L. victorae* in a patchy environment. The dynamics of this host–parasitoid interaction is described by a behavioral state-independent model (May 1978) that captures the broad effects of spatial heterogeneity. (Black line: *D. ananassae*, Gray line: *L. victorae*).

Evidence from the interaction between *Drosophila ananassae* and its parasitic wasp *Leptopilina victorae*, suggests that the local behavioral activity of the wasp on the population dynamics can be understood from model averaging approaches and comparison of competing explanations (Burnham & Anderson 2002). Replicated population dynamic data of *Drosophila* and *Leptopilina* are shown in Fig. 13.5 and, in a range of ecological models incorporating different behavioral attributes, were compared to the data (Table 13.1). In these patchy environments, the most appropriate model for the dynamics of the *Drosophila–Leptopilina* interaction is

$$\begin{cases} \dfrac{dH}{dt} = r \cdot H(t) - H(t) \cdot k \cdot log\left(1 + \dfrac{\alpha \cdot P(t)}{k}\right) \\[4mm] \dfrac{dP}{dt} = H(t) \cdot k \cdot log\left(1 + \dfrac{\alpha \cdot P(t)}{k}\right) - d_p \cdot P(t) \end{cases} \qquad (13.14)$$

where *r* is the intrinsic rate of increase of the host population (*H*), α is the parasitoid searching efficiency, *k* is the clumping parameter of the negative binomial distribution describing how the skew in the distribution of parasitism (small *k* implies highly heterogeneous pattern of parasitism), and d_p is the parasitoid death rate. The best fitting candidate model and its nearest rival both include aspects of parasitoid (rather than host) biology as a significant determinant of the dynamical patterns in this host–parasitoid interaction. These results suggest that, although the broad population level description of behaviors such as mutual

Table 13.1 Candidate functions used to describe different parasitoid behaviors in the interaction between *D. ananassae* and *L. victorae*. Models were fitted (using a Gaussian likelihood) collectively to all the time series data (Fig. 13.5). The best description of the functional response underpinning the population dynamics is based on evaluating AIC (Burnham & Anderson 2002). Model comparisons are made from the difference between the most parsimonious model (a state-independent description of behavior – function 3) and all other functions (ΔAIC).

	Functional response	Likelihood	ΔAIC
1	$\alpha \cdot P_t$	11644.221	1986.342
2	$\dfrac{\alpha \cdot P_t}{1 + T_h \cdot H_t}$	11663.110	2026.120
3	$k \cdot log\left(1 + \dfrac{\alpha \cdot P_t}{k}\right)$	10651.050	–
4	$\dfrac{\beta \cdot P_t^{-m}}{1 + T_h \cdot \beta \cdot P_t^{-m} \cdot H_t}$	11727.706	2155.312
5	$\dfrac{\alpha \cdot P_t}{1 + v \cdot (P_t - 1)}$	10717.060	133.690

interference (Hassell & Varley 1969, Beddington 1975) or resource distribution models (Sutherland 1983) provide insight into the theoretical predictors of population stability, the specific behavioral differences are not distinguished at the population level and the details of the particular interaction remain important. Further, it is possible to determine the effects of parasitoid foraging on the population dynamics without understanding the mechanistic details of the behavior (Ives 1995). Simple descriptions of the effects of spatial heterogeneity on parasitoid foraging (May 1978) can provide a robust description of the population dynamic interactions between hosts and parasitoids, suggesting that a state-independent approach could be used to describe how individual behaviors to patchily distributed hosts translate to affect the population dynamics.

In approaching the problem of linking behavior and dynamics, Ives (1995) suggests that it might be more feasible to characterize variability in parasitoid behavior in terms of the number of hosts attacked and the time parasitoids stay in patches. Predictions from the simple queuing models (see above) and the mechanistic leaving rule approaches (Waage 1979, Haccou et al. 1991, Driessen et al. 1995, Driessen & Bernstein 1999) may provide a solution to this dilemma. Queuing models provide a stochastic distribution to patch residence times and leaving-rule approaches aim to explain this variability in terms of the time between encounters, the frequency with which hosts are rejected, and patch-leaving tendencies. These aspects of parasitoid foraging are not directly explained by the simple relationship between the number of hosts in a patch and the pattern of parasitism as originally proposed by Hassell and May (1973, 1974). There is unexplained variability in the distribution of parasitism with respect to host density and, as mentioned above, this contributes to the stability and persistence of host–parasitoid interactions (Chesson &

Murdoch 1986, Hassell et al. 1991). It is likely to have important effects on the dynamical persistence and stability of many host–parasitoid interactions.

13.7 Evolutionary dynamics

Notwithstanding the effects of taking a generic approach in linking behavior to population dynamics, incorporating aspects of foraging into the evolution of host–parasitoid interactions is a developing area of parasitoid behavioral ecology (van Baalen & Sabelis 1993, 1999, Bernstein et al. 1999, Schreiber et al. 2000). One way to explore how different behaviors affect the evolutionary dynamics of host–parasitoid interactions is to use 'adaptive dynamic' game theory (Metz et al. 1992, Vincent & Brown 2005) and ask how a parasitoid (or host), which plays a different strategy, affects the population and evolutionary dynamics of a resident host–parasitoid interaction. To motivate the idea, we present a structured host–parasitoid model:

$$
\begin{cases}
\dfrac{dH}{dt} = r \cdot H(t-\tau) \cdot \sigma_H - f(P_i,H) \cdot H(t) - \mu_H \cdot H(t) \\[2ex]
\dfrac{dJP_i}{dt} = H(t) \cdot f(P_i,H) - g(P_j,JP_i) \cdot JP(t)_i \\[2ex]
\dfrac{dP_i}{dt} = (1-\pi_{ij}) \cdot g(P_j,JP_i) \cdot JP(t)_i + \pi_{ji} \cdot g(P_i,JP_j) \cdot P(t)_i - dp \cdot P(t)_i
\end{cases}
\qquad (13.15)
$$

where r is the host, (H) intrinsic rate of increase, σ_H is the survival of the pre-susceptible host stage, $f(P_i,H)$ is the attack rate of the parasitoid (P_i) on the host, and μ_H is the mortality rate of the host. Hosts already attacked and harboring a developing parasitoid (JP_i) may suffer further $g(P_j,JP_i)$ attack by heterospecific parasitoids (P_j). Free living parasitoids of type i emerge from hosts that avoid subsequent attack $(1-\pi_{ij} \cdot g(P_j,JP_i))$ and from successful multiparasitism events ($\pi_{ji} \cdot g(P_i,JP_j)$). π_{ij} is the probability of a successful multiparasitism event by parasitoid j on parasitoid i and can govern competitive ability. For instance, if the probability of a successful multiparasitism event is related to attack rate, then an appropriate description of this competition (Bonsall et al. 2004) is

$$
\pi_{ij} = \frac{1}{1 + exp(\alpha_i - \alpha_j)}
\qquad (13.16)
$$

where α_i and α_j are the attack rates of the ith and j^{th} parasitoid, respectively. To explore how different behaviors affect the evolutionary dynamics of host–parasitoid interactions, we ask how different functions for parasitism affect the probability of coexistence between different parasitoids. For instance, under simple mass action function for parasitism ($f(P_i,H) = \alpha_i \cdot P_i$), the fitness (per capita growth rate) of an invading parasitoid strategy j is

$$
\frac{1}{P_j}\frac{dP_j}{dt} = \alpha_j \cdot (1-\pi_{ji}) \cdot JP_i^* - dp_j
\qquad (13.17)
$$

where JP_i^* is the equilibrium density of host harboring developing parasitoids determined from the resident host–parasitoid interaction. Positive fitness is ensured if the per capita

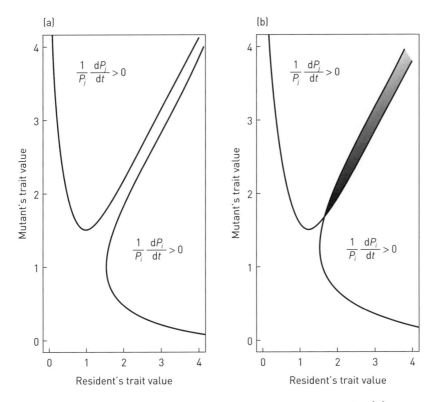

Fig. 13.6 Boundaries of coexistence for competing parasitoids under (a) mass-action (Type I) foraging and (b) mutual interference. In the presence of behaviors such as interference, the coexistence of alternative strategies is feasible and the community dynamics are shaped by evolution. Shaded region denotes areas of coexistence.

growth rate is greater than zero. It is well-known that this strategy does not promote niche differentiation (Briggs 1993, Bonsall et al. 2002): there is no possibility for long-term coexistence (Fig. 13.6a). In contrast, by incorporating behaviors such as mutual interference, the function for parasitism is (Beddington 1975):

$$f(P_i, H) = \frac{\alpha_i \cdot P_i}{1 + v \cdot (P_i - 1)} \tag{13.18}$$

where v is the time wasted interfering with conspecific parasitoids. The per capita fitness function is

$$\frac{1}{P_j} \frac{dP_j}{dt} = \frac{1}{1 + v_j \cdot (P_i{}^* - 1)} \left[\alpha_j \cdot (1 - \pi_{ji}) \cdot \frac{dp \cdot (1 + v_i \cdot (P_i{}^* - 1))}{(1 - \pi_{ij}) \cdot \alpha_i} \right]. \tag{13.19}$$

Here, provided the interference strategies are different amongst the parasitoid strategies, there is the possibility of coexistence (Fig. 13.6b). This occurs through a life-history trade-off in competitive ability and attack rate. Other aspects of parasitoid behavior (superparasitism, multiparasitism) have recently been shown to affect diversity and diversification in host–parasitoid interactions and provide broader interpretation to the debate on neutral versus niche structured assemblages (Bonsall et al. 2004, Gravel et al. 2006, Scheffer & van Nes 2006). The invasion and coexistence of alternative strategies is possible as the behavioral effects of interference operating within any particular parasitoid strategy outweigh the competitive effects operating between strategies. While these behaviors may not necessarily affect the population dynamics, population level interactions provide the stage within which evolutionary games are played out and the link between the evolution of behavior, population, and adaptive dynamics is only starting to be realized. For instance, Abrams and Kawecki (1999) show that adaptive parasitoid foraging across multiple host types – such that the dynamical interaction is of an apparent competition (Holt 1977, Bonsall & Hassell 1997) – frequently destabilizes the dynamics and can affect host abundances and distributions. Asking how behaviors, such as multiple attacks on hosts by conspecific and heterospecific wasps, also has implications for understanding how communities of parasitoids are assembled through evolutionary processes (Bonsall et al. 2004). Further, parasitoids are likely to constantly redistribute to preserve equal gains in all patches. Such intra-generational effects are often not accounted for in the descriptions of adaptive evolutionary dynamics of foraging (Cressman & Křivan 2006). Understanding the effects of movement and foraging on the evolutionary dynamics of host–parasitoid interactions thus requires further exploration.

13.8 Coda

As outlined throughout this chapter, the main aim of behavioral ecology is to interpret behavior in terms of the fitness consequences of an animal adopting a given strategy. As argued here, and in other chapters of this book, many behavioral phenomena (i.e. searching for food, social foraging, state-dependent processes, patch residence times) affect the time a parasitoid can dedicate to host searching and oviposition. Changes in the time allocated to different strategies not only affect forager fitness but also the broader-level dynamics. For instance, details on how optimal decisions affect the distribution of parasitoids are also likely to affect the evolutionary dynamics of host–parasitoid interactions. As a consequence, asking what the effects of different foraging strategies are on host–parasitoid interactions is, in part, a central endeavor of behavioral ecology.

As argued here, understanding the difference in behaviors (mediated through the population dynamics) has evolutionary consequences for the modification of foraging by natural selection. Understanding the mechanistic details of foraging, such as the different timescales between individual-level and population-level phenomena, requires appropriate consideration of the cognitive and perceptual abilities of parasitoids, their responses to host densities, and abilities to move between patches (Křivan 1997, Bernstein et al. 1999, Cressman & Křivan 2006). As such, it can be argued that most aspects of the behavior of host–parasitoid interactions can be encapsulated in relative simple expressions (such as aggregation, interference, pseudo-interference) and this approach has been developed upon since the original models proposed by Thompson, Nicholson, Lokta, and Volterra (Hassell 1978, 2000). Here, we have used these models to explore how dynamical data from host–parasitoid interactions might be used to infer the underlying mechanisms of behavior. We have shown that the best fitting model is one which simply captures the effects of spatial heterogeneity and non-linear foraging in a relatively general way. These results suggest that the information needed to describe the population dynamics of a host and its parasitoid does not require a detailed knowledge of behavior. However, this is not the end as science is more than simply describing patterns. It is about explaining the detailed processes in terms of the inter-relationships between its constitutive elements. In terms of host–parasitoid interactions, this leads naturally to ask how the interaction amongst individuals affects the population dynamics.

Although our simple population models can describe the dynamics of the host–parasitoid interaction without knowledge of the details of the behavior, it is critical to ask if these models (and approaches) are robust enough to retain their descriptive strengths and provide explanatory power when the conditions radically change. For instance, none of the models explicitly consider spatial processes nor do they consider the effects of alternative food sources in terms of refuges or metabolic resources. As we lack the appropriate scientific evidence, it is difficult to claim the higher explanatory powers of the link between behavioral and population models. Currently, we need robust, rigorous, and replicated individual and population level experiments to test these ideas properly. Once we have appropriately designed experiments encapsulating both individual and population level patterns and processes, we will be able to argue which (if any) behaviors or strategies are most influential. Until then, the issues of what drives host–parasitoid dynamics are far from closed. Understanding how different aspects of behavior translate to affect the population dynamics and evolution of parasitoid life-history strategies and coevolution of host–parasitoid interactions remains a challenging, exciting, and fruitful area for future research in behavioral ecology.

References

Abrams, P.A. and Kawecki, T.J. (1999) Adaptive host preference and the dynamics of host-parasitoid interactions. *Theoretical Population Biology* **56**: 307–24.

Bailey, V.A., Nicholson, A.J. and Williams, E.J. (1962) Interaction between hosts and parasites when some host individuals are more difficult to find than others. *Journal of Theoretical Biology* **3**: 1–18.

Beddington, J.R. (1975) Mutual interference between parasites or predators and its effect on searching efficiency. *Journal of Animal Ecology* **44**: 331–40.

Beddington, J.R., Free, C.A. and Lawton, J.H. (1978) Characteristics of successful natural enemies in model of biological control of insect pests. *Nature* **273**: 513–19.

Bernstein, C., Kacelnik, A. and Krebs, J.R. (1988) Individual decisions and the distribution of predators in a patchy environment. *Journal of Animal Ecology* **57**: 1007–26.

Bernstein, C., Kacelnik, A. and Krebs, J.R. (1991) Individual decisions and the distribution of predators in a patchy environment II. The influence of travel costs and the structure of the environment. *Journal of Animal Ecology* **60**: 205–25.

Bernstein, C., Auger, P. and Poggiale, J.C. (1999) Predator migration decisions, the ideal free distribution, and predator-prey dynamics. *American Naturalist* **153**: 267–81.

Bonsall, M.B. and Hassell, M.P. (1997) Apparent competition structures ecological assemblages. *Nature* **388**: 371–3.

Bonsall, M.B., Hassell, M.P. and Asefa, G. (2002) Ecological trade-offs, resource partitioning, and coexistence in a host-parasitoid assemblage. *Ecology* **83**: 925–34.

Bonsall, M.B., Jansen, V.A.A. and Hassell, M.P. (2004) Life history trade-offs assemble ecological guilds. *Science* **306**: 111–14.

Briggs, C.J. (1993) Competition among parasitoid species on a stage-structured host and its effect on host suppression. *American Naturalist* **141**: 372–97.

Burnham, K.P. and Anderson, D.R. (2002) *Model Selection and Multi-Model Inference: A Practical Information-Theoretic Approach.* Springer-Verlag, Berlin.

Charnov, E.L. (1976) Optimal foraging: the marginal value theorem. *Theoretical Population Biology* **9**: 129–36.

Chesson, P.L. and Murdoch, W.W. (1986) Aggregation of risk: Relationships among host-parasitoid models. *American Naturalist* **127**: 696–715.

Clark, C.W. and Mangel, M. (2000) *Dynamic State Variable Models in Ecology. Methods and Applications.* Oxford University Press, Oxford.

Comins, H.N. and Hassell, M.P. (1979) The dynamics of optimally foraging predators and parasitoids. *Journal of Animal Ecology* **48**: 335–51.

Cook, R.M. and Hubbard, S.F. (1977) Adaptive searching strategies in insect parasites. *Journal of Animal Ecology* **46**: 115–25.

Cressman, R. and Křivan, V. (2006) Migration dynamics and the ideal free distribution. *American Naturalist* **168**: 384–97.

Driessen, G. and Bernstein, C. (1999) Patch departure mechanisms and optimal host exploitation in an insect parasitoid. *Journal of Animal Ecology* **68**: 445–59.

Driessen, G., Bernstein, C., van Alphen, J.J.M. and Kacelnik, A. (1995) Count down mechanism for host search in the parasitoid *Venturia canescens. Journal of Animal Ecology* **64**: 117–25.

Emlem, J.M. (1973) *Ecology: An Evolutionary Approach.* Addison-Wesley, New York.

Free, C.A., Beddington, J.R. and Lawton, J.H. (1977) On the inadequacy of simple models of mutual interference for parasitism and predation. *Journal of Animal Ecology* **46**: 543–54.

Fretwell, S.D. and Lucas, J.H.J. (1970) On territorial behaviour and other factors influencing habitat distribution in birds. *Acta Biotheoretica* **19**: 16–36.

Gravel, D., Canham, C.D., Beaudet, M. and Messier, C. (2006) Reconciling niche and neutrality: the continuum hypothesis. *Ecology Letters* **9**: 399–409.

Gross, K. and Ives, A.R. (1999) Inferring host-parasitoid stability from patterns of parasitism among patches. *American Naturalist* **154**: 489–96.

Haccou, P. and Meelis, E. (1994) *Statistical Analysis of Behavioural Data: An Approach Based on Time-Structured Models.* Oxford University Press, Oxford.

Haccou, P., Devlas, S.J. van Alphen, J.J.M. and Visser, M.E. (1991) Information processing by foragers – effects of intra-patch experience on the leaving tendency of *Leptopilina heterotoma. Journal of Animal Ecology* **60**: 93–106.

Harper, D.G.C. (1982) Competitive foraging in mallards – ideal free ducks. *Animal Behaviour* **30**: 575–84.

Hassell, M.P. (1978) *The Dynamics of Arthropod Predator-Prey Systems.* Princeton University Press, Princeton, NJ.

Hassell, M.P. (1980) Foraging strategies, population models and biological control. *Journal of Animal Ecology* **49**: 603–28.

Hassell, M.P. (1982) Patterns of parasitism by parasitoids in patchy environments. *Ecological Entomology* **7**: 365–77.

Hassell, M.P. (2000) *The Spatial and Temporal Dynamics of Host-Parasitoid Interactions.* Oxford University Press, Oxford.

Hassell, M.P. and Huffaker, C.B. (1969) Regulatory processes and population cyclicity in laboratory populations of *Anagasta kuhniella* (Zeller) (Lepidoptera: Phycitidae). III. *The development of population models. Researches on Population Ecology* **11**: 186–210.

Hassell, M.P. and Varley, G.C. (1969) New inductive population model for insect parasites and its bearing on biological control. *Nature* **223**: 1133–6.

Hassell, M.P. and May, R.M. (1973) Stability in insect host-parasite models. *Journal of Animal Ecology* **42**: 693–726.

Hassell, M.P. and May, R.M. (1974) Aggregation of predators and insect parasites and its effect on stability. *Journal of Animal Ecology* **43**: 567–94.

Hassell, M.P. and Pacala, S.W. (1990) Heterogeneity and the dynamics of host-parasitoid interactions. *Philosophical Transactions of the Royal Society of London B, Biological Sciences* **330**: 203–20.

Hassell, M.P., Pacala, S.W., May, R.M. and Chesson, P.L. (1991) The persistence of host-parasitoid associations in patchy environments. I. A general criterion. *American Naturalist* **138**: 568–83.

Holling, C.S. (1959) Some characteristics of simple types of predation and parasitism. *Canadian Entomologist* **91**: 395–8.

Holt, R.D. (1977) Predation, apparent competition and the structure of prey communities. *Theoretical Population Biology* **12**: 197–229.

Hubbard, S.F. and Cook, R.M. (1978) Optimal foraging by parasitoid wasps. *Journal of Animal Ecology* **47**: 593–604.

Ives, A.R. (1995) Spatial heterogeneity and host-parasitoid population dynamics: Do we need to study behavior? *Oikos* **74**: 366–76.

Iwasa, Y., Higashi, M. and Yamamura, N. (1981) Prey distribution as a factor determining the choice of optimal foraging strategy. *American Naturalist* **117**: 710–23.

Krebs, J.R. (1978) Optimal foraging. In: Krebs, J.R. and Davies, N.B. (eds.) *Behavioural Ecology: An Evolutionary Approach*. Blackwell Scientific Publications, Oxford, pp. 23–63.

Křivan, V. (1997) Dynamical ideal free distribution: effects of optimal patch choice on predator-prey dynamics. *American Naturalist* **149**: 164–78.

Lessells, C.M. (1985) Parasitoid foraging: should parasitism be density dependent? *Journal of Animal Ecology* **54**: 27–41.

Lotka, A.J. (1925) *Elements of Physical Biology*. Williams and Wilkins, Baltimore.

MacArthur, R.H. and Pianka, E.R. (1966) On optimal use of a patchy environment. *American Naturalist* **100**: 603–9.

Mangel, M. and Clark, C.W. (1988) *Dynamic Modeling in Behavioral Ecology*. Princeton University Press, Princeton, NJ.

May, R.M. (1978) Host-parasitoid systems in patchy environments: a phenomenological model. *Journal of Animal Ecology* **47**: 833–43.

May, R.M. and Hassell, M.P. (1981) The dynamics of multiparasitoid-host interactions. *American Naturalist* **117**: 234–61.

Metz, J.A.J., Nisbet, R.M. and Geritz, S.A.H. (1992) How should we define fitness for general ecological scenarios? *Trends in Ecology & Evolution* **7**: 198–202.

Murdoch, W.W. (1977) Stabilizing effects of spatial heterogeneity in predator-prey systems. *Theoretical Population Biology* **111**: 2523–73.

Murdoch, W.W. and Oaten, A. (1975) Predation and population stability. *Theoretical Population Biology* **12**: 263–85.

Murdoch, W.W., Briggs, C.J. and Nisbet, R.M. (2003) *Consumer-Resource Dynamics*. Princeton University Press, Princeton.

Newell, G.F. (1971) *Applications of Queuing Theory*. Chapman & Hall, London.

Nicholson, A.J. (1933) The balance of animal populations. *Journal of Animal Ecology* **2**: 131–78.

Nicholson, A.J. and Bailey, V.A. (1935) The balance of animal populations. I. *Proceedings of the Zoological Society of London* **3**: 551–6.

Oaten, A. (1977) Optimal foraging in patches – case for stochasticity. *Theoretical Population Biology* **12**: 263–85.

Pacala, S.W. and Hassell, M.P. (1991) The persistence of host-parasitoid associations in patchy environments. II. Evaluation of field data. *American Naturalist* **138**: 584–605.

Reeve, J.D., Kerans, B.L. and Chesson, P.L. (1989) Combining different forms of parasitoid aggregation: effects on stability and patterns of parasitism. *Oikos* **56**: 233–9.

Rogers, D.J. (1972) Random searching and insect population models. *Journal of Animal Ecology* **41**: 369–83.

Scheffer, M. and van Nes, E.H. (2006) Self-organized similarity, the evolutionary emergence of groups of similar species. *Proceedings of the National Academy of Sciences* **103**: 6230–5.

Schreiber, S.J., Fox, L. and Getz, W.M. (2000) Coevolution of contrary choices in host-parasitoid systems. *American Naturalist* **155**: 637–56.

Sirot, E. and Bernstein, C. (1996) Time sharing between host searching and food searching in parasitoids: State-dependent optimal strategies. *Behavioral Ecology* **7**: 189–94.

Solomon, M.E. (1949) The natural control of populations. *Journal of Animal Ecology* **18**: 1–35.

Stephen, D.W. and Krebs, J.R. (1986) *Foraging Theory*. Princeton University Press, Princeton, NJ.

Stiling, P.D. (1987) The frequency of density dependence in insect host-parasitoid systems. *Ecology* **68**: 844–56.

Sutherland, W.J. (1983) Aggregation and the ideal free distribution. *Journal of Animal Ecology* **52**: 821–8.

Thompson, W.R. (1924) La théorie Mathématique de l'action des parasites entomophages et le facteur de hasard. *Annales de la Faculté des Sciences de Marseille* **2**: 69–89.

Tregenza, T., Parker, G.A. and Thompson, D.J. (1996a) Interference and the ideal free distribution: models and tests. *Behavioral Ecology* **7**: 379–86.

Tregenza, T., Parker, G.A. and Thompson, D.J. (1996b) Interference and the ideal free distribution: oviposition in a parasitoid wasp. *Behavioral Ecology* **7**: 387–94.

van Alphen, J.J.M., Bernstein, C. and Driessen, G. (2003) Information acquisition and time allocation in insect parasitoids. *Trends in Ecology & Evolution* **18**: 81–7.

van Baalen, M. and Sabelis, M.W. (1993) Coevolution of patch selection strategies of predators and prey and the consequences for ecological stability. *American Naturalist* **142**: 646–70.

van Baalen, M. and Sabelis, M.W. (1999) Nonequilibrium population dynamics of 'ideal and free' prey and predators. *American Naturalist* **154**: 69–88.

Vincent, T.L. and Brown, J.S. (2005) *Evolutionary Game Theory, Natural Selection, and Darwinian Dynamics*. Cambridge University Press, Cambridge.

Volterra, V. (1926) Variazioni e fluttuazioni del numero d'individui in specie animali conviventi. *Memorie della Accademia Nazionale dei Lincei* **2**: 31–13.

Waage, J.K. (1979) Foraging for patchily distributed hosts by the parasitoid *Nemeritis canescens*. *Journal of Animal Ecology* **48**: 353–71.

Wäckers, F.L. (1994) The effects of food deprivation on the innate visual and olfactory preferences in the parasitoid *Cotesia rubecula*. *Journal of Insect Physiology* **40**: 641–9.

Wajnberg, É. (2006) Time allocation strategies in insect parasitoids: from ultimate predictions to proximate behavioral mechanisms. *Behavioral Ecology & Sociobiology* **60**: 589–611.

Wajnberg, É., Rosi, M.C. and Colazza, S. (1999) Genetic variation in patch time allocation in a parasitic wasp. *Journal of Animal Ecology* **68**: 121–33.

Wajnberg, É., Curty, C. and Colazza, S. (2004) Genetic variation in the mechanisms of direct mutual interference in a parasitic wasp: consequences in terms of patch-time allocation. *Journal of Animal Ecology* **73**: 1179–89.

Wajnberg, É., Bernhard, P., Hamelin, F. and Boivin, G. (2006) Optimal patch time allocation for time-limited foragers. *Behavioural Ecology & Sociobiology* **60**: 1–10.

Walde, S.J. and Murdoch, W.W. (1988) Spatial density dependence in parasitoids. *Annual Review of Entomology* **33**: 441–66.

Williams, I.S., Jones, T.H. and Hartley, S.E. (2001) The role of resources and natural enemies in determining the distribution of an insect herbivore population. *Ecological Entomology* **26**: 204–11.

14

Linking behavioral ecology to the study of host resistance and parasitoid counter-resistance

Alex R. Kraaijeveld and H. Charles J. Godfray

Abstract

Parasitoids provide excellent model systems for exploring a variety of questions in both behavioral and evolutionary ecology. However, work in the two fields does not always overlap. Here, we review recent studies of the evolutionary ecology of resistance and counter-resistance in host–parasitoid systems, and explore how this work may impact upon behavioral ecology and vice versa. We review briefly the mechanistic basis of the two traits, and then focus in particular on work involving *Drosophila* and its parasitoids. Here, studies have shown costs to the evolution of higher resistance, and possibly higher counter-resistance. But little is known about how information on levels of resistance is incorporated into parasitoid decision making. We look more briefly at other systems where resistance and counter-resistance has been studied, and speculate how behavioral ecological and evolutionary ecological studies may converge in the future.

14.1 Introduction

Scientists study parasitoids for at least two reasons. First, they explore the biology of a group of insects that are intrinsically interesting and which are major players in many terrestrial ecosystems including those involving agriculture and forestry. Second, they are concerned about what they can tell us about general issues in biology that are of importance and interest beyond parasitoid biology, and indeed beyond entomology. In truth, most parasitoid researchers are motivated by both drivers. In the second category, work on parasitoids has illuminated the behavioral ecology agenda, since the foundation of the subject in the 1970s. More recently, parasitoids have been used to explore questions concerning the coevolution of resistance and counter-resistance in resource-consumer systems, areas of research that are more evolutionary ecology than behavioral ecology.

Hosts have evolved numerous stratagems to escape the attentions of parasitoids. Natural selection for enemy-free space may modify the niche of a species such that parasitoid mortality is reduced, though sometimes the parasitoid may itself evolve to chase its host through niche space (Jeffries & Lawton 1984). At the level of the individual, the host may evolve mechanisms of concealment, or more active means of defense that can include physically attacking a potential parasitoid. Even when you have been successfully parasitized, it is still possible to defend yourself, at least if you are attacked by a koinobiont parasitoid. These species do not kill or permanently paralyze their hosts at oviposition, but typically allow them to resume growth and to increase in size until they are large enough to support better parasitoid development (Quicke 1997). It is during this period, when the parasitoid suspends development as an egg, or more typically a first instar larva, that the host can retaliate and destroy the parasitoid through an immune response. Of course, the parasitoid is under strong selection pressure to evolve counter-measures to host resistance (Godfray 1994).

Until recently, immunological defenses against parasites have been studied chiefly from a physiological or genetic perspective. In the last decade, there has been a growing trend to treat immunity-related traits as part of an organism's life history, a field that has become known as ecological immunity (Rolff & Siva-Jothy 2003, Schmid-Hempel 2003, Schmid-Hempel & Ebert 2003). Investment in defense against parasites and pathogens is considered as one more call on an organism's limiting resources, with trade-offs determining the optimum investment in resistance versus growth, reproduction, etc. Exactly the same considerations apply to the natural enemy where investment in counter-resistance strategies again is subject to trade-offs with other fitness components. For both, the optimum investment may depend on the state of the individual or the state of local environment. It may also depend on the investment decisions made by the antagonistic. This approach to host–parasite interactions clearly owes a huge intellectual debt to behavioral ecology. Moreover, while some investment decisions are fixed components of the host or parasite's life histories that set the boundary conditions for behavioral ecological interactions, some decisions will be phenotypically plastic and be intimately linked with the behavioral interaction between the two organisms.

The interaction between hosts and parasitoids is purely antagonistic and, hence, we can expect each to exert significant evolutionary pressures on the other. Hosts will be under strong selection to evolve immune defenses against parasitoids and, in turn, parasitoids will be under strong selection pressure to evolve counter-resistance mechanisms to prevent their eggs and/or larvae from being eliminated by the host's immune system. Each combatant's strategy will be influenced by that of the other, though also by the nature and magnitude of the trade-offs between resistance or counter-resistance and the other trophic and reproductive processes that affect fitness. Of course, these may vary spatially and through time. Moreover, there is an asymmetry: parasitoids must overcome host defenses, while hosts may be selected to 'gamble' on not being discovered by their parasitoids. The odds on the gamble paying off are determined by local population dynamics, leading to feedbacks between ecological and evolutionary processes (Frank 1994, Doebeli 1997).

In nature, it is rare for a host to be attacked by only one species of parasitoid. Several species of parasitoids, possibly using different counter-resistance mechanisms, typically parasitize the same host species (Hawkins & Sheehan 1994). Also, the host may be under attack from other (true) parasites and microbial pathogens. Depending on the details of the physiological mechanisms underlying resistance against each natural enemy, investment in defense against one species may lead to cross-resistance against another species, or there may be trade-offs with increased resistance against one antagonist resulting in a decrease in resistance against another (Poitrineau et al. 2003, 2004). Similarly, hosts may evolve traits which conceal them from particular searching parasitoids and this may or may not lead to correlated benefits through protection from other species (Hochberg & Holt 1995, Hochberg 1997).

The Geographic Mosaic Coevolution hypothesis (Thompson 1994, 2005) predicts that different populations within a species are likely to experience different sets of biotic and abiotic selection pressures so that the outcome of coevolution varies spatially. For host–parasitoid interactions, differences in the local spectrum of parasitoids attacking a particular host, and the variety of hosts available to a particular parasitoid, will be important sources of amongst-population variation in selection. These community-ecology differences may be amplified by their effects on local population dynamics, and on how the different species of hosts and parasitoids respond behaviorally to the local conditions. We can thus expect the reciprocal selection pressures that hosts and parasitoids exert on each other to vary across populations and for ideas from the Geographic Mosaic theory to be relevant for interpreting the resulting spatial patterns.

The aims of this chapter are twofold. It is first to introduce to a behavioral ecological audience some of the recent research on host resistance and parasitoid counter-measures that may be relevant to thinking about how the two subjects abut or overlap. We make no pretence at a comprehensive review, and in particular focus on the interaction between *Drosophila* and its parasitoids. Our second aim is to flag up areas of parasitoid ecological immunity where there are already, or are potentially, links with behavioral ecology. We begin with a brief review of the mechanistic basis of resistance and counter-resistance mechanisms, and then go on to the *Drosophila*-parasitoid system. We then discuss more briefly some other systems and finish by looking ahead to how evolutionary and behavioral ecological studies of parasitoids may interact in the future.

14.2 Resistance and counter-resistance mechanisms

Insects, like most metazoans, have both cellular and non-cellular (humoral) immune defense mechanisms (Strand & Pech 1995). For understandable reasons, studies of the mechanistic basis of invertebrate immunity have lagged behind the equivalent vertebrate subjects. However, the last decade has seen a remarkable advance in our knowledge of insect humoral immunity, largely based on studies of *Drosophila*, but strongly stimulated by the unexpected homologies of invertebrate and vertebrate innate immunity (Hoffmann et al. 1999). Though much remains to be discovered, many details of the signal transduction pathways that are triggered by the recognition of different categories of microorganisms are now known, as are the particular defense strategies they trigger, including the production of antimicrobial peptides, reactive oxygen compounds, and the deposition of melanin around wounds and nodules (DeGregorio et al. 2002, Agaisse & Perrimon 2004).

Resistance against parasitoids does involve the humoral immune system, but much more importantly the cellular arm of the immune system. At a macroscopic level, hemocytes – cells floating free in hemocoel or body cavity – aggregate around the parasitoid (or indeed any other foreign body) and form a multicellular capsule (Nappi 1975, 1981). The cell walls break down and the black pigment melanin is deposited. The parasitoid egg or larva is killed, either by asphyxiation or by the direct effects of necrotizing compounds such as oxidizing free radicals, quinones, and semi-quinones from the capsule (Nappi & Vass 1993, Nappi et al. 1995).

Further insight at the cellular level has been hampered by the fact that, until recently, different classes of hemocytes could only be identified morphologically and not by the type of molecular markers that are used so widely in vertebrate immunology. For example, in *Drosophila*, three major hemocyte types are distinguished, all of which are involved in capsule formation (Meister & Lagueux 2003):

1 plasmatocytes, which recognize immune challenges and carry out phagocytosis;
2 lamellocytes, which form the bulk of the capsules surrounding foreign objects; and
3 crystal cells, which have crystalline inclusions containing the enzymes necessary for humoral and cellular melanization.

When a parasitoid egg is injected into the host, it is recognized as foreign by plasmatocytes that adhere to the chorion, spreading across the egg, so isolating it from the hemocoel. There then occurs a rapid response in the hemopoietic organ, a multi-lobed lymph organ situated behind the brain, with a massive differentiation of lamellocytes (from pro-hemocytes) and an increase in numbers of crystal cells (Sorrentino et al. 2002). Hemocytes are also recruited from a reserve population underneath the cuticle (Lanot et al. 2001). The lamellocytes join the plasmatocytes to form the capsule (Nappi 1975, 1981, Rizki & Rizki 1984), with the crystal cells helping to lay down melanin, a process involving the prophenoloxidase (PPO) system.

At the molecular level, the mechanism of capsule formation is still poorly understood, though some of the transcription factors and signaling pathways involved in hemocyte differentiation have been identified. A recent study has shown that the putative signal released by the plasmatocytes in *Drosophila* is recognized by certain cells in the posterior signaling center of the hemopoietic organ (Crozatier et al. 2004). This recognition requires the expression of the transcription factor 'collier'. Interestingly, the vertebrate ortholog of this gene is also involved in blood cell differentiation (Evans et al. 2003, Crozatier et al. 2004). In the presence of collier, a further currently unknown signal is produced that triggers lamellocyte and to a lesser extent crystal cell differentiation.

Encapsulation has also been closely studied in Lepidoptera, especially in the noctuid *Pseudoplusia includens* (Lavine & Strand 2002). Here, foreign bodies such as parasitoids are recognized by a class of hemocytes called granulocytes, which recruit a second class called plasmatocytes to form the body of the capsule. A cytokine called the plasmatocyte spreading peptide is involved in triggering plasmatocyte adhesion and spreading. As the capsule grows, granulocytes are again recruited and they form an outer layer, which probably resembles the basement membrane lining the hemocoel and prevents further growth of the capsule (Pech & Strand 2000). A further type of hemocyte, the oenocytoid (equivalent to the *Drosophila* crystal cell), is involved in melanization.

Parasitoids attempt to disrupt or avoid the cellular immune response in many different ways. In some species, the parasitoid egg is placed carefully in an organ where it is not exposed to circulating hemocytes, for example, the brain. Other species obtain at least partial protection by having eggs that adhere to insect fat body. As will be detailed below, in the *Drosophila* parasitoid *Asobara tabida*, the degree to which a wasp egg adheres to host tissue (e.g. fat body) is correlated with its probability of escaping parasitism (Kraaijeveld et al. 2001a). In many cases, adapted parasitoids elicit no response from the host, and this is presumably because they mimic host tissue so that they are recognized as self (Asgari & Schmidt 1994, Hayakawa & Yazaki 1997). It has been noted that if the surface of the parasitoid is abraded or chemically treated, or compromised by the attack of a competing parasitoid larva, the invader suddenly becomes visible to the host, eliciting an encapsulation response. In the ichneumonid *Venturia canescens*, the larval camouflage is provided by a layer of virus-like protein particles produced in the female reproductive system.

In addition to these passive means of avoidance, many parasitoid species take more active measures to counter host defenses. For example, some larger parasitoids can wriggle vigorously, and so disrupt capsule formation. Many parasitoids inject toxins and other chemicals at oviposition that disrupt host defense. For example, venom injected by the ichneumonid *Pimpla hypochodriaca* attacks the hemocytes of its noctuid host (*Lacanobia oleracea*) and renders them unable to form a capsule (Richards & Edwards 1999, Richards & Parkinson 2000). Substances injected at oviposition may also modify host physiology to the parasitoid's benefit. The venom of the aphid parasitoid *Aphidius ervi* castrates the host and other injected products may modify host physiology to make resources more available to the parasitoid (Digilio et al. 2000).

The developing parasitoid also releases substances that help regulate the host environment for its own benefit. These include insect hormones that can allow it to manipulate host development. It is known that some secreted material affects the host's immune response, though this type of counter-measure has been relatively little studied.

Several groups of parasitoids, particular braconids, produce a type of cell called a teratocyte that floats free in the host hemolymph. The origin of the cells is the embryonic membranes and, once free, they do not divide but increase in size, sometimes dramatically. The cells are covered with microvilli and have other adaptations for both absorption and secretion. The major function of teratocytes seems to be to assist the parasitoid in efficiently consuming host resources, but they may also release compounds that interfere with host resistance, though sometimes this also requires venom and other products injected at oviposition (Beckage & Gelman 2004).

One of the most interesting types of counter-measures is found in two groups of ichneumonoid wasps, which inject 'viruses' (we explain below why they may not be true viruses) into their hosts to compromise the immune system. The viruses are unique in that they are composed of typically 15–35 double-stranded DNA circles that are permanently integrated in the wasp's genome. Because of this genome arrangement, they have been called *Polydnaviridae* (poly-DNA-viruses, PDVs) containing the genera *Ichnovirus* (IV) largely restricted to Ichneumonidae (Campopleginae), and *Bracovirus* (BV) restricted to Braconidae (Microgastrinae and Cardiochilinae). In the calyx organ of the female reproductive system, the DNA is replicated and packaged into protein capsids, one DNA molecule per capsid in the BVs, more and possibly the whole genome in the IVs. Capsid morphology also differs between IVs and BVs. Once in the host, the virus DNA is expressed but not replicated, and the gene products affect host immunocompetence by such tricks as inhibiting cytoskeletal dynamics and blocking immune signal transduction pathways (Turnbull & Webb 2002, Beckage & Gelman 2004, Webb et al. 2006). PDV-encoded proteins may also disrupt the normal pattern of protein storage in the host to make more resources available for parasitoid growth, and in some species trigger premature metamorphosis (Johner et al. 1999). Proper action of some BVs also requires substances injected during oviposition.

Three PDVs have now been fully sequenced (Espagne et al. 2004, Webb et al. 2006): from the braconids *Cotesia congregatus* (CcBV) and *Micoplitus demolitor* (MdBV), and the ichneumonid *Campoletis sonorensis* (CsIV). These provide a wealth of detail about a strange type of virus. First, their genome is amongst the largest recorded for viruses. Second, the fraction of coding DNA is low for a virus, between 15 and 30%. Third, coding sequences typically exist as related families of genes. Fourth, genes containing introns are not uncommon. This 'flabbiness' might be explained as a reduction in selection pressures for a streamlined genome after the domestication of the virus and the loss of independent replication. Genes with introns might suggest transfer from the eukaryotic host, while gene families might suggest proteins of similar function but with subtle cell, tissue, or development stage specialization (Webb et al. 2006). The different DNA segments are produced in different relative quantities, and this may be a type of dosage control as there is some evidence to suggest that genes for proteins required at high concentrations occur on high copy number segments (Webb et al. 2006). A number of genes with virulence factors that might cause the phenotypic effects of parasitism were identified, but no genes involved in replication.

There is no sequence homology between the IVs and BVs, strongly supporting the hypothesis that they have evolved independently and that their striking similarities are convergent rather than inherited (Turnbull & Web 2002). There are also many differences between the two BVs, and amongst other PDVs for which more limited sequence information is available, suggesting an active shuffling of genes into and out of the PDV genome over evolutionary time. The ultimate origin of the two PDV clades is not yet certain. They may originally have been viruses captured by the host as suggested by their virus-like protein capsid. Alternatively, as their many non-virus features might suggest, they could be an alternative means of delivering substances to disable the host immune response, independently invented in the two wasp subfamilies. Possibly both explanations may apply, with PDVs being assembled from a few transposable element or viral components (e.g. the capsid) but otherwise with host-derived genes.

14.3 *Drosophila* and its parasitoids

Most of the common *Drosophila* species that feed on fermenting substrates in the field in Europe belong to the *melanogaster* and *obscura* species groups, with *D. melanogaster*, *D. simulans*, *D. subobscura*, and *D. obscura* being the most abundant. Parasitism by hymenopteran parasitoids represents a significant cause of mortality for both larvae and pupae (Carton et al. 1986). The larvae are attacked by parasitoids from the families Braconidae and Figitidae (= Eucoilidae), of which the braconid *A. tabida* and the figitids *Leptopilina heterotoma* and *L. boulardi* are the most important. The pupae face attack by species from the families Pteromalidae and Diapriidae, with the pteromalid *Pachycrepoideus vindemiae* being the single most important species (Carton et al. 1986).

Larval hosts defend themselves against parasitoids chiefly through the cellular immune response described above. Parasitoids are not defenseless against the host's immune reaction and different counter-defense mechanisms have evolved in the two main families of larval parasitoids. *A. tabida* has 'sticky' eggs (Kraaijeveld & van Alphen 1994, Eslin et al.

1996): proteinaceous filaments on the egg chorion cause the egg to become attached to and embedded in host tissue, in particular the fat body. This prevents the host's hemocytes from completely surrounding and melanizing the egg. In consequence, the parasitoid embryo can still develop and the larva emerges from the incomplete capsule. Parasitoids of the genus *Leptopilina* use a different approach. They inject virus-like particles (VLPs) into the host together with the egg (Rizki & Rizki 1990, Dupas et al. 1996). The details differ between the two species but the VLPs enter the host's hemocytes and there cause apoptosis (Rizki & Rizki 1990). This disrupts the initial aggregation of hemocytes around the parasitoid egg and thus essentially blocks any subsequent steps in the immune reaction.

14.3.1 Host resistance

Across Europe, there is substantial variation among *D. melanogaster* populations in their ability to encapsulate parasitoid eggs (Kraaijeveld & van Alphen 1995a). In particular, there is a cline from north-western Europe to the central Mediterranean of increasing encapsulation ability against *A. tabida*. We believe that this cline can largely be explained by the relative abundance of *D. subobscura*, an alternative host species for the parasitoid. This is a species that appears completely to lack the ability to encapsulate the eggs of any parasitoid (Kraaijeveld & van der Wel 1994, Eslin & Doury 2006) and which is more common at higher latitudes. In the north of Europe, the majority of hosts attacked by *A. tabida* are *D. subobscura* and this, we think, has led to little investment in counter-resistance by the parasitoid. Northern populations of *D. melanogaster* can thus defend themselves without having to invest a large amount in resistance. In the south, *D. melanogaster* is the main host of *A. tabida*, leading to greater investment in counter-resistance by the parasitoid, and consequently in resistance by the host. This contrasts to the pattern of resistance against the more specialized *L. boulardi*, which is much more a mosaic, without any clear clines (Kraaijeveld & van Alphen 1995a). It is not yet clear why *D. subobscura* does not try to defend itself. Perhaps the population dynamics and trade-offs it experiences favors individuals of this species to redirect resources away from resistance to other functions and to 'gamble' on not being found by a parasitoid.

Resistance and counter-resistance may be graded traits where investment in defense is one-dimensional and always improves fitness, or multidimensional with interactions between specific host and parasitoid genotypes. Our analysis of host resistance against *A. tabida* suggests that it is most easily explained as a graded trait (Kraaijeveld & Godfray 2001). The situation with *L. boulardi* is more complicated with isofemale lines derived from different geographic populations of hosts and parasitoid clearly showing genotype-genotype interactions (Carton 1984, Carton & Nappi 1991, Carton & Nappi 1997, Carton & Dupas 1999). It would be interesting to know whether the additive genetic variation segregating in particular field populations of *L. boulardi* and its hosts is also of this type.

In addition to variation in resistance across host populations, there is also substantial genetic variation within populations (Carton & Boulétreau 1985). As parasitoid mortality is such an important determinant of fitness, this variation is surprising. One explanation is that resistance is costly and that there are trade-offs with other fitness components. Temporal variation in the relative importance of defense against parasitoids and these other fitness components might then explain this standing genetic variation. The temporal variation could arise from abiotic drivers such as the weather, or biotically through population fluctuations caused by the host–parasitoid interaction. It might also arise through spatial variation in the importance of different fitness components in populations connected by migration.

There are two different types of costs of resistance: the costs of actually defending yourself after attack, and the costs of having the ability to mount an immune response irrespective of challenge. Mounting an immune response after being parasitized clearly has costs in *D. melanogaster*: larvae which have successfully encapsulated a parasitoid egg suffer from reduced competitive ability (Tiën et al. 2001). After pupation, their puparial wall is thinner, making them potentially more susceptible to mortality factors such as mechanical damage, desiccation, and attack by pupal parasitoids (Fellowes et al. 1998b). If they survive, the adult flies are smaller, resulting in lower mating success as a male or lower fecundity as a female (Carton & David 1983, Fellowes et al. 1999a). However, because successful parasitoid attack inevitably leads to death, the host will always be selected to pay the cost of trying to defend itself after attack. However, if successful defense leads to hosts with much reduced fitness, this will influence the benefits of investing in the machinery of the immune system.

Perhaps the more interesting cost, from an evolutionary point of view, is that of having the ability to mount an immune reaction: the cost that is paid irrespective of whether parasitism actually takes place. High levels of resistance will evolve when the benefits outweigh the costs. Artificial selection experiments for higher rates of encapsulation of parasitoid eggs have shown that greater resistance is associated with reduced larval competitive ability in *D. melanogaster* (Kraaijeveld & Godfray 1997, Fellowes et al. 1998a). Interestingly, this means that costs of resistance are density-dependent, as high larval competitive ability is only important when food is limited.

There are several possible mechanistic explanations for the trade-off between resistance to parasitoids and larval competitive ability in *D. melanogaster*. The improved survival after parasitoid attack probably arises because of an increase in the density of circulating hemocytes in selected lines (Kraaijeveld et al. 2001b), while the reduction in competitive ability is associated with a decrease in feeding rate (Fellowes et al. 1999b). It is possible that selection leads to a redirection of limiting resources from feeding to defense (see also Chapter 6 by Strand and Casas). Alternatively, the higher density of hemocytes may make the hemolymph more viscous and less efficient at supplying energy to, or removing waste products from, the mouthpart musculature. Finally, the hemopoietic organ (which produces the hemocytes) and the cephalopharyngeal muscles share the same embryonic origins (Fullilove et al. 1977, Tepass et al. 1994) and possibly the trade-off may occur at this early stage of development (Kraaijeveld et al. 2001b).

Comparisons across species in the *melanogaster* group appear to show the same patterns as those found within *D. melanogaster*. Using six species that varied widely in ability to encapsulate the eggs of *A. tabida* (*D. melanogaster, D. simulans, D. sechellia, D. mauritiana, D. yakuba,* and *D. teissieri*), Eslin and Prévost (1996, 1998) found a strong positive correlation between encapsulation ability and numbers of circulating hemocytes. Carton and Kitano (1981) suggest that the ability to encapsulate the eggs of *L. boulardi* is negatively correlated with competitive ability among four *Drosophila* species (*D. melanogaster, D. simulans, D. yakuba,* and *D. teissieri*).

While selection for increased resistance to parasitoids results in a reduction in competitive ability, this does not appear to be a simple symmetric trade-off. Selection experiments in which larvae were reared under crowded conditions resulted in an increase in both larval competitive ability and encapsulation ability, even though larvae had not been exposed to parasitoids at any time during the experiment (Sanders et al. 2005). Competition among larval *D. melanogaster* is of the scramble type, with feeding rate being an important determinant of competitive superiority (Joshi & Mueller 1988, Mueller 1988). It is possible that under crowded conditions larvae are more likely to wound each other with their mouth hooks (Sanders et al. 2005). Crowding and the evolutionary response to crowding may thus select for better wound healing, and as the latter also involves hemocytes and the PPO cascade, better parasitoid defense may occur as a correlated response to selection for traits that improve fitness under crowding.

As stated above, *Pachycrepoideus vindemiae* is the most important pupal parasitoid of *Drosophila*. Parasitoids attacking *Drosophila* pupae lay their eggs inside the puparium, but outside of the actual pupa (Carton et al. 1986). This means that the eggs of pupal parasitoids are not subject to any kind of direct immunological attack. One strategy to resist pupal parasitoid attack might be to increase the thickness of the puparium through which the parasitoids have to drill in order to lay an egg (Zareh et al. 1980). However, puparial thickness does not appear to influence oviposition success (Kraaijeveld & Godfray 2003). Pupal size is positively correlated with the risk of parasitism, either because of a higher chance that large pupae are discovered or because of active rejection of smaller pupae by the searching parasitoid female (Kraaijeveld & Godfray 2003). If parasitoids reject small pupae because of a behavioral ecological 'optimal diet' decision, then this might lead to selection for smaller hosts. Optimal diet and optimal host range decisions are of course made in the context of the spectrum of host qualities available. If all hosts become smaller, then the response of the parasitoid would be to reduce its threshold of host size selection and accept smaller hosts for oviposition, which would vitiate any benefits. But *Pachycrepoideus* has a broad host range (Nøstvik 1954) and it is possible that one species of *Drosophila* could evolve to be outside the optimum host set. Such a strategy would only be favored if the benefits of reduced parasitoid attack were greater than the costs of being smaller as an adult in terms of lower mating success, fecundity, dispersal ability, etc.

Besides parasitoids, *Drosophila* faces potential attack from a range of other natural enemies. For instance, parasitic nematodes (*Parasitylenchus diplogenus*, *Howardula aeronymphium*), which attack larvae and then remain in the host's body until it reaches adulthood, cause a severe reduction in fitness, essentially rendering parasitized females sterile (Jaenike 1992, Jaenike et al. 1995). Entomopathogenic fungi such as *Beauveria bassiana* kill adults 5–10 days after infection (Fytrou et al. 2006), while entomopathogenic bacteria such as *Pseudomonas entomophila* cause mortality in both larvae and adults 1–4 days after infection (Vodovar et al. 2005). The microsporidian *Tubulinosema kingi* causes a reduction in fitness, mostly due to a negative effect on fecundity (Futerman et al. 2006). Infection with *Wolbachia* leads to a reduction in several fitness parameters such as size, fecundity, sperm competitiveness and, interestingly, encapsulation of parasitoid eggs (Champion de Crespigny & Wedell 2006, Fytrou et al. 2006). *Drosophila* can defend itself against bacteria and fungi by means of anti-microbial peptides, some of which are used against a wide range of microbes, whereas the action of others is more specialized (Lemaitre et al. 1997). Whether significant additive genetic variation in anti-microbial resistance exists in *Drosophila* is as yet unknown, and virtually nothing is known about defense mechanisms of *Drosophila* against nematodes, microsporidia, and intracellular bacteria such as *Wolbachia*.

Given that *Drosophila* faces attack from a variety of pathogens and parasitoids, an interesting question is to what degree resistance to these natural enemies is correlated.

Cross-resistance is likely to occur when the same defense mechanism is used against relatively similar natural enemies, whereas trade-offs are more likely when different resistance mechanisms are involved. Cross-resistance does indeed occur amongst larval parasitoids, but it is asymmetric. Selection for resistance to *L. boulardi* leads to an increase in resistance to *A. tabida*, but not the other way around (Fellowes et al. 1999c). However, selection for resistance to either *A. tabida* or *L. boulardi* leads to an increase in resistance to *L. heterotoma* (Fellowes et al. 1999c). A possible explanation for this is that selection for an increase in resistance to *A. tabida* requires a greater density of circulating hemocytes that assists in defense against *L. heterotoma*. But to defend yourself against the relative specialist *L. boulardi*, you need both an increase in hemocytes (leading to cross-resistance against generalists) and some other, currently unknown defense.

14.3.2 Host behavior

The benefits of investing in defense against parasitoids clearly depend on the risk of attack, and this may be influenced by host behavior. Genetic variation in mobility, perhaps linked to foraging behavior, exists among *Drosophila* larvae. 'Rovers' move more frequently than 'sitters', a polymorphism associated with allelic differences at a locus (dg2) that codes for a protein kinase (Sokolowski 1980, Osborne et al. 1997) involved in, among other things, muscle contraction and nerve impulse transmission. This behavioral polymorphism influences the risk of parasitoid attack, as rovers are more likely to be parasitized by species that use vibrational cues to find their host, such as *A. tabida*, whereas sitters are more frequently attacked by species that randomly probe the substrate with their ovipositor, for example *Leptopilina* spp. (Carton & Sokolowski 1992, Kraaijeveld & van Alphen 1995b). As predicted, the relative frequency of rovers and sitters in wild populations is correlated with the local frequency of parasitoids using the two types of search strategies (Sokolowski et al. 1986). The two types of larva do not differ in encapsulation frequency (Kraaijeveld & van Alphen 1995b), which suggests either that they experience similar levels of parasitoid attack and hence the same benefits of investing in defense, or that sufficiently close linkage between the genes involved in mobility and defense traits has not arisen.

The rover-sitter polymorphism may also influence the risk of pupal parasitism as rovers pupate further away from the larval feeding site than sitters (Sokolowski et al. 1986), and risk of discovery by parasitoids declines with distance (Kraaijeveld & Godfray 2003). Pupae cannot defend themselves immunologically from parasitoid attack and so this type of behavioral avoidance mechanism is one of their few means of reducing pupal parasitoid mortality.

It is possible that there are sex-specific differences in the costs and benefits of investment in defense. For example, the two sexes may differ in behavior, leading to different risks of parasite or pathogen discovery or infection. Sex differences in the behavior of *Drosophila* larvae and in the juvenile stages of other holometabolous insects may not be great, but this effect is likely to be more important in hemimetabolous insects. It may also be particularly significant for defense against parasitoids that attack the adult stage. Thus, bumblebee (*Bombus*) workers (sterile females) suffer higher rates of parasitism by conopid flies than males (Schmid-Hempel et al. 1990). There may also be sex-specific differences in the costs of investing in defense. The variance in mating success is typically far greater in males than in females (Bateman 1948, Rolff 2002), with the highest quality males getting a disproportionate share of matings. Other things being equal, males may invest relatively less than females in defense if, by investing in other functions, this increases their probability of becoming a top-quality individual (Rolff 2002). In the adult stage, males may be selected to invest in an intense period of reproductive effort at the expense of longevity. In contrast, females may be selected to invest more in defense, both because relative quality compared with others of their sex is less important and because, by living longer, they are more likely to encounter a pathogen or parasite. Female larvae do have a higher rate of encapsulation of parasitoid eggs than male larvae, but no sex difference have been found in adult resistance to a fungal pathogen (Kraaijeveld et al. 2007).

During courtship, a female decides whether a potential mate is of the right species and then, to degrees that differ across species, assesses his quality as a potential sperm-donor. Might a female *Drosophila* try to assess whether a mate carries genes that would help her offspring defend themselves against natural enemies? And, if so, what cue might she use? In *Drosophila* and some other Diptera, if the larva successfully encapsulates a parasitoid, then the melanized capsule is usually visible through the abdominal wall. Hamilton and Zuk (1982) suggested that females might prefer to mate with males showing they have genes which make them more resistant to parasites. A capsule-bearing *Drosophila* male would indeed advertise the ability to defend against parasitism. However, when offered the choice between males carrying an encapsulated parasitoid egg and unparasitized males, females showed no preference for either (Kraaijeveld et al. 1997). Perhaps any advantage of carrying genes for successful defense is counterbalanced by carrying genes that allowed the host to be discovered and parasitized in the first place. Interestingly, females do show a preference for mating with males from lines selected for high resistance over males from unselected control lines (Rolff & Kraaijeveld 2003). Whether this preference is caused by differences in resistance *per se*, or by differences in associated variation in traits associated with mate choice, such as the form of the mating song or cuticular hydrocarbons, is not known.

14.3.3 Parasitoid counter-resistance

The geographic variation in parasitoid counter-resistance mirrors to a large degree that of host resistance. The ability of southern European populations of *A. tabida* to avoid encapsulation by *D. melanogaster* is much higher than that of northern European populations (Kraaijeveld & van Alphen 1994). This may be explained by the absence of the non-encapsulating *D. subobscura* in southern Europe, leading to strong selection pressure for counter-resistance against the encapsulating *D. melanogaster* (see above). The mechanism underlying this is that eggs from southern European wasps are more likely to become embedded in host tissue than those from northern European populations (Kraaijeveld & van Alphen 1994). Geographic variation in counter-resistance, on a global scale, has been found for *L. boulardi* (Dupas et al. 1996, Dupas & Boscaro 1999). Analogous to the situation for *A. tabida* in Europe, geographic variation in counter-resistance in *L. boulardi* appears to be linked to geographic variation in the relative abundance of two of its host species (in this case *D. melanogaster* and *D. yakuba*), which differ markedly in their ability to encapsulate *L. boulardi* eggs (*D. yakuba* has a higher level of resistance against *L. boulardi* than *D. melanogaster* (Dupas & Carton 1999)).

As with host resistance, we can also ask if there are costs associated with parasitoid counter-resistance. *A. tabida* can be artificially selected for increased counter-resistance against *D. melanogaster* but, compared with the response of the host, this is much harder and the additive genetic variance for counter-resistance is perhaps an order of magnitude less than that for resistance. Because of this, statistical analysis of the selection lines is much more difficult, but there does seem to be a correlation between increased counter-resistance and the degree to which eggs are embedded in host tissue (Kraaijeveld et al. 2001a). There is no evidence that producing such 'sticky' eggs has a cost to the female parasitoid herself. However, there is a suggestion that these eggs take slightly longer (2.5 hours on average) to hatch inside the host (Kraaijeveld et al. 2001a), perhaps because, by being embedded in host tissue, they receive a slower rate of supply of nutrients or oxygen. After hatching from the egg, an A. tabida larva delays development until the host pupates. As this typically happens several days after hatching, a 2.5-hour delay is unlikely to have any fitness consequences in itself. However, when hosts are parasitized more than once, which is common in the field (van Strien-van Liempt & van Alphen 1981, Janssen 1989), the parasitoid larvae fight for possession of the host, which only contains enough resources for the successful development of one parasitoid. The parasitoid that enters the larval stage first has a greatly increased chance of winning the competition for the host (van Strien-van Liempt 1983, Visser et al. 1992c) and thus, even a 2.5-hour head start could have a significant effect on parasitoid fitness.

As with host resistance, the costs of parasitoid counter-resistance seem to be density-dependent. The admittedly provisional evidence collected so far suggests that counter-resistance has few or no costs at low parasitoid densities, but becomes more expensive as parasitoid densities increase and the risks of superparasitism grow. Populations with high rates of superparasitism may thus be selected to invest less in counter-resistance. Behavioral ecological thinking has been pivotal in helping understand the conditions under which solitary wasps are selected to lay an egg in an already parasitized host (Visser et al. 1990, 1992a,b) and may be helpful in further studies of this potential trade-off.

14.3.4 Parasitoid behavior

Parasitoids of *Drosophila*, searching in the field, will encounter a variety of hosts that differ in quality as oviposition sites (Janssen 1989). Differences in the host species' investment in resistance may be a major component of this variation. The relative preference of A. tabida females for larvae of different species of Drosophila matches the probability that parasitoids will successfully complete development in that host species (van Alphen & Janssen 1982). Moreover, this preference shows intra-specific variation, in line with the geographic variation in counter-resistance. As detailed above, parasitoids from northern parts of Europe have a low level of counter-resistance against the encapsulation system of D. melanogaster, in contrast to parasitoids from southern Europe. When offered a choice between D. melanogaster and the non-encapsulating D. subobscura, parasitoids from northern European populations reject D. melanogaster, whereas parasitoids from southern European populations accept both host species for oviposition (Kraaijeveld et al. 1995).

We do not yet know whether parasitoids are able to detect at the time of oviposition how much their hosts have invested in defense. If they could, they might use this information in deciding whether to place an egg in that particular individual. They might also use the information to update their estimation of local patch quality, modifying their host rejection or patch-leaving thresholds accordingly.

There is some evidence that host preference evolves in tandem with counter-resistance. Females from lines selected for increased counter-resistance against D. melanogaster have a higher level of acceptance of this host species than females from the unselected control lines (Rolff & Kraaijeveld 2001). We do not know whether host preference evolves in some way as a simple correlated response to selection on counter-resistance, or whether there is simultaneous direct selection on this trait.

14.4 Other systems

We briefly mention some results from non-*Drosophila* systems that bear on the themes developed above. Some of the first studies of host–parasitoid coevolution involved house flies (*Musca domestica*) and the pupal parasitoid, *Nasonia vitripennis* (Pimentel & Al-Hafidh 1965, Pimentel & Stone 1968, Olson & Pimentel 1974). The original motivation for this work was to study 'population homeostasis', but it also provides some data on resistance/counter-resistance coevolution. As discussed above, the puparial wall of an insect represents a potentially important barrier to pupal parasitoids. House flies cultured with *Nasonia* evolve heavier puparia and a shorter pupal stage (Zareh et al. 1980). Both of these adaptations can be regarded as resistance mechanisms. Heavier puparia are probably the result of a thicker puparial wall, which may make it harder for the parasitoid female to drill through and oviposit. Shortening of the pupal period reduces the time window in which the pupa is vulnerable to parasitism. A possibly cost of becoming more resistant to pupal parasitoids is a reduction in female fecundity (Zareh et al. 1980).

Due to their clonal nature, aphids, and in particular the well-studied pea aphid (*Acyrtosiphon pisum*), provide a valuable model system for investigating genetic variation in resistance against parasitoids. The exact nature of the mechanism that pea aphids use to defend themselves against parasitoids is largely unknown, though what is clear is that cellular encapsulation, as used by *Drosophila* and most insects, plays no role. In the field, there is substantial variation among pea aphid clones in their resistance to different species of *Aphidius* (Braconidae), their main parasitoid (Henter & Via 1995, Ferrari et al. 2001). However, costs of resistance have so far not been identified in the pea aphid. Resistant clones do not differ from susceptible clones in fecundity on either good-quality or poor-quality plants, and there are no negative correlations between resistance to two species of *Aphidius* (*A. ervi* and *A. eadyi*) or to a fungal pathogen (*Pandora neoaphidis*, Ferrari et al. 2001).

One problem in trying to search for possible costs of resistance is the complicated population structure of the pea aphid and its association with symbiotic bacteria. *A. pisum* consists of a series of host-adapted populations, which are specialized on particular host plant species (Via 1999, Ferrari & Godfray 2006, Ferrari et al. 2006). Host-adapted populations differ in resistance to parasitoids (Ferrari & Godfray 2006) and this may, at least partly, be caused by secondary bacterial endosymbionts. In addition to the essential bacterial endosymbiont *Buchnera*, pea aphid clones also harbor several other bacterial endosymbionts. Unlike the case with *Buchnera*, there is variation among clones in the presence or absence of these secondary endosymbionts. Oliver et al. (2003) showed that aphids carrying one of two endosymbionts species had increased resistance to *Aphidius*, while the presence of a third improved resistance to the fungus *Pandora* (Scarborough et al. 2005).

Less work has been done on counter-resistance in the parasitoids of the pea aphid, though Henter (1995) showed considerable additive genetic variation amongst parasitoids in their ability to successfully parasitize pea aphids. The amount of genetic variation thus appears to be much higher than in the *Drosophila* parasitoids described above. It would be interesting to know the degree to which parasitoid populations have become adapted to different pea aphid host races. It would also be nice to know whether differences in host race, or differences in the spectrum of bacteria carried by pea aphids, influence the behavioral ecology of their parasitoids.

We described above how *Drosophila* larvae evolved higher levels of resistance under crowded conditions (Sanders et al. 2005). In other systems, especially armyworm Lepidoptera and locusts, it has been shown that individuals can up-regulate their immune system under crowded conditions because, it is hypothesized, crowding is associated with increased exposure to pathogens. This phenomenon has been called density-dependent prophylaxis (Reeson et al. 1998). The most importance pathogens of these insects are microbial but in the case of the noctuid *Spodoptera exempta*, gregarious forms are also more resistant to parasitoids (Wilson et al. 2001). It would be interesting to know whether *S. exempta* suffers density-dependent parasitism in the field and how the host response affects parasitoid foraging and host choice behavior.

14.5 Conclusions

Parasitoids have proven to be excellent models for exploring some of the classical questions in behavioral ecology, in particular patch use (see also Chapter 8 by van Alphen and Bernstein) and diet breadth (as host range), and in evolutionary ecology for investigating co-evolutionary interactions in consumer-resource systems. The two fields of parasitoid research have overlapped so far by very little. One reason for this is that behavioral ecology largely involves the study of relatively plastic traits, which can be studied by experimentally manipulating the parasitoid's environment. Studies of resistance and counter-resistance involve traits that are less plastic and that cannot be manipulated directly. In consequence, more genetic approaches, in particular artificial selection, have had to be employed.

How might this change in the future? First, host resistance is a major determinant of host quality for a parasitoid. Typically, behavioral ecologists treat the physiological aspect of host quality as a given (see also Chapter 6 by Strand and Casas), while evolutionary ecologists seek to make parasitoid behavior as uniform as possible. It would be interesting to consider feedbacks between the physiological and behavioral aspects of the parasitoid's life history. We already know, at least in theory, that there are interesting feedbacks between the evolution of resistance and counter-resistance and host parasitoid population dynamics (Sasaki & Godfray 1999), and including a behavioral ecological component would also be fascinating.

Second, both behavioral and evolutionary ecologists are becoming more interested in the explicit genetic basis of the traits they study. For a long time, the rover/sitter polymorphism in *D. melanogaster* was very rare in being a behavioral trait in which both the genetic and molecular basis was understood (Osborne et al. 1997). With the increasing use of high throughput '-omics' molecular techniques, many more such loci influencing behavior will be discovered, as is already happening with the genes involved in resistance (DeGregorio et al. 2001, Irving et al. 2001, Roxström-Lindquist et al. 2004, Wertheim et al. 2005). Unfortunately, we do not yet have a well-studied system where both the genomes of the host and parasitoid are known. Perhaps this is an area in which parasitoids scientists involved in behavioral, evolutionary, and physiological research need to come together to agree priorities.

References

Agaisse, H. and Perrimon, N. (2004) The roles of the JAK/STAT signalling in *Drosophila* immune responses. *Immunological Reviews* **198**: 72–82.

Asgari, S. and Schmidt, O. (1994) Passive protection of eggs from the parasitoid *Cotesia rubecula* in the host *Pieris rapae. Journal of Insect Physiology* **40**: 789–95.

Bateman, A.J. (1948) Intra-sexual selection in *Drosophila. Heredity* **2**: 349–68.

Beckage, N.E. and Gelman, D.B. (2004) Wasp parasitoid disruption of host development: Implications for new biologically based strategies for insect control. *Annual Review of Entomology* **49**: 299–330.

Carton, Y. (1984) Analyse expérimentale de trois niveaux d'interaction entre *Drosophila melanogaster* et le parasite *Leptopilina boulardi* (sympatrie, allopatrie, xénopatrie). *Génétique Sélection Evolution* **16**: 417–30.

Carton, Y. and Kitano, H. (1981) Evolutionary relationships to parasitism by seven species of the *Drosophila melanogaster* subgroup. *Biological Journal of the Linnean Society* **16**: 227–41.

Carton, Y. and David, J.R. (1983) Reduction of fitness in *Drosophila* adults surviving parasitization by a cynipid wasp. *Experientia* **39**: 231–3.

Carton, Y. and Boulétreau, M. (1985) Encapsulation ability of *Drosophila melanogaster*: A genetic analysis. *Developmental & Comparative Immunology* **9**: 211–19.

Carton, Y. and Nappi, A.J. (1991) The *Drosophila* immune reaction and the parasitoid capacity to evade it: Genetic and coevolutionary aspects. *Acta Oecologica* **12**: 89–104.

Carton, Y. and Sokolowski, M.B. (1992) Interactions between searching strategies of *Drosophila* parasitoids and the polymorphic behaviour of their hosts. *Journal of Insect Behavior* **5**: 161–75.

Carton, Y. and Nappi, A.J. (1997) *Drosophila* cellular immunity against parasitoids. *Parasitology Today* **13**: 218–27.

Carton, Y. and Dupas, S. (1999) Genetics of resistance and virulence in host-parasitoid systems. In: Wiesner, A., Dunphy, G.B., Marmaras, V.J. et al. (eds.) *Techniques in Insect Immunology*. SOS Publishing, Fair Haven, pp. 279–88.

Carton, Y., Boulétreau, M., van Alphen, J.J.M. and van Lenteren, J.C. (1986) The *Drosophila* parasitic wasps. In: Ashburner, M. (ed.) *The Genetics and Biology of Drosophila*, vol. 3e. Academic Press, London, pp. 347–94.

Champion de Crespigny, F.E. and Wedell, N. (2006) *Wolbachia* infection reduces sperm competitive ability in an insect. *Proceedings of the Royal Society of London Series B Biological Science* **273**: 1455–8.

Crozatier, M., Ubeda, J., Vincent, A. and Meister, M. (2004) Cellular immune response to parasitisation in *Drosophila* requires the EBF orthologue collier. *PLoS Biology* **2**: 196.

DeGregorio, E., Spellman, P.T., Rubin, G.M. and Lemaitre, B. (2001) Genome-wide analysis of the *Drosophila* immune response by using oligonucleotide microarrays. *Proceedings of the National Academy of Sciences USA* **98**: 12590–5.

DeGregorio, E., Spellman, P.T., Tzou, P., Rubin, G.M. and Lemaitre, B. (2002) The Toll and Imd pathways are the major regulators of the immune response in *Drosophila. EMBO Journal* **21**: 2568–79.

Digilio, M.C., Isidoro, N., Tremblay, E. and Pennacchio, F. (2000) Host castration by *Aphidius ervi* venom proteins. *Journal of Insect Physiology* **46**: 1041–50.

Doebeli, M. (1997) Genetic variation and the persistence of predator-prey interactions in the Nicholson-Bailey model. *Journal of Theoretical Biology* **188**: 109–20.

Dupas, S. and Boscaro, M. (1999) Geographic variation and evolution of immunosuppressive genes in a *Drosophila* parasitoid. *Ecography* **22**: 284–91.

Dupas, S. and Carton, Y. (1999) Two non-linked genes for specific virulence of *Leptopilina boulardi* against *Drosophila melanogaster* and *D. yakuba*. *Evolutionary Ecology* **13**: 211–20.

Dupas, S., Brehelin, M., Frey, D.F. and Carton, Y. (1996) Immune suppressive virus-like particles in a *Drosophila*-parasitoid: Significance of their intraspecific morphological variations. *Parasitology* **113**: 207–12.

Eslin, P. and Prévost, G. (1996) Variation in *Drosophila* concentration of haemocytes associated with different ability to encapsulate *Asobara tabida* larval parasitoid. *Journal of Insect Physiology* **42**: 549–55.

Eslin, P. and Prévost, G. (1998) Hemocyte load and immune resistance to *Asobara tabida* are correlated in species of the *Drosophila melanogaster* subgroup. *Journal of Insect Physiology* **44**: 807–16.

Eslin, P. and Doury, G. (2006) The fly *Drosophila subobscura*: A natural case of innate immunity deficiency. *Developmental & Comparative Immunology* **30**: 977–83.

Eslin, P., Giordanengo, P., Fourdrain, Y. and Prévost, G. (1996) Avoidance of encapsulation in the absence of VLP by a braconid parasitoid of *Drosophila* larvae: An ultrastructural study. *Canadian Journal of Zoology* **74**: 2193–8.

Espagne, E., Dupuy, C., Huguet, E. et al. (2004) Genome sequence of a polydnavirus: Insights into symbiotic virus evolution. *Science* **306**: 286–9.

Evans, C.J., Hartenstein, V. and Banerjee, U. (2003) Thicker than blood: Conserved mechanisms in *Drosophila* and vertebrate hematopoiesis. *Developmental Cell* **5**: 673–90.

Fellowes, M.D.E., Kraaijeveld, A.R. and Godfray, H.C.J. (1998a) Trade-off associated with selection for increased ability to resist parasitoid attack in *Drosophila melanogaster*. *Proceedings of the Royal Society of London Series B Biological Science* **265**: 1553–8.

Fellowes, M.D.E., Masnatta, P., Kraaijeveld, A.R. and Godfray, H.C.J. (1998b) Pupal parasitoid attack influences the relative fitness of *Drosophila* that have encapsulated larval parasitoids. *Ecological Entomology* **23**: 281–4.

Fellowes, M.D.E., Kraaijeveld, A.R. and Godfray, H.C.J. (1999a) The relative fitness of *Drosophila melanogaster* (Diptera, Drosophilidae) that have successfully defended themselves against the parasitoid *Asobara tabida* (Hymenoptera, Braconidae). *Journal of Evolutionary Biology* **12**: 123–8.

Fellowes, M.D.E., Kraaijeveld, A.R. and Godfray, H.C.J. (1999b) Association between feeding rate and defence against parasitoids in *Drosophila melanogaster*. *Evolution* **53**: 1302–5.

Fellowes, M.D.E., Kraaijeveld, A.R. and Godfray, H.C.J. (1999c) Cross-resistance following artificial selection for increased defence against parasitoids in *Drosophila melanogaster*. *Evolution* **53**: 966–72.

Ferrari, J. and Godfray, H.C.J. (2006) The maintenance of intraspecific biodiversity: The interplay of selection on resource use and on natural enemy resistance in the pea aphid. *Ecological Research* **21**: 9–16.

Ferrari, J., Müller, C.B., Kraaijeveld, A.R. and Godfray, H.C.J. (2001) Clonal variation and covariation in aphid resistance to parasitoids and a pathogen. *Evolution* **55**: 1805–14.

Ferrari, J., Godfray, H.C.J., Faulconbridge, A.S., Prior, K. and Via, S. (2006) Population differentiation and genetic variation in host choice among pea aphids from eight host plant genera. *Evolution* **60**: 1574–84.

Frank, S.A. (1994) Coevolutionary genetics of hosts and parasites with quantitative inheritance. *Evolutionary Ecology* **8**: 74–94.

Fullilove, S.L., Jacobson, A.G. and Turner, F.R. (1977) Embryonic development: Descriptive. In: Ashburner, M. and Wright, T.R.F. (eds.) *The Genetics and Biology of Drosophila*, vol. 2c. Academic Press, London, pp. 106–209.

Futerman, P.H., Layen, S.J., Kotzen, M.L., Franzen, C., Kraaijeveld, A.R. and Godfray, H.C.J. (2006) Fitness effects and transmission routes of a microsporidian parasite infecting *Drosophila* and its parasitoids. *Parasitology* **132**: 479–92.

Fytrou, A., Schofield, P.G., Kraaijeveld, A.R. and Hubbard, S.F. (2006) *Wolbachia* infection suppresses both host defence and parasitoid counter-defence. *Proceedings of the Royal Society of London Series B Biological Science* **273**: 791–6.

Godfray, H.C.J. (1994) *Parasitoids: Behavioural and Evolutionary Ecology*. Princeton University Press, Princeton.

Hamilton, W.D. and Zuk, M. (1982) Heritable true fitness and bright birds: A role for parasites? *Science* **218**: 384–7.

Hawkins, B.A. and Sheehan, W. (1994) *Parasitoid Community Ecology*. Oxford University Press, Oxford.

Hayakawa, Y. and Yazaki, K. (1997) Envelope protein of parasitic wasp symbiont virus, polydnavirus, protects the wasp eggs from cellular immune reactions by the host insect. *European Journal of Biochemistry* **246**: 820–6.

Henter, H.J. (1995) The potential for coevolution in a host-parasitoid system. II. Genetic variation within a population of wasps in the ability to parasitize an aphid host. *Evolution* **49**: 439–45.

Henter, H.J. and Via, S. (1995) The potential for coevolution in a host-parasitoid system. I. Genetic variation within an aphid population in susceptibility to a parasitic wasp. *Evolution* **49**: 427–38.

Hochberg, M.E. (1997) Hide or fight? The competitive evolution of concealment and encapsulation in parasitoid-host systems. *Oikos* **80**: 342–52.

Hochberg, M.E. and Holt, R.D. (1995) Refuge evolution and the population dynamics of coupled host-parasitoid associations. *Evolutionary Ecology* **9**: 633–61.

Hoffmann, J.A., Kafatos, F.C., Janeway, Jr J.A. and Ezekowitz, R.A.B. (1999) Phylogenetic perspectives in innate immunity. *Science* **284**: 1313–18.

Irving, P., Troxler, L., Heuer, T.S. et al. (2001) A genome-wide analysis of immune responses in *Drosophila*. *Proceedings of the National Academy of Sciences USA* **98**: 15119–24.

Jaenike, J. (1992) Mycophagous *Drosophila* and their nematode parasites. *American Naturalist* **139**: 893–906.

Jaenike, J., Benway, H. and Stevens, G. (1995) Parasite-induced mortality in mycophagous *Drosophila*. *Ecology* **76**: 383–91.

Janssen, A.R.M. (1989) Optimal host selection by *Drosophila* parasitoids in the field. *Functional Ecology* **3**: 469–79.

Jeffries, M.J. and Lawton, J.H. (1984) Enemy free space and the structure of ecological communities. *Biological Journal of the Linnean Society* **23**: 269–86.

Johner, A., Stettler, P., Gruber, A. and Lanzrein, B. (1999) Presence of polydnavirus transcripts in an egg-larval parasitoid and its lepidopterous host. *Journal of General Virology* **80**: 1847–54.

Joshi, A. and Mueller, L.D. (1988) Evolution of higher feeding rate in *Drosophila* due to density-dependent selection. *Evolution* **42**: 1090–3.

Kraaijeveld, A.R. and van Alphen, J.J.M. (1994) Geographical variation in resistance of the parasitoid *Asobara tabida* against encapsulation by *Drosophila melanogaster*: The mechanism explored. *Physiological Entomology* **19**: 9–14.

Kraaijeveld, A.R. and van der Wel, N.N. (1994) Geographic variation in reproductive success of the parasitoid *Asobara tabida* in larvae of several *Drosophila* species. *Ecological Entomology* **19**: 221–9.

Kraaijeveld, A.R. and van Alphen, J.J.M. (1995a) Geographical variation in encapsulation ability of *Drosophila melanogaster* larvae and evidence for parasitoid-specific components. *Evolutionary Ecology* **9**: 10–17.

Kraaijeveld, A.R. and van Alphen, J.J.M. (1995b) Foraging behaviour and encapsulation ability of *Drosophila melanogaster* larvae: Correlated polymorphisms? *Journal of Insect Behavior* **8**: 305–14.

Kraaijeveld, A.R. and Godfray, H.C.J. (1997) Trade-off between parasitoid resistance and larval competitive ability in *Drosophila melanogaster*. *Nature* **389**: 278–80.

Kraaijeveld, A.R. and Godfray, H.C.J. (2001) Is there local adaptation in *Drosophila*-parasitoid interactions? *Evolutionary Ecology Research* **3**: 107–16.

Kraaijeveld, A.R. and Godfray, H.C.J. (2003) Potential life history costs of parasitoid avoidance in *Drosophila melanogaster*. *Evolutionary Ecology Research* **5**: 1251–61.

Kraaijeveld, A.R., Nowee, B. and Najem, R.W. (1995) Adaptive variation in host-selection behaviour of *Asobara tabida*, a parasitoid of *Drosophila* larvae. *Functional Ecology* **9**: 113–18.

Kraaijeveld, A.R., Emmett, D.A. and Godfray, H.C.J. (1997) Absence of direct sexual selection for parasitoid encapsulation in *Drosophila melanogaster*. *Journal of Evolutionary Biology* **10**: 337–42.

Kraaijeveld, A.R., Hutcheson, K.A., Limentani, E.C. and Godfray, H.C.J. (2001a) Costs of counter-defenses to host resistance in a parasitoid of *Drosophila*. *Evolution* **55**: 1815–21.

Kraaijeveld, A.R., Limentani, E.C. and Godfray, H.C.J. (2001b) Basis of the trade-off between parasitoid resistance and larval competitive ability in *Drosophila melanogaster*. *Proceedings of the Royal Society of London Series B Biological Science* **268**: 259–61.

Kraaijeveld, A.R., Barker, C.L. and Godfray, H.C.J. (2007) Stage-specific sex differences in *Drosophila* immunity to parasites and pathogens. *Evolutionary Ecology* (in press).

Lanot, R., Zachary, D., Holder, F. and Meister, M. (2001) Postembryonic hematopoiesis in *Drosophila*. *Developmental Biology* **230**: 243–57.

Lavine, M.D. and Strand, M.R. (2002) Insect hemocytes and their role in immunity. *Insect Biochemistry & Molecular Biology* **32**: 1295–309.

Lemaitre, B., Reichhart, J.-M. and and Hoffman, J.A. (1997) *Drosophila* host defense: Differential induction of antimicrobial peptide genes after infection by various classes of microorganisms. *Proceedings of the National Academy of Sciences USA* **94**: 14614–19.

Meister, M. and Lagueux, M. (2003) *Drosophila* blood cells. *Cell Microbiology* **5**: 573–80.

Mueller, L.D. (1988) Evolution of competitive ability in *Drosophila* by density-dependent natural selection. *Proceedings of the National Academy of Sciences USA* **85**: 4383–6.

Nappi, A.J. (1975) Parasite encapsulation in insects. In: Maramorosch, K. and Shope, R. (eds.) *Invertebrate Immunity*. Academic Press, New York, pp. 293–326.

Nappi, A.J. (1981) Cellular immune response of *Drosophila melanogaster* against *Asobara tabida*. *Parasitology* **83**: 319–24.

Nappi, A.J. and Vass, E. (1993) Melanogenesis and the generation of cytotoxic molecules during insect cellular immune reactions. *Pigment Cell Research* **6**: 117–26.

Nappi, A.J., Vass, E., Frey, F. and Carton, Y. (1995) Superoxide anion generation in *Drosophila* during melanotic encapsulation of parasites. *European Journal of Cell Biology* **68**: 450–6.

Nøstvik, E. (1954) Biological studies of *Pachycrepoideus dubius* Ashmead (Chalcidoidea: Pteromalidae), a pupal parasite of various Diptera. *Oikos* **5**: 195–204.

Oliver, K.M., Russell, J.A., Moran, N.A. and Hunter, M.S. (2003) Facultative bacterial symbionts in aphids confer resistance to parasitic wasps. *Proceedings of the National Academy of Sciences USA* **100**: 1803–7.

Olson, D. and Pimentel, D. (1974) Evolution of resistance in a host population to attacking parasite. *Environmental Entomology* **3**: 621–4.

Osborne, K.A., Robichon, A., Burgess, E. et al. (1997) Natural behavior polymorphism due to a cGMP-dependent protein kinase of *Drosophila*. *Science* **277**: 834–6.

Pech, L.L. and Strand, M.R. (2000) Plasmatocytes from the moth *Pseudoplusia includens* induce apoptosis of granular cells. *Journal of Insect Physiology* **46**: 1565–73.

Pimentel, D. and Al-Hafidh, R. (1965) Ecological control of a parasite population by genetic evolution in the parasite-host system. *Annals of the Entomological Society of America* **58**: 1–6.

Pimentel, D. and Stone, F.A. (1968) Evolution and population ecology of parasite-host systems. *Canadian Entomologist* **100**: 655–62.

Poitrineau, K., Brown, S.P. and Hochberg, M.E. (2003) Defence against multiple enemies. *Journal of Evolutionary Biology* **16**: 1319–27.

Poitrineau, K., Brown, S.P. and Hochberg, M.E. (2004) The joint evolution of defence and inducibility against natural enemies. *Journal of Theoretical Biology* **231**: 389–96.

Quicke, D. (1997) *Parasitic Wasps*. Chapman & Hall, London.

Reeson, A.F., Wilson, K., Gunn, A., Hails, R.S. and Goulson, D. (1998) Baculovirus resistance in the noctuid *Spodoptera exempta* is phenotypically plastic and responds to population density. *Proceedings of the Royal Society of London Series B Biological Science* **265**: 1787–91.

Richards, E.H. and Edwards, J.P. (1999) Parasitization of *Lacanobia oleracea* (Lepidoptera: Noctuidae) by the ectoparasite wasp, *Eulophus pennicornis* – effects of parasitisation, venom and starvation on host hemocytes. *Journal of Insect Physiology* **45**: 1073–83.

Richards, E.H. and Parkinson, N.M. (2000) Venom from the endoparasitic wasp *Pimpla hypochondriaca* adversely affects the morphology, viability, and immune function of hemocytes from larvae of the tomato moth, *Lacanobia oleracea*. *Journal of Invertebrate Pathology* **76**: 33–42.

Rizki, T.M. and Rizki, R.M. (1984) The cellular defense system of *Drosophila melanogaster*. In: King, R.C. and Akai, H. (eds.) *Insect Ultrastructure*. Plenum Publishing, New York, pp. 579–603.

Rizki, R.M. and Rizki, T.M. (1990) Parasitoid virus-like particles destroy *Drosophila* immunity. *Proceedings of the National Academy of Sciences USA* **87**: 8388–92.

Rolff, J. (2002) Bateman's principle and immunity. *Proceedings of the Royal Society of London Series B Biological Science* **269**: 867–2.

Rolff, J. and Kraaijeveld, A.R. (2001) Host preference and survival in selected lines of a *Drosophila* parasitoid, *Asobara tabida*. *Journal of Evolutionary Biology* **14**: 742–5.

Rolff, J. and Kraaijeveld, A.R. (2003) Selection for parasitoid resistance alters mating success in *Drosophila*. *Proceedings of the Royal Society of London Series B Biological Science* **270**: 154–5.

Rolff, J. and Siva-Jothy, M.T. (2003) Invertebrate ecological immunity. *Science* **301**: 472–5.

Roxström-Lindquist, K., Terenius, O. and Faye, I. (2004) Parasite-specific immune response in adult *Drosophila melanogaster*: A genomic study. *EMBO Reports* **5**: 207–12.

Sanders, A.E., Scarborough, C., Layen, S.J., Kraaijeveld, A.R. and Godfray, H.C.J. (2005) Evolutionary change in parasitoid resistance under crowded conditions in *Drosophila melanogaster*. *Evolution* **59**: 1292–9.

Sasaki, A. and Godfray, H.C.J. (1999) A model for the coevolution of resistance and virulence in coupled host-parasitoid interactions. *Proceedings of the Royal Society of London Series B Biological Science* **266**: 455–63.

Scarborough, C., Ferrari, J. and Godfray, H.C.J. (2005) Aphid protected from pathogen by endo-symbiont. *Science* **310**: 1781.

Schmid-Hempel, P. (2003) Variation in immune defence as a question of evolutionary ecology. *Proceedings of the Royal Society of London Series B Biological Science* **270**: 357–66.

Schmid-Hempel, P. and Ebert, D. (2003) On the evolutionary ecology of specific immune defence. *Trends in Ecology & Evolution* **18**: 27–32.

Schmid-Hempel, P., Müller, C., Schmid-Hempel, R. and Shykoff, J.A. (1990) Frequency and ecological correlates of parasitism by conopid flies (Conopidae, Diptera) in populations of bumblebees. *Insectes Sociaux* **37**: 14–30.

Sokolowski, M.B. (1980) Foraging strategies of *Drosophila melanogaster*: A chromosomal analysis. *Behavior Genetics* **10**: 291–2.

Sokolowski, M.B., Bauer, S.J., Wai-Ping, V., Rodriguez, L., Wong, J.L. and Kent, C. (1986) Ecological genetics and behaviour of *Drosophila melanogaster* larvae in nature. *Animal Behaviour* **34**: 403–8.

Sorrentino, R.P., Carton, Y. and Govind, S. (2002) Cellular immune response to parasite infection in the *Drosophila* lymph gland is developmentally regulated. *Developmental Biology* **243**: 65–80.

Strand, M.R. and Pech, L.L. (1995) Immunological basis for compatibility in parasitoid-host relationships. *Annual Review of Entomology* **40**: 31–56.

Tepass, U., Fessler, L.I., Aziz, A. and Hartenstein, V. (1994) Embryonic origin of hemocytes and their relationship to cell death in *Drosophila*. *Development* **120**: 1829–37.

Thompson, J.N. (1994) *The Coevolutionary Process*. University of Chicago Press, Chicago.

Thompson, J.N. (2005) *The Geographic Mosaic of Coevolution*. University of Chicago Press, Chicago.

Tiën, N.S.H., Boyle, D., Kraaijeveld, A.R. and Godfray, H.C.J. (2001) Competitive ability of parasitized *Drosophila* larvae. *Evolutionary Ecology Research* **3**: 747–57.

Turnbull, M. and Webb, B. (2002) Perspectives on polydnavirus origins and evolution. *Advances in Virus Research* **58**: 203–54.

van Alphen, J.J.M. and Janssen, A.R.M. (1982) Host selection by *Asobara tabida* Nees (Braconidae: Alysiinae), a larval parasitoid of fruit inhabiting *Drosophila* species. II. Host species selection. *Netherlands Journal of Zoology* **32**: 215–31.

van Strien-van Liempt, W.T.F.H. (1983) The competition between *Asobara tabida* Nees von Esenbeck, 1834 and *Leptopilina heterotoma* (Thomson, 1862) in multiparasitized hosts. I. The course of competition. *Netherlands Journal of Zoology* **33**: 125–63.

van Strien-van Liempt, W.T.F.H. and van Alphen, J.J.M. (1981) The absence of interspecific host discrimination in *Asobara tabida* Nees and *Leptopilina heterotoma* (Thomson), coexisting larval parasitoids of *Drosophila* species. *Netherlands Journal of Zoology* **31**: 701–12.

Via, S. (1999) Reproductive isolation between sympatric races of pea aphids. I. Gene flow restriction and habitat choice. *Evolution* **53**: 1446–57.

Visser, M.E., van Alphen, J.J.M. and Nell, H.W. (1990) Adaptive superparasitism and patch time allocation in solitary parasitoids: The influence of the number of parasitoids depleting a patch. *Behaviour* **114**: 21–36.

Visser, M.E., van Alphen, J.J.M. and Hemerik, L. (1992a) Adaptive superparasitism and patch time allocation in solitary parasitoids: An ESS model. *Journal of Animal Ecology* **61**: 93–101.

Visser, M.E., van Alphen, J.J.M. and Nell, H.W. (1992b) Adaptive superparasitism and patch time allocation in solitary parasitoids: The influence of pre-patch experience. *Behavioural Ecology & Sociobiology* **31**: 163–71.

Visser, M.E., Luyckx, B., Nell, H.W. and Boskamp, G.J.F. (1992c) Adaptive superparasitism in solitary parasitoids: Marking of parasitized hosts in relation to the pay-off from superparasitism. *Ecological Entomology* **17**: 76–82.

Vodovar, N., Vinals, M., Liehl, P. et al. (2005) *Drosophila* host defense after oral infection by an entomopathogenic *Pseudomonas* species. *Proceedings of the National Academy of Sciences USA* **102**: 11414–19.

Webb, B.A., Strand, M.R., Dickey, S.E. et al. (2006) Polydnavirus genomes reflect their dual roles as mutualists and pathogens. *Virology* **347**: 160–74.

Wertheim, B., Kraaijeveld, A.R., Schuster, E. et al. (2005) Genome-wide gene expression in response to parasitoid attack in *Drosophila*. *Genome Biology* **6**: 94.

Wilson, K., Cotter, S.C., Reeson, A.F. and Pell, J.K. (2001) Melanism and disease resistance in insects. *Ecology Letters* **4**: 637–49.

Zareh, N., Westoby, M. and Pimentel, D. (1980) Evolution in a laboratory host-parasitoid system and its effects on population kinetics. *Canadian Entomologist* **112**: 1049–106.

Part 3

Methodological issues in behavioral ecology

15

State-dependent problems for parasitoids: case studies and solutions

Bernard Roitberg and Pierre Bernhard

Abstract

Many problems that parasitoids face in nature are state-dependent. For example, the optimal decision regarding encounters with already-parasitized hosts often depends upon the egg load (egg state) of the focal organism. Further, many biologically relevant events (e.g. host encounters) occur with some probability. Thus, an appropriate model for analyzing such problems is a stochastic state variable model, with Dynamic Programming (DP) as the appropriate tool to derive an optimal policy (together known as SDP). In this chapter, we consider two case studies where SDP models provide solutions to parasitoid foraging problems. These include: (i) optimal attacks on target and non-target hosts in the context of biological control, as well as (ii) optimal attacks on hosts of known size but unknown instar. Using these case studies, we show a general form of DP, and its infinite horizon form for which we discuss algorithmic solution methods, as the DP equation is not constructive in that case. The application to lifetime processes induces a slight generalization of classic infinite horizon DP. We also stress that, using Kolmogorov's equation, essentially the same computer code that lets us compute an optimal policy can be used to compare its efficiency against that of other policies, avoiding lengthy forward simulations.

15.1 Introduction

It is becoming increasingly clear that many parasitoid behaviors are state-dependent. As a sample of many such relationships, Ueno (1999) and Sirot et al. (1997) showed that oviposition decisions (superparasitism) depend upon egg load (egg state), Desouhant et al. (2005) showed that feeding decisions depend on nutrition state, Roitberg et al. (1993) demonstrated that host discrimination can be a function of information state, while Wajnberg et al. (2006) showed that patch leaving decisions are a function of age.

Of course, we could go on citing many other examples, but the important question that emerges is: How does one incorporate such states into predictions of parasitoid behavior?

The answer can be found in the form of Stochastic Dynamic Programming (SDP) models (Bellman 1957). Though these models were first developed to solve problems in human economics, they have now been adapted to address issues in natural non-human systems, including parasitoids, thanks to the efforts of Clark and Mangel (2000), Houston and McNamara (1999), and others. The models take many forms but, in general, optimal decisions are calculated based upon maximizing expected lifetime reproductive success (the most common fitness proxy, see Roitberg et al. 2001) as a function of current state(s). As noted above, this state-dependant theory has shown great utility for predicting a variety of parasitoid behaviors. In particular, dynamic programming (DP) models have been employed to solve a number of parasitoid problems, including:

1. patch time allocation (Wajnberg 2006, see also Chapter 8 by van Alphen and Bernstein);
2. egg versus time limitation (Rosenheim 1999, Ellers et al. 2000);
3. oviposition decisions (Mangel 1989b); and
4. host feeding (Collier 1995, see also Chapter 7 by Bernstein and Jervis).

DP models have been reviewed elsewhere and, as such, we refer readers to Houston et al. (1988) for background reading. Here, we will use this opportunity to show readers in detail how to formulate and analyze DP models. To do so, we have chosen two classic problems in parasitoid behavioral ecology: optimal fidelity and optimal instar discrimination. These problems provide a kind of window into state variable challenges that parasitoids typically face. We shall also show how the same formulation, in terms of dynamic state models, lets us replace Monte Carlo simulations by a faster and more precise approach, namely Kolomogorov's equation, which turns out to have a close kinship with Bellman's (1957) equation.

15.1.1 Fidelity

In biological control, one of the main questions that researchers face is the possibility of attacks on non-target hosts (Bigler et al. 2006). Thus, the key issue here is fidelity, i.e. will the agent restrict its attacks to targets, as chosen by the biocontrol practitioner? As noted above, there already exists considerable theory on optimal host selection and this question neatly falls into that category, i.e. are there conditions where the parasitoid will attack (presumably) the lower-quality, non-target host? It is difficult to provide a general answer for biological control practitioners because this answer is likely to be context-dependent, with the context here being life history. However, as a simple starting point, and drawing from Roitberg (2000), we will consider two major categories of parasitoids: those that are proovigenic versus those that are synovigenic. The former types are comprised of individuals who begin adult life with a full complement of eggs and then deplete those eggs over time through parasitizing hosts. The latter individuals continually mature new eggs following eclosion. In the former case, each time an egg is laid, the parasitoid will move into the future with one less egg available than before oviposition, compared to the latter whereupon future egg availability will depend upon egg maturation rates and ovarioles. Further, we must consider gregarious parasitoids, i.e. individuals not only have to 'decide' whether to accept or reject hosts but also, upon acceptance, they must decide how many eggs to lay in a given host (Hoffmeister et al. 2005). Here we consider the scenario of an individual foraging within a large non-depleting universe of hosts. When hosts are encountered, target hosts are accepted with requisite clutch size decision but, when non-target hosts are encountered, acceptance/rejection decisions are required. Below, we develop a theory for non-target host acceptance decisions.

15.1.2 Size discrimination

Many parasitoids choose hosts on the basis of size (Wang & Messing 2004). Size is a useful criterion because it often correlates well with nutrient content (Vinson & Iwantsch 1980) and host defensive ability (Gerling et al. 1990). However, when parasitoids are koinobionts, wherein their larvae exploit developing hosts, then the host instar is also extremely important due to differences in growth rates and time limitations. This should not pose a problem for parasitoids that are host specialists because size and instar are highly correlated. However, when parasitoids are generalists, they may encounter a vast range of different sized individuals within the same instar. So, how should koinobiont parasitoids choose hosts?

This question arises from work being conducted by Henry et al. (2005, 2006). This author works with the parasitoid *Aphidius ervi*, a generalist wasp that attacks a variety of aphids. It turns out that, regardless of host species, that the second and third instars are the best hosts while the first and fourth instars are often unsuitable. In addition, independent of host species, the larger the aphid, the more dangerous that aphid is to *A. ervi* via its defensive maneuvers (e.g. kicking). These two factors are interesting but they become complicated in that there is no obvious manner by which the parasitoid can determine instar across host species, especially if size is the criterion employed. For example, for the pea aphid *Acyrthosiphon pisum*, the high-quality second instar is approximately the same size as the low-quality fourth instar of the foxglove aphid *Aulacorthum solani* and, as noted above, Henry et al. (2005) suggests that *A. ervi* cannot identify the instar directly and probably uses sized-based cues. As such, there are six distinct size classes when we combine foxglove and pea aphids of different instars. Let as(i) and ap(j) denote the ith instar and the jth instar of foxglove and pea aphids respectively; the six classes are [as(1)], [as(2)], [*as(3), ap(1)*], [*as(4), ap(2)*], [ap(3)], and [ap(4)]. The italicized classes above indicate possible misidentification of species, and could thus be problematic. So, what should the parasitoid do? Of course, the problem is complicated by the parasitoid state in that identification errors are more costly when egg load is low than when it is high. We develop a model that specifically incorporates possible misidentification into host acceptance policy.

15.2 Egg-laying decision problems

15.2.1 Fidelity: proovogenic gregarious parasitoids

The model
A state model is classically composed of:

1 a sequence of 'stages', here the successive host encounters;
2 a dynamical equation describing the evolution of the state variables from one stage to the next, as a function of the decision variables and some random variables; and

3 a criterion to be maximized through the choice of the decision variables at each stage.

This criterion must be the sum of a function of final state and stage-wise benefits, or of some of these, each stage-wise benefit being a function of current state, control variable, and disturbance (or of some of these).

We now describe the fidelity problem in the formalism of a state variable dynamic system and a criterion to be maximized. Classically, the decision variables themselves may depend on some current information. Here, the available information at any stage will be the current state, a considerable simplification of the theory.

DYNAMICS: Let

- $t \in \{t,\infty\} \subset \mathbb{R}_+$ the current time, a state variable;
- $x \in \{0,1,\ldots,X\} \subset \mathbb{N}$ the number of eggs available, a state variable;
- $c \in \{0,1,\ldots,x\}$ the clutch size in a given host, the decision variable;
- $h(c)$ the 'handling time' it takes to lay a clutch of size c (with $h(0) = 0$);
- T, a random variable, the travel time to the next host encounter.

The variables t, x, c, and T take on new values for each of the encounters of the parasitoid with a host. Let t_k, x_k, c_k, and T_k be the values they have at the encounter number k, $k \in \mathbb{N}$. The whole sequence of the c_k's will be written as $\{c_k\}$ and likewise for other variables.

Trivially, these variables obey the following dynamical equations:

$$\begin{aligned} t_{k+1} &= t_k + h(c_k) + T_k, \\ x_{k+1} &= x_k - c_k \end{aligned} \tag{15.1}$$

CRITERION: Let

- $\omega \in \{\boldsymbol{7}, \boldsymbol{\mathcal{N}7}\}$ the nature of the host encountered, *7arget* or *7on-7arget*, a random variable (there could be an arbitrary number of host types in this model);
- $\lambda_{\boldsymbol{7}}$ and $\lambda_{\boldsymbol{\mathcal{N}7}}$ the probabilities that $\omega = \boldsymbol{7}$ or $\omega = \boldsymbol{\mathcal{N}7}$, respectively;
- $\xi(t) \in \{0,1\}$ the life status indicator, equal to 1 if the parasitoid is alive at time t, 0 if it is dead, a random variable;
- $P(t)$ the probability that $\xi(t) = 1$, or the expectation of ξ;
- $f(\omega,c)$ the fitness gained, say the expected number of offspring, as a function of the nature of the host and the clutch size. It is assumed to be increasing and concave, with $f(\omega,0) = 0$, and bounded by some positive $f_{\max}(\omega)$.

Again, we shall consider sequences of values of the variables $\{\omega_k\}$. The 'decision' to lay c eggs at a given stage may bring less fitness than $f(\omega,c)$, because the parasitoid may die before it has laid its c eggs. Therefore, we consider the conditional expected utility of that decision, given that it is alive at time t. To do so, we introduce the ratio ρ defined as

$$\rho(t,\tau) = \frac{R(t + \tau)}{P(t)} \tag{15.2}$$

which is the probability of the parasitoid surviving until the time $t + \tau$, conditioned on it being alive at time t. The conditional expected utility sought is then

$$g(\omega,t,c) = \sum_{l=1}^{c} [f(\omega,l) - f(\omega,l-1)]\rho(t,h(l)) \tag{15.3}$$

Obviously, we have

$$\forall \omega,t,c, \ 0 \le g(\omega,t,c) \le f(\omega,c) \tag{15.4}$$

and therefore also

$$\forall \omega,t, \ g(\omega,t,0) = 0 \tag{15.5}$$

The foraging activity ends when the parasitoid dies, a random variable, so that we do not know *a priori* the number of encounters to be considered, nor even a limit on that number (although for a realistic set of data, a limit could be found for any practical purpose). We therefore take the criterion to be maximized as

$$J(t_0,x_0;\{c_k\}) = \mathbb{E} \sum_{k=1}^{\infty} \xi(t_k)g(\omega_k,t_k,c_k) \tag{15.6}$$

Parallel with Charnov's (1976) patch leaving problem
The problem considered here is equivalent to Charnov's (1976) classic patch leaving problem for a solitary parasitoid, up to the exact shape of the fitness function, which may differ (Chavnov 1976). As a matter of fact, we have the correspondence shown in Table 15.1 (the function $\tilde{f}$ is defined hereafter).

The number of eggs laid in Charnov's (1976) problem is classically given as a function, $\tilde{f}$, of the time t spent on the patch. To relate it to the clutch size c, noting that $h(c)$ is increasing, for any given time t there is formally a clutch size $c = h^{-1}(t)$ (the largest clutch size c such that $h(c) \le t$). Thus, we let $\tilde{f}(\omega,t) = f(\omega,h^{-1}(t))$. Hence, the analytic form commonly used in the patch leaving problem has no reason to hold here, but the only important point in the theory is that this function be concave. This is insured in the patch leaving problem by search efficiency decrease due to patch depletion, and in the fidelity problem by the law of decreasing return of eggs laid. As a consequence, the work presented here may be viewed largely as an extension of that of Wajnberg et al. (2006) to proovigenic parasitoids, and we treat it in a similar way.

Table 15.1 Parallels between the fidelity and patch leaving problems.

Fidelity	Patch leaving
Host	Patch
Time $h(c)$ to lay c eggs	Time t spent on the patch
Type of host ω	Quality of the patch
Fitness gain $f(\omega,c)$	Cumulative number of eggs laid $\tilde{f}(\omega,t)$
Travel time	Travel time

Dynamic programming (DP)

From the above description, we can derive Bellman's (1957) DP equation. We let

$$G(t_0, x_0) = \max_{\{c_k\}} J(t_0, x_0; \{c_k\}) \tag{15.7}$$

and notice that the problem is stationary in the sense that G does not depend on the index of the initial stage.

The decision variable c_k may depend on x_k and ω_k, and on $\xi(t_k)$, which is known as a consequence of Descartes' law '*cogito, ergo sum*', but not on T_k, which is not known at time t_k. Hence, we get

$$G(t, x) = \mathbb{E}_{\omega, \xi} \max_{c \leq x} \mathbb{E}_T[G(t + T + h(c), x - c) + \xi(t)g(\omega, t, c)] \tag{15.8}$$

Notice, first, that there is no loss of performance in requesting that the decision taken be the same whether the parasitoid is dead or alive, since our formulation allows for a decision taken by a dead parasitoid but with no effect on the performance. Hence, we may shift the expectation operator in ξ to the right of the max operator and inside the square bracket, resulting in

$$G(t, x) = \mathbb{E}_\omega \max_{c \leq x} [\mathbb{E}_T G(t + T + h(c), x - c) + P(t)g(\omega, t, c)] \tag{15.9}$$

In this equation, $G(t, x)$ is the maximal expected fitness gained from time t, if $x(t) = x$. The equation means that the forager has to balance the benefit expected from laying more eggs, i.e. increasing c and thus $g(\omega, t, c)$, against the loss in future ability to gain fitness, measured through the decrease in the function G due to the decreasing egg load $x - c$ that is available.

Now, $G(t, x)$ is an *a priori* expectation. If, say, the probability of being alive at time t is very small, then $G(t, x)$ will accordingly be very small. It turns out to be preferable, for some purposes, to work with the conditional expectation, given that the parasitoid is alive at time t. Let that conditional expectation be

$$F(t, x) = \frac{1}{P(t)} G(t, x) \tag{15.10}$$

Dividing through Equation (15.10) by $P(t)$, we get

$$F(t, x) = \mathbb{E}_\omega \max_{c \leq x} [\mathbb{E}_T \rho(t, T + h(c)) F(t + T + h(c), x - c) + g(\omega, t, c)] \tag{15.11}$$

Notice that, since ω assumes only two values, the expectation operator in ω can be decomposed as

$$F(t, x) = \lambda_7 \max_{c \leq x} [\mathbb{E}_T \rho(t, T + h(c)) F(t + T + h(c), x - c) + g(7, c)]$$
$$+ \lambda_{\mathbf{n7}} \max_{c \leq x} [\mathbb{E}_T \rho(t, T + h(c)) F(t + T + h(c), x - c) + g(\mathbf{n7}, c)] \tag{15.12}$$

This form displays the fact that different decisions may be taken for different host types, since there is a $\max_c$, hence a separate choice of c, for each type.

Now, the test of fidelity is that, $\forall c \in \{1, \ldots, x\}$,

$$\mathbb{E}_T \rho(t, T + h(c))F(t + T + h(c), x - c) + g(\mathbf{\mathit{N7}}, c) < \mathbb{E}_T \rho(t, T)F(t + T, x) \qquad (15.13)$$

which means that, for any non-zero clutch size, the expected fitness gained from laying eggs in a non-target host is less than that gained by bypassing this host. However, the measure of the fitness gained is in terms of the function F, the solution of Bellman's (1957) equation, so that performing this test requires us to solve that equation.

Solving Bellman's equation

The process of solving Bellman's equation is usually a numerical one, since there is no reason why there should exist a closed form solution for realistic data. This is a functional equation, therefore not easy to solve. There are commonly two algorithms used to solve it, both iterative, called 'iteration on the value' (or 'Howard's algorithm') and 'iteration on the policies'. The second one is known to converge faster, but it is so much more complex to program that we use the simpler Howard's algorithm here. To a mathematician, this is a simple Picard iteration for the fixed point Equations (15.9) or (15.11). It can also be understood as the limit of a finite horizon Bellman iteration process as the horizon becomes very large. It looks simpler written for G, but the fact that G may take very small values may make it numerically ill-behaved. If there is a possibility of computing $\rho(t, \tau)$ with reasonable accuracy, even for 'large' t, then the following algorithm is preferred. We denote F^n, $n = 0, 1, \ldots$ the successive approximations of F.

$$F^0(t, x) = 0,$$
$$F^{n+1}(t, x) = \mathbb{E}_\omega \max_{c \le x} [\mathbb{E}_T \rho(t, T + h(c))F^n(t + T + h(c), x - c) + g(\omega, t, c)] \qquad (15.14)$$

The expectation in ω can be decomposed as in Equation (15.12). Since c ranges over a finite and presumably rather small set, the maximization can be performed by simply computing the bracket for each possible value of c and keeping the best results. As for the expectation in T, it is with respect to a scalar variable, so that the most efficient method is to use a simple integration scheme such as the trapezoidal rule.

We notice that this algorithm presents the desirable feature that all approximations F^n satisfy $F^n(t, 0) = 0$ for all t, thus, so does the limit. We offer the following theorem, which is a slight extension of a classic result for Bellman's stationary equation:

THEOREM 1

If there exists $r < 1$ such that $\rho(t, t + T)$ is less or equal to r, then:

1 Equations (15.9) and (15.11) have a unique bounded solution G or F, respectively.
2 These solutions coincide with the definitions (15.7) and (15.10), respectively, and the maximization operator in (15.9) or, equivalently (15.11), yields the optimal strategy.
3 The sequence F^n converges uniformly to the unique solution of the corresponding Bellman equation, and so does the equivalent recursion for G.

A proof of this theorem is provided in the Appendix.

An hypothesis that insures that $\rho(t, T)$ be bounded is the following:

HYPOTHESIS 1

1 T is certainly larger than or equal to some positive T_0.
2 $P(t + T_0)/P(t) \leq r$.

The first of the above two properties suffices to insure convergence of Howard's algorithm in another realistic case:

THEOREM 2

If the lifetime of the parasitoid is less than a fixed positive number L, and if the first hypothesis above holds, then we have the same conclusions as in Theorem 1.

A proof of this theorem is provided in the Appendix.

Numerical experiments

We have conducted a limited set of numerical experiments to validate the methodology, and draw some early biological conclusions. We have used the following parameters, which are not specific to any particular system but are well within the range of many parasitoid species:

- computational horizon: 36 h;
- handling time: 1 min per egg (i.e. proportional to c);
- initial egg load ≤ 144;
- fitness law $f(\omega,c) = f_{max}(\omega)(1 - exp(-c/\pi))$ with $\pi = 2.5$;
- $c_{max} = 10$ (level beyond which superparasitism would reduce the fitness gain by threatening the survival of the host, or severely decreasing survivorship of parasitoid larvae via competition);
- relative fitness value of target versus non-target hosts for a clutch of any given size: 5;
- law of travel time: exponential with an expectation of 30 min, truncated at $T \geq 1$ min;
- survival law: Weibull law with scale parameter: 12 h, shape parameter: 4;
- frequency of target hosts versus non-targets: 4.

We stopped the algorithm when the precision reached 0.01. The results are depicted in Figs 15.1 and 15.2.

There are several aspects of these results that are relevant to biological concerns regarding non-target effects, when the target (the pest insect) is common relative to non-targets and would be particularly pertinent to situations where non-targets are rare, threatened species (Lynch et al. 2002). First, note that, as expected, the optimal clutch size varies as a function of age and egg load. By contrast, and similar to that of solitary parasitoids (Roitberg 2000), there are conditions under which the optimal response to non-target hosts will be rejection. These are the conditions that biological control workers should strive to create. Second, the optimal clutch size for attacks on target hosts:

1 largely tracks egg load up to the maximum;
2 is nearly independent of age for young but not old parasitoids; and
3 almost always exceeds that of optimal clutch size on non-targets, egg load constraints notwithstanding.

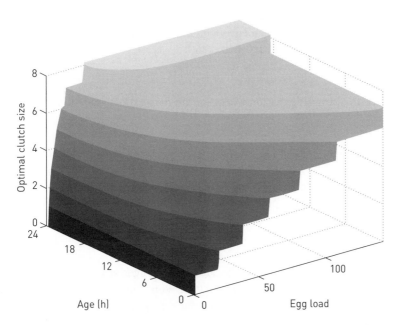

Fig. 15.1 The optimal clutch size from attacks on target hosts by proovigenic parasitoids as a function of egg load and age.

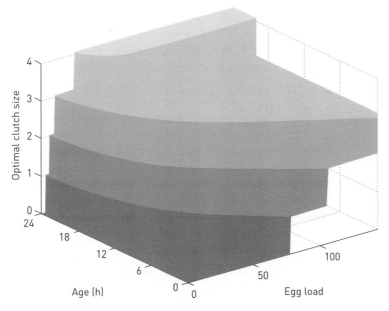

Fig. 15.2 The optimal clutch size from attacks on non-target hosts by proovigenic parasitoids as a function of egg load and age.

This is good from a biological control perspective, especially if damage to hosts (i.e. via mortality or fecundity) is clutch-size dependent. What the figures also tell us is that parasitoid rejection of non-targets can be accomplished by releasing specialist biocontrol agents that have relatively low egg loads or are long-lived, relative to egg load (Louda et al. 2003). An important caveat is that once the parasitoid has attacked some hosts, the problem becomes both frequency-dependent and strongly linked to population dynamics (see also Chapter 13 by Bonsall and Bernstein).

The stationary case

We turn now to a simpler situation where we can derive more analytical results from DP. Assume that:

1 the parasitoids are not egg-limited, having a large enough initial egg load;
2 the survival law is exponential, with a fixed intensity ρ, i.e. $P(t) = exp(-\rho t)$ and, to simplify the calculations, the less critical hypotheses that;
3 the time to lay eggs is proportional to the number of eggs laid, i.e. $h(c) = hc$, h is a positive constant;
4 the random variable T has a bounded support $[0, T_{max}]$.

Because of the first hypothesis, the egg state is no longer a factor, and F is no longer a function of x. But also, $\rho(t,\tau) = exp(-\rho\tau)$ for some positive constant ρ, and is thus independent from t. It is convenient to introduce the parameter

$$\Theta = \mathbb{E}\,e^{-\rho T} \tag{15.15}$$

Then F will also be constant, satisfying the stationary Bellman equation

$$F = \mathbb{E}_\omega \max_c [\Theta e^{-\rho hc}F + g(\omega,c)] \tag{15.16}$$

If we accept the hypothesis of decreasing return of the eggs laid in a single host, the maximum in Equation (15.16) is reached at

$$c^*(\omega) = min\{c|\Theta e^{-\rho hc}F + g(\omega,c) - \Theta e^{\rho h(c+1)}F + g(\omega,c + 1) \leq 0\} \tag{15.17}$$

Or, using the definition of g above, and after a simple calculation,

$$c^*(\omega) = min\{c|f(\omega,c + 1) - f(\omega,c) \leq \Theta F(e^{\rho h} - 1)\} \tag{15.18}$$

In order to investigate the effect of the mortality rate, we expand the above equations in the region of $\rho = 0$, hence letting $exp(\rho h) \simeq 1 + \rho h$. The above equation then becomes

$$c^*(\omega) = min\left\{c\left|\frac{f(\omega,c + 1) - f(\omega,c)}{h} \leq \Theta\rho F\right.\right\} \tag{15.19}$$

Let the optimal c be $c^*(\omega)$, and write $\bar{c}$ for its expectation, and $\bar{g} = \mathbb{E}_\omega g(\omega,c^*(\omega))$ for the expectation of g. It is a simple matter to see that Equation (15.16) now reads as

$$F = F\Theta(1 - \rho h\bar{c}) + \bar{g} \tag{15.20}$$

Further, the hypothesis that T is bounded by some T_{max} lets us expand $\Theta \simeq 1 - \rho\overline{T}$, where $\overline{T}$ is the expected value of T. As a result, we get

$$\rho F \simeq \frac{\bar{g}}{\overline{T} + h\bar{c}} \tag{15.21}$$

This equation has several consequences. First, notice that, as ρ goes to zero and if we let γ be the limit foraging rate, F converges to $\int_0^\infty exp(-\rho t)\gamma dt = \gamma/\rho$, so that ρF converges to γ, while g converges to f and, of course, Θ converges to 1. Hence, in the limit as $\rho \rightarrow 0$, Equations (15.19) and (15.21) are Charnov's (1976) marginal value theorem (MVT), stating that the parasitoid should leave the resource it is exploiting when the intake rate is about to drop below the optimal overall mean intake rate in the environment (see also Chapter 8 by van Alphen and Bernstein). Thus, for a very small mortality rate, simple rate maximizing is optimal (Mangel 1989a). Second, it shows that, as ρ varies, ρF can only decrease, since the ratio $\bar{f}/(\overline{T} + h\bar{c})$ is maximal at the MVT solution. Hence, since Θ is also decreasing with ρ, in Equation (15.19), the right-hand side decreases as ρ is increased from 0 to a positive value. According to the law of diminishing return, this means that $c^*(\omega)$ has to increase for all types of hosts.

Evaluating the selective pressure
The 'optimal' behavior determined by the above theory is more likely to approximate the real behavior of parasitoids in that it most likely gives a larger fitness superiority over simpler behaviors. It is therefore important to be able to evaluate this difference. The DP theory gives us $G(0,x_0)$, which is the expected fitness acquired under the optimal strategy for an initial egg load of x_0. We need now to evaluate the fitness gained using other strategies (Roitberg & Mangel 1997).

A classical means of evaluating this is via Monte-Carlo simulation runs with the assumed simpler strategy $c(\omega,t,x)$. We note here that there is a simpler, faster, and more precise way of doing this. According to Kolmogorov's equation, the expected efficiency of a given strategy $c(\omega,t,x)$ is a solution of the same Equation (15.9), without the max_c operator, when plugging our $c(\omega,t,x)$ into the right-hand side instead of the optimal c^*. As a consequence, the corresponding Howard's algorithm converges to the desired fitness expectation. The same computer code can be used, replacing the max operation by the choice of the given decision strategy. It is therefore much faster and more accurate in that role. With some more work, we may also obtain the variance of the fitness gained by a similar procedure.

Another valuable check would be to compare the optimal fitness gain with the optimum obtained if the parasitoid does not discriminate target from non-target hosts. This is obtained with almost the same computation, by interchanging the max and $\mathbb{E}$ operators in the right-hand side of Bellman's Equation (15.11). See Section 15.2.3 for more details.

With the parameters chosen in Figs 15.1 and 15.2, the corresponding optimal clutch size, for the 'blind' parasitoids, is shown in Fig. 15.3.

The ratio of fitness obtained in the blind behavior to that of the optimal behavior is above 98%, indicating that the 'optimal' behavior has very little selective benefit. This result, though, is very much dependant on the parameters. Figure 15.4 gives a plot of this dependence. We see that, for example, with a smaller initial egg load of 30, this ratio drops below 90%.

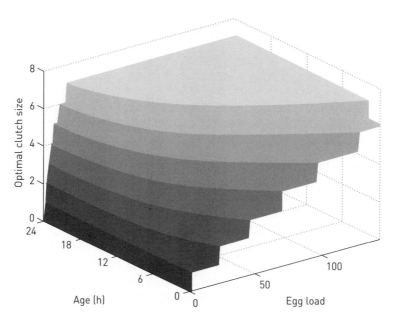

Fig. 15.3 The optimal clutch size from attacks on all hosts by a so-called 'blind' proovigenic parasitoid that cannot discriminate target from non-target hosts. The optimal clutch varies as a function of egg load and age.

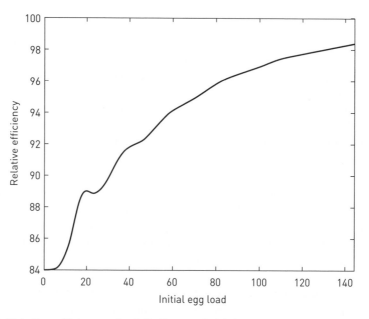

Fig. 15.4 Relative efficiency of a 'blind' parasitoid that cannot discriminate target from non-target hosts as a function of initial egg load.

15.2.2 Fidelity: synovigenic gregarious parasitoids

We turn now to the case of a parasitoid still having a predetermined egg load, but imma-ture. Eggs mature at a fixed rate of m eggs per unit of time, and mature eggs are available to be laid. We assume that egg maturation does not impact on the mortality rate (but see Roitberg (1989)).

Let $y(t)$ denote the number of immature eggs left at time t, and $y_0 = y(0)$. Since we assume that the egg maturation rate is constant, we have

$$y(t) = max\{y_0 - mt, 0\} \tag{15.22}$$

As a simple function of t, we do not need to make it an extra state variable. We shall write the dynamics in terms of the 'state' variables t and x, the control variable c, and the random variable T. Let also, for any τ:

$$\delta(t,\tau) = min\{y(t), m\tau\} = min\{max\{y_0 - mt, 0\}, m\tau\} \tag{15.23}$$

This expression looks awkward, but it can be interpreted as a 'saturation' function:

$$\delta(t,\tau) = \begin{cases} 0 & \text{if } y_0 - mt \leq 0 \\ y_0 - mt & \text{if } 0 \leq y_0 - mt \leq m\tau \\ m\tau & \text{if } y_0 - mt \geq m\tau \end{cases} \tag{15.24}$$

The dynamics are now

$$\begin{aligned} t_{k+1} &= t_k + h(c_k) + T_k, \\ x_{k+1} &= x_k - c_k + \delta(t_k, h(c_k) + T_k) \end{aligned} \tag{15.25}$$

We restrict c_k to be no more than x_k.

Two comments are in order here. On the one hand, x_k is no longer an integer here, since the products mt_k and $m(h(c_k) + T_k)$ are not integers. Yet, the constraint $c_k \leq x_k$ is cor-rect, insisting that c_k is an integer. That way, an egg 'partially matured' is not counted as available. On the other hand, there is a slight error here, since we restrict c_k to be no more than the number of available mature eggs, but this holds at the beginning of the oviposi-tion on the current host only. It neglects eggs that might mature during the oviposition. Yet, this should be a good approximation for most systems (Godfray 1994).

If there is no lower limit on possible values of the random variable T, we cannot give an *a priori* limit on the number of host encounters. So again, we use infinite-stage DP. Using the same notations as before, the DP equation now reads

$$F(t,x) = \mathbb{E}_\omega \max_{c \leq x} \mathbb{E}_T [\rho(t, h(c) + T)F(t + h(c) + T, x - c + \delta(t, h(c) + T)) + g(\omega, t, c)] \tag{15.26}$$

The numerical procedure and convergence theorems also follow as before.

Here again, we have run a few numerical experiments to validate the theory. We have used the same parameters as for the proovigenic case, with a maturation rate of one egg per ten minutes. Results are shown in Figs 15.5 and 15.6. What we now see is that,

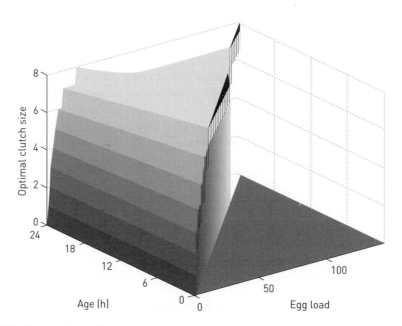

Fig. 15.5 The optimal clutch size from attacks on target hosts by synovigenic parasitoids as a function of egg load and age.

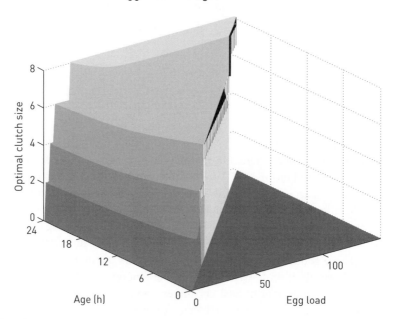

Fig. 15.6 The optimal clutch size from attacks on non-target hosts by synovigenic parasitoids as a function of egg load and age.

qualitatively, the clutch size policies for attacks on target and non-target hosts is similar but optimal clutch size is almost always lower on non-targets. The contrast in policies for these gregarious synovigenic versus proovigenic parasitoids is analogous to the signific-ant differences found by Roitberg (2000) for solitary wasps with different reproductive strategies. Once again, we find that there are abrupt changes or step functions in optimal clutch size, due to the interaction between egg maturation rates and expectation of life for synovigenic parasitoids compared to the nearly monotonic policies when individuals are proovigenic. As before, we caution against release of such organisms without thorough studies of host acceptance policies.

15.2.3 Size acceptance policy

We now turn to a problem with imperfect information, but that is easy to handle. This problem deals with a solitary parasitoid, i.e. there is no clutch size decision to take, only one egg can be laid in each host. Thus, the only decision is to accept or to reject it. The new feature of this problem is that there are several classes of hosts (e.g. size classes) and in each class there are several types of hosts (e.g. different instars of different species with the same size). The classes need not be disjointed, but their union must be the whole set Ω of types. This is because, while every type belongs to at least one class, some may have varying characteristics among individuals that make it possible to belong to one class or another. The parasitoid is assumed to be unable to discriminate the types within the class of a host it encounters. Yet, the expected fitness gained from laying an egg in a host of a given class may vary widely across types within that class. We may notice that the comparison strategy proposed in Section 15.2.1 is of this nature. It amounts to having a single class of hosts for both target and non-target hosts.

   We use similar variables and notations as before, with the following differences:

- Let Ω denote the (finite) set of possible types.
- There are n classes of hosts, which are subsets Ω_i of the set Ω of types.
- The probability of encountering a host of class i is λ_i.
- The probability of a host of class i being of type ω is $\mu_i(\omega)$, with $\mu_i(\omega) > 0$ if, and only if, $\omega \in \Omega_i$.
- The fitness gain expected from laying an egg $f(\omega)$ only depends on the host type ω.

The variables t, x, h (now fixed), and T are as before, as well as the life probability $P(t)$, the conditional probability $\rho(t,\tau) = P(t + \tau)/P(t)$ of surviving from time t to time $t + \tau$, and

$$g(\omega,t) = f(\omega)\rho(t,h) \qquad (15.27)$$

the conditional expected fitness gain of deciding to lay an egg, taking into account the possibility that the parasitoid dies before finishing its oviposition.

   We let $G_i(t,x)$ be the expected future fitness gain when attacking a host of class i at time t with an egg complement of x. The DP equation is now

$$G_i(t,x) = max \left\{ \sum_{j=1}^{n} \lambda_j \mathbb{E}_T G_j(t + T,x), \sum_{\omega \in \Omega_i} \mu_i(\omega)P(t)g(\omega,t) \right.$$

$$\left. + \sum_{j=1}^{n} \lambda_j \mathbb{E}_T G_j(t + h + T,x - 1) \right\} \qquad (15.28)$$

This form clearly displays the fact that the optimal policy, i.e. reject the host if the first term in the max operation is larger, accept it if the second term is larger, depends on the class i of the host, but not on its precise type ω.

As before, it is possible, and may be advisable, to use the conditional expected gain $F_i(t,x) = G_i(t,x)/P(t)$ instead of $G_i(t,x)$. Moreover, by introducing the expected return function $F(t,x) = \sum_i \lambda_i F_i(t,x)$, we easily derive from the above that it satisfies

$$F(t,x) = \sum_{i=1}^{n} \lambda_i max \left\{ \mathbb{E}_T \rho(t,T) F(t+T,x), \sum_{\omega} \mu_i(\omega) g(\omega,t) \right.$$

$$\left. + \mathbb{E}_T \rho(t,h+T) G(t+h+T,x-1) \right\} \qquad (15.29)$$

Again, for each class i, the optimum decision is to lay an egg if the second term in the max operation is larger. We see that, not unexpectedly, the only class dependent term in these equations is the expected utility of deciding to lay an egg knowing the class i of the host,

$$\sum_{\omega} \mu_i(\omega) g(\omega,t) \qquad (15.30)$$

The derivation of Howard's algorithm to numerically solve this equation and its convergence proof are as in the preceding section. Figures 15.5 and 15.6 show results of a special case of that theory (with only one size class).

15.3 Final remarks

Parasitoids are complex organisms that display a great deal of adaptive plasticity to circumstances (Godfray 1994). In many cases, those circumstances, or conditions, are state-dependent. In this chapter, we have shown that this plasticity can be predicted by way of DP models. How might we use such a tool when working with parasitoids? When it comes to their use in biological control, the key is to elucidate the aforementioned plasticity of individual insects in the light of their effects at the population or community level (e.g. impact of parasitoid clutch size on pest population dynamics (Heimpel 2000)). This scaling up from the individual to the population level is made more difficult by the extraordinary range of behaviors that parasitoids display. However, the task is manageable if the individual behaviors can be assembled into a kind of policy that is embedded within population-level models (Mangel & Roitberg 1992). Roitberg (2007) provides a list of parasitoid-based state-dependent opportunities for applied behavioral ecologists working in agricultural settings that range from mass production of parasitoids to pre-release conditioning.

We close by issuing a plea for theoreticians and experimental lab and field parasitoid biologists to work together on parasitoids. Our models clearly show that parasitoid behavior can be greatly affected by physiological constraints and environmental conditions, thus the utility for such models in biological control relies on a good understanding of those constraints and conditions. Likewise, our models point to novel experiments and approaches for experimental biologists, which might otherwise be ignored.

Acknowledgments

All numerical computations, those shown here and others, were performed by Frédéric Hamelin, of the I3S Laboratory, France, whose contribution we gratefully acknowledge. Bernard Roitberg was funded by NSERC, Canada and Pierre Bernhard was supported by the ECOGER project of INRA. We thank the editors of this book for offering us the opportunity to explore new concepts in the behavioral ecology of parasitoid insects.

References

Bellman, R. (1957) *Dynamic Programming*. Princeton University Press, Princeton.

Bigler, F., Babendreier, D. and Kuhlmann, U. (eds.) (2006) *Environmental Impact of Invertebrates for Biological Control of Arthropods: Methods and Risk Assessment*. CABI Publishing, Wallingford.

Charnov, E.L. (1976) Optimal foraging: the marginal value theorem. *Theoretical Population Biology* **9**: 129–36.

Clark, C. and Mangel, M. (2000) *Dynamic State Variable Models in Ecology: Methods and applications*. Oxford University Press, New York.

Collier, T. (1995) Adding physiological realism to dynamic state variable models of parasitoid host feeding. *Evolutionary Ecology* **9**: 217–35.

Desouhant, E., Driessen, G., Amat, I. and Bernstein, C. (2005) Host and food searching in a parasitic wasp *Venturia canescens*: A trade-off between current and future reproduction? *Animal Behaviour* **70**: 145–52.

Ellers, J., Sevenster, J., and Driessen, G. (2000) Egg load evolution in parasitoids. *American Naturalist* **156**: 650–65.

Gerling, D., Roitberg, B.D. and Mackauer, M. (1990) Instar-specific defense of the pea aphid, *Acyrthosiphon pisum*: Influence on oviposition success of the parasite *Aphelinus asychis* (Hymenoptera: Aphelmidae). *Journal of Insect Behaviour* **3**: 501–14.

Godfray, H.C.J. (1994) *Parasitoids: Behavioral and Evolutionary Ecology*. Princeton University Press, Princeton.

Heimpel, G. (2000) Effects of parasitoid clutch size on host-parasitoid population dynamics. In: Hochberg, M.E. and Ives, A.R. (eds.) *Parasite Population Biology*. Princeton University Press, Princeton, pp. 27–40.

Henry, L., Gillespie, D. and Roitberg, B. (2005) Does mother always know best? Oviposition decisions in a parasitoid wasp. *Entomologia Experimentalis et Applicata* **116**: 167–74.

Henry, L., Roitberg, B. and Gillespie, D. (2006) Covariance of phenotypically plastic traits induces an adaptive shift in host selection behaviour. *Proceedings of the Royal Society of London Series B Biological Science* **273**: 2893–9.

Hoffmeister, T., Roitberg, B. and Vet, L. (2005) Evolution of solitary parasitoids: A spatially dependent life history approach. *American Naturalist* **166**: 62–74.

Houston, A. and McNamara, J. (1999) *Models of Adaptive Behaviour*. Cambridge University Press, Cambridge.

Houston, A., Clark, C., McNamara, J. and Mangel, M. (1988) Dynamic models in behavioral and evolutionary ecology. *Nature* **332**: 29–34.

Louda, S.M., Pemberton, R.W., Johnson, M.T. and Follett, P.A. (2003) Nontarget effects – The Achilles' heel of biological control? Retrospective analyses to reduce risk associated with biocontrol introductions. *Annual Review of Entomology* **48**: 365–96.

Lynch, L.D., Ives, A.R., Waage, J.K., Hochberg, M.E. and Thomas, M.B. (2002) The risks of biocontrol: Transient impacts and minimum nontarget densities. *Ecological Applications* **12**: 1872–82.

Mangel, M. (1989a) Evolution of host selection in parasitoids – does the state of the parasitoid matter? *American Naturalist* **133**: 688–705.

Mangel, M. (1989b) An evolutionary interpretation of the motivation to oviposit. *Journal of Evolutionary Biology* **2**: 157–72.

Mangel, M. and Roitberg, B. (1992) Behavioural stabilization of parasite-host dynamics. *Theoretical Population Biology* **42**: 308–20.

Messing, R., Roitberg, B. and Brodeur, J. (2006) Measuring and predicting indirect impacts of biological control: Competition, displacement, and secondary interactions. In: Bigler, F., Babendreier, D. and Kuhlmann, U. (eds.) *Environmental Impact of Invertebrates for Biological Control of Arthropods: Methods and Risk Assessment*. CABI Publishing, Wallingford, pp. 64–77.

Roitberg, B.D. (1989) The cost of reproduction in rosehip flies: Eggs are time. *Evolutionary Ecology* **3**: 183–8.

Roitberg, B. (2000) Threats, flies and protocol gapes: Can behavioral ecology save biological control? In: Hochberg, M.E. and Ives, A.R. (eds.) *Parasite Population Biology*. Princeton University Press, Princeton, pp. 254–65.

Roitberg, B.D. (2007) Why pest management needs behavioral ecology and vice versa. *Journal of Entomological Research* **37**: 14–18.

Roitberg, B. and Mangel, M. (1997) Individuals on the landscape: Behavior can mitigate differences among habitats. *Oikos* **80**: 234–40.

Roitberg, B., Sircom, J., Roitberg, C., van Alphen, J. and Mangel, M. (1993) Life expectancy and reproduction. *Nature* **364**: 18.

Roitberg, B., Boivin, G. and Vet, L. (2001) Fitness, parasitoids and biological control: An opinion. *Canadian Entomologist* **133**: 429–38.

Rosenheim, J.A. (1999) Characterizing the cost of oviposition in insects: A dynamic model. *Evolutionary Ecology* **13**: 141–65.

Sirot, E., Ploye, H. and Bernstein, C. (1997). State dependent superparasitism in a solitary parasitoid: Egg load and survival. *Behavioural Ecology* **8**: 226–32.

Ueno, T. (1999) Host-feeding and acceptance by a parasitic wasp (Hymenoptera: Ichneumonidae) as influenced by egg load and experience in a patch. *Evolutionary Ecology* **13**: 33–44.

Vinson, S.B. and Iwantsch, G. (1980) Host suitability for insect parasitoids. *Annual Review of Entomology* **25**: 397–419.

Wajnberg, É. (2006) Time allocation strategies in insect parasitoids: From ultimate predictions to proximate behavioral mechanisms. *Behavioral Ecology & Sociobiology* **60**: 589–611.

Wajnberg, É., Bernhard, P., Hamelin, F. and Boivin, G. (2006) Optimal patch time allocation for time-limited foragers. *Behavioral Ecology & Sociobiology* **60**: 1–10.

Wang, X. and Messing, R. (2004) Fitness consequences of body-size-dependent host species selection in a generalist ectoparasitoid. *Behavioral Ecology & Sociobiology* **56**: 513–22.

Appendix: proofs of theorems 1 and 2

Proof of theorem 1

We work with Equations (15.11) and (15.14). By its very definition, F is a limited function. We view Equation (15.11) as a fixed point equation in the space C^0 of bounded continuous functions, and Equation (15.14) as a Picard sequence for this equation. Items 1 and 3 of the theorem will therefore be a consequence of Banach's fixed point theorem, if we can prove that the right-hand side of Equation (15.11) is a contraction in a norm for which C^0 is complete. As is traditional, we choose the norm of the uniform convergence $\| F \| = \sup_{(t,x) \in \mathbb{R}^2_+} |F(t,x)|$. Item 2 is then a consequence of DP theory.

We are left with the task of proving that the application $F \mapsto \mathcal{K}(F)$ with

$$\mathcal{K}(F)(t,x) = \mathbb{E}_\omega \max_{c \le x} [\mathbb{E}_T \rho(t, T + h(c)) F(t + T + h(c), x - c) + g(\omega, t, c)] \qquad (A15.1)$$

is a contraction in C^0. Also let

$$\mathcal{H}(F)(\omega, t, x, c) = \mathbb{E}_T \rho(t, T + h(c)) F(t + T + h(c), x - c) + g(\omega, t, c) \qquad (A15.2)$$

Therefore, let V and W be two limited continuous functions. We seek a uniform bound to

$$\Delta(t,x) = |\mathcal{K}(V)(t,x) - \mathcal{K}(W)(t,x)| \qquad (A15.3)$$

We have

$$\begin{aligned}
\Delta(t,x) &= |\mathbb{E}_\omega \max_c \mathcal{H}(V)(\omega, t, x, c) - \mathbb{E}_\omega \max_c \mathcal{H}(W)(\omega, t, x, c)| \\
&\le \mathbb{E}_\omega \max_c |\mathcal{H}(V)(\omega, t, x, c) - \max_c \mathcal{H}(W)(\omega, t, x, c)|
\end{aligned} \qquad (A15.4)$$

For any two limited continuous functions $v(c)$ and $w(c)$, it always holds that $|max_c\ v(c) - max_c\ w(c)| \le max_c|v(c) - w(c)|$. Using this and the monotonicity of the mathematical expectation operator, and then the monotonicity of the max operator, we get

$$\begin{aligned}
\Delta(t,x) &\le \mathbb{E}_\omega \max_c |\mathcal{H}(V)(\omega, t, x, c) - \mathcal{H}(W)(\omega, t, x, c)| \\
&= \mathbb{E}_\omega \max_c |\mathbb{E}_T \rho(t, T + h(c)) V(t + T + h(c), x - c) - \\
&\qquad \mathbb{E}_T \rho(t, T + h(c)) W(t + T + h(c), x - c)| \\
&\le \mathbb{E}_\omega \max_c \mathbb{E}_T \rho(t, T + h(c)) |V(t + T + h(c), x - c) - \\
&\qquad\qquad W(t + T + h(c), x - c)|
\end{aligned} \qquad (A15.5)$$

We have $T + h(c) \ge T \ge T_0$ almost surely by hypothesis and therefore $\rho(t, T + h(c)) \le \rho(t + T_0) \le r$. By definition of the norm, we have endowed C^0 with $|V(t + T + h(c), x - c) - W(t + T + h(c), x - c)| \le \|V - W\|$. Hence, we end up with $\Delta(t,x) \le r\|V - W\|$ and, taking the supremum in (t,x):

$$\|\mathcal{K}(V) - \mathcal{K}(W)\| \le r\|V - W\| \qquad (A15.6)$$

showing that $\mathcal{K}$ is indeed a contraction, and the theorem is proved.

Proof of theorem 2

We assume now that for $t \ge L$, $P(t) = 0$, and also that $T \ge T_0 > 0$. We now work with the recursion for G: $G^0(t,x) = 0$,

$$G^n(t,x) = \mathbb{E}_\omega \max_{c \le x} [\mathbb{E}_T G^{n-1}(t + T + h(c), x - c) + P(t) g(\omega, t, c)] \qquad (A15.7)$$

We then substitute for G^{n-1} on the right-hand side in terms of G^{n-2}, using the same recursion to obtain

$$\begin{aligned}
G^n(t,x) = \mathbb{E}_{\omega_1} \max_{c_1 \le x} \{ &P(t) g(\omega_1, t, c_1) + \\
\mathbb{E}_{T_1} \mathbb{E}_{\omega_2} \max_{c_2 \le x - c_1} \{ &P(t + T_1 + h(c_1)) g(\omega_2, t + T_1 + h(c_1), c_2) + \\
\mathbb{E}_{T_2} G^{n-2}(&t + T_1 + T_2 + h(c_1) + h(c_2), x - c_1 - c_2)] \}
\end{aligned} \qquad (A15.8)$$

We introduce the notation, for all integer k,

$$\theta_k = t + \sum_{i=1}^{k} (T_i + h(c_i)) \tag{A15.9}$$

and carry out this substitution recursively for l steps, to obtain

$$
\begin{aligned}
G''(t,x) = \mathbb{E}_{\omega_1} \max_{c_1 \le x} \{ P(t) g(\omega_1, t, c_1) + \\
\mathbb{E}_{T_1} \mathbb{E}_{\omega_2} \max_{c_2 \le x - c_1} [P(\theta_1) g(\omega_2, \theta_1, c_2) + \\
\mathbb{E}_{T_2} \mathbb{E}_{\omega_3} \max_{c_3 \le x - c_1 - c_2} (P(\theta_2) g(\omega_3, \theta_2, c_3) + \\
\vdots \\
\mathbb{E}_{T_{l-1}} \mathbb{E}_{\omega_l} \max_{c_l \le x - \sum_{k=1}^{l-1} c_k} (P(\theta_{l-1}) g(\omega_l, \theta_{l-1}, c_l) + \\
\mathbb{E}_{T_l} G^{n-l}(\theta_l, x - c_1 - \ldots - c_l)] \ldots)] \}
\end{aligned}
\tag{A15.10}
$$

It follows, from the hypothesis, that $T \ge T_0$, so with respect to T, we may restrict the range to $T \ge T_0$. But then, for l such that $lT_0 \ge L$:

$$P(t + T_1 + \ldots + T_l + h(c_1) + \ldots + h(c_l)) = 0 \tag{A15.11}$$

and thus

$$G^{n-l}(t + T_1 + \ldots + T_l + h(c_1) + \ldots + h(c_l), \, x - c_1 - \ldots - c_l) = 0 \tag{A15.12}$$

so that the above expression can be written without its last term. But then, the right-hand side is independent of n, showing that the recursion reaches a steady state in no more than l steps. Therefore, the solution of the DP equation is

$$
\begin{aligned}
G(t,x) = \mathbb{E}_{\omega_1} \max_{c_1 \le x} \{ P(t) g(\omega_1, t, c_1) + \\
\mathbb{E}_{T_1} \mathbb{E}_{\omega_2} \max_{c_2 \le x - c_1} [P(\theta_1) g(\omega_2, \theta_1, c_2) + \\
\mathbb{E}_{T_2} \mathbb{E}_{\omega_3} \max_{c_3 \le x - c_1 - c_2} (P(\theta_2) g(\omega_3, \theta_2, c_3) + \\
\vdots \\
\mathbb{E}_{T_{l-1}} \mathbb{E}_{\omega_l} \max_{c_l \le x - \sum_{k=1}^{l-1} c_k} [P(\theta_{l-1}) g(\omega_l, \theta_{l-1}, c_l)] \ldots)] \}
\end{aligned}
\tag{A15.13}
$$

Although not needed, it may be a useful exercise to place this back in the right-hand side of Equation (15.9), and check that, upon shifting all mute indices by one, we recover the identical expression. This proves the theorem, except for the uniqueness of the solution of Equation (15.9). This uniqueness follows from the statement that using $\rho(t, t + T_0) < 1$ suffices to show that the mapping of $\mathcal{K}$ in the previous proof still satisfies a weak contraction property: for all V, W in C^0, $\| \mathcal{K}(v) - \mathcal{K}(w) \| < \| V - W \|$, which suffices to get the uniqueness (if not the existence) of the fixed point.

16

A Bayesian approach to optimal foraging in parasitoids

Jean-Sébastien Pierre and Richard F. Green

Abstract

A Bayesian approach provides quantitative models for animals foraging in a patchy environment. Oaten (1977) proposed a stochastic Bayesian model, which assumes that foragers know how their prey are distributed in patches. It also describes how a forager should use information obtained from prey encounters in a patch to decide how long to stay in that patch. Parasitoids present two problems that have no place in Oaten's (1977) Bayesian model. Unlike predators, parasitoids do not remove hosts after attacking them, leaving the problem of how to deal with already attacked hosts. The other problem is that parasitoids can use chemical mechanisms to detect the presence of hosts before they are actually encountered. Green (1990) has shown how to incorporate additional information into a model such as Oaten (1977), but this problem has not been treated in a fully systematic way. This chapter will address this problem.

There are several benefits of using Bayesian methods to study parasitoid behavior. Bayesian models contribute to the study of parasitoid behavior by drawing attention to host distribution and information use. They also make predictions about behavior and provide results that can be used in host–parasitoid population models. The Bayesian framework also helps to understand how parasitoids do achieve Bayesian decisions. It can be shown, for example, that some plausible mechanisms, such as those described by Waage (1979) and Driessen et al. (1995), lead to patch-leaving rules close to what is predicted by Bayesian models. Adapting Bayesian models to parasitoids would broaden and strengthen foraging theory by providing new questions to answer (e.g. how best to use chemical information).

16.1 Introduction

16.1.1 General view

Optimal foraging theory was one of the earliest and most prominent parts of behavioral ecology. Over time, the treatment of foraging changed from optimization models to the description of behavioral mechanisms. In other words, there has been a change in the treatment of foraging from functional explanations, which characterized behavioral ecology in its early days, to causal explanations, which are more common today. In this chapter, we suggest that behavioral ecologists should think again about optimal foraging theory, particularly using the Bayesian approach seen in the model of Oaten (1977) and in the work of his followers. Most general work on foraging refers to 'predators' searching for 'prey', whether the foragers are predators searching for prey, seed-eaters searching for seeds, or nectar-feeders searching for nectar. Sometimes we will refer to parasitoids searching for hosts, but at other times we will refer to predators searching for prey, especially when we are describing work that uses these terms.

Oaten (1977) modeled the problem of a forager that searches an environment in which patches vary in quality (see also Chapter 8 by van Alphen and Bernstein). Such a forager must decide when to leave a patch, based on its prior knowledge of patch quality and its experience in the patch. Studies of parasitoid behavior have done a great deal of work on the problem of how a parasitoid decides when to leave a host patch (Godfray 1994, van Alphen et al. 2003, Wajnberg 2006). Some works on the patch residence time problem are based on Oaten's (1977) model, for example, that of Iwasa et al. (1981). Others are not (Waage 1979) but could provide behavioral mechanisms underlying Bayesian foraging. We think that all of these works on patch residence time, and all of the behavior of parasitoids that they model, can be understood best if viewed in terms of Oaten's (1977) Bayesian foraging model.

First, we will clarify what we mean by a Bayesian approach and show that there is evidence that animals are actually Bayesians in a general sense. Then, we will look at the disciplinary background of our particular problem as part of behavioral ecology in general and optimal foraging theory in particular. We will describe Oaten's (1977) stochastic model as an alternative to Charnov's (1976b) deterministic model for the patch residence time problem. We will summarize the results of the analysis of Oaten's (1977) model. Among these results are:

1 the form of the best rule varies for different distributions of hosts in patches;
2 a Bayesian forager can sometimes find hosts at a much higher rate than can a parasitoid that ignores information about the environment, but occasionally knowledge of the environment is of little value;
3 the form of the best rule can sometimes be very simple: and
4 by calculating the rate achieved by an optimal forager we can study the stability of parasitoid-host population dynamics models.

We will also theoretically discuss two alternative approaches to the patch residence time problem, one proposed by Waage (1979) and the other suggested be Iwasa et al. (1981). We will finally discuss more empirical works on patch-leaving rules in parasitoids that have centered on two ideas. One is an extension of Waage's (1979) model proposed by Driessen et al. (1995). The other, originally developed by Haccou et al. (1991), looks at the giving-up time (i.e. the time between the last encounter with a host and departure from a patch) and describes it in terms of Cox's (1972) proportional hazard model. We will mention how these two ideas may combine nicely with the theory of Bayesian foraging. We conclude by discussing some other applications of the Bayesian approach.

16.1.2 What is a Bayesian approach?

A Bayesian decision-maker has to decide what to do in a particular case. The decision is based on a prior opinion about the present case, perhaps based on experience of other cases, and on experience obtained while observing the present case. For example, imagine that there are two kinds of coins, one that comes up heads two-thirds of the time, the other that comes up tails two-thirds of the time, and that the two kinds of coins are equally common. If a coin is chosen at random, the probability that it will be a heads-favoring coin is one-half. This is the prior probability that the coin favors heads. Now, if the coin is tossed and it comes up heads, the posterior probability that it is a heads-favoring coin is two-thirds. If the coin is tossed again, and it comes up heads again, the probability that it is a heads-favoring coin is four-fifths, etc. In this example, the problem is to decide whether a particular coin favors heads or tails. Our decision arises from Bayes' theorem usage that we give in the Appendix.

The structure of the foraging problems that we consider, and the calculations necessary to solve them, are similar to those described above. We are interested in parasitoids that search for hosts distributed in patches. Some patches have many hosts, some have few. A parasitoid has to decide when to leave one patch and search for another (see also Chapter 8 by van Alphen and Bernstein). That decision may depend on experience in a patch. The problems that foragers face are more difficult than the simple example described above. Patches are not simply of two types, good or bad. Some patches may be very good, some pretty good, some only fair, and some poor. It may be best for the forager to leave a patch soon after entering it, having obtained little information.

Weis (1983) studied an example of parasitoid foraging that invites a Bayesian interpretation. The problem is complicated, but the solution is simple. Van Alphen and Vet (1986) discussed this example, but without mentioning Bayes' theorem. Weis (1983) studied the foraging behavior of a torymid wasp that attacks the larvae of a midge that forms blister-shaped galls on the leaves of goldenrod plants. The parasitoid inserts its ovipositor from the edge of a gall and lays an egg on a host if one is encountered. The parasitoid may insert its ovipositor several times, but eventually it leaves the gall and searches for another. The problem facing the parasitoid is to decide when to leave a gall based on its success in finding hosts in the gall. Each insertion of the ovipositor is either successful in finding a host or not, and the chance of success depends on the number of midge larvae in the gall. Weis (1983) observed that some galls contain only one larva, some contain two, some three, and some four. With a Bayesian approach, the optimal patch-leaving rule may be found by using information about how the numbers of potential hosts are distributed in galls. The optimal patch-leaving rule is simple: leave a gall if no larvae have been found by some fixed time, or as soon as one larva has been found.

When using a Bayesian approach, we do not claim that animals actually calculate probabilities and use them to decide what to do. It is human theorists who do the calculations, and find the best foraging strategies and the rates of finding hosts that they achieve. Then, an experimenter can compare the patch-leaving rules that animals use and the foraging rate that they achieve with the optimal patch-leaving rule and the success that an optimal forager would achieve.

16.1.3 Are animals Bayesians?

At a meeting on 'Bayesian foraging' at Lund University, Sweden in 2003 (whose proceedings were published in the February 2006 issue of *Oikos*) some participants asked the question: 'Are animals Bayesians?' This question invites a facile answer, 'Of course not. Animals cannot calculate probabilities.' A better way to ask the question might be: 'Do animals face problems that can be thought of in Bayesian terms, and do they solve the problems using behaviors that can be understood by comparing them with the solution of the corresponding Bayesian problem?' Discussion of the question of whether animals are Bayesians or not can be found in McNamara et al. (2006). More generally than Bayesian foraging, the Lund meeting was about how animals use, or might use, information in making decisions. Biologists, particularly theoretical biologists, like to think about how they would solve the problems that they imagine animals face. For Bayesians, the question is how animals should use past and present information to decide what to do now. But do animals decide what to do in the present situation based on information both about the present situation and about the past? It seems clear that animals indeed do this. Another question is how do they do? We will give the sketch of an answer in Section 16.3 below.

A well-known parasitoid example, not involving foraging but sex allocation, is an experiment performed by Chewyreuv (1913) and discussed by Charnov (1982). Chewyreuv (1913) offered hosts of two different sizes to a solitary parasitoid. The experiment was performed twice, using two different-sized hosts in each case (the hosts were actually of three different sizes). When the parasitoid was offered small- or medium-sized hosts, it laid more fertilized eggs (producing daughters) on the relatively larger (i.e. the medium-sized hosts), but when it was offered medium-sized or large hosts, it laid more unfertilized eggs (producing sons) on the smaller (i.e. the medium-sized hosts). Whether the medium-sized hosts received more fertilized or unfertilized eggs depended on the size of the other hosts with which the medium-sized hosts were paired. This observation, that was a surprise in 1913, is now well understood in the light of Bayesian decision theory.

16.1.4 The framework of behavioral ecology

Behavioral ecology has always been more about behavior than about ecology. As it developed in the 1950s and 1960s, behavioral ecology began by following two different lines of work. One line attempted to use behavior to understand ecological questions. Works relating to this include optimal foraging theory, which began in 1966. The other line used ecology to help understand behavior. This line included the effect of nest-site location on the nesting behavior of kittiwakes (Cullen 1957) and the influence of resource distribution on flocking behavior of weaver finches (Crook 1964). The book by Klopfer (1962) pursued both lines, studying both the ecological context and the ecological consequences of behavior. A new method, i.e. optimization, was also introduced to study behavior in the 1960s. Optimization was introduced to the study of foraging by MacArthur and Pianka (1966) and Emlen (1966). It was also used by Hamilton (1967) in his study of 'extraordinary sex ratios', inspired by an interest in the female-biased sex ratios often seen in parasitoids (see also Chapter 12 by Ode and Hardy).

Behavioral ecology separated itself from ethology in the 1970s. Tinbergen (1963) had argued that ethology used four different approaches to behavior: (i) causal; (ii) developmental; (iii) functional, and (iv) evolutionary, but actually ethology concentrated on causal explanations (behavioral mechanisms, proximate causes), and tended to neglect functional explanations (ultimate causes). Behavioral ecology switched the emphasis to functional explanations and also focused attention on some questions that had been ignored by ethologists, including foraging. One of the most important differences between behavioral ecology and ethology was the use of optimization models by behavioral ecologists to generate predictions about animal behavior.

Behavioral ecology was announced to the world by the readable collection of papers edited by Krebs and Davies (1978). It became popular because it offered a new point of view. It invited us to think of behavior as a strategy that an animal uses to solve a problem posed by nature. It also proposed to model the problem and find an optimal solution. However, as time went on, behavioral ecology paid more attention to mechanisms of behavior and causal explanations and less attention to optimization models and functional explanations (Krebs & Davies 1991). We do not propose to ignore behavioral mechanisms in order to concentrate on optimization models. We do believe, however, that parasitoid foraging would be better understood if optimization theory and behavioral observations were treated together.

Here we describe an idealization of the behavioral ecology approach: (i) An animal faces a problem posed by nature, then (ii) a theorist devizes a mathematical model of the problem that the animal faces, leading to (iii) a strategy to be found, which is the optimal solution of the mathematical problem, and (iv) the animal's behavior is observed in an experiment and compared with the optimal behavior, and finally (v) the ecological consequences of the behavior are studied. Most pieces of work in behavioral ecology include only one or two of these parts, but as a whole, Bayesian foraging includes all five.

16.1.5 Optimal foraging theory

Optimal foraging theory was begun by mathematically-inclined ecologists in 1966. MacArthur and Pianka (1966) and Emlen (1966) used rather crude mathematical models to investigate the consequences of optimal foraging for patch choice and food choice. Over the next ten years, a number of models were developed, tending to become more mathematical and less ecological as time went on. By the mid- to late-1970s, there were clearly stated mathematical models, especially by Charnov (1976a,b), and a review by Pyke et al. (1977) that described four problems that foragers face. The two most important of these were prey (host) choice and patch residence time.

Much experimental work was done in the late 1970s and early 1980s, by testing predictions of optimal foraging theory. This work was summarized in a review by Krebs et al. (1983) and in chapter 9 of Stephens and Krebs (1986). Questions asked were, for example, to see whether animals stay longer in patches if the travel time between patches is longer. Actually, they often do. This sort of result might be more interesting if experimenters thought in Bayesian terms. Does what an animal does in this patch depend on what happens in the rest of the environment? If animals stay longer in a patch of a particular quality when travel time is longer, are they responding to something more than just the particular patch? What exactly do they know about their environment?

After the review of foraging theory by Stephens and Krebs (1986), interest in testing optimal foraging theory seemed to decline, although interest in foraging behavior remains high. Concerning insect parasitoids, the review of foraging in Godfray (1994) bucked the trend away from optimal foraging theory by seriously considering both empirical and theoretical work on foraging. However, even in works on parasitoid foraging, there has been a tendency to do descriptive work rather than work based on optimization models. In this chapter, we suggest that some of the observational work, including the Waage-like 'incremental' and 'decremental' patch-leaving mechanisms and the Cox (1972) proportional hazard model interpretation of giving-up times (see Chapter 8 by van Alphen and Bernstein), which were reviewed by van Alphen et al. (2003) and by Wajnberg (2006), might be understood better in terms of optimal foraging theory. Indeed, they might be strongly related to the Bayesian point of view.

16.2 The patch residence-time problem: a Bayesian approach

16.2.1 The problem

Here, we are interested in the problem of how long a parasitoid should spend searching a patch of hosts or, more precisely, what rule a parasitoid should use to decide when to leave a patch (see also Chapter 8 by van Alphen and Bernstein). Charnov's (1976b) famous 'marginal value theorem' (MVT) offered a solution to the patch residence-time problem. Charnov (1976b) assumed that the environment contains several easily distinguished types of patches, and that each type of patch has a fixed 'gain function'. As a forager searches a patch, the gain rate declines smoothly. An optimal forager should leave a patch when the gain rate (the 'marginal value') in that patch has fallen to a rate equal to the long-term average rate of finding prey (or hosts) in the environment as a whole. Charnov's (1976b) theorem specified when a forager should leave a patch, but it did not say how a forager should decide when to leave a patch.

Krebs et al. (1974) suggested that the MVT might hold if a forager were to use a fixed giving-up-time rule, i.e. leave a patch if no prey have been found within some fixed time, the giving-up time (GUT). Krebs et al. (1974) performed an experiment with black-capped chickadees in an aviary to decide which of three patch-leaving rules a forager should use. They compared the fixed GUT rule with two other rules:

1 a forager should stay a fixed time in each patch (the fixed-time rule); or
2 a forager should stay until a fixed number of prey have been found (the fixed-number rule).

Evidence indicated that the fixed GUT rule was more likely what the birds used than were the other two rules. Krebs et al.'s (1974) paper was important because it suggested comparison between different patch-leaving rules and because it tested some possibilities experimentally. Obviously, the authors did not foresee the results of the studies published later by Oaten (1977) and Iwasa et al. (1981), which showed that the form of the optimal strategy depends on how prey are distributed in patches. Their paper purportedly tested Gibb's (1958) 'hunting by expectation' hypothesis, which suggests that foragers might use the fixed-number rule. Actually, the fixed-number rule is best under a particular prey distribution, i.e. all patches contain the same number of prey. They did not offer this kind of distribution, so it is not surprising that their birds did not use this rule.

The problem with Charnov's (1976b) MVT is that it is based on a deterministic model, i.e. all patches of a particular type are exactly the same and a forager would have exactly the same experience in each patch. Energy would be obtained constantly in a patch and the rate of energy gain would decrease steadily as long as the forager remained in a patch. Charnov (1976b) claimed that his model could easily be made stochastic, but did not explain what was meant by this. Oaten (1977) disputed this claim, pointing out that if the optimal patch-leaving rule was based on expectations rather than deterministic rates, then it would be necessary to know what rule is used to find the expectations. That is, the optimal patch-leaving rule would have to be known before that rule could be found. Oaten (1977) proposed his own model, which was explicitly stochastic. Here, we list the seven assumptions of Oaten's (1977) model, and then we add an eighth, which, if it holds, makes the calculation of optimal patch-leaving rules much simpler.

16.2.2 Assumptions of Oaten's (1977) model

1 Prey are found in patches and the predator knows the distribution of the number of prey in each.
2 The predator knows the joint distribution of the capture times, given the number of prey in a patch.
3 Prey are not replaced as they are captured.
4 The predator knows the time τ that it takes to travel from patch to patch.
5 The predator decides when to leave a patch based on its knowledge of assumptions 1, 2, and 4 and on its experience in the patch.
6 Given a strategy we can calculate the long-term average rate of finding prey:

$$R = \frac{E(G)}{E(S) + \tau} \qquad (16.1)$$

where $E(G)$ is the expected number of prey caught in each patch and $E(S)$ is the expected length of time spent searching each patch.
7 The predator uses the strategy that maximizes R.
8 The number of prey found in a patch by time t is a sufficient statistic for the number of prey remaining in the patch.

16.2.3 Advantages and disadvantages of Oaten's (1977) model

Oaten's (1977) model has some advantages over Charnov's (1976b). Oaten's first two assumptions require that attention be paid to how prey (or hosts) are distributed among patches, and on how a forager searches a patch. There is no single, simple solution to the problems that Oaten's (1977) model foragers face, as there is for Charnov's (1976b) model. Charnov's (1976b) model seems more general than Oaten's (1977), but Oaten's model makes a point of considering factors that are of ecological and behavioral importance, factors that Charnov's model ignores.

On the other hand, Oaten's (1977) model also has some disadvantages. One of these is that Oaten ignores the possibility that a forager can distinguish among patch types, while

Charnov (1976b) does consider patch types. Stephens and Charnov (1982) distinguished the models of Oaten (1977) and Charnov (1976b) by saying that Charnov considers patch 'types', which a forager can distinguish immediately without search, while Oaten considers patch 'subtypes', which a forager can distinguish only with search. Stephens and Krebs (1986) criticized 'Oaten and his followers' for not dealing with patch types. This is an important criticism, but it is not difficult to deal with patch types along with subtypes, as Green (1990) showed. We simply find the best strategy for dealing with patch subtypes for each patch type. Unfortunately, this idea of dealing with types and subtypes simultaneously has not been pursued. This is particularly important for parasitoids because they do, at least sometimes, assess patch quality without search. We might say that the idea of a patch type in the sense of Stephens and Charnov (1982) is equivalent to the use of chemical information by parasitoids to assess patch quality. It would be worthwhile to try to incorporate this idea in Bayesian models of parasitoid foraging.

Two other disadvantages of Oaten's (1977) model are the computational difficulty of finding the optimal strategy and the fact that studies of predator behavior seldom know how prey are distributed in the field. The first difficulty is not so serious if we use the relatively simple method of calculation illustrated by Green (2006). The second difficulty is less serious for studies on parasitoids because host patches, which may be a leaf if the hosts are aphids, or a gall, if the hosts are gall-forming flies, are often small enough that host numbers can be determined fairly easily.

16.2.4 Results and predictions

What the best patch-leaving rule looks like
If assumption 8 is added to the other assumptions of Oaten's (1977) model, then the best patch-leaving rule depends only on the time that a forager has spent in a patch and the number of hosts found up to that time. Therefore, the best patch-leaving rule may be characterized by a set of points (t,x), whose coordinates are time in a patch, t, and the number of hosts, x, found by time t. The forager should leave a patch when one of these points is reached.

We can illustrate a forager's experience in a patch by a trajectory specified by a function, $x(t)$, which is the number of hosts found in the patch by time t. Such a trajectory is illustrated in Fig. 16.1. The form of the trajectory is determined by the times at which hosts are discovered.

This optimal patch-leaving rule was found using dynamic programming (see also Chapter 15 by Roitberg and Bernhard). Green (1980) gave the first example of Oaten's (1977) model that was solved using dynamic programming. Mathematical details were included. The calculations needed to find the best rule are fairly complicated, but the resulting rule seen in Fig. 16.1 has a simple form. Roughly speaking, the forager should stay for some time without finding a prey and, if a prey is found before that time, remain in the patch for as long as the average rate of finding prey in the patch is above some particular value. A simpler, more general method of using dynamic programming to find optimal patch-leaving rules was given by Green (2006). The example illustrated in Fig. 16.1 was given in Green (1987). More recently, Olsson and Holmgren (1998, 2000) and Olsson (2006) have used dynamic programming to find optimal foraging strategies. Olsson and Holmgren (1998) changed Oaten's (1977) assumption that foragers are rate maximizers (his assumptions 6 and 7) and found instead the strategy that maximizes survival.

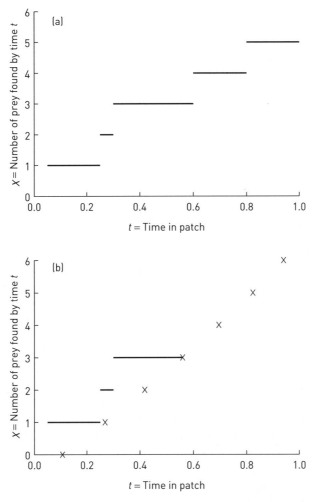

Fig. 16.1 A forager's experience in a patch. (a) The number of prey encountered, x, plotted against time in the patch, t. (b) Same as (a) except that the stopping points are indicated by X. Note that a forager leaves a patch when one of the stopping points is reached. Such a patch-leaving rule is optimal for a negative binomial host distribution with a shape parameter $k = 1$ and an average of five hosts per patch. Search is assumed to be systematic, and travel time between patches is assumed to equal $\tau = 0.1$, while the time to search a patch completely is 1.

The form of the best patch-leaving rule depends on the distribution of hosts and the pattern of search by the parasitoid

In one of the most important papers on Bayesian foraging, Iwasa et al. (1981) showed that different prey distributions call for different patch-leaving rules. They used two approaches. First, they considered the three particular patch-leaving rules that had been suggested by Krebs et al. (1974):

1 fixed time rule;
2 fixed number rule; and
3 fixed giving-up-time rule.

Then, they looked at two different prey distributions:

1 each patch has exactly five prey; and
2 half the patches have no prey and half have exactly ten prey.

Of the three rules that they considered, the fixed number rule was best when each patch has the same number of prey, and the fixed GUT rule was best when half the patches are empty and half have exactly ten prey. Iwasa et al. (1981) did a quantitative analysis of these examples, which showed that for a rule of a given type, the rate of finding prey was not sensitive to the exact form of the rule used. For example, if a predator used a fixed-time rule, the rate of finding prey is not sensitive to the fixed time.

The second approach used by Iwasa et al. (1981) was based on the expected number of prey remaining in a patch when a particular number of prey have been found by a particular time. This expected number is calculated using Bayes' theorem. The idea is that a forager should leave a patch if the expected number of hosts remaining falls below a certain value. The rule found by this method, which Pyke (1978) suggested as a way to implement Charnov's (1976b) MVT in the stochastic case, does not always yield the optimal patch-leaving rule. This method yields the best rule in three of the four cases that Iwasa et al. (1981) treated, i.e. when the distribution of prey is regular, binomial, and Poisson, but it does not yield the best rule for the negative binomial distribution, which is perhaps of most interest. The optimal patch-leaving rules are illustrated in Fig. 16.2, for the four prey distributions considered by Iwasa et al. (1981).

The form of the optimal patch-leaving rule depends on the pattern of search as well as the distribution of prey in patches. Assuming that the search is random, Fig. 16.2(d) shows the optimal patch-leaving rule for the same negative binomial prey distribution that requires the patch-leaving rule shown in Fig. 16.1, when the search is systematic. In both cases, an optimal forager should leave a patch when some time has been spent in the patch and no prey, or few prey, have been found. The difference between the optimal rules for random and for systematic search is that the set of leaving points tends to curve up more for random search. This difference illustrates the point that there are two different reasons to leave a patch. One is that the forager discovers that the patch is poor, which is the only reason why a systematic forager should leave a patch before searching it completely. The other reason to leave a patch is that it has been depleted. In Charnov's (1976b) model, for which the MVT gives the solution, foragers gain no information about patch quality while searching a patch, and patch depletion is the only reason to leave. For Oaten's (1977) model, if we assume that prey have a negative binomial distribution and that the search is random, then a forager should leave a patch if it is found to be poor, or if it becomes depleted.

16.2.5 Advantages of a quantitative treatment of foraging theory

Analysis of particular examples of Oaten's (1977) stochastic model, for the patch residence-time problem, shows that the optimal patch-leaving rules sometimes have a simple form (Iwasa et al. 1981, Green 1980, 1987). This is a qualitative result, but a quantitative treatment (Iwasa et al. 1981, Green 1984, 1987, 2006) shows that the performance of the best

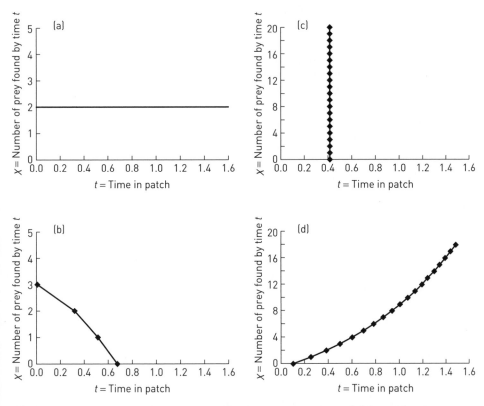

Fig. 16.2 Stopping rules for four different prey distributions: (a) Regular prey distribution where each patch has the same number of prey ($x = 5$); (b) Binomial prey distribution with parameters $n = 10$ and $p = 0.5$; (c) Poisson distribution with parameter $\lambda = 5$; (d) Negative binomial distribution with parameters $k = 1$ and $m = 5$ (or $p = 1/6$). In all cases, the expected number of prey in a patch is $\mu = 5$ and search time is $\tau = 0.1$. Search is assumed to be random. The best rule for a regular prey distribution is the fixed number rule, while the best rule for a Poisson prey distribution is the fixed time rule. In all cases, the patch leaving decision is achieved when the trajectory (t, x) gets outside of the domain defined by the stopping rule.

rule is not sensitive to the exact rule used. Together, these results show that in some cases it is easy for an animal to forage optimally, and even if the animal does not use the best possible rule, it still can do very well.

A quantitative treatment of a number of different kinds of patch-leaving rules shows how much better an optimal forager can do than a sub-optimal forager. In particular, we can compare the rate of finding prey for an optimal forager, which uses information optimally, with the rate achieved by a 'naïve' forager that ignores information gained while searching a patch and leaves at some time chosen independently of experience in the patch (Green 1980, 1987). This is important because sometimes information matters and sometimes it does not. In particular, if we wish to test experimentally whether an animal uses information about patch quality obtained while foraging in order to decide when to leave a patch, it is necessary to use patches with prey distributions that make information use important to the animal. An example in which this was not done was an experiment on rats performed by Mellgren et al. (1984) and Mellgren and Brown (1987).

Perhaps the most important use of a quantitative treatment of foraging is found in the use of individual behavior to study population ecology. There have been a number of attempts to use foraging behavior to study the stability of predator-prey models (Hassell & May 1974, Murdoch & Oaten 1975). There have also been exhortations to use individual behavior to understand population dynamics (Hassell & May 1985, Schoener 1986). More recently, there has been the development of 'individual-based' models, promoted especially by DeAngelis and Gross (1992), in which population dynamics are determined by the inter-action of many individuals whose behavior is modeled separately. These attempts to include behavior in the study or population dynamics are an improvement over the original population models of Lotka and Volterra (Lotka 1925, Volterra 1926) and of Nicholson and Bailey (1935), which treat the individuals that make up populations as randomly inter-acting particles, but they are still flawed because of their failure to treat foraging beha-vior, and particularly information use by foragers, carefully enough.

A serious attempt to put foraging theory into a predator-prey model was made by Murdoch and Oaten (1975), who investigated whether a forager using a fixed GUT patch-leaving rule would stabilize the interaction between predators and prey, if prey distribution follows:

1 a Poisson distribution;
2 a negative binomial distribution with the shape parameter k proportional to density; or
3 a negative binomial distribution with a constant shape parameter k.

They found that foragers stabilized the system in each of these three cases. Green (1990) re-analyzed this problem and concluded that, if the predators use the foraging strategy that is optimal when prey density is at equilibrium, stability would only be found in case (3). The general conclusion of this re-analysis is that the conditions under which forag-ing can stabilize a predator-prey model are even more limited than had been believed. This conclusion is consistent with that of Bernstein (2000) in that there are no convinc-ing mathematical treatments that produce stability for host–parasitoid models. A particu-lar implication of Green's (1990) re-analysis of Murdoch and Oaten's (1975) model is that the stability of the system does not depend simply on the degree of aggregation of the prey at equilibrium, but also depends on the way that the degree of aggregation changes with prey density. This point is of interest because there was, for a time, the belief that the stability of some population models depended on the degree of aggregation of the numbers of prey, or hosts, or competitors, or the degree of attack. This was summarized in what was known as the CV^2 rule, i.e. the system had a stable equilibrium if the squared coefficient of variation was greater than one (see references in Hassell et al. 1991). The critical factor was not the degree of aggregation at equilibrium, but rather how the degree of aggregation changes with density.

16.3 Bayesian foraging and proximal mechanisms

We will now address the question of 'how do they do?' We have seen the interest of Bayesian foraging theory. Animals, and especially parasitoids, sample their environment while foraging, and seem to update their evaluation of the patch quality (see Chapter 8 by van Alphen and Bernstein). Obviously, we rejected the idea that parasitoids calculate probabilities but saw that Bayesian rules are robust to errors. It is also difficult to admit that parasitoids use statistics, even as simple as the number of hosts found at time *t*, and compare it to a stopping optimal policy. However, the rules have to be implemented, and we have to find plausible decision mechanisms. For that, we will return to the work of Iwasa et al. (1981), which we already discussed from another point of view, and relate it to giving-up rules suggested by several other authors, especially Waage (1979).

16.3.1 The expected-number-left estimator of Iwasa et al. (1981)

The first mention of parasitoids in a Bayesian context was found in Iwasa et al. (1981). Although their work is mainly theoretical, the authors briefly mentioned Waage (1979) in their discussion. They attempted to draw a link between the evolution through time of a Bayesian estimator of the number of hosts remaining in the patch, and the mechanism proposed by Waage (1979), which we discuss below. Assuming random search, Iwasa et al. (1981) provides several important results that are applicable to parasitoids in a patch-leaving decision context. First, the decision rule for leaving a patch should be sequential, according to decision theory. Second, the number of hosts found and the time spent in the patch are sufficient statistics and this is precisely one of the simplifications of Oaten's (1977) model used by Green (1980, 2006). Third, the estimate of the number of remaining hosts in the patch suffers sudden jumps when a host item is found. Iwasa et al. (1981) show that these jumps go upward when the distribution of items among patches is clumped (negative binomial distribution), and downward when the distribution of items is more uniform (binomial distribution). The estimate decreases with time in both cases during all intervals where no items are found. The authors mention the fact that the curve representing the estimate of the remaining host in the patch is a function of time, which looks like the curve used in the patch leaving mechanism proposed by Waage (1979) (Figs 16.3a,b).

16.3.2 Waage's (1979) patch-leaving mechanism and Bayesian decision

Waage (1979), referring to the parasitoid *Nemeritis canescens*, proposed a very influential model of the patch-leaving decision mechanism, independently of any consideration of sampling or optimality (see also Chapter 8 by van Alphen and Bernstein). The model is not Bayesian in any way, but mechanistic. Its advantage was that, instead of being an *ad hoc* 'rules of thumb', as previously considered, it aimed to represent some plausible psychological mechanism of patch leaving decision. The forager was assumed to enter the patch with a basic tendency to stay in it. In the original article, this initial tendency, noted *aP*, is proportional to the amount of kairomones emitted by the hosts in the patch. That was biologically sound in the context of *N. canescens* but can be more generally considered as the state of the forager when entering the patch, not necessarily depending on the amount of chemical cues that can be detected in the patch. The tendency to remain in a patch was supposed to decrease monotonously (linearly in Waage 1979) during search, when no host is found. Any host discovery results in a sudden increment in this tendency, and this feature is now referred as an incremental mechanism. The forager leaves the patch when the tendency falls to a given threshold (Fig. 16.3a). As a consequence, the more often

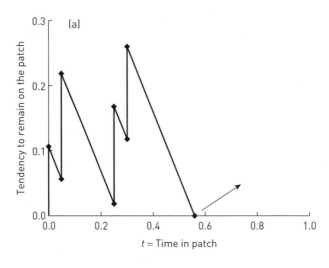

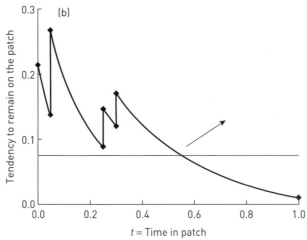

Fig. 16.3 A patch-leaving rule like Waage's (1979) model, when host follows a negative binomial distribution with $k = 1$ and $m = 5$, for an animal that has the same experience as the one illustrated in Fig. 16.1. (a) Responsiveness (i.e. the tendency to remain in a patch) is plotted against time. Search is assumed to be systematic and a travel time of $\tau = 0.1$ is used to find the best patch-leaving rule. The forager leaves the patch when the responsiveness falls under a threshold (here zero, black arrow). The figure is drawn supposing increments to be constant for each encounter with a host. (b) Evolution in time of the Iwasa et al. (1981) estimator of the number of hosts remaining in the patch. The expected number of remaining hosts is plotted against time. The forager leaves the patch when the estimation of remaining hosts falls under a given value (black arrow). Note the similarity between the two processes.

hosts are found, the longer will the parasitoid remain in the patch. The tendency to stay in the patch evolves as

$$m(t) = aP - bt + \sum_{i=0}^{n(t)} I(t_i) \tag{16.2}$$

where aP is the initial tendency, b the slope of its decrease, $n(t)$ the number of hosts met at time t, and $I(t_i)$ the sudden increment triggered by the discovery of a host at time t_i (Wajnberg 2006). The time dependence of the increments is a complication justified by the observation that the discovery of a host had little effect when it immediately followed a previous one, and Waage (1979) proposed a functional form for this dependence. However, Fig. 16.3a assumes increments that are constant.

The tendency to stay in the patch, as described by Waage (1979), is typically what psychologists call a motivational variable. In psychology, motivation (or 'drive') refers to what causes the initiation, direction, intensity, and persistence of a behavior (Geen 1995). The similarity between changes in motivation in Waage's (1979) model and the evolution of Iwasa et al.'s (1981) Bayesian estimator of the number of prey remaining in the patch was so striking (Fig. 16.3b) that Iwasa et al. (1981), in their conclusion, interpreted Waage's (1979) tendency or motivation as the estimator itself. They stated that the animal was actually estimating the number of remaining hosts in the patch, assimilating, without formal proof, their own Bayesian model with a non-Bayesian mechanism. Despite this, their intuition was excellent and should have been more widely mentioned. We propose to reverse this explanation and to assume that natural selection has tailored simple behavioral processes or rules, which mimic closely a Bayesian estimator, because Bayesian estimation leads to optimal decisions in an uncertain world. The evidence of estimation is only provided by the observation of animal decisions. In other words, the decision appears to be taken 'just as if' the forager was applying a Bayesian decision procedure. As McNamara et al. (2006) pointed out: 'We emphasize the distinction between the mathematical procedure that can be used to find the optimal solutions and the mechanism an animal might use to implement such solutions.' We have then to assess whether Waage's (1979) mechanism, or similar mechanisms proposed by other authors, are good candidates to perform the function of Bayesian analogical mechanisms. Motivational mechanisms would then achieve a kind of 'emerging computation' (Forrest 1990) at the individual level.

The process described by Waage (1979) has several interesting characteristics. In a virtual experiment that we call 'the infinite or undepletable patch model', the nature of the process, i.e. an infinite series of random shocks, produces a decision sampler. If the random shocks occur as a Poisson process with a constant rate λ, which obviously depends on the patch density, there is a threshold value λ_c, at which the process outcome switches. Below λ_c, the probability of still being in the patch at long term is lower than the probability of leaving it. Inversely, when $\lambda > \lambda_c$, the probability of leaving the patch is lower than the probability of still being in it when t tends to infinity. This means that the forager is able to assess whether the rate of encounters with hosts is above or below a critical density. Above that density, the forager has a non-zero probability of still being in the patch when t tends to infinity. Below that density, it will almost surely have left the patch. The uncertainty of the decision and the 'errors' of the process are greater in the neighborhood of λ_c (Pierre et al. submitted). Figure 16.4 represents the probability of still being in the patch at time t, depending on the rate of encounters with hosts λ.

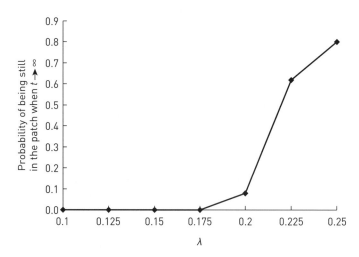

Fig. 16.4 Probability of still being in the patch when t tends to infinity, obtained asymptotically by Monte-Carlo simulation, for very long times and for various values of the encounter rate with hosts λ. The conditions assumed are those of Waage's (1979) process in an infinite or non-depletable patch (λ constant, incremental model). For very low values of λ, all animals should leave the patch when t tends to infinity. For very high values of λ, almost no animal should leave the patch. In intermediate cases, there is a proportion of animals that should never leave the patch. In all cases, $aP = 10$, $b = 1$, $I = 5$.

The similarity of the class of processes, which we will call 'Waage-like processes', with the Iwasa et al. (1981) Bayesian estimator, was reinforced by evidence of decremental processes, which they predicted. Iwasa et al. (1981) maintained that the jumps of the Bayesian estimator of the number of hosts remaining in the patch should drop suddenly instead of rise when meeting a host item, if the spatial distribution of the hosts within the patches tends to be uniform. This was found both experimentally and theoretically by Driessen et al. (1995), who presented the first evidence of a decremental process in the parasitoid *Venturia canescens*. This work was regarded as strongly supporting Iwasa et al.'s (1981) views, and therefore the Bayesian sampling idea, although the Driessen et al. (1995) model was not inspired by the Bayesian theory.

The analogy is strongly reinforced when the stopping rule implied by Waage's (1979) incremental or Driessen' et al. (1995) decremental processes are explicated. Let us go back to Equation 16.2 and see what relation between n, the number of hosts found, and t, the time elapsed since the entry into the patch, arises when the parasitoid leaves the patch. From Equation 16.2, setting $m(t)$ to 0, and considering I as constant, we get

$$n = -\frac{aP}{I} + \frac{b}{I}t \tag{16.3}$$

In the incremental case ($I > 0$), we obtain a set of stopping points on a straight line whose slope is positive, intersecting the t-axis at aP/b (Fig. 16.5a). In the decremental case ($I < 0$), the set of stopping points is on a straight line with a negative slope, intersecting the

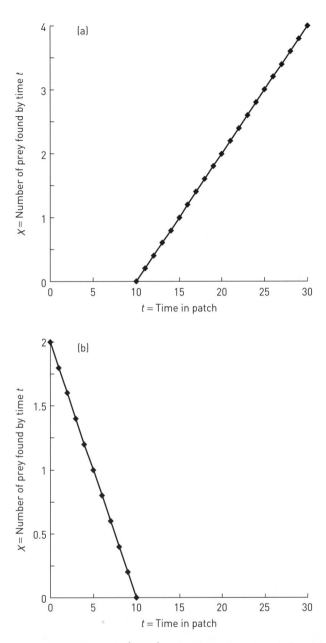

Fig. 16.5 Stopping rule of Waage's (1979) process in the plan defined by t, the time spent in the patch, and n, the number of hosts found since the entry in the patch. (a) Incremental process; (b) decremental process. Parameters for both figures are $aP = 10$, $b = 1$, $I = 5$, and $I = -5$ for (a) and (b), respectively. Note the respective similarities with Figs 16.2d and b (Iwasa et al. (1981), model stopping rules), respectively.

n-axis at *aP/I*, and the *t*-axis at *aP/b* (Fig. 16.5b). Qualitatively, those figures are similar to those of Iwasa et al. (1981) (Figs 16.2d and b, respectively). The incremental model corresponds to a situation when hosts follow a negative binomial distribution, and the decremental one corresponds to a binomial distribution. This means that, although the model itself is not Bayesian, the decision rule is consistent with the Bayesian model.

Pierre et al. (2003) were the first to design an experiment to demonstrate that parasitoids used some sort of Bayesian sampling rule. The species used was a mealybug parasitoid, *Leptomastyx dactylopii*, which perfectly discriminates healthy hosts from parasitized hosts. They placed the foraging females in artificial patches of varying quality containing from 50 to 100% of parasitized hosts. Using Wald's (1947) theory of sequential sampling, they predicted that the stopping points, defined by the pair (n, T), where *n* is the number of parasitized hosts found after having met *T* hosts, should be organized on a straight line whose parameters depended on the patch composition. They also produced a simulation showing that a Waage-like process was suitable to produce a pattern of patch-leaving similar to that observed experimentally. Such an approach has not been much followed in the field of parasitoids, but many experiments support the existence of Waage-like mechanisms describing parasitoid patch-leaving rules.

16.3.3 Evidence of Waage's-like processes

Incremental and decremental processes were recently detected in insect parasitoids using techniques of survival analysis, particularly the Cox's (1972) proportional hazard model (Therneau & Grambsch 2000, see also Chapter 8 by van Alphen and Bernstein and also Chapter 18 by Wajnberg and Haccou). This model is essentially a regression model with multiple independent variables. With such methods, the incremental or decremental effect of each encounter with a host, of each host handling, and of each oviposition, can be detected and estimated in terms of increases or decreases in the hazard rate (Wajnberg 2006). In the patch residence-time problem, the hazard rate is the chance (rate) of leaving a patch at a given time, assuming that the forager is still in the patch at that time. The Cox proportional hazard model assumes that the hazard rate is a function of time on which certain events can have a multiplying effect. For example, a parasitoid may have a particular hazard rate function, but when an unparasitized host is found, the hazard rate is reduced by a constant proportion.

Wajnberg and Haccou (Chapter 18, this volume) treats these methods comprehensively, and therefore we will only report some results related to our Bayesian point of view. In parasitoids, one of the first uses of such a method was proposed by Haccou et al. (1991), on *Leptopilina heterotoma*. Using this method, the authors interpreted the tendency to leave the patch as the baseline of a Cox proportional hazard model. This implies that their decision process is not a Waage-like process. There is indeed some discrepancy between the incremental or decremental processes found by Cox proportional hazard models and Waage-like models (Wajnberg 2006). In Waage-like models, everything, except the times at which hosts are found, is deterministic. Thus, conditioned on a set of host discoveries occurring at given times $t_1, t_2, \ldots, t_k$, the patch residence time is determined absolutely. On the contrary, a Cox regression supposes a random survival function and only a multiplicative effect of any covariate, here the times of host discoveries.

However, the method was often used (Hemerik et al. 1993, van Roermund et al. 1994, Wajnberg et al. 1999, 2000, Outreman et al. 2005), and evidence accumulates that incremental, decremental, and even combined (Outreman et al. 2005) processes are real (Wajnberg 2006). Van Alphen et al. (2003), and more recently Wajnberg (2006), gave a comprehensive review of these works. We can thus assert that some decision processes act by way of sudden increments and decrements of the tendency to stay in the patch, and mimic closely a Bayesian estimator as defined by Iwasa et al. (1981). We suggest thinking about such processes as 'quasi-estimation' in the sense that the process 'estimates' the remaining quality of the patch. A kind of analogical computer realizes mechanically the updating of the information and triggers the decision. In Waage-like models, we can think of the monotonous decrease in the motivation as an updating of the time elapsed since the entry into the patch, and of the motivation increments or decrements as an updating of the number of hosts discovered. It is also important to note that neither memory nor learning are assumed in such processes.

This aspect of Bayesian decision theory in parasitoids and in other animals is important, but works linking psychological processes to the success of decisions are still scarce. Such work would answer Kacelnik and Todd's (1992) call for the introduction of psychological mechanisms in behavioral ecology.

16.3.4 Humans and parasitoids

The relationship between neuropsychological processes and behavioral ecology becomes a growing concern in higher animals and humans, and the pioneer works realized in parasitoids still have some influence. Hutchinson and Gigerenzer (2005), for instance, assert that much human decision-making can be described by simple algorithmic process models (heuristics). Referring to Iwasa et al. (1981), but not to Waage (1979), he and his collaborators proved recently that, in a foraging task, humans use an incremental rule, as if we were adapted to forage for aggregated resource items. Further, they showed that humans cannot switch easily from an incremental to a decremental rule when needed, i.e. when facing a more or less uniform distribution. In this particular domain, humans do worse than some parasitoids (Outreman et al. 2005).

Unfortunately, the tiny brain of a parasitoid is not convenient for neurophysiological investigation. Nevertheless, let us consider briefly what is becoming known in more complex animals such as birds or mammals. Under the name of 'neuroeconomics', a field of investigations is growing in importance. Glimcher (2002) mentions that 'in the last decade, evidence has been accumulating that the brains of complex animals like mammals perform operations which closely correspond to the optimization problems behavioral ecologists describe as the ultimate causes of behavior.' Glimcher (2002) claims that little attention has been paid by neurologists to the variables that behavioral ecologists study and that, as a result, many classical physiological studies have almost entirely ignored the variables that behavioral ecologists identify as critical at the level of ultimate causation. Glimcher (2002), Glimcher and Rustichini (2004), and Glimcher et al. (2005) conducted a series of experiments to see how different parts of the brain were activated in tasks involving decision making. In macaques, they showed that activity in the posterior parietal cortex is correlated with the relative subjective desirability of action. When each new item occurs in a decision task, bursts of groups of neurons are observed, depending on the kind of problem that must be solved by the subject. This kind of phenomenon is what might be suspected to be behind the increments, decrements, and general decrease of motivation described by Waage-like models.

16.3.5 Can Bayesian foraging be tailored by natural selection or individual adaptation?

The proximal mechanisms discussed above can be simply tuned by natural selection as well as by epigenetic processes. Consider the original process proposed by Waage (1979) (Equation 16.2). Genetic and epigenetic variation can act on:

1 the initial motivation aP;
2 the slope of decrease b; and
3 the sign and size of increments (or decrements) l.

This is a rather simple tuning and, providing that genetic variability exists, the mechanism described by the equation is able to lead to optimality. Memory and learning capacities can also tune these three parameters more accurately if the environment characteristics are highly variable in space or time. Both genetic variability and experience effects were shown at least on the size of the increments. Wajnberg et al. (1999) found a significant genetic variability in the reaction to the discovery of an attacked host in a patch. Outreman et al. (2005) found a behavioral plasticity in an *Aphidius* species, able to switch from incremental to decremental process as a function of its state. Although still scarce, these studies indicate that genetic and epigenetic variability exist in those traits and give room to natural selection and individual adjustment (Wajnberg et al. 2004).

Another piece of evidence is given by Tenhumberg et al. (2001), who devized a Stochastic Dynamic Programming (SDP) model (see also Chapter 15 by Roitberg and Bernhard) to find the optimal patch leaving strategy inspired by the parasitoid *Cotesia rubecula*. After optimization, they ran simulations of patch use and analyzed the residence times obtained using a Cox proportional hazard model. They found that the optimal policies defined by SDP can be summarized by a modification of the tendency to leave the patch at each oviposition, either in an incremental or in a decremental way. This suggests strongly that foraging behavior might be improved by tuning the size of increments or decrements in motivation, and that these parameters are likely to be a target for natural selection.

16.3.6 Bayesian foraging and learning

Parasitoids, like almost all animals, show learning capabilities. Many works considering learning in parasitoids are cited in van Alphen et al. (2003). Learning is often included in the category of cognitive capacities and the consideration of its importance led to the development of a new concept called 'Cognitive Ecology' (Real 1993, Dukas 1998). Learning itself is a growing point in insect ecology (Papaj & Lewis 1993). The influence of learning on foraging strategy has often been modeled (McNamara & Houston 1985, Stephens 1993). Although learning is a way of incorporating information, there is no direct link between learning models and Bayesian models. Learning implies information use, but Bayesian procedures do not imply learning directly. For instance, in the case of Iwasa's et al. (1981) model, a parasitoid can be rigidly hardwired for a rule adapted to a negative binomial distribution, providing that its hosts are always distributed in this way. Its type of procedure will always be incremental. The Bayesian nature of its decision making requires only some sort of memory. Parasitoids have to remember how many hosts were found in a period of time. Let us note that this memory does not need to be cognitive. Any process depending on the past experience can suit such a requirement.

Learning is more than that. For instance, it may be necessary to learn the characteristics of a new habitat, in order to be able to switch adequately from an incremental to a decremental process. McNamara and Houston (1985) give a nice definition of learning: 'a rule can be said to learn about an environmental parameter when the value of a state variable converges to the parameter value in question.' They give, as an example, the optimal marginal gain of the Charnov (1976b) model, which can be unknown and learned in that sense by the forager. This definition obviously does not apply to Bayesian foraging. Of course, the estimate of the host distribution can be considered as a state variable, but Bayesian processes do not modify durably the state of the forager. As soon as a decision is taken, the process is renewed. The McNamara and Houston (1985) model does not rely on a stochastic flow of information, but is mainly deterministic. The state variable is not probabilistic. At the beginning of its foraging, the naïve forager starts with an initial value of γ^*, the optimal marginal gain in the environment. Its reward leads it to correct this value until the gain is maximized. This is an iterative optimization process. Of course, learning and Bayesian estimation can be combined, but we are not aware of such an approach.

16.4 Contributions of Bayesian foraging theory to other questions than optimal patch time allocation

16.4.1 Superparasitism

Many solitary parasitoids are able to recognize hosts already parasitized by themselves, by conspecifics or by allospecifics. Such recognition ability makes them able to avoid superparasitism, i.e. ovipositing in an already parasitized host. Superparasitism is generally counter-adaptive as the second egg laid in a host has a low probability of winning the larval competition to produce an adult. However, superparasitism may be adaptive in some cases (Visser et al. 1990). The interesting point is that the proportion of parasitized hosts in a patch is a key component of patch quality. In some cases, the discovery of parasitized hosts should be incorporated in a Bayesian foraging model.

We cited previously the work on *Leptomastix dactylopii* by Pierre et al. (2003), who showed that the patch-leaving decision mimicked a Bayesian rule assessing the proportion of parasitized hosts in the patch. More recently, Kolss et al. (2006) used explicitly the Bayesian framework of Iwasa et al. (1981) to assess whether encounters with parasitized hosts should be taken into account or not in deciding when to leave a patch. They found that the advantage of assessing host status is larger when the patches are highly variable. They note that these conclusions are consistent with those for predators by Green (1980), Iwasa et al. (1981), and Stephens (1993).

16.4.2 Bayesian foraging and population dynamics

Several attempts have been made to link individual behavior to population dynamics, beginning with Hassell and May (1974) and Murdoch and Oaten (1975) (Section 16.2.5 and also Chapter 13 by Bonsall and Bernstein). In particular, Murdoch and Stewart-Oaten (1989) concluded that the optimal foraging theory, combined with the widely spread character of negative binomial distribution of hosts, resulted in a Holling type II functional response (Holling 1959). In the Lotka–Volterra models (Lotka 1925, Volterra 1926), as in Nicholson and Bailey's (1935) model, a type II functional response is likely to have a destabilizing effect on the dynamics of both hosts and their parasitoids. These authors concluded that optimal foraging behavior did not lead straightforwardly to the stability of host–parasitoid systems. This way of investigation seems to be almost abandoned nowadays.

In our opinion, it could be worth reconsidering the question. Individual foraging is a master key to understanding the exploitation of patches of hosts by parasitoids. The Bayesian point of view incorporates stochasticity and patchiness of the environment and is therefore fully in tune with modern population dynamics. The possibility of selection on foraging traits (Wajnberg et al. 1999, Wajnberg 2004, Wajnberg et al. 2004) is also encouraging for finding applications to biological control.

16.5 Conclusions

What is the substance of Bayesian foraging theory? Obviously it lies in the use of past information in order to improve the decisions. This use of information has several characteristics. It implies that the external world is treated as a stochastic flow of events and not as a predictable series of events. Then, Bayesian foraging has to be included in a stochastic dynamic game. It also puts in perspective the need of being omniscient. Indeed, it is not necessary to know all the characteristics of the environment, but only some statistical characteristics of the stochastic flow. It is more likely that the first two moments of the distributions (i.e. mean and variance) are enough to take optimal decisions. These latter conclusions are suggested by Oaten (1977) and Iwasa et al. (1981). This gives a place to think about sampling procedures and especially about sequential ones. Let us note that in the deterministic Charnov (1976b) model, the so-called omniscience of the forager was limited to the knowledge of the distribution of patch types in the environment as well as the mean travel time between patches. From this deterministic point of view, only some statistics on the environment were supposed to be known and not, for instance, the precise location of each patch of each type.

The link with optimality, mentioned by the Oaten (1977) and Green (2006) models, is now obvious and their emphasis on the need for taking stochasticity into account is relevant. As Oaten (1977) pointed out, it is unrealistic to try to calculate optimal solutions by relying on laws of large numbers alone and ignoring the information foragers gain about patch quality while foraging. SDP has widely confirmed that stochasticity and the forager's state strongly determine the optimal solutions (see also Chapter 15 by Roitberg and Bernhard). The simplicity of observing the foraging behavior of parasitoids makes them good candidates for further investigation linking both approaches.

We can now summarize the steps for building a Bayesian foraging model:

1 Determine the set of distributions of hosts in the environment;
2 Calculate the optimal policy as a function of these distributions;
3 Determine a reasonable prior distribution for the parasitoid;
4 Establish the Bayesian estimators at each step of the foraging process, and calculate the posterior distribution;

5 Apply optimal policy to the posterior;
6 Infer the decision rules;
7 Compare with actual data.

Steps 1 and 2 are not Bayesian. They require optimization methods. Steps 3 to 5 are. The interplay with optimality techniques is quite natural.

The Waage (1979) model, and similar mechanisms, has provided good insights into the relationship between adaptation and proximal mechanisms. The recent development of 'neuroeconomics' (Glimcher 2002) gives hope to the elucidation of how the computations are done by natural neural networks. Behavioral ecologists interested in foraging in parasitoids should collaborate with neurophysiologists to study animals with higher cognitive capacities in the course of foraging tasks.

Our survey of the literature showed that references to Bayesian theory are not frequent in the field of parasitoid studies. Further, almost all of them concern the patch residence time problem. Although parasitoids can be considered to be 'smart' enough for that, we think that other behaviors could benefit from study using the Bayesian approach. Understanding of mate choice, sex ratio, and host choice might all be improved by knowledge of characteristics of the environment and the pattern of encounter with hosts or potential mates. As females of most parasitoid species can control the sex of their offspring, sex ratio should be sensitive to estimates of host quality distribution (see also Chapter 12 by Ode and Hardy). Although we could mention some works, these questions have not been addressed frequently from such a point of view.

A disappointing aspect of Bayesian foraging theory concerns applications to biological control. Actually, no direct link has been shown between the use of information by parasitoids and their capacity to control the density of their host. It only may help by participating to our general knowledge of the patch exploitation strategies in these insects.

References

Bernstein, C. (2000) Host-parasitoid models: The story of a successful failure. In: Hochberg, M.E. and Ives, A.R. (eds.) *Parasitoid Population Biology*. Princeton University Press, Princeton.

Charnov, E.L. (1976a) Optimal foraging: Attack strategy of a mantid. *American Naturalist* **110**: 141–51.

Charnov, E.L. (1976b) Optimal foraging: The marginal value theorem. *Theoretical Population Biology* **9**: 129–36.

Charnov, E.L. (1982) *The Theory of Sex Allocation*. Princeton University Press, Princeton.

Chewyreuv, I. (1913) Le rôle des femelles dans la determination du sexe de leur descendance dan le groupe des Ichneumonides. *Compte Rendu des Séances de la Société de Biologie de Paris* **74**: 695–9.

Cox, D.R. (1972) Regression models and life tables. *Journal of the Royal Statistical Society B* **74**: 187–220.

Crook, J.H. (1964) The evolution of social organization and visual communication in the Weaverbirds (Ploceinae). *Behaviour Supplement* **10**: 1–178.

Cullen, E. (1957) Adaptations in the kittiwake to cliff-nesting. *Ibis* **99**: 275–302.

DeAngelis, D.L. and Gross, L.J. (eds.) (1992) *Individual-based Models and Approaches in Ecology: Populations, Communities and Ecosystems*. Chapman & Hall, London.

Driessen, G., Bernstein, C. and van Alphen, J.J.M. (1995) A count-down mechanism for host search in the parasitoid *Venturia canescens*. *Journal of Animal Ecology* **64**: 117–25.

Dukas, R. (1998) *Cognitive Ecology: The Evolutionary Ecology of Information Processing and Decision Making*. The University of Chicago Press, Chicago.

Emlen, J.M. (1966) The role of time and energy in food preference. *American Naturalist* **100**: 611–17.

Forrest, S. (1990) Emergent computation: Self-organizing, collective, and cooperative phenomena in natural and artificial computing networks. Introduction to the Proceedings of the Ninth Annual CNLS Conference. *Physica D – Nonlinear Phenomena* **42**: 1–11.

Geen, R.G. (1995) *Human Motivation: A Social Psychological Approach*. Wadsworth Publishing, New York.

Gibb, J.A. (1958) Predation by tits and squirrels on the eucosmid *Ernarmonia conicolana* (Heyl.). *Journal of Animal Ecology* **27**: 375–96.

Glimcher, P.W. (2002) Decisions, decisions, decisions: Choosing a biological science of choice. *Neuron* **36**: 323–32.

Glimcher, P.W. and Rustichini, A. (2004) Neuroeconomics: The consilience of brain and decision. *Science* **306**: 447–52.

Glimcher, P., Dorris, M. and Bayer, H.M. (2005) Physiological utility theory and the neuroeconomics of choice. *Games & Economic Behavior* **52**: 213–56.

Godfray, H.C.J. (1994) Parasitoids: *Behavioral and Evolutionary Ecology*. Princeton University Press, Princeton.

Green, R.F. (1980) Bayesian birds: A simple example of Oaten's stochastic model of optimal foraging. *Theoretical Population Biology* **18**: 244–56.

Green, R.F. (1984) Stopping rules for optimal foragers. *American Naturalist* **123**: 30–43.

Green, R.F. (1987) *A stochastic model of optimal foraging: Systematic search for negative-binomially distributed prey*. University of Minnesota Duluth, Computer Science, Mathematics and Statistics, Technical report No. 87-2.

Green, R.F. (1990) Putting ecology back into optimal foraging theory. *Comments on Theoretical Biology* **1**: 387–410.

Green, R.F. (2006) A simpler, more general method of finding the optimal foraging strategy for Bayesian birds. *Oikos* **112**: 274–84.

Haccou, P., De Vlas, S.J., van Alphen, J.J.M. and Visser, M.E. (1991) Information processing by foragers: Effects of intra-patch experience on the leaving tendency of *Leptopilina heterotoma*. *Journal of Animal Ecology* **60**: 93–106.

Hamilton, W.D. (1967) Extraordinary sex ratios. *Science* **156**: 477–88.

Hassell, M.P. and May, R.M. (1974) Aggregation of predators and insect parasites and its effect on stability. *Journal of Animal Ecology* **43**: 567–94.

Hassell, M.P. and May, R.M. (1985) From individual behaviour to population dynamics. In: Sibly, R.M. and Smith, R.H. (eds.) The British Ecological Society Symposium, Volume 25: *Behavioural Ecology*. Blackwell Scientific Publications, Oxford, pp. 3–32.

Hassell, M.P., May, R.M., Pacala, S.W. and Chesson, P.L. (1991) The persistence of host-parasitoid associations in patchy environments. I. A general criterion. *American Naturalist* **138**: 568–83.

Hemerik, L., Driessen, G. and Haccou, P. (1993) Effects of intra-patch experiences on patch time, search time and searching efficiency of the parasitoid *Leptopilina clavipes*. *Journal of Animal Ecology* **62**: 33–44.

Holling, C.S. (1959) Some Characteristics of simple types of predation and parasitism. *Canadian Entomologist* **91**: 385–98.

Hutchinson, J.M.C. and Gigerenzer, G. (2005) Simple heuristics and rules of thumb: Where psychologists and behavioural biologists might meet. *Behavioural Processes* **69**: 97–124.

Iwasa, Y., Higashi, M. and Yamamura, N. (1981) Prey distribution as a factor determining the choice of optimal foraging strategy, *The American Naturalist* **117**: 710–23.

Kacelnik, A. and Todd, I.A. (1992) Psychological mechanisms and the marginal value theorem: Effect of variability in travel time on patch exploitation. *Animal Behaviour* **43**: 313–22.

Klopfer, P.H. (1962) *Behavioral Aspects of Ecology*. Prentice-Hall, Englewood Cliffs.

Kolss, M., Hoffmeister, T.S. and Hemerik, L. (2006) The theoretical value of encounters with parasitized hosts for parasitoids, *Behavioral Ecology & Sociobiology* **61**: 291–304.

Krebs, J.R. and Davies, N.B. (1978) *Behavioural Ecology*. 1st edn. Blackwell Scientific Publications, Oxford.

Krebs, J.R. and Davies, N.B. (1991) *Behavioural Ecology*. 4th edn. Blackwell Scientific Publications, Oxford.

Krebs, J.R., Ryan, J.C. and Charnov, E.L. (1974) Hunting by expectation or optimal foraging? *Animal Behaviour* **22**: 953–64.

Krebs, J.R., Stephens, D.W. and Sutherland, W.J. (1983) Perspectives in optimal foraging. In: Brush, A.H. and Clark, J.R. (eds.) *Perspective in Ornithology*. Cambridge University Press, Cambridge, pp. 165–216.

Lotka, J. (1925) *Elements of Physical Biology*. Williams and Wilkins Co, Baltimore.

MacArthur, R.H. and Pianka, E.R. (1966) On optimal use of a patchy environment. *American Naturalist* **100**: 603–9.

McNamara, J. and Houston, A. (1985) A simple model of information use in the exploitation of patchily distributed food. *Animal Behaviour* **33**: 553–60.

McNamara, J.M., Green, R.F. and Olsson, O. (2006) Bayes' theorem and its applications in animal behaviour. *Oikos* **112**: 243–51.

Mellgren, R.L. and Brown, S.W. (1987) Environmental constraints on optimal-foraging behavior. In: Commons, M.L., Kacelnik, A. and Shettleworth, S.J. (eds.) *Quantitative Analysis of Behavior*, Volume VI: *Foraging*. Lawrence Erlbaum Associates, Hillsdale, pp. 133–51.

Mellgren, R.L., Misani, L. and Brown, S.W. (1984) Optimal foraging theory: Prey density and travel requirements in *Rattus norvegicus*. *Journal of Comparative Psychology* **98**: 142–53.

Murdoch, W.W. and Oaten, A. (1975) Predation and population stability. *Advances in Ecological Research* **9**: 1–131.

Murdoch, W.W. and Stewart-Oaten, A. (1989) Aggregation by parasitoids and predators: Effects on equilibrium and stability. *American Naturalist* **134**: 288–301.

Nicholson, A.J. and Bailey, V.A. (1935) The Balance of animal populations. *Proceedings of the Zoological Society of London* **3**: 551–98.

Oaten, A. (1977) Optimal foraging in patches: A case for stochasticity. *Theoretical Population Biology* **12**: 263–85.

Olsson, O. (2006) Bayesian foraging with only two patch types. *Oikos* **112**: 285–97.

Olsson, O. and Holmgren, N.M.A. (1998) The survival-rate-maximizing policy for Bayesian foragers: Wait for good news. *Behavioral Ecology* **9**: 345–53.

Olsson, O. and Holmgren, N.M.A. (2000) Optimal Bayesian foraging policies and prey population dynamics: Some comments on Rodriguez-Girones and Vasquez. *Theoretical Population Biology* **57**: 137–54.

Outreman, Y., Le Ralec, A., Wajnberg, É. and Pierre, J.S. (2005) Effects of within- and among-patch experiences on the patch-leaving decision rules in an insect parasitoid. *Behavioral Ecology & Sociobiology* **58**: 208–17.

Papaj, D.R. and Lewis, A.C. (eds.) (1993) *Insect Learning: Ecological and Evolutionary Perspectives*. Chapman & Hall, New York.

Pierre, J.S., van Baaren, J. and Boivin, G. (2003) Patch leaving decision rules in parasitoids: Do they use sequential decisional sampling? *Behavioral Ecology & Sociobiology* **54**: 147–55.

Pyke, G.H. (1978) Optimal foraging in hummingbirds: Testing the marginal value theorem. *American Zoologist* **18**: 739–52.

Pyke, G.H., Pulliam, H.R. and Charnov, E.L. (1977) Optimal foraging: A selective review of theory and tests. *Quarterly Review of Biology* **52**: 137–54.

Real, L.A. (1993) Toward a cognitive ecology. *Trends in Ecology & Evolution* **8**: 413–17.

Schoener, T.W. (1986) Mechanistic approaches to community ecology: A new reductionism? *American Zoologist* **26**: 81–106.

Stephens, D.W. (1993) Learning and behavioural ecology: Incomplete information and environmental predictability. In: Papaj, D.R. and Lewis, A.C. (eds.) *Insect Learning: Ecological and Evolutionary Perspectives*. Chapman & Hall, New York, pp. 195–218.

Stephens, D.W. and Charnov, E.L. (1982) Optimal foraging: Some simple stochastic models. *Behavioral Ecology & Sociobiology* **10**: 251–63.

Stephens, D.W. and Krebs, J.R. (1986) *Foraging Theory*. Princeton University Press, Princeton.

Tenhumberg, B., Keller, M.A. and Possingham, H.P. (2001) Using Cox's proportional hazard models to implement optimal strategies: An example from behavioural ecology. *Mathematical & Computer Modelling* **33**: 560–97.

Therneau, T.M. and Grambsch, P.M. (2000) *Modelling Survival Data, Extending the Cox Model*. Springer-Verlag, New York.

Tinbergen, N. (1963) On aims and methods of ethology. *Zeitschrift für Tierpsychologie* **20**: 410–33.

van Alphen, J.J.M. and Vet, L.E.M. (1986) An evolutionary approach to host finding and selection. In: Waage, J.K. and Greathead, D. (eds.) *Insect Parasitoids*. Academic Press, London, pp. 23–61.

van Alphen, J.J.M., Bernstein, C. and Driessen, G. (2003) Information use and time allocation in insect parasitoids. *Trends in Ecology & Evolution* **18**: 81–7.

van Roermund, H.J.W., Hemerik, L. and van Lenteren, J.C. (1994) Influence of intrapatch experiences and temperature on the time allocation of the whitefly parasitoid *Encarsia formosa* (Hymenoptera: Aphelinidae). *Journal of Insect Behaviour* **7**: 483–501.

Visser, M.E., van Alphen, J.J.M. and Nell, H.W. (1990) Adaptive superparasitism and patch time allocation in solitary parasitoids: The influence of the number parasitoids depleting a patch. *Behaviour* **114**: 21–36.

Volterra, V. (1926) Variazioni E Fluttuazioni Del Numero D'individui In Specie Animali Conviventi. *Memoria Academia Linnei di Roma* **2**: 31–113.

Waage, J.K. (1979) Foraging for patchily-distributed hosts by the parasitoid, *Nemeritis canescens*. *Journal of Animal Ecology* **48**: 353–71.

Wajnberg, É. (2004) Measuring genetic variation in natural enemies used for biological control: Why and how? In: Ehler, L., Sforza, R. and Mateille, T. (eds.) *Genetics, Evolution and Biological Control*. CABI Publishing, Wallingford, pp. 19–37.

Wajnberg, É. (2006) Time-allocation strategies in insect parasitoids: From ultimate predictions to proximate behavioural mechanisms. *Behavioral Ecology & Sociobiology* **60**: 589–611.

Wajnberg, É., Rosi, M.C. and Colazza, S. (1999) Genetic variation in patch time allocation in a parasitic wasp. *Journal of Animal Ecology* **68**: 121–33.

Wajnberg, É., Fauvergue, X. and Pons, O. (2000) Patch leaving decision rules and the Marginal Value Theorem: An experimental analysis and a simulation model. *Behavioral Ecology* **11**: 577–86.

Wajnberg, É., Curty, C. and Colazza, S. (2004) Genetic variation in the mechanisms of direct mutual interference in a parasitic wasp: Consequences in terms of patch-time allocation. *Journal of Animal Ecology* **73**: 1179–89.

Wald, A. (1947) *Sequential Analysis*. John Wiley & Sons, New York.

Weis, A.E. (1983) Patterns of parasitism by *Torymus capite* on hosts distributed in small patches. *Journal of Animal Ecology* **52**: 867–78.

Appendix: Bayes' theorem

Bayesian inference is a logical process involving the updating of the probability of a hypothesis about the state of nature. This process uses Bayes' theorem and, historically, was based on a subjective conception of probability. From the Bayesian point of view, probability is not interpreted as the limit of a frequency but rather as the numeric translation of a state of knowledge about nature, or as the degree of confidence that can be attributed to a hypothesis. Bayes' theorem is derived from the definition of conditional probability, the product rule for joint probability, and the addition rule for the probability of a union of events. The probability of an unknown state of nature S, conditioned on some information (the realization of an event) B, is given by

$$p(S|B) = \frac{p(S \cap B)}{p(B)} = \frac{p(S)p(B|S)}{p(B)} \tag{A16.1}$$

where $p(S)$ is the prior probability of S. $p(S \cap B)$ is the probability of the simultaneous occurrence of S and B. $p(S|B)$ is called the posterior probability of S conditioned on the occurrence of B. On this background, a decision theory (Bayesian decision theory) can be built by defining a decision criterion on $p(S|B)$ and by attributing costs to decision errors. A sequential decision procedure is achieved by reiterating Equation A16.1 for a series of successive events $\{B_1, B_2, \ldots, B_n\}$ until a given criterion is reached. A typical sequential procedure is that of Wald's (1947) decisional sampling based on the likelihood ratio of two hypotheses.

The Bayesian approach has been criticized by classical statisticians (frequentists), mainly because the prior did not seem in any way definable. As McNamara et al. (2006) pointed out, this is not a problem when applying Bayesian theory to animal behavior. The reason given is that evolutionary history and previous experience are likely to have determined well-defined priors. Genes are influenced by the main characteristics of the environment and they drive adapted behaviors. Individual experience and memory tune the behavior more accurately to the actual environmental characteristics. Another reason, which is more pragmatic, is that most decision processes are sequential, the posterior probability takes the place of the prior at the next iteration, and the sequence of posteriors tends toward a good estimate of environmental quality.

Equation A16.1 can be extended to probability density functions. If $f(x)$ is the density of the continuous variable x, then

$$f(x|B) = \frac{f(x \cap B)}{p(B)} = \frac{f(x)p(B|x)}{p(B)} \tag{A16.2}$$

where $f(x)$ is the prior density of x and $f(x|B)$ is the posterior density subject to the knowledge of B. This formula, widely used in Bayesian decision theory, is easily derived by definition of the density.

The analogy with animal decisions is straightforward in the case of an animal foraging for discrete prey or host items. The forager meets items at times $\{t_1, t_2, \ldots, t_n\}$, evaluates by some mechanism whether the patch is valuable or not, and decides to leave the patch when some criterion warns that patch profitability is too low. A useful metaphor is then to consider the forager as sampling its environment, and as making use of Bayesian decision. Looking for optimality in the decision criterion is a tool for understanding the design of foraging behavior. A question is to understand how Bayesian decision theory can lead to the optimization of foraging. This question involves the use of information by animals. If some fit appears between what this theory predicts and the actual behavior of the animal, this raises the question of the underlying mechanisms. Obviously, animals do not compute conditional probabilities explicitly, and thus we must investigate the underlying mechanisms and evaluate whether these mechanisms can mimic the result of the human calculation. By doing this, behavioral ecology tackles the question of the relations between ultimate and proximate causes of behavior.

17

Finding optimal behaviors with genetic algorithms

Thomas S. Hoffmeister and Éric Wajnberg

Abstract

Optimality models used in behavioral ecology, and especially on insect parasitoids, have taken a variety of approaches, from classical analytical tools to individual-based simulations. The increasing awareness that much of the observable behavior in parasitoids depends on the state of the insect (be it the physiological or informational state) has led to the increasing use of stochastic dynamic programming models. However, optimal behaviors of one individual often depend upon the behavior of conspecifics, further complicating the issue. While classical game theory may be applied when behavior is not state dependent, genetic algorithms (GA) provide a powerful way in finding optimal behaviors for situations where such optimal behavior depends upon an animal's state and on the frequency of alternative behaviors of conspecifics. More generally, GAs can be used when there is a need to find the optimal strategy among a number of different alternative behaviors whose number is far too great to be exhaustively checked. GAs are search algorithms that proceed in a fashion analogous to natural selection. They are individual-based simulations that identify optimal solutions by searching the enormous space of potential solutions mimicking the process of evolution in biological systems.

In this chapter, we discuss research questions in parasitoid behavioral ecology that might benefit from applying GA as a research tool for optimality models. Examples will be developed and explained. Among others, using the foraging problem of estimating habitat quality, we show how fast GAs find optimal solutions for parameter spaces that are impossible to solve numerically otherwise.

17.1 Introduction

Behavioral ecology studies the ecological and evolutionary basis of animal behavior and the roles of behavior in enabling an animal to adapt to its environment. Often, the aim is to try to understand what sort of optimal behavioral strategies animals should adopt, in a particular environmental situation, in order to maximize their production of progeny (i.e. their so-called 'fitness'). Along with other areas of evolutionary biology, behavioral ecology has incorporated a number of techniques, which have been borrowed from optimization theory. An optimized strategy is a strategy that offers the highest fitness return to an animal, given all the different factors and constraints it is facing. Such optimal behavioral rules are then tested experimentally on real animals, and this has especially been done several times on insect parasitoids, producing important and interesting results (see Godfray 1994, for several significant examples).

Several different methods are available to find optimal theoretical predictions, depending on the behavioral ecological problem to be solved. Most of the time, and especially for early work, models were sufficiently simple to be solved using standard maximization techniques. This is the case for most static, optimal foraging theoretical models in which the state of the animal remains unchanged during its foraging time and when there are no competitors trying simultaneously to maximize their own fitness in the environment (e.g. the marginal value theorem (MVT) of Charnov 1976). However, behavioral ecologists, and especially those working on insect parasitoids, are progressively addressing questions that are closer to real situations and that are correspondingly becoming increasingly complex and difficult to solve analytically. For example, if changes in the state of the animal during the foraging process are able to change the optimal behavioral decisions it should adopt, then stochastic dynamic programming models (SDPs) should be used to find the optimal trajectory of the foraging animal within the state-space, leading to maximal lifetime fitness gain (Clark & Mangel 2000, and see also Chapter 15 by Roitberg and Bernhard). If animals need to learn the different features of their habitat to behave in an optimal way and if they can do that by updating some prior information with current foraging experience, then so-called Bayesian approaches can be used (see also Chapter 16 by Pierre and Green). Also, if foraging animals are trying to maximize their fitness gain in the presence of competitors trying to reach the same goal, then models from game theory (Maynard-Smith 1982, Giraldeau & Caraco 2000) can be used to find the best behavioral strategy they should adopt, eventually leading to Evolutionarily Stable Strategies. Finally, these different approaches can be combined when the problem to be solved involves several of the corresponding situations they can address.

Finding optimal behavioral strategies with one or several of these methods is certainly the most elegant way to solve the questions being addressed but, in an increasing number of cases, the problems to be solved appear to be too complicated to find optimal solutions that way. This will be the case, for example, if the number of parameters (and/or their combination) is leading to an enormously large state-space of possible solutions, making it infeasible to search the state-space of these so-called NP-complete problems entirely (i.e. without using heuristics, see Garey and Johnson (1979)). In these cases, the optimal behavioral strategies animals should adopt can still be found by means of numerical methods. Among these methods, Genetic algorithms (GA) appears to be both the simplest one to be implemented on a computer and one of the most efficient to find a rapid and supposedly optimal solution. The aim of this chapter is to present what this numerical method is and how it works, but also, by means of several working examples, how it can be used to solve a variety of research questions that can be raised on the behavioral ecology of parasitoids.

GAs were invented in the early 1970s to mimic some of the processes observed in natural evolution. GAs, based on the observation that life with such a level of complexity was able to evolve in a relatively short time, use the power of evolution to solve optimization problems. The idea was invented and then initially explored by Holland (1975), with the idea of using mathematically-based artificial evolution as a method to conduct a structured search for solutions to complex problems. As stated by another pioneer in GA optimization, Goldberg (1989): 'Three billion years of evolution can't be wrong. It's the most powerful algorithm there is.' Hence, the design of the GA is inspired by Darwinian natural selection and biological genetic evolution. Thus, GAs include common ingredients present in real organisms that allow them to evolve, such as chromosomes, genetic recombination through crossing-over, and mutation, as well as forces acting on them such as fitness and selection.

Biologists willing to study in a formalized way the biological systems they are working on have used, since Leonardo da Pisa in the 12th century – and are still using – methods and concepts than have been designed by mathematicians. GAs are one of the few examples (with neural networks, Haykin 1999) where, in turn, biological concepts were used as a template for designing methods used by optimization theoreticians. However, despite being biologically-inspired, applications of GAs are mainly outside the field of biology, and it is only recently that such optimization methods started to be used to solve ecological research problems.

In the following, we will outline how a GA works, illustrating this with the example of the traveling salesman problem (TSP), a well-known example for a simple problem with a huge search space. Further, we will provide three examples from parasitoid behavioral ecology, in order to illustrate various technical approaches on different research topics.

17.2 Outline of a GA

Although a GA of the TSP contains some special features that differ from many other forms of a GA, we use this example since it represents a widely known optimization problem with a relatively huge state space of solutions that is almost impossible to solve by enumeration. In the TSP, there are N cities with a given distance between each of them. A traveling salesman starting at one of the cities, has to visit all of the N cities and then return to the first city. However, his task is to find a sequence of cities to minimize the distance traveled in total, without visiting a city more than once. To search the state space of possible solutions, we need to define some solution and calculate the traveled distance. In order to compare solutions, we define a given set of solutions and evaluate the performance of each of them. We will drop poor solutions and create new solutions. Since we create new solutions by altering existing ones, the new solutions only differ in small parts and we search the space around good solutions only. This process is called hill-climbing and is extremely efficient in finding optimal solutions.

This is essentially how a GA works. As you will see, the outline of the basic GA is very general. There are many parameters and settings that can be implemented differently in various problems. A GA typically comprises a population of individuals (or solutions to a problem) that exhibit heritable characters. These characters are numerically coded on a 'string' (Goldberg 1989), which is often called a 'chromosome' because of its intuitive conformity with the evolutionary coding of characters in organisms. Different 'traits' of an individual are coded on 'genes', and different 'character states' are coded as alternative 'alleles' of a gene. The numerical coding of character states may either be in binary code, or represented by integer or real values (Fig. 17.1).

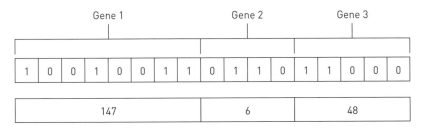

Fig. 17.1 Coding of 'genes' on a 'chromosome' for a GA, in binary (upper row) and integer (lower row) coding. Each 'gene' codes for a character state of the organism.

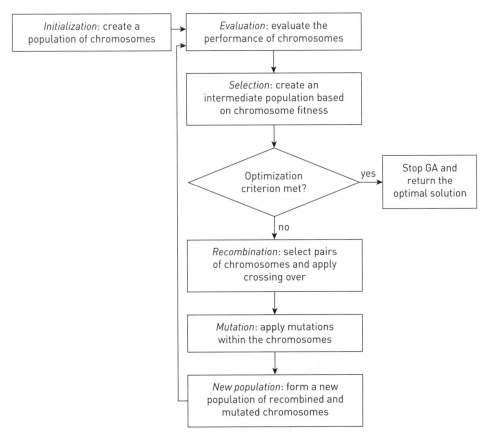

Fig. 17.2 Flow chart of a GA. After initialization, the process is iterated until the optimization criterion is met.

The first step in implementing a GA is the initialization of a population by generating a random population of n chromosomes (Fig. 17.2). Obviously, n should be suitable for the research question to be tackled. Obitko and Slavík (1999) suggest that a population of 20 to 30 chromosomes often performs best (Kumsawat et al. 2005), but some GA problems might require populations of 50 or 100 chromosomes. Generally, the number of chromosomes in the population should be related to the length of the chromosome, i.e. longer chromosomes require larger populations of chromosomes. Note that each chromosome represents an individual that can be tested against a fitness function or, more generally, against the problem defined in the GA. In our example of the TSP, the chromosome holds N genes with the number-code of one of N cities (each city occurring a single time) and the sequence of visits to cities is given by the sequence of genes on the chromosome. This type of coding is called permutation encoding.

The next step in a GA is the evaluation of each chromosome j using a fitness function $f(j)$, or – as is often the case in game theoretical or other frequency and density dependent approaches – a simulation where chromosome j plays against all other chromosomes in the population and the performance of each chromosome j is evaluated at the end of the simulation (Fig. 17.2). If a stochastic simulation (or so-called Monte Carlo, Rubinstein 1981) is used, the values reached in the evaluation of chromosome j will depend upon stochastic events and might vary noticeably between simulation runs. In order to achieve reliable values, the evaluation process may be replicated sufficiently often (Barta et al. 1997, Hoffmeister & Roitberg 1998, Reinhold et al. 2002) to receive a stable average performance value for a chromosome. In the case of the TSP, the performance of a chromosome does not depend on the coded behavior of alternative chromosomes and thus the total distance traveled on the journey between cities only depends upon the sequence of cities that are visited and the distances between cities. Typically, the position of cities would be coded as the x/y-position on a grid, the sequence of cities that are visited is read from the chromosome, and the distances traveled are summed up across all N cities visited.

While the performance values of chromosomes in the evaluation process are absolute values, in the next step of the GA, a selection process among chromosomes is performed that is based upon relative fitness values, rather than the absolute performance of a chromosome. In this step an intermediate population is formed from the original population of chromosomes, where the number of copies at which a given chromosome is represented depends upon the fitness of the chromosome.

The next question is how to select parents for the next generation. The general idea of the selection process is that chromosomes with low fitness will be eliminated and chromosomes with high fitness will spread in the population by bestowing their code to multiple copies. This can be done in many ways. The most basic approach is called 'roulette wheel selection'. Imagine all genotypes (represented by chromosomes) are arranged in a pie chart in slices of a size proportional to their relative fitness. Around that pie chart, a roulette wheel is placed. For population size n, n spins of the roulette wheel are performed and the chromosome at which the pointer stops is selected with replacement and introduced to the intermediate population (Fig. 17.3a). That way, chromosomes with high fitness values will be represented with numerous copies, while poorly performing chromosomes will be represented with only a few copies or no copy at all.

However, due to the stochastic nature of roulette wheel selection, the representation of chromosomes might not match exactly the relative fitness (= (r/sum of fitnesses) × 100, with r being the fitness of chromosome i) of each chromosome. An improvement to this selection process is the stochastic remainder selection (Goldberg 1989). Imagine a population of $n = 100$. When the total sum for the fitness is set to 100, a chromosome j_1 with relative fitness 1.4 would receive 1 copy in the intermediate population, a chromosome j_2 with relative fitness 5.9 would receive 5 copies, and so on. In a first step of selection, these

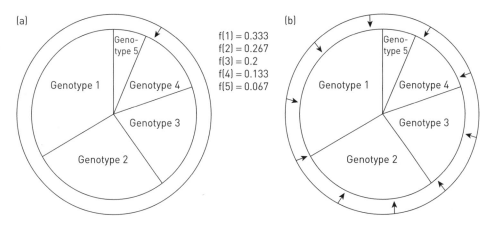

Fig. 17.3 Selection process of a GA by (a) roulette wheel selection or (b) stochastic remainder selection. Values in the center of the graph represent relative fitness values of the five genotypes present in the parental generation. The pointers show selected genotypes. The wheel is spun n times in (a) (with n = population size), whereas n equally spaced pointers are spun once in (b) (Aarsen 1983).

integer parts of the fitness are allocated. Only the non-integer parts of the relative fitness (0.4 for j_1, 0.9 for j_2 etc.) would enter the roulette wheel selection in a second step of selection as described above. The same gain in precision is reached with the so-called universal selection. Universal selection can be implemented by placing n equally spaced pointers at the roulette wheel on which the genotypes are arranged according to their proportional fitness. A single spin of the wheel determines all selected chromosomes. For each pointer, a single chromosome is created from the genotype the pointer aims at (Fig. 17.3b).

If there are huge differences between fitness values, few other than the best chromosome will have a chance to be selected in the above-mentioned process. Under such conditions, it is advisable to use rank selection rather than a selection based on proportional fitness values. Otherwise, the optimization process might converge to a local optimum. In rank selection, the worst chromosome receives fitness 1, the second worst receives fitness 2, and so on, and the best chromosome receives fitness n. This improves the chances of low-fitness chromosomes being selected at the cost of the speed at which the GA converges to the optimal solution. However, the risk that the GA is caught on a local optimum should be decreased. A further alternative method is steady state selection where a large number of chromosomes with high fitness values are retained and a number of low fitness values are deleted, and the intermediate population is filled up with copies of the retained chromosomes.

One method that distinctly improves the performance of GAs is the introduction of elitism into the selection process. The following step of recombination and mutation in a GA, discussed below, includes the risk that the best chromosomes might be destroyed in the process. Elitism is the fact that the best or a few best chromosomes are retained unaltered in the next generation. If j elite chromosomes are selected, the intermediate population then has the size $n-j$ chromosomes and is produced by one of the above-mentioned selection schemes from the entire set of chromosomes in the population, including those that have been selected as elite chromosomes. In the case of our TSP example, we might use elitism combined with roulette wheel selection. From our population of 30 chromosomes we retain the two chromosomes with the highest fitness values and form an intermediate population of 28 chromosomes, based upon the relative fitness of all 30 chromosomes. This intermediate population of 28 chromosomes now undergoes recombination and mutation, which is the next step in a GA.

Recombination by genetic operators, such as crossing over and mutations, form the core of GA and are applied in a probabilistic fashion. Crossing over combines elements of two chromosomes in a filial chromosome by exchanging substrings and can, for instance, combine adaptive character states from different genes. Mutations alter individual genes on a chromosome and both crossing over and mutation help to search the previously unsearched state space. While crossing over in a population that converges to the optimum will often lead to no changes (two similar chromosomes are paired and exchange sequences), mutations always change the code of a chromosome. Usually, crossing over rates chosen in GAs are generally high, for example $p_{cross} = 0.8$, while the probability that a mutation occurs at a given position of a chromosome should be low, e.g. $p_{mut} = 0.01$. However, there is no fixed value for either crossing over or mutation rate that performs best and optimal rates are a matter of selection by trial and error for each GA. There is a number of different ways of applying crossing over and mutation, and the following description presents only some possibilities.

To employ crossing over, two chromosomes from the intermediate population are randomly drawn. With probability p_{cross}, crossing over occurs at a randomly chosen position on the chromosome. In binary coded chromosomes, the crossing over point may be at any position, i.e. also within genes, while in chromosomes coded in integer or real values, the crossing over point will always be between genes. In crossing over, the sequence of the first chromosome, up to the position of crossing over, is written on the filial chromosome and combined with the sequence of the second chromosome from the position of crossing over until the end of the chromosome, and vice versa (Fig. 17.4).

An exception to this mechanism can be found in chromosomes with permutation encoding, as in our example of the TSP. Since each city can only occur once, the sequence of cities up to the point of crossing over is taken from the first chromosome. Following, the remaining slots on the filial chromosome are filled by selecting cities from the second chromosome in the sequence they occur in this chromosome, if they have not yet been occurring in the sequence on the filial chromosome. For example, let the first and second chromosomes having a length of 10 and the sequences 1, 2, 3, 4, 5, 6, 7, 8, 9, 10 and 1, 3, 5, 7, 9, 2, 4, 6, 8, 10. With crossing over after position 6 we get for the filial chromosome A: 1, 2, 3, 4, 5, 6, 7, 9, 6, 8, 10 (sequence from second parental chromosome underlined). The next step in a GA is to incorporate mutations which are applied with p_{mut}, at each position of a chromosome. In binary encoded chromosomes, a mutation changes the allele from 0 to 1 or 1 to 0. In value encoded chromosomes, a new value is either chosen randomly within the range of possible values or the current value is changed by a small increment or decrement. Whether the random value should be drawn from a uniform or non-uniform distribution depends on the phenotypic behavior the gene is coding for. For example, if residence times of a parasitoid (that are usually skewed to the right) are considered, it makes intuitive sense to draw values from an exponential distribution rather than a uniform distribution. Our example of the TSP again requires a different scheme. Since the number of a given city cannot be changed to another value and since all cities are to be represented a single time on the chromosome, for each mutation, two positions

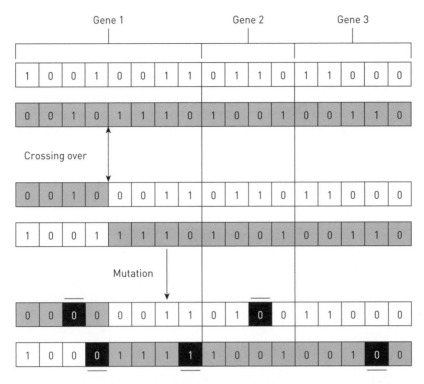

Fig. 17.4 Recombination and mutation in a GA: daughter chromosomes are created from parent chromosomes by crossing over and mutation.

on the chromosome are randomly drawn and the numbers are swapped, thereby altering the sequence of cities to be visited.

After applying crossing over and mutations, a new filial population of chromosomes is created that contains those chromosomes of the parent generation that had high fitness values, but many of these will be slightly altered through crossing over and mutation. By applying selection and the genetic operators of crossing over and mutation, we generate a new set of individuals that have a better than average chance of performing well. By closing the generational loop in a GA (Fig. 17.2) and iterating the process of evaluation, selection, and recombination across many generations, the overall fitness of the population improves as the GA converges to the optimal solution. The obvious next question is when should a GA be stopped? Again, a number of possibilities exist. GAs may either be run for a fixed number of generations (Barta et al. 1997, Bouskila et al. 1998, Reinhold et al. 2002, Perry & Roitberg 2005) or for a fixed computing time. An alternative to these arbitrary stop criteria is to analyze the performance variation of the GA. The GA might be stopped, if a given number of generations has passed with no improvement in the fitness of the best chromosome, or if the improvement in the fitness of the best chromosome has not exceeded a certain increment within a specified number of generations. In our example of the TSP, we stop the GA after 400 generations. As can be seen in Fig. 17.5, the distance traveled decreases most rapidly within the first few generations of the GA runs and quickly converges on a minimum value for the traveled distance that is stable from

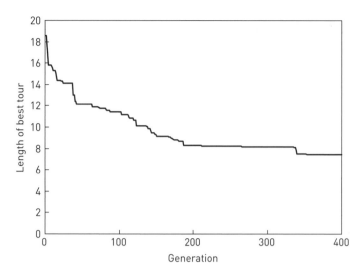

Fig. 17.5 Evolution of the length of the best tour across generations of the GA for the TSP. The most drastic reduction in tour length is achieved in the first couple of generations by the optimizing algorithm.

generation 353 onward. Likewise, the number of genotypes, i.e. chromosomes with different encoded sequences, which is close to the population size in the beginning of the GA, also decreases across generations. Note that always some variability in chromosomes will be retained due to crossing over and mutation events.

17.3 A first example of parasitoid behavioral ecology: Learning and optimal patch visitation

Most of insect parasitoids are exploiting hosts that are patchily distributed in the environment (Godfray 1994). There is now a huge amount of both theoretical and experimental work that have been developed to understand what should be the optimal time female parasitoids should invest in each host patch they are exploiting before leaving to find another patch in the environment (Wajnberg 2006, see also Chapter 8 by van Alphen and Bernstein). For this reason, several optimal foraging theoretical models were built in order to understand what behavioral strategy should be adopted in order to maximize that rate at which hosts are encountered and exploited per time unit.

In this respect, one of the most important rate maximization models is the MVT (Charnov 1976) that predicts the optimal time a foraging animal should remain on a host patch (Wajnberg 2006, see also Chapter 8 by van Alphen and Bernstein). This model is based on several assumptions that are probably unrealistic, the most important one being that animals are omniscient, having complete knowledge of the quality of all available patches in the environment and of the mean time needed to reach them (Stephens & Krebs 1986).

This assumption was addressed by several authors, which considered that animals sample patches in the habitat and learn progressively from both their profitability and interpatch travel times. This is the case for the learning process proposed by McNamara and

Houston (1985), which appears to be both simple and efficient. In their model, the foraging animal initially does not know the quality of the available patches in its habitat and the travel times needed to reach them, but is able to collect such information progressively through the following learning process: patches of different quality are visited sequentially, t_n is the time spent on the nth patch, R_n is the fitness acquired on that patch, and τ_n is the time spent traveling between the nth and $n+1^{th}$ patch. Let $T_n = t_n + \tau_n$, so $T_1 + T_2 + \ldots + T_n$ is the total time between arrival on the first patch and arrival on the $n+1^{st}$. It is supposed that the animal has some initial constant values R_0 and T_0 ($R_0, T_0 > 0$) and thus has an initial estimate $\gamma_0 = R_0/T_0$ of the average profitability of its environment. According to the MVT, the animal should leave the first patch encountered when its rate of fitness acquisition drops to γ_0. When it arrives on the next patch, T_1 and R_1 will be known and an updated estimate γ_1 can be computed that will be use to determine when this other patch should be abandoned, and so on. Then, the whole learning process works iteratively according to the two following process: (i) On arriving on the ith patch, and only at that time, the animal computes an estimate of average environment profitability γ_i where

$$\gamma_i = \frac{R_0 + R_1 + \ldots + R_i}{T_0 + T_1 + \ldots + T_i} \qquad (17.1)$$

and (ii) the animal leaves this i^{th} patch when its rate of fitness acquired on that patch reaches γ_i.

McNamara and Houston (1985) demonstrated that animals using such a simple learning rule asymptotically (i.e. when $n \to \infty$) converge to the optimal patch allocation strategy predicted by the MVT (i.e. γ_i converges to the true average profitability of the environment γ^*), even if there are different patch qualities in the habitat. The rate of convergence is lower if there is an important variation in patch quality. However, in practical situations, female parasitoids are not sampling an infinite number of patches in their lifetime and might not reach an optimal behavioral strategy when adopting such a learning rule. In this case, as noted by McNamara and Houston (1985), the order in which the different patch types are visited influences the estimates γ_i and thus influences both the time animals will spend on each visited patch and their total fitness gain.

Insect parasitoids are short-lived animals and this is known to influence their patch time allocation foraging strategy (Wajnberg et al. 2006). Thus, they are obviously not visiting an infinite number of patches in their lifetime and, if we assume that they are using the learning rule proposed by McNamara and Houston (1985) in an environment in which several patch qualities can be found, we can ask the question of what should be the optimal order in which all patches should be visited in order for them to maximize their fitness gain. When just a low number of patches are visited, a stochastic dynamic programming model (SDP) can be used to find the optimal visiting order (see also Chapter 15 by Roitberg and Bernhard). However, when the number of patches the animal can visit increases, and especially if there is a large number of patch qualities in the habitat, the number of possible combinations will be too large and a GA should be used instead to find the optimal solution. As an example, suppose that a parasitoid female is able to visit 20 patches in her lifetime, and that the 20 patches are all of different quality. There are thus 20! (i.e. about 2.43×10^{18}) possible orders in which the patches can be visited. If a computer is able to evaluate the fitness gained for 100 of these orders per second, it will need more than 770 millions of years to perform all of the computations!

Thus, we developed a GA to find the optimal visiting sequence of 20 patches of different quality. The GA we used for this is similar to the one used to solve the classical TSP presented above. The differences are that: (i) it is the optimal order of patches, and not cities, that we look for: and (ii) the fitness of each genotype does not correspond to the total distance traveled between cities, but to the final γ_i after the 20th patch visited is left. In our example, the quality (in terms of number of hosts to be attacked) of the ith patch is $10 \times i$, and the female is traveling exactly 100 time steps between each patch before reaching each of them. We designed a GA with 50 chromosomes, each bearing 20 genes indicating in what order the different patches should be visited (i.e. so-called permutation encoding). In order to compute the fitness of each chromosome, we used the learning process proposed by McNamara and Houston (1985) with $\gamma_0 = 0.1$, and the fitness gain on each patch was computed as in Wajnberg et al. (2006), using iteratively a Holling (1959) Type II functional response, with attack rate $a = 0.01$ and handling time $Th = 0.1$. At each generation, the two best chromosomes were kept and used for the next generation (i.e. elitism), and each pair of chromosomes experienced a crossing over event at a rate of 80%. In order to always have all patches visited, and after a location is drawn randomly, a crossing over consisted of copying a part of the chromosome of the first parent (i.e. part of the patch order) and the rest of the patches was taken in the same order as in the second parent. Finally, in order to increase the convergence rate of the GA, mutations were allowed using a systematic process: patches were systematically chosen and exchanged if it led to an increase in the resulting fitness. After 300 generations, the best, although not necessary optimal order in which to visit the 20 patches in order to obtain the highest fitness gain rate, is shown in Fig. 17.6.

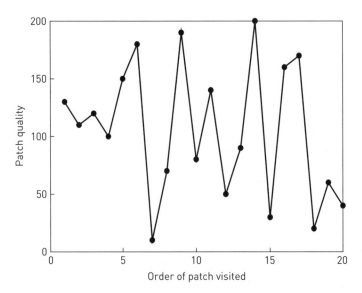

Fig. 17.6 Sequence of 20 visited patches of different quality leading to the higher fitness rate gain, produced by a GA, for female parasitoids learning progressively the quality of their habitat, using the McNamara and Houston's (1985) learning process.

The best visiting sequence thus seems to be to visit first medium-quality patches. Doing this, females are most likely getting an estimate of the average quality of the habitat that is as close as possible to the true average profitability of their environment. Then, females should progressively and alternatively sample patches of higher and lower quality, leading them to refine their first estimate and then progressively approach the true value of average habitat profitability.

17.4 A second example of parasitoid behavioral ecology: optimal patch time allocation for competing parasitoids with a limited time horizon

As already explained above, the question regarding optimal patch time allocation for female parasitoids foraging for hosts has arguably been the most studied problem in behavioral ecology over the past few decades (van Alphen et al. 2003, Wajnberg 2006). We have seen that the optimal time females should spend on each patch of hosts is given by the MVT (Charnov 1976), but we also saw that such a theoretical maximization model suffers from being based on several assumptions that are most likely unrealistic. Besides considering that female parasitoids are omniscient and should thus know the quality of all available patches in the environment, and on the mean time needed to reach them (see the GA we developed above to see how such an assumption can be addressed), this model also supposes: (i) that females never reach the end of their life (i.e. infinite time horizon); and (ii) that there is no competition because females are supposed to forage alone in each patch (Yamamura & Tsuji 1987).

The effect of relaxing the assumption that foraging parasitoid females are living during a never-ending period has been addressed theoretically using an SDP model by Wajnberg et al. (2006). They demonstrated that time-limited females should spend a longer time on each visited patch than what is predicted by Charnov's (1976) MVT and the deviation should increase when females are approaching the end of their life. The effect of competition was addressed using game theoretical models by Sjerps and Haccou (1982), Haccou et al. (1999), Haccou et al. (2003), and Hamelin et al. (2007a,b). These authors demonstrated that, in most cases, females are engaged in a war of attrition type when foraging simultaneously on a host patch (see also Chapter 9 by Haccou and van Alphen).

Building a theoretical deterministic model, while relaxing simultaneously both the assumptions of no time horizon and no competition, will obviously be a difficult task, if even feasible. Thus, we will consider here, as a simplified example, how such a problem could be approached by using a GA. As usual, the main difficulty in developing a GA is to find a way to encode the question to be solved in terms of a number of genes along chromosomes. For the sake of simplicity, we will consider that:

1 all patches in the environment are of the same quality;
2 the travel time between patches is fixed;
3 all foraging females live for the same finite time; and
4 each female parasitoid is foraging in the environment either alone or in the presence of a single competitor.

The genes on chromosomes should code for the behavior of the female in these two different situations. A possible, simplified way to code such a situation can be to have six genes, three coding the phenotype of the female when it is alone, the other three coding the same phenotype when it is with a competitor. In each of these two cases, one gene is coding for the time the female will invest on each patch she has just encountered. A second gene is coding for a switching time after which the (additional) effect of a third gene, in terms of patch residence time, is taken into account. Doing so, the optimal increase in patch residence time, when a time-limited female is approaching the end of her life (Wajnberg et al. 2006), is explicitly considered. The six genes are coding for phenotypes on a continuous scale, with the only restriction being that all values should be positive (or null). So, the mutation of a gene will consist of replacing its value by a value randomly drawn from an exponential or a Weibull distribution (for patch times) or from a uniform distribution (for switching times). Moreover, the foraging female does not know what behavioral strategy its competitor (if any) – who is also a time-limited forager trying to maximize its own fitness – is adopting. She is also unaware of the probability that a competitor will be foraging simultaneously on the same patch. Thus, we are in the above-mentioned situation in which the fitness of each chromosome in the population cannot explicitly be computed, so Monte Carlo simulation should be used instead. For this, at each generation, the fitness of each chromosome will be estimated by its average fitness computed in several (say a hundred) simulated situations. For each of them, another chromosome is drawn randomly in the population and used to code for the behavioral strategy played by the competitor. Also, for each sequentially visited patch, the presence of the competitor is drawn randomly with a given, fixed probability, which should ideally be around 0.5 in order to have both parts of the chromosome evolving at the same speed.

Running such a GA will provide interesting information about both the optimal time females should invest on a patch, with or without the presence of a competitor, and when and what should be the optimal changes in patch residence upon approaching the end of their life, comparatively in the two situations.

17.5 A third example of parasitoid behavioral ecology: optimal mark persistence in a parasitoid threatened by hyperparasitoids

Many animals leave chemical information in the environment, either in order to communicate with conspecifics or to mark sites for future recognition. The latter occurs frequently in herbivorous and parasitoid insects that use small and discrete breeding sites for the production of offspring. While early investigators focused on the deterring effects that such substances might have on conspecifics searching for oviposition sites (Prokopy et al. 1977, Boller 1981, Prokopy 1981, McNeil & Quiring 1983, Lou & Wang 1985), an evolutionary approach determined that the evolutionary benefit of marking depends upon the fact that a female may encounter her own marking and might use it to avoid wasting eggs and time on resources that have already been exploited (Roitberg & Mangel 1988). Conspecifics encountering such a mark are informed about the current quality of the resource and may adaptively decide to superparasitize or to refrain from oviposition (Roitberg & Mangel 1988, van Alphen & Visser 1990, Nufio & Papaj 2001, Hoffmeister & Roitberg 2002). As long as the behavior of an animal in response to encountering a mark is in the interest of the marker, the marking behavior is of mutual benefit. However, marks, and especially external marks that are placed on rather than into the resource, are an example of public information (information derived from the behavior of other individuals, see Danchin et al. 2004, Dall et al. 2005) that can be exploited by lower (Hoffmeister & Roitberg 1997) and higher trophic levels (Prokopy & Webster 1978, Roitberg & Lalonde 1991, Baur & Yeargan 1994, Hoffmeister & Gienapp 1999).

Thus, marking behavior does not necessarily lead only to fitness increments but may in turn incur costs, for example, through an increased mortality risk of the offspring developing in the marked substrate. Given that the marking pheromone can evolve toward higher or lower persistence, we may predict that the detectability of a marking pheromone will be related to the costs and benefits that a mark has for the marking individual. Unfortunately, the persistence of marking pheromones has been analyzed only in a small number of cases, yet variation in persistence may suggest that the residual activity of a marking pheromone is subject to evolutionary selection (Quiring & McNeil 1984, Averill & Prokopy 1987, Höller et al. 1991). This sets the stage for a game between trophic levels, since the cost of marking may depend on the frequency of marking individuals in the population and thus, the density of marked and non-marked resources an organism from a higher trophic level might exploit. Hoffmeister and Roitberg (1998) use such logic for an evolutionary game between a marking herbivore host and a mark-exploiting parasitoid of the herbivore offspring that they modeled using a GA. We can easily translate this problem into a GA for a game between a parasitoid marking the oviposition site externally and its hyperparasitoid exploiting the marks to find the primary parasitoid's offspring.

Let us assume that the offspring of a primary parasitoid is vulnerable against competition with conspecifics. The degree of vulnerability may change over time with older larvae being competitively superior to younger larvae. Thus, the mortality risk through a second oviposition into the host, resulting in competition, might be equal for both contestants (50%) right after the oviposition and show an exponential decay over time for the first larva afterwards. Consequently, the benefit of a marking pheromone to a female's offspring is highest directly after oviposition and decays over time. For the female herself, the benefit of the mark stems not only from the fact that she does not kill her own first offspring in sibling competition, but also from the fact that she can avoid ovipositing into previously used hosts, thereby exposing a second offspring to sibling competition. Thus, the female also benefits from the mark if her first offspring is no longer vulnerable to competition.

However, a female parasitoid might not stay at the same site for her entire life and, over time, the probability of revisiting a host she has parasitized some time ago will also follow an exponentially decreasing function. Consequently, the fitness return from a mark will be highest for a female directly after oviposition and decay over time. How steep the decay function will be depends on the fate of her offspring in competition and her spatial searching behavior. The costs of a marking pheromone might be twofold. First of all, they impose costs related to the production of the chemical, and chemical substances less vulnerable to oxidation might be more costly to synthesize than others. Second, costs do accrue from mark-exploiting hyperparasitoids, i.e. natural enemies of the parasitoid that search more intensely in areas where they detect marking pheromone of the parasitoid. Again, these costs depend on the window of vulnerability of the primary parasitoid's offspring. For a chemical substance, we may assume that the decay of the marking pheromone follows first-order kinetics. Thus, the detectability of the marking pheromone M_t at time t is a function of the initial detectability β at t_0 and the decay rate τ, thus:

$$M_t = \beta e^{-t/\tau} \qquad (17.2)$$

To allow the analysis of the full state space of possible optimal decay rates of the marking pheromone, we define a gene with a binary character state for marking or non-marking and a second gene for the decay rate τ, with 30 discrete character states in value encoding.

Because the decision whether or not to accept a previously parasitized host for oviposition is a function of a parasitoid's life expectancy and egg load (van Alphen & Visser 1990, Minkenberg et al. 1992, Fletcher et al. 1994), we need two genes for a reaction norm of a parasitoid's tendency to accept previously parasitized and marked hosts as a function of the above-mentioned parameters. While the acceptance rate of marked hosts will generally increase with increased time and egg limitation, the gene may hold a variety of reaction norms that differ in the form (concave or convex) and the steepness of the function (Hoffmeister & Roitberg 1998). Since it is unknown to a parasitoid whether or not she or a conspecific will return to a marked site and at which physiological state, the number of possible combinations, and thus the search space for the benefit of marking is enormous. Further, the probability that a marked site is detected and exploited by hyperparasitoids will be dependent on the rate of marking and the detectability of marks in the parasitoid population. If few marks exist and the hyperparasitoid invests more searching time on marked compared to unmarked sites (Roitberg & Lalonde 1991, Hoffmeister & Gienapp 1999), hyperparasitoid attack will concentrate on marked sites, because more time will be invested in searching for rather than on sites. In contrast, if most genotypes in a population of parasitoids apply marks of high detectability, each marked site will benefit from a dilution effect. Consequently, the parasitoids in our scenario evolve in the frequency-dependent context of an evolutionary game between trophic levels and the search space for optimal solutions is too large to employ a deterministic fitness function. Since we need to use a stochastic fitness function, it will be necessary to employ a repeated evaluation GA, where averaging across 100 Monte Carlo repeats of the same generation leads to reliable fitness estimates for coded strategies.

In the simulation we may use a population of 100 parasitoids that, in the starting condition, represents a random draw from the population of possible genotypes. Each parasitoid has an egg load of 50 eggs and searches and oviposits in a habitat containing 4000, 8000, or 12,000 hosts, thus different scenarios cover a range of conditions from high to low competition for hosts. To evaluate the effect of marking costs we employ scenarios with no, intermediate, and high attack rate of hyperparasitoids, each run for 1000 generations and replicated with 5 different starting conditions, to see whether the GA converges repeatedly on the same strategy. Running such a GA elucidates how environmental characteristics shape the use of a chemical signal in a frequency and density dependent way and points toward important sources that select for the variability of marking systems across a range of ecological scenarios.

17.6 Conclusion

GAs are powerful tools to find optimal behavioral strategies for research questions, whose answers lie within a huge search space. While deterministic models are certainly the most elegant way to elucidate optimal solutions for theoretical approaches in behavioral ecology, their scope is limited to research questions tackling a small-sized search space of solutions. We feel that there is a large array of research questions in parasitoid behavioral ecology where approaches would benefit from including state-dependence of optimal behavioral decisions, as well as the fact that many such decisions are made in a frequency-dependent game context within and between trophic levels. Naturally, both these mechanisms increase the search space of solutions considerably. When investigating optimal life history strategies in herbivores, Bouskila et al. (1998) found importantly different solutions when modeling a dynamic life history problem in a game context compared to a non-game context. However, few papers exist to date that used GAs to solve a research question in parasitoid behavioral ecology (Hoffmeister & Roitberg 1998, McGregor & Roitberg 2000). GAs are certainly not the only approach that can be used to find optimal behavioral solutions in a dynamic game context, but for research questions with a huge search space of possible solutions they represent an efficient means of finding good solutions. As has been described above and can be seen from the examples given here, GAs cannot only be applied to a variety of research questions, but there is also a variety of ways to employ a GA. Such variability in available methods and the fact that optimizing a GA is subject to fine tuning, often by trial and error, may account for the fact that GAs have been rarely used by parasitoid behavioral ecologists. Moreover, GA is simulation-based, and simulations tend to be viewed as less elegant compared to deterministic approaches. Yet, GAs have been shown to be a very powerful method to search for optimal solutions within enormous and often complicated search spaces. By running them repeatedly with different starting conditions, it is relatively easy to demonstrate whether or not the algorithm found the optimum of the search space. Thus, GAs offer a great potential to research questions in parasitoid behavioral ecology and, with an increased complexity of approaches used in experiment and theory, it is likely that they will become a widely used research tool for parasitoid behavioral ecologists.

References

Aarsen, L.W. (1983) Ecological combining ability and competitive combining ability in plants: Toward a general theory of coexistence in systems of competition. *American Naturalist* **122**: 707–31.

Averill, A.L. and Prokopy, R.J. (1987) Residual activity of oviposition-deterring pheromone in *Rhagoletis pomonella* (Diptera: Tephritidae) and female response to infested fruit. *Journal of Chemical Ecology* **13**: 167–77.

Barta, Z., Flynn, R. and Giraldeau, L.A. (1997) Geometry for a selfish foraging group: A genetic algorithm approach. *Proceedings of the Royal Society of London Series B Biological Science* **264**: 1233–8.

Baur, M.E. and Yeargan, K.V. (1994) Developmental stages and kairomones from the primary parasitoid *Cotesia marginiventris* (Hymenoptera: Braconidae) affect the response of the hyperparasitoid *Mesochorus discitergus* (Hymenoptera: Ichneumonidae) to parasitized caterpillars. *Annals of the Entomological Society of America* **87**: 954–61.

Boller, E.F. (1981) Oviposition-deterring pheromone of the European cherry fruit fly: Status of research and potential applications. In: Mitchell, E.R. (eds.) *Management of Insect Pests with Semiochemicals: Concepts and Practice.* Plenum Press, New York, pp. 457–62.

Bouskila, A., Robinson, M.E., Roitberg, B.D. and Tenhumberg, B. (1998) Life-history decisions under predation risk: Importance of a game perspective. *Evolutionary Ecology* **12**: 701–15.

Charnov, E.L. (1976) Optimal foraging: the marginal value theorem. *Theoretical Population Biology* **9**: 129–36.

Clark, C.W. and Mangel, M. (2000) *Dynamic State Variable Models in Ecology – Methods and Applications.* Oxford University Press, New York.

Dall, S.R.X., Lotem, A., Winkler, D.W. et al. (2005) Defining the concept of public information. *Science* **308**: 353–6.

Danchin, E., Giraldeau, L.A., Valone, T.J. and Wagner, R.H. (2004) Public information: From noisy neighbors to cultural evolution. *Science* **305**: 487–91.

Fletcher, J.P., Hughes, J.P. and Harvey, I.F. (1994) Life expectancy and egg load affect oviposition decisions of a solitary parasitoid. *Proceedings of the Royal Society of London Series B Biological Science* **258**: 163–7.

Garey, M.R. and Johnson, D.S. (1979) *Computers and Intractability: a Guide to the Theory of NP-completeness*. W.H. Freeman, San Francisco.

Giraldeau, L.A. and Caraco, T. (2000) *Social Foraging Theory*. Princeton University Press, Princeton.

Godfray, H.C.J. (1994) *Parasitoids. Behavioral and Evolutionary Ecology*. Princeton University Press, Princeton.

Goldberg, D.E. (1989) *Genetic Algorithms in Search, Optimization, and Machine Learning*. Addison Wesley, Reading.

Haccou, P., Sjerps, M. and van der Meijden, E. (1999) To leave or to stay, that is the question: Predictions from models of patch leaving strategies. In: Olff, H., Brown, V.K. and Drent, R.H. (eds.) *Herbivores: Between Plants and Predators*. Blackwell Science, Oxford, pp. 85–108.

Haccou, P., Glaizot, O. and Cannings, C. (2003) Patch leaving strategies and superparasitism: An asymmetric generalized war of attrition. *Journal of Theoretical Biology* **225**: 77–89.

Hamelin, F., Bernard, P., Nain, P. and Wajnberg, É. (2007a) Foraging under competition: Evolutionarily stable patch-leaving strategies with random arrival times. 1. Scramble competition. In: Jørgensen, S., Quincampoix, M. and Vincent, T. (eds.) *Advances in Dynamic Games and Applications*, vol. 9, Annals of the International Society of Dynamic Games. Birkhäuser, Basel, pp. 327–48.

Hamelin, F., Bernard, P., Shaiju, A.J. and Wajnberg, É. (2007b) Foraging under competition: Evolutionarily stable patch-leaving strategies with random arrival times. 1. Interference competition. In: Jørgensen, S., Quincampoix, M. and Vincent, T. (eds.) *Advances in Dynamic Games and Applications*, vol. 9, Annals of the International Society of Dynamic Games. Birkhäuser, Basel, pp. 349–65.

Haykin, S. (1999) *Neural Networks: a Comprehensive Foundation*, 2nd edn. Prentice-Hall, Englewood Cliffs.

Hoffmeister, T.S. and Roitberg, B.D. (1997) Counterespionage in an insect herbivore – parasitoid system. *Naturwissenschaften* **84**: 117–19.

Hoffmeister, T.S. and Roitberg, B.D. (1998) Evolution of signal persistence under predator exploitation. *Ecoscience* **5**: 312–20.

Hoffmeister, T.S. and Gienapp, P. (1999) Exploitation of the host's chemical communication in a parasitoid searching for concealed host larvae. *Ethology* **105**: 223–32.

Hoffmeister, T.S. and Roitberg, B.D. (2002) Evolutionary ecology of oviposition marking pheromons. In: Hilker, M. and Meiners, T. (eds.) *Chemoecology of Insect Eggs and Egg Deposition*. Blackwell Wissenschaftsverlag, Berlin, pp. 319–47.

Holland, J.H. (1975) *Adaptation in Natural and Artificial Systems*. The University of Michigan Press, Ann Arbor.

Höller, C., Williams, H.J. and Vinson, S.B. (1991) Evidence for a two-component external marking pheromone system in an aphid hyperparasitoid. *Journal of Chemical Ecology* **17**: 1021–35.

Holling, C.S. (1959) Some characteristics of simple types of predation and parasitism. *Canadian Entomologist* **91**: 385–98.

Kumsawat, P., Attakitmongcol, K. and Srikaew, A. (2005) A new approach for optimization in image watermarking by using genetic algorithms. *IEEE Transactions on Signal Processing* **53**: 4707–19.

Lou, H.Z. and Wang, X.S. (1985) An oviposition deterrent extracted from the larval frass of *Ostrinia furnacalis*. *Chinese Journal of Biological Control* **1**: 53.

Maynard-Smith, J. (1982) *Evolution and the Theory of Games*. Cambridge University Press, Cambridge.

McGregor, R.R. and Roitberg, B.D. (2000) Size-selective oviposition by parasitoids and the evolution of life-history timing in hosts: Fixed preferences vs frequency-dependent host selection. *Oikos* **89**: 305–12.

McNamara, J.M. and Houston, A.I. (1985) Optimal foraging and learning. *Journal of Theoretical Biology* **117**: 231–49.

McNeil, J.N. and Quiring, D.T. (1983) Evidence of an oviposition-deterring pheromone in the alfalfa blotch leafminer, *Agromyza frontella* (Rondani) (Diptera: Agromyzidae). *Environmental Entomology* **12**: 990–2.

Minkenberg, O.P.J.M., Tatar, M. and Rosenheim, J.A. (1992) Egg load as a major source of variability in insect foraging and oviposition behavior. *Oikos* **65**: 134–42.

Nufio, C.R. and Papaj, D.R. (2001) Host marking behavior in phytophagous insects and parasitoids. *Entomologia Experimentalis et Applicata* **99**: 273–93.

Obitko, M. and Slavík, P. (1999) Visualization of genetic algorithms in a learning environment. In: *Spring Conference on Computer Graphics*, SCCG'99. Comenius University, Bratislava, pp. 101–6.

Perry, J.C. and Roitberg, B.D. (2005) Games among cannibals: Competition to cannibalize and parent-offspring conflict lead to increased sibling cannibalism. *Journal of Evolutionary Biology* **18**: 1523–33.

Prokopy, R.J. (1981) Oviposition-deterring pheromone system of apple maggot flies. In: Mitchell, E.R. (eds.) *Management of Insect Pests with Semiochemicals*. Plenum Publishing Corporation, New York, pp. 477–94.

Prokopy, R.J. and Webster, R.P. (1978) Oviposition-deterring pheromone in *Rhagoletis pomonella*: a kairomone for its parasitoid *Opius lectus*. *Journal of Chemical Ecology* **4**: 481–94.

Prokopy, R.J., Greany, P.D. and Chambers, D.L. (1977) Oviposition-deterring pheromone in *Anastrepha suspensa*. *Environmental Entomology* **6**: 463–5.

Quiring, D.T. and McNeil, J.N. (1984). Intraspecific competition between different aged larvae of *Agromyza frontella* (Rondani) (Diptera: Agromyzidae): Advantages of an oviposition-deterring pheromone. *Canadian Journal of Zoology* **62**: 2192–6.

Reinhold, K., Kurtz, J. and Engvist, L. (2002) Cryptic male choice: Sperm allocation strategies when female quality varies. *Journal of Evolutionary Biology* **15**: 201–9.

Roitberg, B.D. and Mangel, M. (1988) On the evolutionary ecology of marking pheromones. *Evolutionary Ecology* **2**: 289–315.

Roitberg, B.D. and Lalonde, R.G. (1991) Host marking enhances parasitism risk for a fruit-infesting fly *Rhagoletis basiola*. *Oikos* **61**: 389–93.

Rubinstein, R.Y. (1981) *Simulation and the Monte Carlo Method*. John Wiley & Sons, New York.

Sjerps, M. and Haccou, P. (1982) Effect of competition on optimal patch leaving: A war of attrition. *Theoretical Population Biology* **46**: 300–18.

Stephens, D.W. and Krebs, J.R. (1986) *Foraging Theory*. Princeton University Press, Princeton.

van Alphen, J.J.M. and Visser, M.E. (1990) Superparasitism as an adaptive strategy for insect parasitoids. *Annual Review of Entomology* **35**: 59–79.

van Alphen, J.J.M., Bernstein, C. and Driessen, G. (2003) Information acquisition and time allocation in insect parasitoids. *Trends in Ecology & Evolution* **18**: 81–7.

Wajnberg, É. (2006) Time allocation strategies in insect parasitoids: From ultimate predictions to proximate behavioral mechanisms. *Behavioral Ecology & Sociobiology* **60**: 589–611.

Wajnberg, É., Bernhard, P., Hamelin, F. and Boivin, G. (2006) Optimal patch time allocation for time-limited foragers. *Behavioral Ecology & Sociobiology* **60**: 1–10.

Yamamura, N. and Tsuji, N. (1987) Optimal patch time allocation under exploitative competition. *American Naturalist* **129**: 553–67.

18

Statistical tools for analyzing data on behavioral ecology of insect parasitoids

Éric Wajnberg and Patsy Haccou

Abstract

Experiments performed by behavioral ecologists, both in the laboratory and in the field, are producing data that need to be accurately and properly analyzed by powerful statistical methods. Unfortunately, for most of the time and especially for experiments done on insect parasitoids:

1 these data are not following normal distributions (since counts, proportions, time duration, etc. are measured);
2 it is (nearly) impossible to use balanced or independent designs; and/or
3 there is pseudoreplication.

Thus, standard, well-established methods cannot be used and alternative statistical approaches have to be applied. This chapter gives a detailed overview on how the different problems mentioned above can be handled accurately.

18.1 Introduction

Nowadays, cheap computer devices and/or programs make it easy to collect large amounts of information from behavioral observations. As a consequence, behavioral ecologists, and especially those working on insect parasitoids, regularly need to handle a large variety of data. Results obtained from analyses of these data are usually confronted with ultimate predictions derived from theoretical models that tell us what animals should do to behave optimally. To arrive at the right conclusions, it is essential that the data are analyzed with correct and powerful statistical methods. Usually, only a small part of the collected data consists of values that are independent and have a Gaussian distribution. In this case, standard methods can be used, based on the so-called General Linear Model that includes most of the 'classical' methods that most readers will be familiar with, including linear regression, analysis of variance (ANOVA), and analysis of covariance (ANCOVA). There are a large number of textbooks available that present these 'classical' methods (Zar 1999).

However, in an increasing number of cases, behavioral ecologists working on insect parasitoids also collect, and should thus also analyze, data that are either non-normally distributed – which is the case, for example, for time durations data, counts, or percentages – or that are non-independent or come from unbalanced experimental set-ups. Indeed, many of the chapters in this book deal with traits that are not normally distributed, such as time intervals (see also Chapter 8 by van Alphen and Bernstein, and Chapter 9 by Haccou and van Alphen) or sex ratios (see also Chapter 12 by Ode and Hardy). Further, especially when they are collecting data from fieldwork, scientists have to cope with pseudoreplication problems. In this chapter, we will present the basic theoretical framework and the statistical methods that can be used to handle such types of data correctly. Examples dealing with behavioral ecology of insect parasitoids will also be presented throughout.

18.2 An introduction to generalized linear models (GLMs)

All of us are in need of analyzing relations between observed values and a set of explanatory variables. Traditionally, this is done with ANOVA and regression methods. Such statistical tools have several limitations, like the assumption of a normal distribution for errors of measurement. Over the last few decades, several methods have been developed to deal with more general models. However, only a small number of behavioral ecologists working on insect parasitoids feel comfortable with, and have indeed applied these. Here we will provide an introduction to these methods that will hopefully overcome these problems and stimulate researchers in behavioral ecology of parasitoids to apply the methods of analysis most suited to their data.

18.2.1 The general framework

The framework we consider is as follows: We have a set of n observations $y_1, \ldots, y_n$, that are realizations of n random variables $Y_1, \ldots, Y_n$. These are usually called dependent variables. For the moment we assume that they are independently distributed. Further, we have a set of p explanatory variables, $x_1, \ldots, x_p$, with values associated to each of the n observations: $x_{11}, \ldots, x_{1p}; x_{21}, \ldots, x_{2p}; \ldots; x_{n1}; \ldots, x_{np}$. These variables are assumed to be known without error (in practice the methods also work well when these variables are random, with a much lower variance than the Y_i). They are also often called independent variables, or predictor variables. Note that the independent variables can be of several types. For instance binary (e.g. sex), nominal (e.g. colors), ordinal (e.g. sizes small, medium or large), or quantitative (e.g. weight in micrograms). For notational convenience, we will add a variable x_0 that equals one for every observation. This will allow us to formulate a null model of no effect in a notationally efficient way (see below).

We seek to relate the observations y to the set of explanatory variables x. This is done by assuming that the distribution of Y_i is somehow related to a predictor variable η_i, which is a linear combination of the explanatory variables:

$$\eta_i = \sum_{j=0}^{p} \beta_j x_{ij} \qquad (18.1)$$

The coefficients β_j are estimated from the data. For instance, in a regression analysis, these are the regression coefficients. In an ANOVA, they represent the estimated effects of the factors. The way in which this so-called linear predictor affects the distribution of the Y_i is determined by a function that relates a parameter of this distribution to the value of η_i. In general, this parameter is the expectation of Y_i, and here we will only consider such cases. If, for convenience, we denote the expectation by μ_i, then it is assumed that

$$\eta_i = g(\mu_i) \tag{18.2}$$

The function $g(\cdot)$ is called the link function (McCullagh & Nelder 1989). Note that in some texts the link function is the inverse of $g(\cdot)$.

In classical analyses, the Y_i follow normal distributions with an average value of μ_i and common variance σ^2, and:

$$\mu_i = \eta_i = \sum_{j=0}^{p} \beta_j x_{ij} \tag{18.3}$$

In this case, the link function is thus the identity function. The variance σ^2 is an example of a so-called nuisance parameter, which means that this parameter is not related to the question that we are examining and, ideally, we do not want it to affect the outcome of the analysis in any way.

The meaning of Equations (18.1) to (18.3) is clear for the situation in which the explanatory variables are all quantitative, a situation that corresponds to a simple regression analysis. However, in order to express the effects of qualitative variables in this way, explanatory variables have to be assigned a value. Binary variables can, for instance, be coded 0 or 1. Nominal or ordinal variables with k classes can be represented by a set of $k-1$ so-called dummy variables, each taking the value 0 or 1. Table 18.1 gives two possible ways of using such dummy variables to code a nominal trait with three classes: Red, Yellow, and Blue.

For non-quantitative variables, the effects of the different classes can be derived from the coefficients β_j. For method A in Table 18.1, for instance, the coefficients for the two dummy variables give the effects of class Blue and Yellow, respectively, relative to class Red. The effect of Blue relative to that of Yellow is given by the difference between the coefficients of the two variables. If method B is used instead, the coefficient of the first dummy variable gives the effect of Blue relative to Yellow, whereas the effect of Blue

Table 18.1 Two equivalent methods of encoding a three-category nominal variable with two binary variables.

	Method A		Method B	
	Var1	Var2	Var1	Var2
Red	0	0	0	0
Yellow	0	1	0	1
Blue	1	0	1	1

relative to Red is given by the sum of the coefficients of the two dummy variables. Thus, the two methods give the same results for the relative effects, as they should.

When there are only qualitative variables, the classical analysis is the ANOVA. When there are qualitative as well as quantitative variables we are in the context of an ANCOVA. Interactions between the effects of the predictor variables can be examined by adding new predictor variables to the model, that consist of combinations of other variables.

The objectives of analyses are to estimate the coefficients β_j from the data, to select a model by identifying those explanatory variables (and their interactions) that have significant effects, to test goodness-of-fit of the selected model, and finally to make predictions based on the model. Prediction can concern, for example, the outcome of an experiment if the range of values of one of the explanatory variables is changed.

18.2.2 The Gaussian case

Estimation

Here, the Y_i are assumed to be random variables following independent normal distributions $N(\mu_i, \sigma^2)$, where the μ_i are given by Equation (18.3). We will first consider the situation where the common variance σ^2 is known. The probability density functions of the Y_i are thus of the form:

$$f(y_i; \mu_i, \sigma^2) = \frac{1}{\sqrt{2\pi\sigma^2}} \exp\left[-\frac{1}{2}\frac{(y_i - \mu_i)^2}{\sigma^2}\right] \tag{18.4}$$

Thus, the log-likelihood of all observations equals

$$L(\mathbf{y}; \boldsymbol{\mu}, \sigma^2) = \log\left(\frac{1}{\sqrt{2\pi\sigma^2}}\right) - \frac{1}{2}\sum_{i=1}^{n}\frac{(y_i - \mu_i)^2}{\sigma^2} \tag{18.5}$$

Note that we denote vectors by bold face, for example, $\mathbf{y} = y_1, \ldots, y_n$. The estimates of the regression parameters β_j are those values that maximize this expression (Mood et al. 1974). As can easily be seen from Equation (18.5), this implies that they minimize the sum of squares:

$$\sum_{i=1}^{n}(y_i - \mu_i)^2 = \sum_{i=1}^{n}\left(y_i - \sum_{j=0}^{p}\beta_j x_{ij}\right)^2 \tag{18.6}$$

which is a well-known result (Mood et al. 1974). There are a lot of statistical software packages that contain pre-programmed algorithms for finding the least-squares (or, equivalently, the maximum likelihood) estimates of the β_j for different types of models, but we will not go into the details of those methods here.

When the number of observations n is sufficiently large, the maximum likelihood estimators $\hat{\beta}_j$ have asymptotically a multivariate normal distribution with expected values β_j under the hypothesis that the fitted model is true. This result holds in general for maximum likelihood estimators, regardless of the distribution of the Y_i. Most statistical computer programs provide the estimated variance/covariance matrix of this multivariate distribution. Confidence intervals for the estimates can be based on this distribution, or alternatively on the fact that the distribution of

$$(\hat{\boldsymbol{\beta}} - \boldsymbol{\beta})' V(\hat{\boldsymbol{\beta}})^{-1} (\hat{\boldsymbol{\beta}} - \boldsymbol{\beta}) \tag{18.7}$$

where $\boldsymbol{\beta} = (\beta_0, \ldots, \beta_p)'$ is the vector of regression coefficients, $\hat{\boldsymbol{\beta}}$ is the corresponding vector of its estimators, and $V(\hat{\boldsymbol{\beta}})^{-1}$ is the inverse of the variance/covariance matrix of the parameters, tends to a χ^2 distribution with $p + 1$ degrees of freedom (denoted by df). We illustrate this with an example for a situation with two estimated parameters (i.e. with $p = 1$), β_0 and β_1. The estimated values of respectively the parameters and their variance/covariance matrix are

$$\hat{\beta}_0 = 0.9, \hat{\beta}_1 = 2.1,$$

$$V(\hat{\beta}) = \begin{pmatrix} 1.2 & 0.3 \\ 0.3 & 0.6 \end{pmatrix}, \tag{18.8}$$

with inverse:

$$V^{-1}(\hat{\beta}) = \begin{pmatrix} 0.95 & -0.48 \\ -0.48 & 1.90 \end{pmatrix} \tag{18.9}$$

For large sample sizes, we thus find that the following statistic has a χ^2 distribution with 2 df:

$$(0.9 - \beta_0, 2.1 - \beta_1) \begin{pmatrix} 0.95 & -0.48 \\ -0.48 & 1.90 \end{pmatrix} \begin{pmatrix} 0.9 - \beta_0 \\ 2.1 - \beta_1 \end{pmatrix}$$
$$= 7.3341 + 0.95\beta_0^2 + 0.306\beta_0 - 7.116\beta_1 - 0.96\beta_0\beta_1 + 1.9\beta_1^2. \tag{18.10}$$

The 5% critical value of a χ^2 distribution with 2 df equals 5.99. Thus, the equation:

$$7.3341 + 0.95\beta_0^2 + 0.306\beta_0 - 7.116\beta_1 - 0.96\beta_0\beta_1 + 1.9\beta_1^2 = 5.99 \tag{18.11}$$

defines an ellipse that corresponds to the simultaneous 95% confidence interval of the two parameters.

Model selection: the deviance
The maximum possible value for the likelihood is when the model fully describes the observations, which is when the μ_i are equal to the y_i:

$$L(\mathbf{y}; \mathbf{y}, \sigma^2) = \log\left(\frac{1}{\sqrt{2\pi\sigma^2}}\right) - \frac{1}{2} \sum_{i=1}^{n} \frac{(y_i - y_i)^2}{\sigma^2} = \log\left(\frac{1}{\sqrt{2\pi\sigma^2}}\right) \tag{18.12}$$

Note that this model is fully uninformative, since it implies that we have a separate parameter for each observation. The maximum likelihood estimators of the μ_i are denoted by

$$\hat{\mu}_i = \sum_{j=0}^{p} \hat{\beta}_j x_{ij} \tag{18.13}$$

and the likelihood obtained with these estimators equals

$$L(\mathbf{y};\hat{\boldsymbol{\mu}},\sigma^2) = \log\left(\frac{1}{\sqrt{2\pi\sigma^2}}\right) - \frac{1}{2}\sum_{i=1}^{n}\frac{(y_i - \hat{\mu}_i)^2}{\sigma^2} \tag{18.14}$$

Two times the difference between the maximum and the observed likelihoods equals

$$2(L(\mathbf{y};\mathbf{y},\sigma^2) - L(\mathbf{y};\hat{\boldsymbol{\mu}},\sigma^2)) = \sum_{i=1}^{n}\frac{(y_i - \hat{\mu}_i)^2}{\sigma^2} \tag{18.15}$$

When the Y_i have a Gaussian distribution, this statistic has a χ^2 distribution with $n - (p+1)$ degrees of freedom, since there are $p + 1$ estimated parameters. We will denote this distribution by $\chi^2_{n-(p+1)}$. In the context of GLMs, the right-hand side of Equation (18.15) is called the scaled deviance. In this case, the deviance itself equals

$$D(\mathbf{y},\boldsymbol{\mu}) = \sum_{i=1}^{n}(y_i - \hat{\mu}_i)^2 \tag{18.16}$$

and model selection, as well as goodness-of-fit procedures for GLMs, are based on these quantities. In particular, differences between (scaled) deviances are used to compare nested models. For instance, suppose we want to test whether a factor with q different levels has a significant effect (i.e. a one-way ANOVA). As explained above, such a factor is represented by $q - 1$ binary variables. Thus, the corresponding model has q parameters (because of the additional parameter β_0), and the scaled deviance for this model will follow a χ^2_{n-q} distribution. The estimator of β_0 in the null model is $\bar{y}$, the mean value of the y_i, and its scaled deviance therefore equals

$$\frac{D(\mathbf{y},\boldsymbol{\mu}0)}{\sigma^2} = 2(L(\mathbf{y};\mathbf{y},\sigma^2) - (\mathbf{y};\bar{\mathbf{y}},\sigma^2)) = \sum_{i=1}^{n}\frac{(y_i - \bar{y})^2}{\sigma^2} \tag{18.17}$$

which follows a χ^2_{n-1} distribution. A test whether the examined factor has a significant effect is based on the difference between the two scaled deviances:

$$\frac{D(\mathbf{y},\boldsymbol{\mu}0)}{\sigma^2} - \frac{D(\mathbf{y},\boldsymbol{\mu})}{\sigma^2} = \sum_{i=1}^{n}\frac{(y_i - \bar{y})^2}{\sigma^2} - \sum_{i=1}^{n}\frac{(y_i - \hat{\mu}_i)^2}{\sigma^2} \tag{18.18}$$

This statistic has a χ^2 distribution with $(n-1) - (n-q) = q - 1$ degrees of freedom under the null model, and thus the effect of the factor can be statistically tested.

Until now we have assumed that the value of the nuisance parameter σ^2 is known. When this is not true, it can be estimated by

$$\hat{\sigma}^2 = \frac{1}{n-1}\sum_{i=1}^{n}(y_i - \bar{y})^2 \tag{18.19}$$

Substituting this in Equation (18.18) gives the test statistic:

$$\frac{D(\mathbf{y},\boldsymbol{\mu}0)}{\hat{\sigma}^2} - \frac{D(\mathbf{y},\boldsymbol{\mu})}{\hat{\sigma}^2} = \frac{\displaystyle\sum_{i=1}^{n}(y_i - \bar{y})^2 - \sum_{i=1}^{n}(y_i - \hat{\mu}_i)^2}{\dfrac{1}{n-1}\displaystyle\sum_{i=1}^{n}(y_i - \bar{y})^2} \tag{18.20}$$

For large n, this statistic is known to follow a χ^2_{q-1} distribution. An alternative is to use the following test statistic:

$$\frac{\left(\sum_{i=1}^{n}(y_i - \bar{y})^2 - \sum_{i=1}^{n}(y_i - \hat{\mu}_i)^2\right)\frac{1}{q-1}}{\frac{1}{n-1}\sum_{i=1}^{n}(y_i - \bar{y})^2} \tag{18.21}$$

which follows a $F_{(q-1,n-1)}$ distribution under the null-hypothesis of no effect. This distribution is exact, i.e. it also holds for small n, and so provides a better test for small sample sizes. Note that the right-hand side of Equation (18.19) equals $D(\mathbf{y},\boldsymbol{\mu}\mathbf{0})/(n-1)$, so we can write Equation (18.21) as

$$\frac{(D(\mathbf{y},\boldsymbol{\mu}\mathbf{0}) - D(\mathbf{y},\boldsymbol{\mu}))\frac{1}{q-1}}{D(\mathbf{y},\boldsymbol{\mu}\mathbf{0})\frac{1}{n-1}} \tag{18.22}$$

This demonstrates that an ANOVA is equivalent to an 'analysis of deviance' for the model considered, i.e. the linear model with independent normally distributed error terms and an identity link function. The same can be shown for a regression analysis or multi-factor ANCOVA, where tests are also based on ratios of sums of squares.

Note that tests of effects and concomitant model selection can also be based on the asymptotic distribution of the β_j (Section 18.2.2). Asymptotically, the two procedures are equivalent. For small n, however, they can lead to different conclusions.

The two-way layout: orthogonality of effects
The multifactor ANOVA has a large advantage that most other analyses do not have, namely so-called orthogonality of effects. This means that the outcomes of tests for a certain (combination of) variables(s) do not depend on whether or not other variables are included in the null-hypothesis model. For example, consider the two-way layout, with two factors f_1 and f_2. We consider four nested models:

1 M_0: the null-model where none of the factors are included;
2 M_1: the model with only f_1;
3 M_{12}: the model with both f_1 and f_2; and
4 M_{1*2}: the model with f_1, f_2 and an interaction effect between f_1 and f_2.

Suppose also that f_1 has four levels and f_2 has three. This means that there are six more predictor variables, taking 0/1 values: three for f_1 and two for f_2. Table 18.2 gives a possible way of coding all these predictor variables. The expressions for μ_i corresponding to the different models are summarized in column 2 of Table 18.3. We also introduce some notation (in column 3) to be able to distinguish the models in the subsequent equations. Note that the (estimators for the) values of the coefficients β_j will differ for the different models. The fourth column gives the expression for the sum of squares to be minimized to estimate the coefficients β_j. The deviance for the null model M_0 is given in Equation (18.17). The other deviances are calculated by substituting the estimates of the β_j in the

Table 18.2 One example of two ordinal variables that are encoded by means of binary variables. The number of binary variables needed for one ordinal variable equals the number of levels of the ordinal variable minus one. Further, to represent the interactions between the two factors we need six more variables: $x_6 = x_1 \times x_4$, $x_7 = x_2 \times x_4$, $x_8 = x_3 \times x_4$, $x_9 = x_1 \times x_5$, $x_{10} = x_2 \times x_5$, $x_{11} = x_3 \times x_5$. For example the interaction between the fourth level of f_1 and the second level of f_2 is represented by $x_6 = 1$, $x_7 = 1$, $x_8 = 1$, $x_9 = 0$, $x_{10} = 0$, $x_{11} = 0$.

Levels of f_1	Variables for f_1			Levels of f_2	Variables for f_2	
	x_1	x_2	x_3		x_4	x_5
1	0	0	0	1	0	0
2	1	0	0	2	1	0
3	1	1	0	3	1	1
4	1	1	1	–	–	–

Table 18.3 Overview of the different models used as examples for testing the effect of two factors on the average value of a Gaussian trait, their parameters, and corresponding degrees of freedom of their residual sums of squares. See text for a detailed explanation.

Model	μ_i	Notation	Minimization of	Degrees of freedom
M_0	$\beta_0 x_{i0} (= \beta_0)$	$\mu 0$	$\sum_{i=1}^{n} (y_i - \mu 0)^2$	$n - 1$
M_1	$\sum_{j=0}^{3} \beta_j x_{ij}$	$\mu 1_i$	$\sum_{i=1}^{n} \left(y_i - \sum_{j=0}^{3} \beta_j x_{ij} \right)^2$	$n - 4$
M_{12}	$\sum_{j=0}^{5} \beta_j x_{ij}$	$\mu 12_i$	$\sum_{i=1}^{n} \left(y_i - \sum_{j=0}^{5} \beta_j x_{ij} \right)^2$	$n - 6$
M_{1*2}	$\sum_{j=0}^{11} \beta_j x_{ij}$	$\mu 1*2_i$	$\sum_{i=1}^{n} \left(y_i - \sum_{j=0}^{11} \beta_j x_{ij} \right)^2$	$n - 12$

expressions in column 4. Degrees of freedom corresponding to the deviances are equal to the number of observations minus the number of estimated parameters. These are given in column 5.

To test whether inclusion of an additional term in the model has a significant effect, we can use an F-test, just as in the one-way ANOVA, as explained above. For instance, to calculate the numerator of the F-statistic for testing model M_1 versus M_0, we divide the difference between the deviances of the two models by the proper degrees of freedom. With the notation for the deviances introduced before (Equation 18.16), this gives

$$\frac{(D(\mathbf{y},\boldsymbol{\mu 0}) - D(\mathbf{y},\boldsymbol{\mu 1}))\frac{1}{3}}{D(\mathbf{y},\boldsymbol{\mu 0})\frac{1}{n-1}} \tag{18.23}$$

which follows a $F_{(3,\, n-1)}$ distribution. The outcome of this test tells us whether factor f_1 has a significant effect or not. A subsequent test of M_{12} versus M_1 provides the same information about the other factor. Note, however, that there is no *a priori* reason to include the two factors in the model in this specific sequence. We might as well do it the other way round. This would lead to a sequence of models M_0, M_2, M_{12}, and M_{1*2} (where M_2 denotes the model with only factor 2 included). In this case, a test as to whether the effect of f_1 is significant is performed by using the differences between the deviances of models M_2 and M_{12}.

In classical multifactor ANOVAs, the two expressions turn out to be equal. Thus, the sequence in which different factors are included in the model has no effect on their significance. This is a pleasant feature of these analyses, but it is unfortunately not usually the case, even in classical analyses beyond the ANOVA. A well-known example where this no longer holds true is, for example, in multiple regression, where there are several quantitative variables. There, it does make a difference if different ways of successively including predictor variables in the models are being used. In GLMs, other than the classical case, the effects of different factors are also usually non-orthogonal. Therefore, it is recommended not to consider just one deviance table, but several, corresponding to different model sequences.

Model selection: some further considerations
Finally, it has to be noted that model selection is usually not just a question of blindly performing a sequence of tests. It also involves knowledge of the biology behind the data, and often plain common sense. To give an example, tests may show that none of the main effects (or only one) are significant, whereas their interaction is. However, it is not a good idea to use a model with only interactions and no main effects (or only one main effect), since such a model is usually difficult to interpret. Another example is that any function of an independent variable may be included in the models. For instance, we may choose $\log(x_j)$, or its square root, as one of the predictor variables. Such variables may even give significant results. That does not mean, however, that it always makes sense to include them. In general, we would advise the use of only well-interpretable transformations. For instance, a multiplicative effect rather than an additive one might sometimes be more realistic. This can be realized by using log-transformations of the original variables. If we denote the original variables by z_j, then $x_j = \log(z_j)$ will be used in the model. In the Gaussian case, i.e. with an identity link function, this means that in terms of the original variables:

$$\mu_i = \sum_{j=1}^{p} \beta_j x_{ij} = \sum_{j=1}^{p} \beta_j \log z_{ij} = \log\left(\prod_{j=1}^{p} z_{ij}^{\beta_j}\right) \tag{18.24}$$

Biological reasoning might also bring us to include predictor variables in the model, even though their effects are not significantly different from zero, simply because we know that they should have an effect. Not finding the effect can be due to the fact that, with the current number of observations, the power of the test is too small compared to its magnitude. The opposite may also occur: if the number of observations is very large, effects may be significant, even though they are very small. In this case, it is advisable to consider the relevance of the effect as a decision factor for including it in the model.

Goodness of fit: residuals
Once a model has been selected, we wish to get an impression of how good it describes the current data set. In the Gaussian case, this is usually done by studying the residuals corresponding to differences between observed values and those predicted by the model:

$$r_i = y_i - \hat{\mu}_i \tag{18.25}$$

If the fitted model is accurate, these residuals should follow a normal distribution with variance equal to σ^2. The so-called Pearson residual uses the estimated variance to scale this to a standard normal distribution:

$$\tilde{r}_i = \frac{y_i - \hat{\mu}_i}{\sqrt{\hat{\sigma}^2}} \tag{18.26}$$

Note that the deviance equals the residual error sum of squares (Equation 18.16). In general we can write:

$$D(y,\mu) = \sum_{i=1}^{n} D(y_i,\mu_i) \tag{18.27}$$

where $D(y_i,\mu_i)$ indicates the contribution of observation i to the total deviance. This has a positive value. For GLMs, we can thus define a 'deviance residual':

$$r_i = sign(y_i - \hat{\mu}_i)\sqrt{D(y_i,\mu_i)} \tag{18.28}$$

In the Gaussian case this is equal to Equation (18.25). Unfortunately, for non-Gaussian models the distribution of these residuals is not always known.

18.2.3 The non-Gaussian case

The models in the previous section are widely applied. Their applicability, however, is restricted in a number of ways. One of them is that the data are continuous quantities that can assume any real value and that follow normal distributions. In practice, we often have to deal with different types of data. GLMs extend the framework of the classical analyses to such situations.

Binary data
An obvious model for binary data is based on the binomial distribution. The observations Y_i correspond to the number of times a 'success' is observed in conjunction with the covariate values $x_{i1}, \ldots, x_{ip}$. They are assumed to follow independent binomial distributions with parameters m_i and p_i, so we have

$$Pr(Y_i = y_i) = \binom{m_i}{y_i} p_i^{y_i}(1 - p_i)^{m_i - y_i}, \, y_i = 0, \, 1, \ldots, \, m_i \tag{18.29}$$

The expectation μ_i equals $m_i \times p_i$, and thus we can write this expression as

$$\Pr(Y_i = y_i) = \binom{m_i}{y_i}\left(\frac{\mu_i}{m_i}\right)^{y_i}\left(1 - \frac{\mu_i}{m_i}\right)^{m_i - y_i}, \; y_i = 0, 1, \ldots, m_i \tag{18.30}$$

This leads to the log-likelihood:

$$L(\mathbf{y};\boldsymbol{\mu}) = \sum_{i=1}^{n} \log\binom{m_i}{y_i} + \sum_{i=1}^{n} y_i \log\left(\frac{\mu_i}{m_i}\right) + \sum_{i=1}^{n} (m_i - y_i) \log\left(1 - \frac{\mu_i}{m_i}\right) \tag{18.31}$$

There are several link functions that can be used to specify the relationship between the expectations and the explanatory variables. We will consider the logistic function:

$$\eta_i = \log\left(\frac{\mu_i}{m_i - \mu_i}\right) \tag{18.32}$$

where η_i is the linear predictor given in Equation (18.1). In terms of the expected values, this gives

$$\frac{\mu_i}{m_i - \mu_i} = e^{\eta_i} \Rightarrow \mu_i = m_i \frac{e^{\eta_i}}{1 + e^{\eta_i}} \tag{18.33}$$

When the explanatory variables are all quantitative, this analysis is called a logistic regression (McCullagh & Nelder 1989). Maximization of Equation (18.31), with these expressions substituted, gives the estimators of the β_j. The deviance equals

$$2(L(\mathbf{y};\mathbf{y}) - L(\mathbf{y};\hat{\boldsymbol{\mu}})) = 2\left(\sum_{i=1}^{n} y_i \log\left(\frac{y_i}{\hat{\mu}_i}\right) + \sum_{i=1}^{n} (m_i - y_i) \log\left(\frac{m_i - y_i}{m_i - \hat{\mu}_i}\right)\right) \tag{18.34}$$

It has to be noted that the distribution of this statistic is usually not well-approximated by a χ^2 distribution (McCullagh & Nelder 1989). For large n, however, the differences between the deviances for different models approximately follow χ^2 distributions, with the numbers of degrees of freedom equal to the differences between the numbers of parameters in the different models.

In some cases, however, the binomial distribution does not adequately describe the data, due to so-called overdispersion. Such a phenomenon corresponds to the case in which observed variances are larger than the binomial variance $mp(1 - p)$. This can occur, for instance, when the observations are in fact mixtures of independent, non-identified binomially distributed variables with (slightly) different success probabilities. Underdispersion can also occur, but its mechanistic explanation is not totally clear. In both cases, we may use a model where

$$E(Y_i) = m_i p_i, \; Var(Y_i) = \sigma^2 m_i p_i (1 - p_i) \tag{18.35}$$

with σ^2 being an unknown nuisance parameter, which is then also estimated from the data. A nuisance parameter is a parameter whose precise value is not of interest, and conclusions from a statistical analysis should preferably be independent of this value. The test statistics now follows approximately a χ^2 distribution multiplied by $\hat{\sigma}^2$. The asymptotic variance/covariance matrix of the estimators is also multiplied by this value.

Counts
Sometimes observations can take any discrete value larger than or equal to zero. Examples are the size of egg batches laid by a gregarious pro-ovigenic parasitoid, the number of host encounters within a certain time interval, the number of females attacking simultaneously a patch of hosts (see also Chapter 9 by Haccou and van Alphen), etc. In this case, a model based on the Poisson distribution might be appropriate:

$$\Pr(Y = y) = e^{-\mu} \frac{\mu^y}{y!}, \, y = 0, 1, \ldots \tag{18.36}$$

with log-likelihood:

$$L(\mathbf{y};\boldsymbol{\mu}) = \sum_{i=1}^{n} (y_i \log \mu_i - \mu_i - \log y_i!) \tag{18.37}$$

The usual link function for this model is

$$\eta_i = \log \mu_i \Rightarrow \mu_i = e^{\eta_i} \tag{18.38}$$

The deviance equals

$$2(L(\mathbf{y};\mathbf{y}) - L(\mathbf{y};\hat{\boldsymbol{\mu}})) = 2 \sum_{i=1}^{n} \left(y_i \log \frac{y_i}{\hat{\mu}_i} - (y_i - \hat{\mu}_i) \right) = 2 \sum_{i=1}^{n} y_i \log \frac{y_i}{\hat{\mu}_i} \tag{18.39}$$

and the model is called a log-linear model (McCullagh & Nelder 1989).

It is well-known that the expectation of a Poisson distribution equals it variance. As for the Binomial model, however, overdispersion may inflate the variance. When observations correspond to numbers of occurrences in time intervals, overdispersion may be caused by random variation of the length of the intervals. Another possibility is inter-individual variability. The model may be generalized to account for such possibilities, by assuming:

$$Var(Y_i) = \sigma^2 E(Y_i) \tag{18.40}$$

where the nuisance parameter σ^2 can also be estimated from the data. Effects on tests and distribution of the estimators are accounted for in the same way as in the Binomial model.

Nominal or ordinal observations
Sometimes, the observations are counts of numbers in certain classes. Measurements can be on a nominal scale, for example, different substrates of a host species, or an ordinal scale, for example, we might measure the size of daughters emerging from a host as small, medium, or large. We then might be interested in effects of explanatory variables on the

probability that observations fall into a certain class. For instance, the host size might affect the proportions of small- and medium-sized daughters, compared to large ones. In this case, the appropriate model for the observations is a multinomial distribution:

$$\Pr(Y_1 = y_1, \dots Y_r = y_r) = \frac{m!}{\prod\limits_{k=1}^{r} m_k!} \prod_{k=1}^{r} p_k^{y_k} \tag{18.41}$$

with

$$\sum_{k=1}^{r} m_k = m, \ \sum_{k=1}^{r} p_k = 1 \tag{18.42}$$

where y_k denotes the number in class k. The log-likelihood becomes

$$L(\mathbf{y};\mathbf{m},\mathbf{p}) = \sum_{i=1}^{n} \log\left(\frac{m!}{\prod\limits_{k=1}^{r} m_k!}\right) + \sum_{i=1}^{n} \sum_{k=1}^{r} y_{i,k} \log p_{i,k} \tag{18.43}$$

If we denote the expectation of the number in class k for the ith observation by $\mu_{i,k}$, we can write this as

$$L(\mathbf{y};\mathbf{m},\boldsymbol{\mu}) = \sum_{i=1}^{n} \log\left(\frac{m!}{\prod\limits_{k=1}^{r} m_k!}\right) + \sum_{i=1}^{n} \sum_{k=1}^{r} y_{i,k} \log \frac{\mu_{i,k}}{m_k} \tag{18.44}$$

and the deviance equals

$$D(y,m) = 2 \sum_{i=1}^{n} \sum_{k=1}^{r} y_{i,k} \log\left(\frac{y_{i,k}}{\hat{\mu}_{i,k}}\right) \tag{18.45}$$

The linear predictor now becomes a vector with elements:

$$\eta_k = \sum_{j=0}^{p} \beta_{j,k} x_{ij,k}, \ k = 1, \dots, r-1 \tag{18.46}$$

where $x_{ij,k}$ denotes the value of covariate j for category k of the ith observation and, analogous to the previous models, $x_{i0,k} = 1$ for all i and k. The length of η is the number of categories minus one, due to the restrictions in Equation (18.42). To take these restrictions into account, the link function contains the ratio of expectations, relative to a chosen reference category. Here, we will use category r for that purpose. A logistic link function gives

$$\log\left(\frac{\mu_{i,k}}{\mu_{i,r}}\right) = \eta_k \Rightarrow \frac{\mu_{i,k}}{\mu_{i,r}} = e^{\eta_k}. \tag{18.47}$$

When explanatory variables are only quantitative, this again gives a logistic regression (as in the situation with binary variables (Section 18.2.3 above)). It should be noted that the ratio can be written in many different ways:

$$\frac{\mu_{i,k}}{\mu_{i,r}} = \frac{p_{i,k}}{p_{i,r}} = \frac{p_{i,k}}{1 - \sum_{k=1}^{r-1} p_{i,r}} = \frac{m_i p_{i,k}}{m_i\left(1 - \sum_{k=1}^{r-1} p_{i,r}\right)} = \frac{\mu_{i,k}}{m_i - \sum_{k=1}^{r-1} \mu_{i,k}}. \tag{18.48}$$

From this, it can be seen that the model is the multi-category analog of the one used for binary data (Equation 18.32).

It can be seen from Equation (18.46) that the number of parameters in a full model equals $(r-1)(p+1)$. An alternative special case of the model, often used, is

$$\eta_k = \beta_{0,k} + \sum_{j=1}^{p} \beta_j x_{ij,k} \tag{18.49}$$

where it is assumed that the covariates have the same effects on the different categories. This model only has $r + p - 1$ parameters.

When the data are measured on an ordinal scale, models can also be formulated in terms of the cumulative probabilities:

$$\gamma_{i,k} = \sum_{l=1}^{k} p_{i,l}, k = 1, \dots, r-1 \tag{18.50}$$

In this case, a logistic model would be, for instance:

$$\log\left(\frac{\gamma_{i,k}}{1 - \gamma_{i,k}}\right) = \beta_{0,k} + \sum_{j=1}^{p} \beta_j x_{ij,k} \tag{18.51}$$

However, it has to be noted that, in such models, the parameters are constrained since the cumulative probabilities $\gamma_{i,k}$ must increase with k. It can be deduced from Equation (18.51) that this implies that

$$\beta_{0,1} \leq \dots \leq \beta_{0,r-1} \tag{18.52}$$

and any method used for estimation will have to take such constraints into account.

As in previous models, overdispersion may occur here, due to the same causes as mentioned before, and can be dealt with in the same way.

Survival data
In many situations, we have to deal with data that are time intervals until occurrence of certain events, or time intervals between events. For instance, behavioral ecologists often

record and analyze time intervals until the onset or termination of certain behaviors. Another example is residence time in host patches (see also Chapter 8 by van Alphen and Bernstein). Normal distributions usually do not give an accurate description of such data. Further, the data need special treatment due to the occurrence of so-called censoring. Indeed, sometimes the event in question is not observed during the experiment. For instance, a parasitoid may not leave a patch of hosts before the end of the observation period. When this occurs, we can only say that the patch-leaving time would have been longer than the observation time. Such observations cannot be simply treated as missing values, since they do contain information about the patch-leaving time. Bressers et al. (1991) give several examples of analyses that lead to erroneous conclusions due to the wrong treatment of censors. Survival analysis methods are especially designed to deal with this problem. Many of the models used in such analyses are, in fact, GLMs.

To illustrate this, we first consider the exponential regression model. In this model the Y_i are assumed to be exponentially distributed, with probability density function:

$$f(y;\lambda) = \lambda e^{-\lambda y} \tag{18.53}$$

where λ denotes the so-called hazard rate. In the exponential model this is a constant, but in more general models the hazard rate may depend on the time y. The hazard rate $\lambda(y)$ is the probability per time unit that an event happens at time y, given that it has not yet occurred. This is an important parameter in survival analysis, and most models are formulated in terms of this parameter rather than the expectation of Y. The interpretation of the hazard rate depends on the biological problem that is examined. For instance, if the observations are times until a patch of hosts is left, the hazard rate is the patch leaving tendency. In the case of, for example, grooming bouts, the hazard rate can be the tendency to start (or stop) grooming.

In the exponential regression model it is assumed that

$$\lambda_i = e^{\eta_i} \tag{18.54}$$

where η_i is the linear predictor of Equation (18.1). The null model parameter e^{β_0} is called the baseline hazard, and is denoted by λ_0, so an alternative way to denote the previous expression (which is the usual notation in survival analysis literature) is

$$\lambda_i = \lambda_0 e^{\sum_{j=1}^{p} \beta_j x_{ij}} \tag{18.55}$$

The expectation of Y_i equals

$$\mu_i = \frac{1}{\lambda_i} \tag{18.56}$$

and thus, we find that, in this case:

$$\mu_i = e^{-\eta_i} \Rightarrow \eta_i = -\log \mu_i \tag{18.57}$$

Thus, the exponential regression model is a generalized linear model with a logarithmic link function.

The probability of observing a time interval larger than some fixed value y equals

$$\Pr(Y > y) = e^{-\lambda y} \tag{18.58}$$

Censoring is taken into account by including this probability in the log-likelihood. For instance, if $y_1, \ldots, y_k$ are the time intervals observed in uncensored cases and $y_{k+1}, \ldots, y_n$ are censor times of the other observations, the log-likelihood equals

$$L(\mathbf{y}; \boldsymbol{\lambda}) = \sum_{i=1}^{k} \log \lambda_i - \sum_{i=1}^{n} \lambda_i y_i \tag{18.59}$$

and the deviance equals

$$2(L(\mathbf{y};\mathbf{y}) - L(\mathbf{y};\hat{\boldsymbol{\lambda}})) = 2 \left(\sum_{i=1}^{k} \log \frac{\hat{\lambda}_i}{y_i} + \sum_{i=1}^{n} \hat{\lambda}_i y_i - n \right) \tag{18.60}$$

and the residuals:

$$y_i - \frac{1}{\hat{\lambda}_i} \tag{18.61}$$

are known to have an exponential distribution with parameter 1 under the hypothesis that the model is correct (Kalbfleisch & Prentice 2002). Thus, the fit of a model can be tested by means of goodness-of-fit tests for an exponential distribution, where residuals of censored observations are treated as censors (see Haccou & Meelis 1994, for a description of several such tests).

The exponential regression model is a special case of the so-called Cox regression model, which is also called the proportional hazards model. In the more general formulation, the baseline hazard rate can be any positive function of the time interval y. This model assumes the following relationship between hazard rate and explanatory variables:

$$\lambda_i(y) = \lambda_0(y) e^{\sum_{j=1}^{p} \beta_j x_{ij}} \tag{18.62}$$

Hence, the variables are assumed to have a multiplicative effect on the baseline hazard. For instance, suppose that x_{11} equals A and x_{21} equals B, and that the other explanatory variables are all equal, i.e. $x_{1j} = x_{2j} = x_j$ for $j = 2, \ldots p$, then the model states that the ratio of the hazard rates is constant:

$$\frac{\lambda_1(y)}{\lambda_2(y)} = \frac{\lambda_0(y) e^{\beta_1 A + \sum_{j=2}^{p} \beta_j x_j}}{\lambda_0(y) e^{\beta_1 B + \sum_{j=2}^{p} \beta_j x_j}} = e^{\beta_1 (A-B)} \tag{18.63}$$

This is called the proportionality assumption. This assumption can be tested for any of the explanatory variables by fitting a model that allows a different baseline hazard for the parameter in question, and then testing whether the estimated baseline hazards are indeed proportional. For instance, to test this for an explanatory variable X_1, which can assume two values, A and B, we use the model:

$$\lambda_i(y;A) = \lambda_{0A}(y)e^{\sum_{j=2}^{p} \beta_j x_{ij}} , \quad \lambda_i(y;B) = \lambda_{0B}(y)e^{\sum_{j=2}^{p} \beta_j x_{ij}} \qquad (18.64)$$

This is an example of a so-called stratified model. Proportionality tests of the baseline hazards can be formal or simply graphical (Kleinbaum & Klein 2005).

The probability density function is related to the hazard rate in the following way:

$$f(y; \lambda(y)) = \lambda(y)e^{-\int_{0}^{y} \lambda(s)ds} \qquad (18.65)$$

and the probability that Y is larger than y equals

$$\Pr(Y > y) = e^{-\int_{0}^{y} \lambda(s)ds} \qquad (18.66)$$

With these expressions we can write the log-likelihood, which is then maximized to estimate the β_j ($j = 1, \ldots, p$), as well as the baseline hazard as a function of time. There are many computer programs available to do this. Tests of models can be based on the deviances or on the asymptotic multinormal distribution of the $\hat{\beta}_j$, as before. It can be shown that, here too, the residuals have an exponential distribution with parameter 1 if the model is correct (Kalbfleisch & Prentice 2002), which can be used to derive goodness-of-fit tests.

There are several generalizations of this model. An important one is to allow time-dependence of explanatory variables. Then the x_{ij} become functions of y also. For example, Hemerik et al. (1993) examined the effect of the number of rejected hosts after the most recent oviposition on the patch-leaving tendency. This covariate changes in time during a patch visit and was, therefore, modeled as a time-dependent variable.

Another generalization is to allow for repeated events. For instance, a patch might be left and revisited several times. The successive times on the patch can, for instance, be considered as separate observations, and we can include an extra covariate that counts the previous number of patch visits. Additional covariates may be, for example, durations of previous visits, or the durations of time intervals spent off the patch (see Wajnberg 2006, for a detailed recent survey of all parameters that were taken into account in studies on patch time allocation in insect parasitoids).

Finally, more than one type of event might be considered. For instance, a wasp on a patch might perform several types of behavior like walk, groom, rest, or examine a host. A walking bout can thus be followed by three different types of event, and we might be interested in studying effects of covariates on the hazard rates of each of these different types. These are called cause-specific hazard rates, and they can be studied separately, by considering other events as censors. For instance, to study the behavioral tendency to stop walking and start grooming, walking intervals that are followed by resting or host encounters are treated as censored observations. Methods for analyzing complete continuous time records of behavior can be found in Haccou and Meelis (1994).

The proportional hazards model assumes a multiplicative relationship between the explanatory variables and the hazard rate. An alternative is to assume such an effect on the time until occurrence of an event Y. This leads to the so-called accelerated failure time model. In the case of a constant hazard rate (i.e. the exponential model described previously), the models are equivalent. In the general case, however, this is not true. Accelerated failure time models are more difficult to interpret than proportional hazard models, and as a consequence are much less popular. Therefore, we will not describe these models here, but refer to the textbooks by Kalbfleisch and Prentice (2002), and Kleinbaum and Klein (2005) for details.

18.3 Non-independent data

The methods discussed so far assume that measures are all independent, which means that each data point has been collected from a different subject to ensure that the sample size reflects all independent replicates. However, some experimental set-ups used in behavioral ecology are explicitly based on so-called repeated measure designs in which subject are measured several times. This is, for example, the case for experiments done to estimate the learning ability of parasitoid females in which the same individuals are tested before and after a series of repeated experiences with hosts (van Baaren & Boivin 1998). As a general rule, repeated measures are produced in so-called longitudinal studies in which measures are collected at successive times. However, repetition can also sometimes correspond to data collected at different locations in space. Concerning behavioral studies done on insect parasitoids, this is especially the case, for example, in experiments done on four-way olfactometers (Vet et al. 1983) in which the behavior of the same individuals is recorded and compared between the different odorized fields of the device.

Using a standard ANOVA in this case is not appropriate because the data violate the assumption of independence and the analysis will fail to take proper account of correlations between the repeated measures. This can easily be seen from the following example based on simulated data: A normally-distributed behavioral trait is measured at three different times on the same individuals that were previously submitted to a specific treatment or a control group. We want to test both the effect of the treatment and any significant change in average values between the three recording times. This is clearly a repeated measure design but let us try to analyze it with a standard two-way ANOVA (Section 18.2.2). For this, values for the 3 times for 20 individuals were randomly drawn from 3 normal distributions, all having an average value of 0.0 and a standard deviation of 1.0. The first 10 individuals were supposed to experience the treatment, while the 10 remaining ones did not. In order to accurately simulate the repeated design, values corresponding to the three simulated times were randomly drawn from distributions and were correlated according to the following correlation matrix:

$$\begin{pmatrix} 1 & \rho & \rho^2 \\ \rho & 1 & \rho \\ \rho^2 & \rho & 1 \end{pmatrix}. \tag{18.67}$$

Different values of ρ, ranging from 0.0 to 1.0, were used and for each of them the whole simulation design was replicated 500 times. Each replicate was then analyzed with

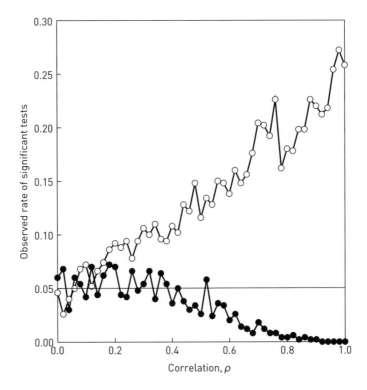

Fig. 18.1 Observed rate of significant tests for the treatment (open circles) and time (closed circles) effects after a two-way standard ANOVA on data simulated with different correlation values of the repeated design. Rates of significant tests are computed for a nominal level of 5%, which is indicated in this figure.

a standard two-way ANOVA to test both the treatment and the time effect, and the observed frequency of tests significant at a 5% level was computed in each case. Figure 18.1 gives the results obtained.

Since there were no differences between treatments and times in the simulated data, about 5% of the computed tests should be significant at a 5% level. As can be seen in Fig. 18.1, this appears to be the case only when the correlation ρ is close to 0.0, which corresponds to a situation of independent data. When the correlation ρ increases, however, the significance level of the test drops, which results in a larger number of falsely positive tests for the treatment effect. For the test of a time effect on such autocorrelated data, increasing values of the correlation ρ progressively lead to more conservative tests with a corresponding lack of power. At the extreme, when $\rho = 1.0$, the data remain unchanged over the course of time, and there cannot be a time effect.

Thus, standard ANOVA leads to wrong statistical conclusions, as it generally does when there are repeated measures, and so other methods should be used instead. One possible way to analyze data coming from a repeated design can be to consider the repeated measures as different variables (see Fig. 18.2 for a graphical meaning of this). Doing this, the correlation between the repeated measures among individuals is explicitly taken into account and standard multi-dimensional ANOVA (MANOVA) can then be used to test

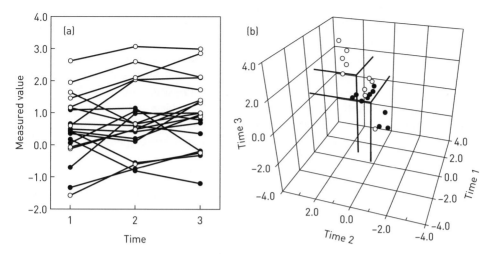

Fig. 18.2 Two graphical representations of the same data set showing how data coming from a repeated design can be considered as a multi- (here three-) dimensional data set. Data were simulated as in Fig. 18.1 (with $\rho = 0.75$), except that 1.0 was added to each value of individuals that were previously submitted to a specific treatment (open circles). (a) Values recorded at three different times are plotted as for a standard longitudinal study. (b) Data are plotted in a three-dimensional space and lines are indicating average values for individuals that were previously submitted or not to a specific treatment.

the different effects studied. Multivariate ANOVAs are simple generalizations of their standard univariate counterparts; the main difference is that inter- and intra-group variance-covariance matrices are used instead of the simple inter- and intra-group variances. Several tests can be used, for example, the Wilks' lambda (Johnson & Wichern 2002), to test, globally, the effect of factors that are not repeated, like the treatment effect in the simulated example above. The repeated factor (e.g. time) can also be tested, but after some data transformation. In the simulated example presented above, this will correspond to the test of the hypothesis that the three expected values μ_1, μ_2 and μ_3, corresponding to the three recording times, are equal. This is equivalent to testing simultaneously the two related hypothesis $\mu_1 - \mu_2 = 0$ and $\mu_2 - \mu_3 = 0$, which can be done with a Hotteling's T^2 test (Johnson & Wichern 2002). For this, each difference between the successive times, and their corresponding average values, should first be computed. Then, the following statistic:

$$\frac{(k - p + 1)}{kp} n \, \bar{Y}' \, \hat{V}^{-1} \bar{Y} \tag{18.68}$$

where n is the total number of individuals observed, p is the number of repeated measures (here 3), $\bar{Y}$ the mean average vector of the differences, $\hat{V}$ the estimated intra-group variance-covariance matrix, and k its corresponding df in a multivariate ANOVA done on the differences, is known to follow approximately an F-distribution with p and $k - p + 1$ df.

However, such an approach will be difficult to apply when the experimental set-up involves more than one factor, and their potential interactions, and/or when we are interested in testing interactions between the repeated factor (e.g. time) and others. Actually, a related method, so-called repeated measures ANOVA, is nowadays available in all software packages. It is based on the fact that, as opposed to data usually analyzed with multivariate analyses, repeated designs consist of collecting repeated measures of the same parameter. In repeated measures ANOVA, an individual is called a subject and the repeated factor is called a within-subjects factor, which represent different trials. Other factors are called between-subjects factors and are constituted of different groups. The method enables us to statistically test:

1 the within-subject main effect to know whether average values are changing between different trials;
2 between-subject main and interaction effects to estimate, globally, influences of the different corresponding factors; and also
3 within-subject-by-between-subject interaction effects to see whether changes among the different trials are influenced by any other factors.

For tests that involve only between-subjects effects, computations are simply based on simple ANOVA done on the sum of values obtained during the repeated trials divided by the square root of their number. Tests involving within-subjects effects are usually based on multivariate approaches similar to those described above (Davis 2003).

Repeated measures ANOVA carries the standard set of assumptions associated with an ordinary ANOVA, extended to the matrix case: multivariate normality for the within-subject factor, homogeneity of covariance matrices, and independence among groups for between-subject factors. Repeated measures ANOVA is robust to violations of the first two assumptions. Violations of independence among groups produce a non-normal distribution of the residuals, which results in invalid F tests. The most common violations of independence occur when either random selection or random assignment is not used. Some additional assumptions should also be verified, depending on the statistical test used to test the within-subject effect (see Davis 2003, for a thorough discussion of this).

Finally, as with fully independent data, there are now methods to analyze repeated designs in which the trait studied is not normally distributed. More accurately, the GLMs presented in Section 18.2 above can be extended to non-independent repeated designs by means of the so-called Generalized Estimating Equation of Liang and Zeger (1986), leading to methods for analyzing traits that follow a binomial (i.e. percentages) or a Poisson (i.e. counts) distribution (Hardin & Hilbe 2003).

18.4 Pseudoreplication

We have seen that experiments done on the behavioral ecology of insect parasitoids can sometimes produce non-independent data. This has been mainly presented to appear when each individual is measured several times (repeated measure designs), and we saw that specific statistical approaches are needed in this case (Section 18.3). For repeated measure designs, the experimental set-up has supposedly been built to intentionally produce non-independent data. However, some experimental set-ups, that are not correctly designed, can sometimes also unintentionally and insidiously produce non-independent data that are then wrongly analyzed statistically. This is mainly, but not exclusively, observed in field experiments (Hurlbert 1984).

The problem was originally recognized by Hurlbert (1984), who analyzed 176 field experiments from 156 papers published in the ecological literature during 1960–1980. Among these, an alarming rate of 27% was guilty of so-called pseudoreplication. Considering only the 101 studies using statistical analyses, 48% were pseudoreplicated. Pseudoreplication is defined as the use of statistical methods to test for treatment effects with data from experiments where either treatments are simply not replicated (though samples may be) or experimental units are not statistically independent (Hurlbert 1984). Since the original paper of Hurlbert (1984), pseudoreplication has been widely reported in environmental, ecological, and behavioral studies (Steward-Oaten & Murdoch 1986, Searcy 1989, Hurlbert & White 1993, Heffner et al. 1996, Ramirez et al. 2000).

Potential problems might arise when treatments are spatially or temporally segregated. Ramirez et al. (2000) proposed a hypothetical, although classic, experimental example. The experiment aims at studying the ability of an insect parasitoid to be attracted from a plant attacked by one of its herbivorous hosts. Using an olfactometer, individual wasp females are offered a choice between two odorous areas: one receives volatiles from an attacked plant, the other volatiles from an unattacked plant. Fifty observations are performed and the attacked and unattacked plants are changed every five observations. The olfactometer is cleaned after each observation. With such an experimental set-up, using a simple t-test for paired data, for example, to compare walking parameters of the females in both areas over the 50 replicates, would be wrong for two reasons. The first one is due to the fact that volatiles coming for the attacked and unattacked plants are always released in the same areas of the olfactometer and some undetectable differences between these locations could generate differences in the recorded behavioral parameters that are not necessarily related to plant volatiles. This is an example of treatments that are spatially segregated, and this is the reason why it is usually proposed to rotate the olfactometer by 90 or 180° after each replicate. The other reason, that corresponds to a temporally segregation of treatments, is more insidious. It comes from the fact that plants are not changed after each observation but only after every fifth observation. Therefore, observations are not independent within groups. The ten groups of five observations, however, constitute legitimate replicates on which a simple t-test for paired data could be applied, and each is best represented by the mean value of the five observations performed in it (Ramirez et al. 2000). Another way to analyze the full data set is to use repeated measures ANOVA, as explained in Section 18.3 above (Ramirez et al. 2000).

Other potential problems are related to experiments that are actually interconnected. For example, an experiment performed on insects that were reared in four different climatic chambers will produce data that are not totally independent due to any (maybe even non-detectable) variation between the different climatic chambers. Finally, as we have seen in Section 18.3 above, pseudoreplication can also be generated when several observations are done repeatedly on the same individual (repeated designs).

Pseudoreplication is not truly a problem of experimental design itself. Rather, it is often the result of a combination of experimental design and an inappropriate statistical analysis (Ramirez et al. 2000). It is interesting to see that the problem is usually not understood and it is sometimes claimed by researches that all experiments, at the extreme, are dependent since all of them are run on the same planet Earth. This is maybe the reason why poorly designed or incorrectly analyzed experimental work is literally flooding the ecological literature (Hurlbert 1984). Actually, pseudoreplication is the result of non-independent experimental units at the specific scale of analysis for the hypothesis being tested (Ramirez et al. 2000).

In any experimental field or laboratory work, it is known on first principles that, due to stochasticity, two or more objects are different whatever the trait measured. Then, if we increase the number of samples taken from each unit and statistically compare them (e.g. with a t-test or a simple ANOVA) with a nominal risk of, for example, 5%, the chance of finding a significant difference will increase with increase in the number of samples per unit. However, increasing the number of independent experimental units per treatment will not increase the chance of finding a significant difference under the null hypothesis. This has been proposed as a possible criterion for distinguishing pseudoreplication from true replication (Hurlbert 1984). Such a result can be more accurately understood using the following simulated example.

We wish to compare average values of a behavioral trait, related to wasp females foraging activity, between two wasp populations. The two populations are each constituted by several different families, each family originating from a mated female (i.e. isofemale lines). In the two populations, the trait studied is known to follow a normal distribution with an average value of 0.1 and a standard deviation of 1.0, but there are known differences in average values between the families. The first experimental design, using pseudoreplicates, is based on only two families, one taken in each population, and having an average value of 0.0 and 0.2, respectively. Samples of the same size are randomly drawn from each family. The second experimental design, using true replicates, is based on only one sample taken from each family and the same number of families is sampled in each population. For the two designs, the sampling protocol is repeated 500 times with different sample sizes and the two populations are compared in each case with a simple t-test. Figure 18.3 gives the observed frequency of tests significant at a 5% level, which was computed in both cases for different sample sizes. As expected, with a properly designed experiment based on true replicates, the probability of judging an effect as significant when there is no effect (i.e. an error of the first kind) remains constant and does not depend on sample sizes used. However, with a wrongly designed experiment using pseudoreplicates, the risk of wrongly declaring a non-existing effect as being significant clearly increases with sample size. The reason for this is that, in the later case, the null hypothesis tested is actually no longer that of no difference between the two populations compared, but that of no difference between the two families from which samples were taken.

Pseudoreplication is usually considered as an insidious bane. Even if some occurrences are clear-cut, like in the simulated example above, others can be more subtle and difficult to detect and require an accurate understanding of the system under study if the problem is to be avoided (Heffner et al. 1996). Common sense is usually needed along with biological knowledge, and sometimes intuition should be applied (Hurlbert 1984). Sometimes, replications are simply impossible or not desired, for example, if the cost in time and/or money of each of them is great. Experiments involving unreplicated treatments can sometimes be the only or best option (Hurlbert 1984). In such a case, erroneous use of statistical tests can lead to wrong conclusions, and it would be far better to recognize weaknesses in the experimental set-up used and to use descriptive statistics rather than formal tests of hypotheses.

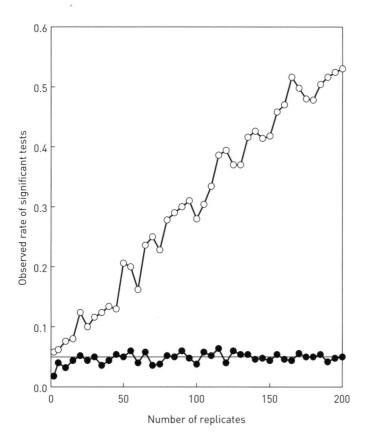

Fig. 18.3 Observed rate of significant tests comparing two wasp populations when true replicates (black circles) or pseudoreplicates (white circles) are used on simulated data with different sample sizes (see text). Rates of significant tests are computed for a nominal level of 5%, which is indicated in this figure.

18.5 Unbalanced set-ups

Many experiments to analyze the behavioral ecology of insect parasitoids are aimed at studying the effect of one or several independent qualitative factors on the average value of a behavioral trait. We already saw some examples of possible tested factors above. They usually correspond to different treatments whose effects have to be quantified and tested. When the effect of more that one factor is studied, their potential interactions are also of interest. In order to analyze the data obtained in such experiments, standard two-way (or more) ANOVAs (for normally-distributed traits) or GLMs (Sections 18.2.2 and 18.2.3) can be used. Usual statistical procedures are designed to handle so-called balanced data, which is data with equal numbers of observations for every combination of the different classification factors tested (Sokal & Rohlf 1981). However, even with the best of intentions, it is frequently impossible to produce such evenly balanced designs. For instance, in an experiment testing the effect of different treatments on wasp females on a behavioral feature of their offspring, the number of offspring per treated female is likely to differ and will not be a parameter under experimental control. Moreover, even if an experiment is correctly designed, with a balanced number of replicates in each case, some individuals might accidentally die before being measured and the resulting data set will become unbalanced due to the existence of missing values.

Analyzing such unbalanced data is usually considered to be a complicated matter. One approach, if possible, is to avoid such a problem at the expense of losing some information. For example, if there are at least five replicates in every subclass in an analysis, but some of them have six of seven replicates, it would be possible to reduce the sample size of all subclasses to five to acquire a balanced design that may be analyzed with a standard ANOVA or a GLM. Of course, removal of any individuals from a subclass to equalize sample sizes must be done at random. Such a procedure will be legitimate only if the number of replicates in each subclass itself does not affect the trait under study. For example, in a gregarious parasitoid species, the behavior of all individuals that emerged from a single parasitized host might be influenced by their number. Further, reducing subclasses to a common sample size is usually not a good idea, especially if:

1 there is an important variation in subclass sample sizes;
2 the original data were scarce and/or expensive to obtain; or
3 the error (i.e. intra-subclass) variance is too important.

Another possibility, if the design is unbalanced due to missing values, is to estimate those values that are missing. If there are several replicates in each subclass, the missing values can be estimated by the observed average values of each corresponding subclass. If the behavioral trait under study is following a known distribution, missing data can also be approximated by simulated values randomly drawn from this distribution by using the average and variance of each corresponding subclass. Finally, if there is only one replicate per subclass, appropriate and usually easy to compute estimating methods are also available (Sokal & Rohlf 1981, Little & Rubin 1987). Once the missing values have been estimated, a standard ANOVA or a GLM can be computed, but the number of degrees of freedom of the error term has to be reduced by the number of values that have been estimated.

Estimating missing values is, however, not always feasible, especially if too many values are missing. Performing a two-way (or more) ANOVA or a GLM with unbalanced subclass sizes still remains possible, but statistical software packages are needed since computational procedures becomes considerably more complicated. In this case, a so-called type III ANOVA can be computed, which is designed to test the same hypotheses that would be tested if the data were balanced. Indeed, in this case, hypotheses and the associated tests are not functions of the number of replicates per treatment combination. On balanced data, such an ANOVA will give the same results as those obtained using standard ANOVA. However, results of such an analysis with unbalanced data should be interpreted cautiously (Shaw & Mitchell-Olds 1993). Probably the most important and obvious problem that can arise when analyzing unbalanced data by this method can be seen using the following example based on simulated values: A normally-distributed behavioral trait is measured on individuals belonging to four different wasp species and after four different treatments. In each species-treatment combination, three independent replicates were performed, so the full balanced design represents a total of $4 \times 4 \times 3 = 48$ replicates. All values were first drawn from a normal distribution with an average of 0.0

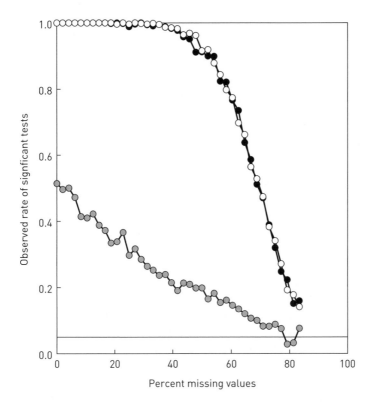

Fig. 18.4 Observed rate of significant tests for the species effect (white circles), the treatment effect (black circles), and their interaction (gray circles) after a two-way type III ANOVA on data simulated with different proportions on missing values (see text). Rates of significant tests are computed for a nominal level of 5%, which is indicated in this figure.

and a standard deviation of 1.0. In order to have significant effects for the two main factors and a significant interaction between them, $0.4 \times i \times j$ was added to each individuals belonging to the ith species and that was treated with the jth treatment (i and $j = 1, 2, 3, 4$), and the simulation design was replicated 500 times. By doing so, and using a nominal level of 5%, both the species and the treatment effect were always declared as being significant and their interaction was declared as being significant in 51.4% of the cases (Fig. 18.4).

The same computation was repeated each time with an increasing number of missing values that were uniformly distributed over the entire data set and the data were analyzed with a two-way type III ANOVA. The observed frequency of tests significant at a 5% level was computed in each case and results are shown in Fig. 18.4. As can be seen in this figure, an increase in the rate of missing values progressively leads to a reduction in the number of tests that are declared as significant. This is true for both the main factors tested and their interaction. For the main effects, the phenomenon seems to appear only when more than half of the data are missing but these factors were initially strongly significant (average p-values for the two factors with the full, balanced design were around 10^{-5} in both cases). On data simulated with less significant factors, the decrease in the power of the corresponding tests would appear much earlier, as this can be seen for the interaction (average p-value for the interaction with the full, balanced design was around 0.112). Thus, the power of the test is reduced when data are unbalanced and it is, therefore, possible to overlook an effect (i.e. judge it as non-significant) when the effect truly exists. So, particular caution should be taken when interpreting failure to reject null hypotheses from unbalanced data (Shaw & Mitchell-Olds 1993).

18.6 Conclusion

Behavioral ecologists, and especially those working on insect parasitoids, regularly fall into standard traps when they analyze results of their experimental work. They indeed often have to deal with unbalanced or non-independent data, or handling pseudoreplications, sometimes even without knowing it. Further, they are in most cases collecting non-normally distributed data. In many cases, standard ANOVA (or regression methods) are applied, but arguments presented in this chapter show that this can lead to wrong conclusions about the effects that are tested. As we have shown, nowadays there is a whole arsenal of more rigorous and efficient methods available, which can be used in most of these cases. When no specific method is available (e.g. for pseudoreplicated data), we have tried to inform the reader about possible misinterpretation of the results obtained from wrongly designed statistical analyses. Thus, the objective of this chapter is to provide the reader with the basic knowledge for analyzing, in a more correct and efficient way, results collected from experimental field or laboratory works on the behavioral ecology of insect parasitoids.

　Most of the statistical procedures presented in this chapter cannot be done by hand but a large number of statistical software packages are now available that incorporate these methods. Examples are:

1 SAS® (http://www.sas.com);
2 STATISTICA® (http://www.statsoft.com/);
3 SPSS® (http://www.spss.com/);
4 SYSTAT® (http://www.systat.com/);
5 Splus® (http://www.insightful.com/); or
6 R (http://www.r-project.org/).

The last one can be recommended because it is a free, efficient, and a simple-to-learn statistical computing and graphic language.

References

Bressers, W.M.A., Meelis, E., Haccou, P. and Kruk, M.R. (1991) When did it really start or stop: the impact of censored observations on ethological analysis of durations. *Behavioural Processes* **23**: 1–20.

Davis, C.S. (2003) *Statistical Methods for the Analysis of Repeated Measurements*. Springer, New York.

Haccou, P. and Meelis, E. (1994) *Statistical Analysis of Behavioural Data*. Oxford University Press, Oxford.

Hardin, J.W. and Hilbe, J.M. (2003) *Generalized Estimating Equations*. Chapman & Hall. CRC, Boca Raton, FL.

Heffner, R.A., Butler, M.J. and Reilly, C.K. (1996) Pseudoreplication revisited. *Ecology* **77**: 2558–62.

Hemerik, L., Driessen, G. and Haccou, P. (1993) The effects of intra-patch experiences on patch leaving tendency, search time and search efficiency of parasitoids of the species *Leptopilina clavipes*. *Journal of Animal Ecology* **62**: 33–44.

Hurlbert, S.H. (1984) Pseudoreplication and the design of ecological field experiments. *Ecological Monographs* **54**: 187–211.

Hurlbert, S.H. and White, M.D. (1993) Experiments with freshwater invertebrate zooplanktivores: quality of statistical analyses. *Bulletin of Marine Science* **53**: 128–53.

Johnson, R.A. and Wichern, D.W. (2002) *Applied Multivariate Statistical Analysis*. Prentice Hall, Englewood Cliffs, NJ.

Kalbfleisch, J.D. and Prentice, R.L. (2002) *The Statistical Analysis of Failure Time Data*, 2nd edn. John Wiley & Sons, New York.

Kleinbaum, D.G. and Klein, M. (2005) *Survival Analysis – a Self-Learning Text*, 2nd edn. Springer, New York.

Liang, K.Y. and Zeger, S.L. (1986) Longitudinal data analysis using generalized linear models. *Biometrika* **73**: 13–22.

Little, R.J.A. and Rubin, D.B. (1987) *Statistical Analysis with Missing Data*. Wiley, New York.

McCullagh, P. and Nelder, J.A. (1989) *Generalized Linear Models*, 2nd edn. Chapman & Hall, London.

Mood, A.M., Graybill, F.A. and Boes, D.C. (1974) *Introduction to the Theory of Statistics*, 3rd edn. McGraw-Hill, Singapore.

Ramirez, C.C., Fuentes-Contreras, E., Rodriguez, L.C. and Niemeyer, H.M. (2000) Pseudoreplication and its frequency in olfactometric laboratory studies. *Journal of Chemical Ecology* **26**: 1423–31.

Searcy, W.A. (1989) Pseudoreplication, external validity and the design of playback experiments. *Animal Behaviour* **38**: 715–17.

Shaw, R.G. and Mitchell-Olds, T. (1993) ANOVA for unbalanced data: an overview. *Ecology* **74**: 1638–45.

Sokal, R.R. and Rohlf, F.J. (1981) *Biometry. The Principles and Practice of Statistics in Biological Research*, 2nd edn. Freeman, San Francisco.

Steward-Oaten, A. and Murdoch, W.W. (1986) Environmental impact assessment: 'pseudoreplication' in time. *Ecology* **67**: 929–40.

van Baaren, J. and Boivin, G. (1998) Learning affects host discrimination behavior in a parasitoid wasp. *Behavioral Ecology & Sociobiology* **42**: 9–16.

Vet, L.E.M., van Lenteren, J.C., Heymans, M. and Meelis, E. (1983) An airflow olfactometer for measuring olfactory responses of hymenopterous parasitoids and other small insects. *Physiological Entomology* **8**: 97–106.

Wajnberg, É. (2006) Time-allocation strategies in insect parasitoids: from ultimate predictions to proximate behavioural mechanisms. *Behavioral Ecology & Sociobiology* **60**: 589–611.

Zar, J.H. (1999) *Biostatistical Analysis*, 4th edn. Prentice Hall, New Jersey.

Index

Note: Page numbers in *italics* refer to Figures; those in **bold** refer to Tables.